Batch 670482LV00001B

670482LVX00107B CASE	9783030431037 6.14X9.21	Real Algebraic Varieties 464	<K>	1.0	MATTE	(3)
670482LVX00108B CASE	9783030597054 6.14X9.21	Linking Sensitive Data : Methods and Tec 492	<L>	1.0625	MATTE	(2)
670482LVX00109B CASE	9783030465032 6.14X9.21	Biodeterioration of Wooden Cultural Herit 556	<N>	1.1875	MATTE	(3)
670482LVX00110B CASE	9783110679601 6.14X9.21	Sprache und Empathie 640	<Q>	1.375	MATTE	(1)
670482LVX00111B CASE	9783030515799 6.14X9.21	The Web of Data 700	<S>	1.5	MATTE	(2)
670482LVX00112B CASE	9781734032727 6.00X9.00	Succeeding as a Management Consultant 462	<K>	1.0	MATTE	(1)
670482LVX00113B CASE	9789493056596 6.00X9.00	Living among the Dead: My Grandmother 178	<B>	0.4375	MATTE	(1)
670482LVX00114B CASE	9781786349279 6.00X9.00	Navigating Uncertainty in the South Chin 144	<A>	0.375	MATTE	(1)
670482LVX00115B CASE	9781922348487 5.50X8.50	Scotland's Story 306	<G>	0.75	MATTE	(1)
670482LVX00116B CASE	9781613321157 5.00X8.00	My Life in 100 Objects 252	<E>	0.625	MATTE	(1)
670482LVX00117B CASE	9783110013160 5.00X8.00	Hystoria Tartarorum C. de Bridia Monachi 60 <AAA>		0.25	MATTE	(1)

Smart Innovation, Systems and Technologies

Volume 196

Series Editors

Robert J. Howlett, Bournemouth University and KES International, Shoreham-by-sea, UK

Lakhmi C. Jain, Faculty of Engineering and Information Technology, Centre for Artificial Intelligence, University of Technology Sydney, Sydney, NSW, Australia

The Smart Innovation, Systems and Technologies book series encompasses the topics of knowledge, intelligence, innovation and sustainability. The aim of the series is to make available a platform for the publication of books on all aspects of single and multi-disciplinary research on these themes in order to make the latest results available in a readily-accessible form. Volumes on interdisciplinary research combining two or more of these areas is particularly sought.

The series covers systems and paradigms that employ knowledge and intelligence in a broad sense. Its scope is systems having embedded knowledge and intelligence, which may be applied to the solution of world problems in industry, the environment and the community. It also focusses on the knowledge-transfer methodologies and innovation strategies employed to make this happen effectively. The combination of intelligent systems tools and a broad range of applications introduces a need for a synergy of disciplines from science, technology, business and the humanities. The series will include conference proceedings, edited collections, monographs, handbooks, reference books, and other relevant types of book in areas of science and technology where smart systems and technologies can offer innovative solutions.

High quality content is an essential feature for all book proposals accepted for the series. It is expected that editors of all accepted volumes will ensure that contributions are subjected to an appropriate level of reviewing process and adhere to KES quality principles.

** **Indexing: The books of this series are submitted to ISI Proceedings, EI-Compendex, SCOPUS, Google Scholar and Springerlink** **

More information about this series at http://www.springer.com/series/8767

Tomonobu Senjyu · Parikshit N. Mahalle ·
Thinagaran Perumal · Amit Joshi
Editors

Information and Communication Technology for Intelligent Systems

Proceedings of ICTIS 2020, Volume 2

 Springer

Editors
Tomonobu Senjyu
Department of Electrical and Electronics Engineering
University of the Ryukyus
Nishihara, Japan

Thinagaran Perumal
Universiti Putra Malaysia
Serdang, Malaysia

Parikshit N. Mahalle
Sinhgad Technical Education Society
SKNCOE
Pune, India

Amit Joshi
Global Knowledge Research Foundation
Udaipur, Rajasthan, India

ISSN 2190-3018   ISSN 2190-3026   (electronic)
Smart Innovation, Systems and Technologies
ISBN 978-981-15-7061-2   ISBN 978-981-15-7062-9   (eBook)
https://doi.org/10.1007/978-981-15-7062-9

© The Editor(s) (if applicable) and The Author(s), under exclusive license to Springer Nature Singapore Pte Ltd. 2021

This work is subject to copyright. All rights are solely and exclusively licensed by the Publisher, whether the whole or part of the material is concerned, specifically the rights of translation, reprinting, reuse of illustrations, recitation, broadcasting, reproduction on microfilms or in any other physical way, and transmission or information storage and retrieval, electronic adaptation, computer software, or by similar or dissimilar methodology now known or hereafter developed.

The use of general descriptive names, registered names, trademarks, service marks, etc. in this publication does not imply, even in the absence of a specific statement, that such names are exempt from the relevant protective laws and regulations and therefore free for general use.

The publisher, the authors and the editors are safe to assume that the advice and information in this book are believed to be true and accurate at the date of publication. Neither the publisher nor the authors or the editors give a warranty, expressed or implied, with respect to the material contained herein or for any errors or omissions that may have been made. The publisher remains neutral with regard to jurisdictional claims in published maps and institutional affiliations.

This Springer imprint is published by the registered company Springer Nature Singapore Pte Ltd.
The registered company address is: 152 Beach Road, #21-01/04 Gateway East, Singapore 189721, Singapore

Organizing Committee

Technical Program Committee Chairs

Tomonobu Senjyu, Editorial Advisory Board, Renewable Energy Focus, University of the Ryukyus, Okinawa, Japan
Thinagaran Perumal, Faculty of Computer Science and Information Technology, University Putra Malaysia, Serdang, Malaysia
Dr. Parikshit N. Mahalle, Singhad Group, Pune, India
Dr. Nilanjan Dey, Techno India Institute of Technology, Kolkata, India
Dr. Nilesh Modi, Chairman, Professor, Dr. Baba Saheb Ambedkar University, Ahmedabad, India

Technical Program Committee Members

Dr. Aynur Unal, Standford University, USA
Prof. Brent Waters, University of Texas, Austin, TX, USA
Er. Kalpana Jain, CTAE, Udaipur, India
Prof. (Dr.) Avdesh Sharma, Jodhpur, India
Er. Nilay Mathur, Director, NIIT Udaipur, India
Prof. Philip Yang, Price water house Coopers, Beijing, China
Mr. Jeril Kuriakose, Manipal University, Jaipur, India
Prof. R. K. Bayal, Rajasthan Technical University, Kota, Rajasthan, India
Prof. Martin Everett, University of Manchester, England
Prof. Feng Jiang, Harbin Institute of Technology, China
Dr. Savita Gandhi, Professor, Gujarat University, Ahmedabad, India
Prof. Xiaoyi Yu, National Laboratory of Pattern Recognition, Institute of Automation, Chinese Academy of Sciences, Beijing, China
Prof. Gengshen Zhong, Jinan, Shandong, China

Prof. A. R. Abdul Rajak, Department of Electronics and Communication Engineering, Birla Institute of Dr. Nitika Vats Doohan, Indore, India
Dr. Harshal Arolkar, Immd. Past Chairman, CSI Ahmedabad Chapter, India
Mr. Bhavesh Joshi, Advent College, Udaipur, India
Prof. K. C. Roy, Principal, Kautaliya, Jaipur, India
Dr. Mukesh Shrimali, Pacific University, Udaipur, India
Mrs. Meenakshi Tripathi, MNIT, Jaipur, India
Prof. S. N. Tazi, Government Engineering College, Ajmer, Rajasthan, India
Shuhong Gao, Mathematical Sciences, Clemson University, Clemson, SC, USA
Sanjam Garg, University of California, Los Angeles, CA, USA
Garani Georgia, University of North London, UK
Chiang Hung-Lung, China Medical University, Taichung, Taiwan
Kyeong Hur, Department of Computer Education, Gyeongin National University of Education, Incheon, Korea
Sudath Indrasinghe, School of Computing and Mathematical Sciences, Liverpool John Moores University, Liverpool, England
Ushio Inoue, Department of Information and Communication Engineering, Engineering Tokyo Denki University, Tokyo, Japan
Dr. Stephen Intille, Associate Professor, College of Computer and Information Science and Department of Health Sciences, Northeastern University, Boston, MA, USA
Dr. M. T. Islam, Institute of Space Science, Universiti Kebangsaan Malaysia, Selangor, Malaysia
Lillykutty Jacob, Professor, Department of Electronics and Communication Engineering, NIT, Calicut, Kerala, India
Dagmar Janacova, Tomas Bata University in Zlín, Faculty of Applied Informatics nám. T. G, Czech Republic, Europe
Jin-Woo Kim, Department of Electronics and Electrical Engineering, Korea University, Seoul, Korea
Muzafar Khan, Computer Sciences Department, COMSATS University, Pakistan
Jamal Akhtar Khan, Department of Computer Science College of Computer Engineering and Sciences, Salman bin Abdulaziz University, Kingdom of Saudi Arabia
Kholaddi Kheir Eddine, University of Constantine, Algeria
Ajay Kshemkalyani, Department of Computer Science, University of Illinois, Chicago, IL, USA
Madhu Kumar, Associate Professor, Computer Engineering Department, Nanyang Technological University, Singapore
Rajendra Kumar Bharti, Assistant Professor, Kumaon Engineering College, Dwarahat, Uttarakhand, India
Prof. Murali Bhaskaran, Dhirajlal Gandhi College of Technology, Salem, Tamil Nadu, India
Prof. Komal Bhatia, YMCA University, Faridabad, Haryana, India
Prof. S. R. Biradar, Department of Information Science and Engineering, SDM College of Engineering and Technology, Dharwad, Karnataka, India

Organizing Committee

A. K. Chaturvedi, Department of Electrical Engineering, IIT Kanpur, India
Jitender Kumar Chhabra, NIT, Kurukshetra, Haryana, India
Pradeep Chouksey, Principal, TIT College, Bhopal, Madhya Pradesh, India
Chhaya Dalela, Associate Professor, JSSATE, Noida, Uttar Pradesh, India
Jayanti Dansana, KIIT University, Bhubaneswar, Odisha, India
Soura Dasgupta, Department of TCE, SRM University, Chennai, India
Dr. Apurva A. Desai, Veer Narmad South Gujarat University, Surat, India
Dr. Sushil Kumar, School of Computer and Systems Sciences, Jawaharlal Nehru University, New Delhi, India
Amioy Kumar, Biometrics Research Lab, Department of Electrical Engineering, IIT Delhi, India
Prof. L. C. Bishnoi, GPC, Kota, India
Dr. Vikarant Bhateja, Lucknow, India
Dr. Satyen Parikh, Dean, Ganpat University, Ahmedabad, India
Dr. Puspendra Singh, JKLU, Jaipur, India
Dr. Aditya Patel, Ahmedabad University, Gujarat, India
Mr. Ajay Choudhary, IIT Roorkee, India
Prashant Panse, Associate Professor, Medi-Caps University, India
Roshani Raut, Associate Professor, Vishwakarma Institute of Information Technology, Pune, India
Rachit Adhvaryu, Assistant Professor, Marwadi University, India
Purnima Shah, Assistant Professor, Adani Institute of Infrastructure and Engineering, India
Mohd. Saifuzzaman, Lecturer, Daffodil International University, Bangladesh
Nandakumar Iyengar, Professor, City Engineering College, India
Nilu Singh, Assistant Professor, SoCA, BBD University, Lucknow, India
Habib Hadj-Mabrouk, Researcher in Artificial Intelligence and Railway Safety, University Gustave Eiffel, France
Ankita Kanojiya, Assistant Professor, GLS University, India
Ripal Ranpara, Assistant Professor and DBA, Atmiya University, India
Santhosh John, Institute Director, Middle East Institute for Advanced Training, Oman
Nageswara Rao Moparthi, Professor, KL University India
Akhilesh Sharma, Associate Professor, Manipal University Jaipur, India
Sylwia Werbińska-Wojciechowska, Associate Professor, Wroclaw University of Science and Technology, Poland
Divya Srivastava, Ph.D. Scholar, Indian Institute of Technology Jodhpur, India
Dr. Debajyoti Mukhopadhyay, Professor, WIDiCoReL Research Lab, India
Dr. Neelam Chaplot, Associate Professor, Poornima College of Engineering, India
Shruti Suman, Associate Professor, K L University, Andhra Pradesh, India
M. V. V. Prasad Kantipudi, Associate Professor, Sreyas Institute of Engineering and Technology, Hyderabad, India
Urmila Shrawankar, Professor, G H Raisoni College of Engineering, Nagpur, India
Dr. Pradeep Laxkar, Associate Professor, Mandsaur University, India
Muhammad Asif Khan, Researcher, Qatar University, Qatar

Dr. Gayatri, Associate Professor, CEPT University, India
Jagadeesha Bhat, Associate Professor, St. Joseph Engineering College, Mangalore, India
Dr. Madhan Kumar Srinivasan, Associate Vice President, Accenture Technology, India
Ajay Vyas, Assistant Professor, Adani Institute of Infrastructure Engineering, India
Mounika Neelam, Assistant Professor, PSCMRCET, India
Prakash Samana, Assistant Professor, Gokula Krishna College of Engineering, India
Dr. Monika Jain, Professor and Head-ECE, I.T.S Engineering College, Greater Noida, India
Dr. K. Srujan Raju, Professor and Dean Student Welfare, CMR Technical Campus, Telengana, India
Dr. Narendrakumar Dasre, Associate Professor, Ramrao Adik Institute of Technology, India
Anand Nayyar, Professor, Researcher and Scientist, Duy Tan University, Vietnam
Sakshi Arora, Assistant Professor, SMVD University, India
Dr. Ajay Roy, Associate Professor, Lovely Professional University, India
Prasanth Vaidya, Associate Professor, Sanivarapu, India
Dr. Jeevanandam Jotheeswaran, Director, Amity University Online, India
John Moses C., Associate Professor, JNTUH, India
S. Vaithyasubramanian, Assistant Professor, Sathyabama University, India
Ashish Revar, Assistant Professor, Symbiosis University of Applied Sciences, Indore, India

Organizing Chair

Mr. Bharat Patel, Chairman, CEDB
Dr. Priyanka Sharma, Raksha Shakti University, Ahmedabad, India
Amit Joshi, Director—Global Knowledge Research Foundation

Organizing Secretary

Mr. Mihir Chauhan, Organizing Secretary, ICTIS 2020

Conference Secretary

Mr. Aman Barot, Conference Secretary, ICTIS 2020

Supporting Chairs

Dr. Vijay Singh Rathore, Professor and Head, JECRC, Jaipur, India
Dr. Nilanjan Dey, Techno India Institute of Technology, Kolkata, India
Dr. Nisarg Pathak, Swarnim Gujarat Sports University, Gandhinagar, India

Preface

This SIST volume contains the papers presented at the ICTIS 2020: Fourth International Conference on Information and Communication Technology for Intelligent Systems. The conference was held during May 15–16, 2020, organized on a digital platform ZOOM due to pandemic COVID-19. The supporting partners were InterYIT IFIP and Knowledge Chamber of Commerce and Industry (KCCI).

This conference aimed at targeting state of the art as well as emerging topics pertaining to ICT and effective strategies for its implementation in engineering and intelligent applications. The objective of this international conference is to provide opportunities for the researchers, academicians, industry persons, and students to interact and exchange ideas, experience, and expertise in the current trend and strategies for information and communication technologies. Besides this, participants will also be enlightened about the vast avenues and current and emerging technological developments in the field of ICT in this era and its applications will be thoroughly explored and discussed. The conference is anticipated to attract a large number of high-quality submissions and stimulate the cutting-edge research discussions among many academic pioneering researchers, scientists, industrial engineers, students from all around the world and provide a forum to researchers; propose new technologies, share their experiences, and discuss future solutions for design infrastructure for ICT; provide a common platform for academic pioneering researchers, scientists, engineers, and students to share their views and achievements; enrich technocrats and academicians by presenting their innovative and constructive ideas; and focus on innovative issues at the international level by bringing together the experts from different countries. Research submissions in various advanced technology areas were received, and after a rigorous peer review process with the help of the program committee members and external reviewers, 75 papers were accepted with an acceptance rate of 0.19 for this Volume.

The conference featured many distinguished personalities like Mike Hinchey—Ph.D., University of Limerick, Ireland, President, International Federation of Information Processing; Bharat Patel—Honorary Secretary General, Knowledge Chamber of Commerce and Industry, India; Aninda Bose—Sr. Editor, Springer, India; Mufti Mahmud—Ph.D., Nottingham Trent University, UK; Suresh Chandra

Satapathy—Ph.D., Kalinga Institute of Industrial Technology, Bhubaneswar, India; Neeraj Gupta—Ph.D., School of Engineering and Computer Science, Oakland University, USA; Nilanjan Dey—Ph.D., Techno India College of Technology, Kolkata, India. We are indebted to all our organizing partners for their immense support to make this virtual conference successfully possible. A total of 23 sessions were organized as a part of ICTIS 2020 including 22 technical and 1 inaugural session. Approximately, 154 papers were presented in 22 technical sessions with high discussion insights. The total number of accepted submissions was 112 with a focal point on ICT and intelligent systems. Our sincere thanks to our Organizing Secretary, ICTIS 2020—Mihir Chauhan and Conference Secretary, ICTIS 2020—Aman Barot and the entire team of Global Knowledge Research Foundation and Conference committee for their hard work and support for the entire shift of ICTIS 2020 from physical to digital modes in these new normal times.

Nishihara, Japan
Pune, India
Serdang, Malaysia
Udaipur, India

Tomonobu Senjyu
Parikshit N. Mahalle
Thinagaran Perumal
Amit Joshi

Contents

Internet of Things: Challenges, Security Issues and Solutions 1
Sweta Singh and Rakesh Kumar

Modeling and Performance Evaluation of Auction Model
in Cloud Computing. 17
Gagandeep Kaur

Framework for Resource Management in Cloud Computing 25
Gagandeep Kaur

Unique Stego Key Generation from Fingerprint Image
in Image Steganography. 33
A. Anuradha and Hardik B. Pandit

Self-powered IoT-Based Design for Multi-purpose Smart
Poultry Farm 43
Tajim Md. Niamat Ullah Akhund, Shouvik Roy Snigdha,
Md. Sumon Reza, Nishat Tasnim Newaz, Mohd. Saifuzzaman,
and Masud Rana Rashel

Phrase-Based Statistical Machine Translation of Hindi Poetries
into English 53
Rajesh Kumar Chakrawarti, Pratosh Bansal, and Jayshri Bansal

A Smart Education Model for Future Learning and Teaching
Using IoT 67
Swati Jain and Dimple Chawla

Internet of Things (IoT)-Based Advanced Voting Machine System
Enhanced Using Low-Cost IoT Embedded Device
and Cloud Platform 77
Miral M. Desai, Jignesh J. Patoliya, and Hiren K. Mewada

IoT: Security Issues and Challenges 87
Er Richa

**Coronary Artery Disease Prediction Using Neural Network
and Random Forest-Based Feature Selection** 97
Aman Shakya, Neerav Adhikari, and Basanta Joshi

**A Collaborative Approach to Decision Making in Decentralized
IoT Devices** 107
Venus Kaurani and Himajit Aithal

**Analysis of MRI Images for Brain Tumor Detection Using Fuzzy
C-Means and LSTM** 117
Sobit Thapa and Sanjeeb Prasad Panday

A Survey on Recognizing and Significance of Self-portrait Images 127
Jigar Bhatt and Mangesh Bedekar

Anomaly Detection in Distributed Streams 139
Rupesh Karn, Suvrat Ram Joshi, Umanga Bista, Basanta Joshi,
Daya Sagar Baral, and Aman Shakya

Envisaging Bugs by Means of Entropy Measures 149
Anjali Munde

**TRANSPR—Transforming Public Accountability Through
Blockchain Technology** 157
P. R. Sriram, N. J. Subhashruthi, M. Muthu Manikandan, Karthik Gopalan,
and Sriram Kaushik

The Scope of Tidal Energy in UAE 167
Abaan Ahamad and A. R. Abdul Rajak

Knowledge Engineering in Higher Education 177
Shankar M. Patil, Vijaykumar N. Patil, Sonali J. Mane, and Shilpa M. Satre

**Building a Graph Database for Storing Heterogeneous
Healthcare Data** 193
Goutam Kundu, Nandini Mukherjee, and Safikureshi Mondal

**Analysis and Optimization of Cash Withdrawal Process Through
ATM in India from HCI and Customization** 203
Abhishek Jain, Shiva Subhedar, and Naveen KumarGupta

**Survey of Mininet Challenges, Opportunities, and Application
in Software-Defined Network (SDN)** 213
Dhruvkumar Dholakiya, Tanmay Kshirsagar, and Amit Nayak

**Some Novelties in Map Reducing Techniques to Retrieve and Analyze
Big Data for Effective Processing** 223
Prashant Bhat and Prajna Hegde

Contents

A Study Based on Advancements in Smart Mirror Technology 233
Aniket Dongare, Indrajeet Devale, Aditya Dabadge, Shubham Bachute, and Sukhada Bhingarkar

Unidirectional Ensemble Recognition and Translation of Phrasal Sign Language from ASL to ISL 241
Anish Sujanani, Shashidhar Pai, Aniket Udaykumar, Vivith Bharath, and V. R. Badri Prasad

Unravelling SAT: Discussion on the Suitability and Implementation of Graph Convolutional Networks for Solving SAT 251
Hemali Angne, Aditya Atkari, Nishant Dhargalkar, and Dilip Kale

CES: Design and Implementation of College Exam System 259
Punya Mathew, Rasesh Tongia, Kavish Mehta, and Vaibhav Jain

ANN-Based Multi-class Malware Detection Scheme for IoT Environment 269
Vaibhav Nauriyal, Kushagra Mittal, Sumit Pundir, Mohammad Wazid, and D. P. Singh

Secure Web Browsing Using Trusted Platform Module (TPM) 279
Harshad S. Wadkar and Arun Mishra

CP-ABE with Hidden Access Policy and Outsourced Decryption for Cloud-Based EHR Applications 291
Kasturi Routray, Kamalakanta Sethi, Bharati Mishra, Padmalochan Bera, and Debasish Jena

Viral Internet Challenges: A Study on the Motivations Behind Social Media User Participation 303
Naman Shroff, G. Shreyass, and Deepak Gupta

Early Flood Monitoring System in Remote Areas 313
John Colaco and R. B. Lohani

A Survey on Securing Payload in MQTT and a Proposed Ultra-lightweight Cryptography 323
Edward Nwiah and Shri Kant

Life Cycle Assessment and Management in Hospital Units Using Applicable and Robust Dual Group-Based Parameter Model 337
Vitaliy Sarancha, Leo Mirsic, Stjepan Oreskovic, Vadym Sulyma, Bojana Kranjcec, and Ksenija Vitale

A Novel Approach to Stock Market Analysis and Prediction Using Data Analytics 347
Sangeeta Kumari, Kanchan Chaudhari, Rohan Deshmukh, Rutajagruti Naik, and Amar Deshmukh

Energy Trading Using Ethereum Blockchain 357
M. Mrunalini and D. Pavan Kumar

**Supply Chain Management for Selling Farm Produce
Using Blockchain** 367
Anita Chaudhari, Jateen Vedak, Raj Vartak, and Mayuresh Sonar

**Design and Construction of a Multipurpose Solar-Powered
Water Purifier** 377
Ayodeji Olalekan Salau, Dhananjay S. Deshpande,
Bernard Akindade Adaramola, and Abdulkadir Habeebullah

Named Entity Recognition for Rental Documents Using NLP 389
Chinmay Patil, Sushant Patil, Komal Nimbalkar, Dhiraj Chavan,
Sharmila Sengupta, and Devesh Rajadhyax

Empirical Analysis of Various Seed Selection Methods 399
Kinjal Rabadiya and Ritesh Patel

**A Comparative Study on Interactive Segmentation Algorithms
for Segmentation of Animal Images** 409
N. Manohar, S. Akshay, and N. Shobha Rani

**Implications of Quantum Superposition in Cryptography:
A True Random Number Generation Algorithm** 419
Dhananjay S. Deshpande, Aman Kumar Nirala,
and Ayodeji Olalekan Salau

**Influences of Purchase Involvement on the Relationship Between
Customer Satisfaction and Loyalty in Vietnam** 433
Pham Van Tuan

**A Study on Factors Influencing Consumer Intention to Use UPI-Based
Payment Apps in Indian Perspective** 445
Piyush Kumar Mallik and Deepak Gupta

**A Proposed SDN-Based Cloud Setup in the Virtualized Environment
to Enhance Security** 453
H. M. Anitha and P. Jayarekha

**A Light-Weight Cyber Security Implementation for Industrial
SCADA Systems in the Industries 4.0** 463
B. R. Yogeshwar, M. Sethumadhavan, Seshadhri Srinivasan,
and P. P. Amritha

**Conventional Biometrics and Hidden Biometric:
A Comparative Study** 473
Shaleen Bhatnagar and Nidhi Mishra

Performance Analysis of Various Trained CNN Models on Gujarati Script 483
Parantap Vakharwala, Riya Chhabda, Vaidehi Painter, Urvashi Pawar, and Sarosh Dastoor

Real-Time Human Intrusion Detection for Home Surveillance Based on IOT 493
Mohith Sai Subhash Gaddipati, S. Krishnaja, Akhila Gopan, Ashiema G. A. Thayyil, Amrutha S. Devan, and Aswathy Nair

Token Money: A Study on Purchase and Spending Propensities in E-Commerce and Mobile Games 507
N. P. Sreekanth and Deepak Gupta

Data Augmentation for Handwritten Character Recognition of MODI Script Using Deep Learning Method 515
Solley Joseph and Jossy George

Improved Automatic Speaker Verification System Using Deep Learning 523
Saumya Borwankar, Shrey Bhatnagar, Yash Jha, Shraddha Pandey, and Khushi Jain

Detection and Prevention of Attacks on Active Directory Using SIEM 533
S. Muthuraj, M. Sethumadhavan, P. P. Amritha, and R. Santhya

Loki: A Lightweight LWE Method with Rogue Bits for Quantum Security in IoT Devices 543
Rahul Singh, Mohammed Mohsin Hussain, Milind Sahay, S. Indu, Ajay Kaushik, and Alok Kumar Singh

Implication of Web-Based Open-Source Application in Assessing Design Practice Reliability: In Case of ERA's Road Projects 555
Bilal Kedir Mohammed, Sudhir Kumar Mahapatra, and Avinash M. Potdar

Influencer Versus Peer: The Effect of Product Involvement on Credibility of Endorsers 565
S. Rajaraman, Deepak Gupta, and Jeeva Bharati

The Influence of Fan Behavior on the Purchase Intention of Authentic Sports Team Merchandise 573
Anand Vardhan, N. Arjun, Shobhana Palat Madhavan, and Deepak Gupta

Purchase Decisions of Brand Followers on Instagram 581
R. Dhanush Shri Vardhan, Shobhana Palat Madhavan, and Deepak Gupta

Deep Learning Based Parking Prediction Using LSTM Approach 589
Aniket Mishra and Sachin Deshpande

Beyond Kirana Stores: A Study on Consumer Purchase Intention for Buying Grocery Online 599
R. Sowmyanarayanan, Gowtam Krishnaa, and Deepak Gupta

Car Damage Recognition Using the Expectation Maximization Algorithm and Mask R-CNN 607
Aseem Patil

Deep Learning Methods for Animal Recognition and Tracking to Detect Intrusions 617
Ashwini V. Sayagavi, T. S. B. Sudarshan, and Prashanth C. Ravoor

VR Based Underwater Museum of Andaman and Nicobar Islands 627
T. Manju, Allen Francis, Nikil Sankar, Tharick Azzarudin, and B. Magesh

Network Performance Evaluation in Software-Defined Networking 633
Shilpa M. Satre, Nitin S. Patil, Shubham V. Khot, and Ashish A. Saroj

Performance Improvement in Web Mining Using Classification to Investigate Web Log Access Patterns 647
Charul Nigam and Arvind Kumar Sharma

Array Password Authentication Using Inhibitor Arc Generating Array Language and Colored Petri Net Generating Square Arrays of Side 2^k 659
S. Vaithyasubramanian, D. Lalitha, A. Christy, and M. I. Mary Metilda

sVana—The Sound of Silence 669
Nilesh Rijhwani, Pallavi Saindane, Janhvi Patil, Aishwarya Goythale, and Sartha Tambe

Coreveillance—Making Our World a "SAFER" Place 681
C. S. Lifna, Akash Narang, Dhiren Chotwani, Priyanka Lalchandani, and Chirag Raghani

Large-Scale Video Classification with Convolutional Neural Networks 689
Bh. SravyaPranati, D. Suma, Ch. ManjuLatha, and Sudhakar Putheti

Saathi—A Smart IoT-Based Pill Reminder for IVF Patients 697
Pratiksha Wadibhasme, Anjali Amin, Pragya Choudhary, and Pallavi Saindane

Predictive Analysis of Alzheimer's Disease Based on Wrapper Approach Using SVM and KNN 707
Bali Devi, Sumit Srivastava, and Vivek Kumar Verma

A Novel Method for Enabling Wireless Communication Technology in Smart Cities 717
Vijay A. Kanade

Electronic Aid Design of Fruits Image Classification for Visually Impaired People 727
V. Srividhya, K. Sujatha, M. Aruna, and D. Sangeetha

Temperature Regulation Based on Occupancy 735
Rajesh Kr. Yadav, Shanya Verma, and Prishita Singh

Score Prediction Model for Sentiment Classification Using Machine Learning Algorithms 745
Priti Sharma and Arvind Kumar Sharma

CFD Analysis Applied to Hydrodynamic Journal Bearing 755
Mihir H. Amin, Monil M. Bhamare, Ayush V. Patel, Darsh P. Pandya, Rutvik M. Bhavsar, and Snehal N. Patel

Author Index 777

About the Editors

Dr. Tomonobu Senjyu received his B.S. and M.S. degrees in Electrical Engineering from the University of the Ryukyus in 1986 and 1988, respectively, and his Ph.D. degree in Electrical Engineering from Nagoya University in 1994. Since 1988, he has been with the Department of Electrical and Electronics Engineering, University of the Ryukyus, where he is currently a Professor. His research interests include stability of AC machines, power system optimization and operation, advanced control of electrical machines and power electronics. He is a member of the Institute of Electrical Engineers of Japan and IEEE.

Dr. Parikshit N. Mahalle holds a B.E. degree in CSE and an M.E. degree in Computer Engineering. He completed his Ph.D. at Aalborg University, Denmark. Currently, he is working as a Professor and Head of the Department of Computer Engineering at STES Smt. Kashibai Navale College of Engineering, Pune, India. He has over 18 years of teaching and research experience. Dr. Mahalle has published over 140 research articles and eight books, and has edited three books. He received the "Best Faculty Award" from STES and Cognizant Technologies Solutions.

Dr. Thinagaran Perumal received his B.Eng., M.Sc. and Ph.D. Smart Technologies and Robotics from Universiti Putra Malaysia in 2003, 2006 and 2011, respectively. Currently, he is an Associate Professor at Universiti Putra Malaysia. He is also Chairman of the TC16 IoT and Application WG National Standard Committee and Chair of IEEE Consumer Electronics Society Malaysia Chapter. Dr. Thinagaran Perumal is the recipient of 2014 IEEE Early Career Award from IEEE Consumer Electronics Society. His recent research activities include proactive architecture for IoT systems; development of the cognitive IoT frameworks for smart homes; and wearable devices for rehabilitation purposes.

Dr. Amit Joshi is currently the Director of the Global Knowledge Research Foundation. An entrepreneur & researcher, he holds B.Tech., M.Tech. and Ph.D. degrees. His current research focuses on cloud computing and cryptography. He is an active member of ACM, IEEE, CSI, AMIE, IACSIT, Singapore, IDES, ACEEE, NPA and several other professional societies. He is also the International Chair of InterYIT at the International Federation of Information Processing (IFIP, Austria). He has published more than 50 research papers, edited 40 books and organized over 40 national and international conferences and workshops through ACM, Springer and IEEE across various countries including India, Thailand, Egypt and Europe.

Internet of Things: Challenges, Security Issues and Solutions

Sweta Singh and Rakesh Kumar

Abstract Today Internet has become a part of our daily life, enabling not only people and the computing component to connect and communicate with each other but also has extended its domain to the objects or the 'things' which are embedded with electronic components, software, communication technologies, sensors and actuators. The main aim of Internet of things (IoT) is to configure a "smart environment" with everything getting smarter providing advanced services, reducing human effort and making use of the technologies to its true potential. With IoT, the computation and communication is now not restricted to computing devices only but now everything around us will get connected, developing a new and advanced era of communication. But as more and more things are getting connected, has given way for more issues, challenges and more security threats. This paper aims at focusing on the key concept of IoT, issues, challenges involved in its deployment, basically emphasizing on the security challenges and also highlighting the proposed solutions by different researchers so far.

1 Introduction

With Internet becoming ubiquitous [1], providing services anytime, anywhere and affecting everyday life of human is now seeking for a technological forward shift where the computation and communication capability is not only confined to computing devices but to the everyday objects too. With this objective of "connecting the unconnected world" has brought up the concept of "Internet of things (IoT)." IoT is defined to be the next era of communication which basically focuses

S. Singh (✉) · R. Kumar
Department of Computer Science Engineering, Madan Mohan Malaviya University of Technology, Gorakhpur, India
e-mail: swetasss22691@gmail.com

R. Kumar
e-mail: rkiitr@gmail.com

© The Editor(s) (if applicable) and The Author(s), under exclusive license to Springer Nature Singapore Pte Ltd. 2021
T. Senjyu et al. (eds.), *Information and Communication Technology for Intelligent Systems*, Smart Innovation, Systems and Technologies 196, https://doi.org/10.1007/978-981-15-7062-9_1

at "how devices or things can be made smarter?" and "how the smart things can be connected through an intelligent network?" Different researcher has a different way of defining IoT. Vermesan et al. [2] define IoT as an interaction between the physical and the digital world. Sethi and Sarangi [3] have defined it as an agglomeration of different technologies working together aiming at a meaningful objective. Authors of [4] define IoT as a fusion of technology which aim at bringing everything under a common infrastructure with an objective to control thing and have knowledge of its state too. Peña-López et al. [5] define it as a paradigm where networking and computation capability is embedded in any type of conceivable thing or object. Chen et al. [6] define it as a smart network which connects everything to the Internet. Madakam et al. [7] define IoT as a comprehensive network of intelligent or smart devices which are capable of sharing data, information, resources have capability to autoconfigure itself and take intelligent action in case of any change in the environment or on encountering certain event. Considering the several definitions for IoT, Internet of things can be summarized as a paradigm which involves convergence of multiple technologies including embedded systems, machine learning, networking, cloud computing, artificial intelligence, big data, etc. Thus, proving a platform for objects embedded with sensors, actuators, electronics, software, computation and communication technologies [8] to connect to each other and communicate with each other to meet certain objective or to develop certain meaningful applications. According to Zhao and Ge [9] by 2015–2020, the devices are considered to be semi-intelligent, whereas targeted to achieve Comprehensively Intelligent Object after 2020. The target is being achieved where around 50 billion of devices have been connected and soon will come an era where the number of connected things will surpass the number of people and will develop a smart and automated world. IoT aims at easing human effort, has control over everything by automating things, configuring a smart environment where system provides advanced services, making use of existing technologies to its true potential, connects everything around us by extending the connectivity, computation and communication to things such as fan, light bulb, microwave, cars, coffee machine, street lights, garbage bins, etc. With various advantages or applications, IoT also has certain drawbacks where it faces certain challenges or issues [6] including technical challenges, architectural challenges, managerial challenges, data management challenges and security challenges. Here, in this paper, aim at focusing the security challenges of IoT which has given way to attackers to exploit the system and hindering its acceptance or advancement. The paper also discusses certain attacks which could be made due to the integration of physical and the cyberdomain [10], the solutions which have been presented by several researchers to overcome the security issues, with several considerations to be kept in mind while proposing a new solution. Generally, the type of application defines the security issue and the complexity in defining the appropriate solutions [9]. Thus, to ensure a more efficient secure solution, one needs to uncover the vulnerability which could be exploited by the attacker. Whatever the solution may be, it must ensure confidentiality, availability, integrity, data privacy and trust, authentication and identity [11].

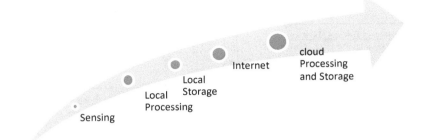

Fig. 1 Working of IoT

IoT is a paradigm which aims at embedding computation and communication capability in the conceivable objects so that the "things" can communicate with each other and exchange data and information. It aims at expanding the interdependence of humans to interact, collaborate and contribute to things. With IoT, the physical world can be transformed to a smart digital world where everything will be connected and can communicate with each other forming a smart environment. Here, the intelligent things are capable of auto-organization, sensing environment, collecting data from the external world, and react to certain changes in the environment. Generally, the devices are constrained devices with limited memory, limited power, low computation power and size. This nature of devices has made the traditional security measure to be unacceptable. Sensors and actuators are the key elements of the IoT which gathers data from outside environment and transferred for further processing and decision making according to which actuators respond or take required action. IoT is basically characterized by four essential features [9, 12], i.e., comprehensive perception or sensing which means detailed information or data can be collected anytime and anywhere from the environment. Reliable transmission which means the data collected by the smart thing has to be transmitted either via wired or wireless medium in real time for further processing. Intelligent processing where the transmitted data is processed and analyzed and last is appropriate action where according to the decision made by the processing system is brought to action by the actuators. The working of IoT is depicted in Fig. 1.

2 Issues and Challenges in IoT

Every technology is developed to reduce human effort, improve performance, maximize and make better use of resources, increase efficiency and achieve an advanced level of service. In simple words, it can be simply stated as the advantage of technology is undisputable with its objective unquestionable but with advantages brings up few concerns or certain challenge. Internet of things (IoT) has its advancement

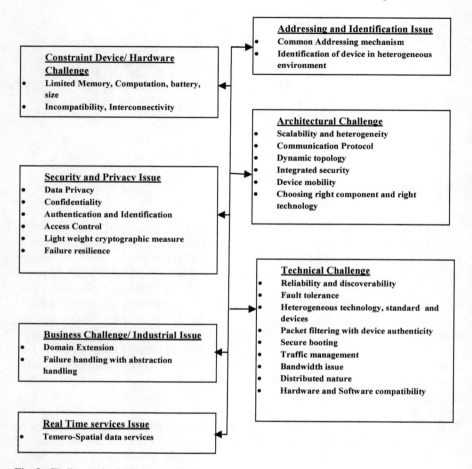

Fig. 2 Challenges in deployment of IoT

in certain applications with advanced services but it also encounters certain issues which serve as a hindrance in meeting out its goal. Few issues in deployment of IoT are depicted in Fig. 2.

2.1 Security Threats Sources in IoT

(1) *Security issues and Attacks at the Perception Layer* [9, 11, 13, 14]: Perception or the sensing layer in the IoT comprises of sensor nodes or the smart devices with local connectivity and the responders or actuators [15]. The sensors could be mechanical sensor, electrical, chemical or any other which aims at accumulating the data from the physical environment. The sensing layer technology includes

RFID, RSN's, WSN etc., to collect the data from external environment and object identification. Since data serves to be the core of IoT, any threat or intended target to these data could lead to severe or unintended consequences. Few threats are described as under.

(a) *Physical tampering or hardware corruption*: The sensors nodes are generally placed as well operate at some external environment or remote areas. Any physical access or intended tampering of the physical device may hinder the real-time capturing of data or even gather incorrect data by making use of smart malicious technical gadgets and connecting it to the device. Hardware failure or corruption of theses smart devices can also hinder the entire system functioning or result in system failure, i.e., no data collection--no processing--no action.

(b) *Node hijacking/capturing*: An IoT application does not comprises a single node but a number of constrained nodes with sensors which could be the prime target of the adversaries. The malicious attacker may attempt to capture the node by taking control over the node or replace part of the device with its own such that it could now act as a part of the system comprising the overall security of the application.

(c) *Frequency interference*: RFID technology is used in this layer to identify object and to collect the data from the device. Any interference in the radio waves may lead to data loss or corrupted data. Thus, ensuring a lightweighted RFID authentication security measure [16, 17] is attempted in ensuring data privacy.

(d) *Sleep deprivation attack*: The devices in perception layer are power constrained. Hence, they are programmed to undergo sleep mode in case of inactiveness. The attacker can target this behavior of the node by sending unnecessary data or running infinite loops using malicious codes which aim to drain the battery of the device. The drainage of battery will make the particular node/device inactive and will result in stopping or denial of service from that particular node.

(e) *Trust gaining attack using node authentication*: Another mechanism which could be opted by the attacker is to win the trustworthiness of the system by claiming to be an authorized node and become part of the IoT system. It is often termed as replay attack. The attacker by claiming to be a new device attempts to be a part of system and attain the authenticity certificate or can capture a packet from the node and send response claiming its genuine identity but with a mollifying intention. In simple words, "first gain trust and then attack."

(f) *Side channel attack*: Side channel attack is a technique used by the attacker to capture the data. In this electromagnetic waves leaking from the sides of the device are analyzed to attain the data. This attack aims at compromising the physical system. The attacker continuously monitors the behavior of the victim node such as the electromagnetic leaks, power consumption, sound monitoring, etc. to gain information which is exploited to compromise the victim system. Cache attack, timing attack electromagnetic attack, power attacks are few examples of side channel attack. Here, time analysis attack is most common where the attacker compromises the victim by analyzing the computation time of the node or analysis of encryption time of the node. Based on this analysis, information is collected and is exploited to compromise the targeted device.

(g) *False data injection attack/malicious coding attack*: An injection of erroneous data or wrong code or fake data may result in uncorrected data processing with undertaking of inappropriate action. This attack targets the overall characteristics of IoT resulting in malfunctioning of the system or false system outcome. In certain application, it could lead to severe consequences.

(h) *Data leakage*: Sensors are connected to local network to communicate with other devices or to push the gathered data to the processing nodes. During this transmission, the real-time collected data is prone to leakage, data copying or confidentiality leakage.

(i) *Booting attack*: IoT framework consists of number of smart devices connected together, working cooperatively accomplishing a specific task. Since size of IoT network varies from few devices to thousand of devices with each device produced by different vendor. When these devices are assembled together to form an IoT system, it leaves a small loophole, i.e., security at the time of booting of these very devices while switching from non-operational mode to operational mode. This is the point where the inbuilt security mechanism lacks. Taking advantage of this the attacker can attempt to target the nodes when they are getting started or restated.

(2) *Security issues and attacks at the network layer* [9, 11, 13, 14, 18]

(a) *Diverse target communication/routing-based attack*: IoT is characterized by scalability, i.e., IoT operates in a geometrically large environment comprising of large number of nodes/devices. IoT involves communication between the devices and forward transmission of data to the processing nodes or the cloud and reverse transmission of the instruction based on the decision made from the cloud to the actuator. This involves routing of data from nodes to cloud and cloud to the devices. Malicious attacker may target this and may attempt redirection of packet to itself or an unauthorized device or aim at making the packets to get trapped in infinite loop. By routing the packet to an unintended location may result in loss of confidentiality and data privacy as data could be either forged or be copied/read by an unauthorized entity. It may also lead to power consumption of the nodes leading to denial-of-service attack. Wormhole attack, Blackhole attack and Sinkhole attack are examples of this type of attack.

(b) *Traditional network attacks*: IoT being interconnection of devices with heterogeneous network and communication protocols, it gives an opportunity to the attacker to stay as a part of the system for relatively large time period launching several traditional network attacks with an aim of traffic diversion, data loss, data forging, data capturing and copying, congestion, resource exhaustion and many more. Few common network attacks include man-in-the-middle attack, denial-of-service attack, snooping attack, sniffing attack, etc.

(c) *Data-in-transit attack*: It is considered that the data in transit is more vulnerable to attack than the data stored at a site/location. Since IoT involves transmission of large number of valuable information exchange between number of devices, to the processing nodes, to gateway and cloud gives a handful opportunity for the adversary to target the data. The data movement in IoT incurs number of communication technologies which are targeted by the attackers for data breaching.

(d) *Network congestion/data traffic* [3]: Data being valuable is not always attacked for access or get monitored by the attacker but the malicious attacker may target the data to get lost in the network resulting in system failure or false/unexpected results. The attacker may get into the system and attempt to send bulk of meaningless or unimportant and unnecessary data in the network such that the entire bandwidth of channel gets occupied and the data from the authorized devices gets dropped due to network congestion. Another possibility due to network congestion is the shutdown or denial of service. When attacker sends enormous number of fake or invalid requests in the network and in general the IoT comprise of lossy network with limited bandwidth, will occupy the entire bandwidth or the communication channel denying the genuine packets to move through the network resulting in denial-of-service attack.

(e) *Identity masking*: Spoofing or identity masking is another common attack where attacker claims to an authorized device by masking itself as an authorized one thus gaining access to the valuable data. Here, instead of data reaching to legitimate devices reaches the attacker who claims to be the authorized device.

(f) *Sybil attack*: An attack where a device/node claims or holds multiple identities thus misleading the data traffic. The attacker can claim to be an authorized/legitimate device by faking itself thus causing the data traffic intended for the authorized device, to reach the attacker. Another possibility due to Sybil attack is, the attacker holding multiple identity being a part of the route through which the data flows to reach the destination can read or capture the data packets while passing through it. Sybil attack also leads to routing table overflows (routing table is filled with invalid entries or entry of the attacker holding multiple identities such that there is no memory space left in the memory restricted nodes for valid entry) and routing table poisoning (resulting in unnecessary route confusion).

(3) *Security issues and attacks at the application layer* [9, 11, 13, 14]

(a) *Data privacy threat*: For IoT application, data plays a significant role. Data in motion is more vulnerable and is highly targeted by the attacker than the data at rest. Targeting these data and launching attack may lead to unintended disclosure or fabrication of data by the attacker compromising the IoT system.

(b) *Phishing attack*: An attack targeting multiple devices with minimum effort. Since all the devices are connected and controlled at a user interface. Any disclosure of the user account Id and password (authenticating credentials) can make the entire IoT system comprising of thousands of devices to get targeted.

(c) *Unauthorized access*: Obtaining the authenticating credentials of the use, the attacker can gain uncontrolled and unauthorized access to data and the user account. Once gaining access the attacker may be able to corrupt or make the system vulnerable to multiple threats or attacks.

(d) *Software vulnerability*: Software attack may attempt at exploiting the system by sending trozans, worms, viruses, malicious scripts, spywares, etc. to attain the confidential credentials and data. They are basically sent using emails or in software updates. Once entering the system exploiting the entire system.

(e) *Data loss*: Due to massive collection of data attained from number of sensors deployed over a large geographical area calls for requirement of data analytics and management. Data needs to be stored or processed at a central site which may be targeted by the attacker. Data storage and management consumes a reasonable system resource which could be a reason for data loss.

(4) *Security issue at the cloud and gateway* [13]

(a) *Malicious code or malware injection attack*: *The* cloud serves as the backend supporting technology in IoT, which manages and processes the data and derive inference on basis of which actuator responds to the environment. In centralized IoT, the attacker may target this central server by sending malicious code or injecting malware using cross site scripting which in turn can paralyze the entire IoT system by malfunctioning or inappropriate service provision by cloud. Once the cloud server is targeted may lead to incorrect data analytics, processing, inappropriate service provision and action.

(b) *Cloud service interruption*: Cloud server provides several types of services such as proving platform, analytics tools, software tools, data storage and access mechanism and different resources according to requirement. It plays a major role in scaling up and down of resources with load balancing, serving as the backbone or support system of IoT. Paralyzing the support can hinder the functioning of IoT and result in failure of IoT application. Cloud also plays an essential role in providing real-time data support.

(c) *Flooding at cloud attack*: With malicious intention of flooding the cloud, the attacker may intend to send a bulk of unnecessary service request over the cloud, which may result in providing the resources or services to number of unauthorized and fake devices, leading to resource exhaustion and blocking the services to the authorized devices. This type of attempt

involves a massive number of devices which are targeted as victims forming a "botnet."
(d) *On-board attack*: This type of attack generally occurs when attacker install itself as a new device and attempt to join the IoT network. Scalability allows any number of devices to be a part of IoT system. This attack targets the gateway as in communication the gateway serving in between the network and the devices is responsible for key distribution and sharing of key. Thus, the attacker may target the gateway to capture the key and get access to the system.
(e) *Encryption-Decryption attack*: End-to-end encryption is done to ensure privacy of data such that only the designated authorized entity can access the valuable data. But with gateway sitting in between, the encrypted data is first decrypted and then re-encrypted to be further relayed to the cloud. This decryption of data at the gateway can make the data susceptible to breach.
(f) *Update attack*: For advanced application, there emerges the need to regularly update the firmware and software. Being resource constrained (low computation and storage), the smart devices are incapable of downloading and installing the updates. Thus, this downloading of firmware updates is made at the gateway which is then applied on the devices. By inserting malicious script or viruses to these updates, the attacker can breach the IoT security.

Table 1 provides a summarized picture of issues with the possible solution.

3 Existing Security Solutions

(a) Blockchain Technology [10, 13, 20]: It is a new technology contributing to a secure and transparent IoT system distributed across a network ensuring confidentiality, integrity and privacy. It serves a backbone technology to secure digital transactions termed digital cryptocurrency. It is a chain of block termed as ledger which includes data, timestamp, hash of current block and the previous block. This hash enables the blocks to be chained together. Here, the hash can simply be considered as a unique identifier similar to fingerprint. For every new block has a new hash. Thus, if anyone tries to forge, the data creates a new block/hash or need to change the entire chain hash value which is just impossible. The security and functioning of blockchain is basically dependent on 'proof of work' and 'distributed hash keys.' The data structure used to hold records of transaction is the Merkle tree. The first block in the ledger is the genesis block. The entire blockchain forms a peer-to-peer network architecture such that the entire ledger is distributed to each node/miner which identifies the block. The distributed concept makes it more secure. The concept of blockchain was initially introduced for secure digital transaction removing the role of third party (bank) but

Table 1 Security issues with the possible solutions

Security issues/challenges	Possible solutions
Physical tampering/hardware corruption	Tamper proof shielding
Node capturing	Access control mechanism
Radio frequency interference	Anti-collision protocol such as binary tree protocol and collision tree protocol [19]
RFID security threat	RFID security using access control, RFID tag chip protection [16, 19]
Sleep deprivation attack	Authentication of nodes such that only authorized node can send data packets or communicate with the nodes
Side channel attack	Channel tracing/complementary duals
False data injection	Access control
Replay attack	Kerberos authentication protocol
Routing attack	Secure routing mechanism/RPL [14]
False data packet	Authentication/intrusion detection system
Communication threat/MITM	Key agreement/node authentication, lightweight cryptographic solution
Unauthorized access	Authentication/encryption/access control
Network congestion	Queuing/priority messaging
Booting attack	Secure booting/bootstrapping device authentication and identity management
Software vulnerability	Firewall/anti-spyware/anti-virus
Data-in-transit attack	End-to-end encryption
Denial-of-service attack	Device authentication
Sybil attack	Key agreement/mutual authentication
Unauthorized access	Access control
Data loss/theft	End-to-end encryption
Gateway encryption-decryption attack	Protocol supporting end-to-end encryption
On-board attack	Encrypted key
Update attack	Signature validation
Architectural issue	Black SDN
Interoperability and heterogenity issue	Middleware solution [13]

now has been introduced in several applicable areas such as automatic cars, in IoT for a secure transparent infrastructure. The key benefits of IoT are data privacy, secure distributed data storage, encrypted data and access control.

(b) Software defined network (SDN) [6, 10, 21]: Major contribution of a software defined network is defined in developing a secure framework for smart cities. It comprises of four basic functional unit named black network [22] that defines a network that provides a secure transmission or carriage of data or metadata associated with each data frame in the IoT network. The main contribution of it is to encrypt the payload and the data held in IoT link layer which is later decrypted by the receipt. Black network contributes to integrity, confidentiality and privacy of data in IoT network, hence contributing to network security. The second component is the SDN controller that manages the communication taking place between the IoT nodes and the entire IoT framework components. It also contributes to routing resolution in the black network by creating random or dynamic routes through the intermediate awaken nodes. The remaining components are the Unified Registry and Key Management component. The Registry contributes in provision of database (holding records on nodes and their attributes), heterogeneous technology, translation and addressing schemes. Key management component ensures a secure communication via exchanging and distributing key between the IoT devices in the entire network.

(c) End-to-end encryption [10, 12, 14, 23]: For secure data transmission and communication between the IoT devices, is it required for the data packets to be encrypted such that only the authorized entity can access the data ensuring data integrity. Traditional encryption mechanism could not be applicable to IoT due to the constrained nature of device. Here, several lightweight and cost-effective cryptographic solutions have been proposed by the researchers [10].

(d) RFID security [16, 17, 24]: In RFID, the data is transmitted in form of radio frequency signals which is interrupted or prone to security breach attack. To protect the data, the RF signals must be encrypted. Several algorithms are defined to ensure lightweight encryption of these signals to prevent any data theft. It is also required to use protective chips, etc. to prevent the data to be read from the RFID tags.

(e) Secure/lightweight booting: A secure bootstrapping measure is defined by [25] to ensure authentication of low powered and constrained device before they operate, i.e., before they switch from non-operational to operational mode. They contribute to physical layer security where the device gets a secret key to reach the server.

(f) Access control: [21, 23, 26, 27]: Access control ensures the integrity, availability and confidentiality of data. Several research works have been proposed defining various access control mechanism to address and resolve the security issues. Some model and mechanism of IoT security are addressed in [21]. Access control for data stream is provided in [not good]. Several access control mechanisms include, role-based access control, ID-based access control, attribute-based, trust-based and several other schemes.

(g) Authentication [10, 14, 23, 27–30]: This mechanism contributes to 60% of IoT security by identifying the devices and the user in the IoT network such that only authorized personal or devices which are not under the control of an unauthorized entity (non-manipulated device) could only access the system or data. It is successful countermeasures for several attacks such as MITM attack, Sybil attack, relay attack, etc. Key establishment and key distribution are the primary tasks in user/device authentication. For device authentication, a mutual authentication scheme is provided where a key is shared between the nodes and both agree and share the key information forehand. A symmetric, asymmetric and hybrid crypto-based authentication scheme has been defined by the authors to ensure authentication between the users/devices. Researchers are motivated in defining lightweight and multi-factor authentication scheme for IoT security. Device identification and authentication in general combine together such that only recognized/authenticated device contributes to the IT heterogeneous system. Authors of [9] have produced a mutual authentication scheme based on hashing and feature extraction. Another work has been contributed by [23] for ID based authentication. By use of shared time-stamped keys that change regularly.

(h) Secure routing [10, 14]: To ensure secure communication of data and ensure availability of data to the authorized user or device, a secure routing protocol must be defined. The Internet Engineering Task Force (IETF) working group for routing over low powered lossy networks has standardized a routing protocol which is based on IPv6 especially designed for constrained devices, RPL [31]. The key consideration while proposing a secure routing solution is the constrained nature of nodes as it has limited memory while cannot store a large routing table or handle complex route calculations and second is the lossy network in IoT system. Several researchers have contributed to target certain attacks and have proposed am effective solution to mitigate and defend such types of attack.

(i) Trust management [10, 14, 21]: Trust management contributes to nearly 20% of IoT security which aims at identifying and removal of fake/malicious nodes. It incurs trust entity to ensure that all security parameters are ensured and a secure communication is established. It involves an automated o dynamic calculation of trust value of each node in the network to detect the malicious node. Another approach is using token or key [14]. Trust management plays a significant role in establishment of trust between the devices/users and during the interaction between them. It involves an agreement between the IoT systems from secure communication [23]. A concept of mutual trust in [4] for inter-model security is defined by creating an access control mechanism.

(j) Federated architecture: Since IoT incurs no well-defined fixed architecture with universal policy and standard from controlling the system and implementation has made it difficult to define a secure IoT framework or solution. At attempt of a federated architecture is made by [14] which is based on access control ensuring flexibility and scalability of IoT system. Another attempt is proposed by [14] proposing concept of Secure Mediation Gateway focusing on distributed architecture.

(k) Designing of attack resistant systems [9]: Designing of tamper proof system could be a solution from physical layer attacks, where the attackers target the nodes placed in the external world to attain the information, or tamper the collected data. It involves a proper node and antenna designing such as hardware design, selecting appropriate security chip, data acquisition unit, circuit design as well antenna design for communication distance and stability. It also should ensure the disposal of garbage data collected such that it does not create a mess or overhead on the cloud.

(l) Auto-immunity through machine learning: Another contribution which can be undertaken in field of IoT security is by introducing several machine learning algorithm, such that the nodes itself draw up pattern and analyze the data before transmission [13]. This feature will further enhance security feature by ensure secure data transmission and undertaking relevant action. Artificial neural network and machine learning may contribute to adaptive learning, fault tolerance, prevent privacy leakage, and a defense against spoofing, eaves dropping and other network attacks. Table 2 aims at highlighting the key areas to which the security solutions are contributing.

Table 2 Security solutions proposed with their area under consideration

Security solutions	Concerned area
Blockchain mechanism	Data security ensuring privacy, data integrity by encrypted data, transparency and confidentiality. Data protection from spoofing attack
Fog computing [32]	Decentralization, data distribution with secure data sharing, real-time service with reduce latency and data security
Authentication and identity management [33]	Device authentication. Prevent unauthorized access
Fuzzy based/lightweight encryption [17, 34, 35]	Secure communication or transmission of data between devices and between different layers
Trust-based solution	Calculate trust value to hold trustworthiness between the devices in layer, between the layers and between the IoT system and the user
Federated architecture	Interoperability and secure information sharing in semi-autonomous and distributed system
Black SDN/black network [22]	Data security, data privacy, integrity and secure data transmission
Edge computing [36]	Prevent data breaches as data is processed at the node itself, data security, solve bandwidth issue, immediate decision making leading to quick service
Firewall/IDS/software solution	Application layer/interface security
RPL	Secure routing
Key agreement	Secure communication between authorized device

4 Conclusion and Future Scope

While working on IoT security and meeting the various challenges incurred in IoT, the researchers need to consider few parameters which should be kept in mind while proposing a solution or working on the security measures. It can be categorized into primary considerations which are the essential or mandatory parameters and must to be considered. Since IoT includes, constraint devices, it is necessary that whatever security measure is chosen or proposed must be lightweight, and compatible with the resource constraint devices consuming low power, low memory and little or no complex computation. The second is the low power lossy network, i.e., the communication network or channel in IoT is generally lossy, thus the security measure must consider the communication channel or transmission network such that there is no or acceptable loss of data. The second category is the secondary factors extending additional coverage which include device authentication, data encryption, lightweight solutions, packet filtering, packet routing in the lossy network, and software update with no additional bandwidth requirement with a reliable and cost-effective solution.

The survey has contributed in highlighting the security aspects, i.e., the possible threats in IoT system. Since the success of any technology and its acceptance basically depends on how quickly the problem is identified and how effectively is resolved. Several works have been proposed by the researchers and still contribution is being made to ensure the adaption and extension of IoT from personal domain to an enterprise level. The major security areas which could be contributed by the researchers are in ensuring data privacy, network security, cloud security, gateway security, contributing a lightweight security solution which is compatible in the constraint, mobile and heterogeneous IoT environment. With new application brings up a new challenge; thus, it is essential to consider all the aspects of IoT and create a secure session of nodes communication. This survey may contribute in serving a valuable resource in further advancements and works of researchers in upcoming application areas and security challenges.

References

1. Roman, R., Najera, P., Lopez, J.: Securing the internet of things. Computer **44**(9), 51–58 (2011)
2. Vermesan, O., Friess, P., Guillemin, P., Gusmeroli, S., Sundmaeker, H., Bassi, A., Jubert, I.S., Mazura, M., Harrison, M., Eisenhauer, M., Doody, P.: Internet of things strategic research roadmap. Int. Things-Glob. Technol. Soc. Trends **1**, 9–52 (2011)
3. Sethi, P., Sarangi, S.R.: Internet of things: architectures, protocols, and applications. J. Electr. Comput. Eng. (2017)
4. Hossain, M.M., Fotouhi, M., Hasan, R.: Towards an analysis of security issues, challenges, and open problems in the internet of things. In 2015 IEEE World Congress on Services, pp. 21–28 (2015)
5. Peña-López, I.: ITU Internet Report 2005: The Internet of Things (2005)
6. Chen, S., Xu, H., Liu, D., Hu, B., Wang, H.: A vision of IoT: Applications, challenges, and opportunities with china perspective. IEEE Internet Things J. **1**(4), 349–359 (2014)

7. Madakam, S., Lake, V., Lake, V., Lake, V.: Internet of Things (IoT): a literature review. J. Comput. Commun. **3**(05), 164 (2015)
8. Goyal, K.K., Garg, A., Rastogi, A., Singhal, S.: A literature survey on Internet of Things (IoT). Int. J. Adv. Netw. Appl. **9**(6), 3663–3668 (2018)
9. Zhao, K., Ge, L.: A survey on the internet of things security. In: 2013 Ninth International Conference on Computational Intelligence and Security, pp. 663–667 (2013)
10. Hassan, W.H.: Current research on Internet of Things (IoT) security: a survey. Comput. Netw. **148**, 283–294 (2019)
11. Lin, J., Yu, W., Zhang, N., Yang, X., Zhang, H., Zhao, W.: A survey on internet of things: Architecture, enabling technologies, security and privacy, and applications. IEEE Internet Things J. **4**(5), 1125–1142 (2017)
12. Gou, Q., Yan, L., Liu, Y., Li, Y.: Construction and strategies in IoT security system. In 2013 IEEE International Conference on Green Computing and Communications and IEEE Internet of Things and IEEE Cyber, Physical and Social Computing, pp. 1129–1132 (2013)
13. Hassija, V., Chamola, V., Saxena, V., Jain, D., Goyal, P., Sikdar, B.: A survey on IoT security: application areas, security threats, and solution architectures. IEEE Access **7**, 82721–82743 (2019)
14. Andrea, I., Chrysostomou, C., Hadjichristofi, G.: Internet of Things: Security vulnerabilities and challenges. In: 2015 IEEE Symposium on Computers and Communication (ISCC), pp. 180–187 (2015)
15. Kelly, S.D.T., Suryadevara, N.K., Mukhopadhyay, S.C.: Towards the implementation of IoT for environmental condition monitoring in homes. IEEE Sens. J. **13**(10), 3846–3853 (2013)
16. Phan, R.C.W.: Cryptanalysis of a new ultralightweight RFID authentication protocol—SASI. IEEE Trans. Dependable Secure Comput. **6**(4), 316–320 (2008)
17. Amendola, S., Lodato, R., Manzari, S., Occhiuzzi, C., Marrocco, G.: RFID technology for IoT-based personal healthcare in smart spaces. IEEE Internet Things J. **1**(2), 144–152 (2014)
18. Sharma, R., Pandey, N., Khatri, S. K.: Analysis of IoT security at network layer. In: 2017 6th International Conference on Reliability, Infocom Technologies and Optimization (Trends and Future Directions) (ICRITO), pp. 585–590 (2017)
19. Jia, X., Feng, Q., Ma, C.: An efficient anti-collision protocol for RFID tag identification. IEEE Commun. Lett. **14**(11), 1014–1016 (2010)
20. Khan, M.A., Salah, K.: IoT security: Review, blockchain solutions, and open challenges. Futur. Gener. Comput. Syst. **82**, 395–411 (2018)
21. Flauzac, O., González, C., Hachani, A., Nolot, F.: SDN based architecture for IoT and improvement of the security. In: 2015 IEEE 29th International Conference on Advanced Information Networking and Applications Workshops, pp. 688–693 (2015)
22. Chakrabarty, S., Engels, D.W.: A secure IoT architecture for Smart Cities. In: 2016 13th IEEE Annual Consumer Communications & Networking Conference (CCNC), pp. 812–813 (2016)
23. Abomhara, M., Køien, G.M.: Security and privacy in the Internet of Things: Current status and open issues. In: 2014 International Conference on Privacy and Security in Mobile Systems (PRISMS), pp. 1–8 (2014)
24. Fan, K., Gong, Y., Liang, C., Li, H., Yang, Y.: Lightweight and ultralightweight RFID mutual authentication protocol with cache in the reader for IoT in 5G. Secur. Commun. Netw. **9**(16), 3095–3104 (2016)
25. Hossain, M.M., Hasan, R.: Boot-IoT: a privacy-aware authentication scheme for secure bootstrapping of IoT Nodes. In ICIOT, pp. 1–8 (2017)
26. Ouaddah, A., Mousannif, H., Elkalam, A.A., Ouahman, A.A.: Access control in the Internet of Things: big challenges and new opportunities. Comput. Netw. **112**, 237–262 (2017)
27. Neto, A.L.M., Souza, A.L., Cunha, I., Nogueira, M., Nunes, I.O., Cotta, L., Gentille, N., Loureiro, A.A., Aranha, D.F., Patil, H.K., Oliveira, L. B.: AOT: Authentication and access control for the entire iot device life-cycle. In: Proceedings of the 14th ACM Conference on Embedded Network Sensor Systems CD-ROM, pp. 1–15 (2016)
28. Ye, N., Zhu, Y., Wang, R.C., Malekian, R., Lin, Q.M.: An efficient authentication and access control scheme for perception layer of internet of things (2014)

29. Hernández-Ramos, J.L., Jara, A.J., Marin, L., Skarmeta, A.F.: Distributed capability-based access control for the internet of things. J. Internet Serv. Inf. Secur. (JISIS) **3**(3/4), 1–16 (2013)
30. Crossman, M.A., Liu, H.: Study of authentication with IoT testbed. In: 2015 IEEE International Symposium on Technologies for Homeland Security (HST), pp. 1–7 (2015)
31. Al-Fuqaha, A., Guizani, M., Mohammadi, M., Aledhari, M., Ayyash, M.: Internet of things: a survey on enabling technologies, protocols, and applications. IEEE Commun. Surv. Tutorials 17(4):2347–2376
32. Chiang, M., Zhang, T.: Fog and IoT: an overview of research opportunities. IEEE Internet Things J. **3**(6), 854–864 (2016)
33. Mahalle, P.N., Anggorojati, B., Prasad, N.R., Prasad, R.: Identity establishment and capability based access control (IECAC) scheme for internet of things. In: The 15th International Symposium on Wireless Personal Multimedia Communications pp. 187–191 (2012)
34. Al Salami, S., Baek, J., Salah, K., Damiani, E.: Lightweight encryption for smart home. In: 2016 11th International Conference on Availability, Reliability and Security (ARES) pp. 382–388 (2016)
35. Mao, Y., Li, J., Chen, M.R., Liu, J., Xie, C., Zhan, Y:. Fully secure fuzzy identity-based encryption for secure IoT communications. Comput. Stand. Interf. **44**, 117–121 (2016)
36. Yu, W., Liang, F., He, X., Hatcher, W.G., Lu, C., Lin, J., Yang, X.: A survey on the edge computing for the Internet of Things. IEEE Access **6**, 6900–6919 (2017)

Modeling and Performance Evaluation of Auction Model in Cloud Computing

Gagandeep Kaur

Abstract Cloud computing addresses various issues in modern technology such as resource allocation, resources adaption, and resource provisioning. To provide better services to users in cloud computing, economic models have been used. This research work provides the definition of game theory and Pareto optimality. The proposed work in cloud computing (CC) is performed for users who are using cloud for the execution of their applications and uses game theory auction model to achieve outcomes and proposes metrics such as resource utilization and profit maximization. The proposed metrics are simulated using Cloudsim simulator for parallel workflow tasks and achieve Nash equilibrium point for non-cooperative users even when users have insufficient knowledge of the environment.

1 Introduction

The most complicated issue in CC is the problem of RA. This problem is modeled to help cloud customer in making a rational decision in competitive market. To design and model RA problem in CC, several consumer characteristics should be highlighted. For example: 1. Users always try to get better service at lesser price. 2. In organization practice, buyers of the cloud resources have more than one behavioral limitation, so they have to make a trade-off of one limitation for another. 3. Cloud uses 'pay-as-you-go' feature and bidding can be modified by each user which corresponds to previous behaviors of the participants. 4. Consumers are unknown to each other because they are distributed globally throughout the cloud. Therefore, Game Theory (GT) auctions are used to answer the problem of RA in clouds.

G. Kaur (✉)
Panjab University, Chandigarh, India
e-mail: gagan381@gmail.com

2 Motivation

There are several economic market models such as tendering model, auction model, posted price model, bid-based proportional resource sharing model, commodity model, and bargaining model. These models are applied in different scenarios. Commodity model balances the price by evaluating demand and supply standards from the market participants [1]. Grid RA market models such as posted price model, states that distributing resources are established on discussing the flat fee, usage interval or period, QoS among the provider and the consumer of grid resource [2]. Modern resource management systems such as grid computing [3], utility computing [4], data mining [5], agent systems [6] are using market mechanisms. These systems use various context such as pricing, control, delivery, and routing. Zong et al. [7] combine the cloud resources' characteristics and describe the combinatorial double auction of cloud service problem as a 0--1 integer programming problem, then match resources and price the cloud products after sorting the resource packages' average price. Also, it improves the utility of buyers and providers in the algorithm implementation, so as to get the CDATM model based on the average price of the cloud service combination. Zhang et al. [8] model the RA process of a mobile CC system as an auction mechanism. The evaluation of the individual rationality and incentive compatibility (truthfulness) properties of the customers is proposed in this method. Jafari [9] provides a wide range of survey and analysis of the auction-based RA method. This work classifies the important cloud RA mechanisms into different categories such as double-sided, one-sided, combinatorial, and other types of auction-based mechanisms.

3 Modeling of Proposed Problem

Definition 1 The game G is represented as a three-tuple vector $<P, S, U>$. P is the players in the allocation game such that $P = \{1, 2, 3, \ldots, m\}$. S are set of players' strategies. Strategy space for player k is $S_k = \{s_k^1, s_k^2, \ldots, s_k^n\}$. U is the utility function of game players. At the outcome of one game, the utility received by a single player is payoff which determines the players' preference.

Definition 2 (*Pareto optimality*). An allocation is Pareto optimal and if there is no other allocation in which some other individual is better off, and no individual is worse off.

The proposed problem in this research work is solved using game theory. Each player intends to select a strategy to maximize its own utility so that the objective of a resource allocation game would be considered as the following optimization problem:

$$\text{Maximize } u_i(a_i) \text{ subject to} \sum_{a_{ij} \in a_i} a_{ij} = k(i), a_{ij} \geq 0$$

where a_{ij} is assignment vector and is represented by selective strategy of player. $k(i)$ is the tasks of user U_i.

The constant elasticity of substitution (CES) function is

$$C = \left[\sum_{i=1}^{n} a_i^{\frac{1}{s}} c_i^{\frac{s-1}{s}}\right]^{\frac{s}{s-1}}$$

$$\ln C = \frac{\ln \sum_{i=1}^{n}\left(a_i^{1-r} c_i^r\right)}{r}$$

Apply Hoptial law

$$\lim_{r \to 0} \ln C = \frac{\sum_{i=1}^{n} a_i \ln c_i}{\sum_{i=1}^{n} a_i}$$

3.1 Allocation Model

To get rapid access for low cost and flexible resources, cloud users no longer need to make large upfront investments due to virtualization. The main aim of virtualization is to get rid of several user concerns such as server upkeep, constraints, and scalability. Different clouds provide different purchasing options for its customers. It is assumed that sum of bids is $\Theta_k = \sum_{i=1}^{N} b_k^i$.

Sum of bids by competitors is

$$\theta_k^{-i} = \sum_{j \neq i}^{N} b_k^j$$

Execution time of task i on resource k is

$$t_k^i = \frac{x_k^i}{C^k q_k^i} = \mu_k^i + \mu_k^i \frac{\theta_k^{-i}}{b_k^i}$$

Execution cost of task k is

$$e_k^i = b_k^i t_k^i = \mu_k^i \theta_k^{-i} + \mu_k^i b_k^i$$

Payoff is the amount of resource received and each player selects their best strategy. To achieve an equilibrium point, in bids offered by different players, following steps are taken:

$$\Theta_k^1 = \sum_n b_k^i$$

$$h_k^i\left(\Theta_k^{(1)}\right).$$

So $b_k^{i(m)} = h_k^i\left(\Theta_k^{(m)}\right)$

$$x_k^{i(m)} = \frac{b_k^{i(m)}}{\Theta_k^{(m+1)}}$$

For all satisfied players, the price flows: $\Theta_k^{(m+1)} = \Theta_k^m$.

$$b_k^{i*} = \max x\left(b_k^i, \theta_k^{-i*}\right)$$

where b_k^{i*} is equilibrium bid

$$\sum_N x_k^{i(m)} = 1.$$

Different resource prices are $\theta_{k1}^* \neq \theta_{k2}^*$
Different values for resource prices are $\sum_N x_k^i$
Equilibrium price θ_k^* that let $\sum_N x_k^i = 1$ is unique and Nash equilibrium exists.

4 Performance Evaluation

4.1 Metrics

To evaluate the proposed model, following metrics are used:

- Execution time = Total task length/(Allocated Vms * Basic Capacity)
- Payoff of the user = Execution time * bid
- Resource Utility: Resource utility is the measure of percentage of satisfied users. Satisfied users are the users whose task execution is completed within deadline.
- Average resource utility = ((successful users + 0.0)/num user) * 100
- Profit: Profit is the measure of total execution time at the server side for the users' tasks.

Profit = [1 − ((Total Execution Time−Actual Time)/Actual Time)] * 100

Table 1 Workloads used to evaluate the auction model

Workload name	No. of users	Jobs
NASA iPSC	69	18,239
LANL CM5	213	122,060
SDSC Par95	98	53,970
SDSC Par96	60	32,135
LLNL T3D (original)	153	22,779
SDSC SP2	437	59,715
DAS2 fs1 (original)	36	40,315
DAS2 fs3 (original)	64	12

4.2 Performance Evaluation of Proposed Model Using Different Workloads

Cloudsim simulator is used for the performance evaluation of proposed model and has been executed for different number of workloads. In this research, the proposed model has been evaluated using PWA workload (Parallel Workload Archive) [10]. There are totally 38 logs but the model has been evaluated using eight logs. The original logs come in different formats [10]. All logs are converted to the standard workload format (SWF). Archives are chosen to different number of users and jobs (Table 1).

The proposed model has been executed for different workloads. Performance has been evaluated for average resource utility and profit with normal distribution (Figs. 1 and 2).

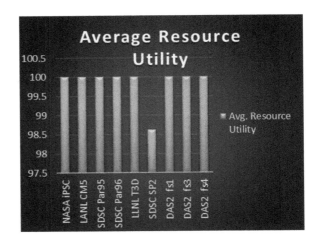

Fig. 1 Performance evaluation of proposed auction model for average resource utility with normal distribution

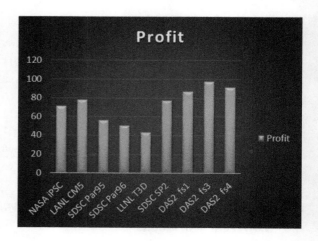

Fig. 2 Performance evaluation of proposed auction model for profit with normal distribution

5 Conclusion

Non-cooperative auction game is proposed to evaluate the performance of proposed model in cloud scenarios. The performance evaluation of proposed auction model gives 100% average resource utility for all parallel workload archive workloads except SDSC SP2. The performance is evaluated for profit for different parallel workload archive and workload DAS2 fs3 gives higher profit while workload LLNL T3D gives lower profit.

References

1. Stuer, G., Vanmechelen, K., Broeckhove, J.: A commodity market algorithm for pricing substitutable grid resources. Futur. Gener. Comput. Syst. **23**(5), 688–701 (2007)
2. Posted price model based on GRS and its optimization using in grid resource allocation. In: 2007 International Conference on Wireless Communications, Networking and Mobile Computing
3. Towards a micro-economic model for resource allocation in grid computing systems. 2002 ieeeexplore.ieee.org
4. Yolken, B., Bambos, N.: Game based capacity allocation for utility computing environments
5. Joita, L., Rana, O.F., Freitag, F., Chao, I., Chacin, P., Navarro, L., Ardaiz, O.: A catallactic market for data mining services. Futur. Gener. Comput. Syst. **23**(1), 146–153 (2007)
6. Shehorya, O., Kraus, S.: Methods for task allocation via agent coalition formation. Artif. Intell. **101** (1998)
7. Zong, X., Zhang, T., Wang, Y., Li, L.: A combinatorial double auction model for cloud computing based on the average price of cloud resource packages. Researchgate.net/publication/311928001
8. Zhang, Y., Niyato, D., Wang, P.: An auction mechanism for resource allocation in mobile cloud computing systems. In: International Conference on Wireless Algorithms, Systems, and Applications WASA 2013

9. Sheikholeslami, F., Nima Jafari, N: Auction-based resource allocation mechanisms in the cloud environments: a review of the literature and reflection on future challenges. 12 Mar 2018. https://doi.org/10.1002/cpe.4456
10. Parallel workloads archive from large scale parallel supercomputers. https://cs.huji.ac.il/lab/parallel/workloads

Framework for Resource Management in Cloud Computing

Gagandeep Kaur

Abstract Cloud computing has become popular in these days as it provides on-demand availability of computer resources such as processor, storage, and bandwidth throughout the world. Virtualization helps in the growth of data and computation centers and makes workload balancing much simpler and easier. Virtualization in modern days makes the things easier for cloud users as they no longer need to make large upfront investments and get rapid access to low-cost and flexible resources. Virtualization aims to reduce end-user apprehensions concerning server upkeep, constraints, and scalability. Cloud computing uses the concept of virtualization and uses IT assets as utilities in todays' world. Cloud computing uses cloud marketplace to access and integrate the services and offerings. Cloud users and cloud providers have different aims when managing the resources, policies, and demand patterns in real-world scenario. In this paper, a new three-tier architecture has been presented to manage the resources in economic cloud market considering deadline and execution time of the tasks for users.

1 Introduction

Cloud providers provide various benefits like scalability, flexibility, reducing IT costs, rapid elasticity, and on-demand self-service but despite these benefits, many users hesitate in moving their IT systems to the cloud. This is due to problems like loss of control, lack of trust, and multitenancy introduced by cloud environments. This paper proposes an architecture for resource management in cloud computing systems. The proposed model is based on three-tier architecture that help cloud service providers in (i) managing their infrastructure resources efficiently (ii) maximizing their revenue. The proposed model meets the demands of consumers and service providers while following various system policies and demand patterns, follows economic method,

---

G. Kaur (✉)
Panjab University, Chandigarh, India
e-mail: gagan381@gmail.com

and provides a fair basis in efficiently managing decentralization and heterogeneity that is present in human economics.

2 Related Work

The concept of utility computing has been inspired and recently combined with the requirements and standards of Web 2.0 [1] to create cloud computing [2, 3]. Lu and Bharghavan [4] propose the structure for resource management in current indoor mobile computing environment. This work investigates methodologies for the organization of crucial networking resources in internal mobile computing situations: adaptively re-adjusting the superiority of service within pre-assigned bounds in request to adjust network dynamic range and consume flexibility. Kailasam et al. [5] state that the objective of resource allocation is to assign appropriate resources to the suitable workloads on time, so that applications can utilize the resources effectively. Islam et al. [6] present a model for resource management in cluster computing that accepts numerous scheduling strategies to survive dynamically. This work built Octopus, an extensible and distributed hierarchical scheduler that implements new space sharing, gang scheduling and load sharing strategies. Czajkowski et al. [7] state that metacomputing systems are expected to help remote and/or simultaneous use of geographically distributed computational resources. Resource management in such systems is difficult by five concerns that do not normally occur in other circumstances: site autonomy and heterogeneous substrates at the resources, and application constraints for policy extensibility, co-allocation, and online control. Raman et al. [8] state that traditional resource management systems use a system model to define resources and a centralized scheduler to manage their allocation. The research created and implemented the classified advertisement (classad) matchmaking structure, common method to resource management in distributed environment with decentralized possession of resources. Jhawar et al. [9] categorize and formalize several requirements of users and service providers with respect to security, reliability, and availability of the service and address the satisfaction of such requirements in the overall problem of resource allocation in infrastructure clouds.

3 Design of the Proposed Model

There are two significant players in the proposed model: provider and consumer/broker. For the trading of resources and deciding service access pricing, both cloud consumer/broker and Cloud Service Provider (CSP) use economic models and interaction protocols. IT assets and services are considered as utilities and it is clear that there is trade-off between service provider and consumer. This allows the use of service by the user under provided SLA.

3.1 Architecture of the Proposed Model

The architecture of the proposed model is three-tier and classifies the requirements of cloud providers and end users. The proposed three-tier architecture system is illustrated in Fig. 1 and shows the partition of Cloud Computing Layer (CCL), Virtual Layer (VL), and Physical Layer (PL).

The CCL and VL give the high-level activities that include the information transferred by several physical agents such as servers.

Cloud Computing Layer In the proposed model, the CCL organizes the cloud activities on top of the VL and is known as CCL. In this layer, the proposed model provides behavior such as cloud service model, cloud deployment model, and various tasks running in various applications make requests on the other layers such as on-demand allocation and reliability. Cloud service model provides services to virtual data center for virtualization within virtualized infrastructure and cloud deployment model provides the isolation of virtualized infrastructure within virtual layer for virtual machines.

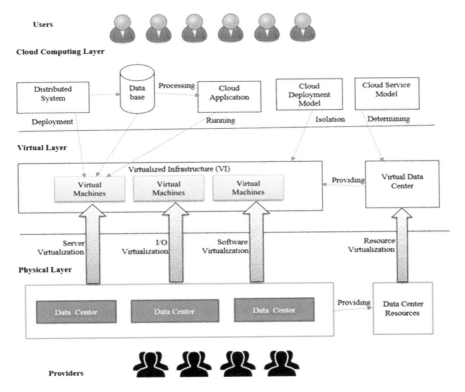

Fig. 1 Three-tier architecture of the proposed model

Virtual Layer In the proposed model, VL provides virtual instances such as virtual machines to the cloud users. This research work utilizes virtual machines for the execution of tasks. The main purpose of developing virtual machines is to achieve the advantage of scalability and to handle the allocation of cloud computing resources. VMs are multiplexed among the resources of a physical machine allowing multitenancy and resource distribution such as server, I/O, and software.

Virtualization offers a layer of abstraction above the physical resource, allowing two methodologies such as users and providers to resource management in cloud implementation model. The proposed model assigns virtual machines according to FCFS policies. The VL consists of following component.

Virtualized Infrastructure (VI) is described as a complex network structure where some or all of the elements such as servers are virtualized. The proposed cloud computing platform consists of VIs which depend on virtualization approaches to distribute resources among users. Virtualization techniques used for VI are:

Server Virtualization (SV): SV resolves the challenge of VM provisioning in cloud computing. SV is a chunk of hardware that is virtualized by applying software or firmware called a hypervisor which separates the equipment into numerous remote and autonomous virtual requests.

Resource Virtualization (RV): RV resolves the VIs needs for resource allocation and controlling of VMs.

I/O Virtualization: IOV is a technology that divides logical and physical resources. IOV is considered as an enabling data center technology that gathers IT infrastructure as a shared pool of resources such as storage, computing, and networking.

The proposed cloud service model achieves the market goals of the CSP that correctly converges to the optimal resource allocation based on maximizing resource utilization and profit and cloud users achieve high performance by maintaining a pool of resources in a cost-effective way by dynamic resource allocation and job scheduling simultaneously in a heterogeneous cloud environment.

Physical Layer The proposed model uses PL to model hardware resources such as CPU, memory, and bandwidth for the execution of tasks. These hardware resources are provided by the cloud providers through data centers. In the proposed model, data centers are used for the execution of large number of tasks and responsible for providing both physical and virtual resources. The proposed research work employs IaaS as a service model in cloud computing. The objective of an IaaS provider in proposed research work is to maximize revenue which involves the maximization of resource utility by the establishment of VMs onto physical machines. The proposed model evaluates the online current state of the cloud while allocating the VMs to a suitably underutilized resources that meets the QoS requirements. Workloads used for the evaluation of the model do not remain static and change during periods of peak user activity. Thus, a VM that is utilizing several resources can have a possibility for QoS influence on additional VMs running in the similar physical machine and must migrate to a resource with appropriate spare capacity.

3.2 Modeling of Resource Allocation Model

The proposed system is evaluated at the Infrastructure as a Service (Iaas) level. In the proposed system, it is assumed that the IaaS provider manages an infrastructure which can provide to his users best combination of VMs and to give the best combination of VMs, providers should have information regarding the execution of tasks. Assume that r resources are allocated to t tasks. Users U_i have a number of similar independent K_i tasks which require equal amount of computing. The sum of total cost of each resource should be minimized and t appropriate tasks are executed in the least time along with minimal cost. In the proposed solution, it is assumed that matrix m consisting of t rows each for one user and r columns each for one resource. Also, m_i is the representative of ith row of the matrix m and m_{ij} signifies the amount of each task of user U_i that is allocated to R_i. The allocation vector of m_i is represented as:

$$\sum_{m_{ij} \in m_i} m_{ij} = K_i$$

The objective of the proposed problem is to minimize expenses and increase utility and the utility for U_i can be evaluated as: $U_i(m_i) = \frac{1}{\text{Total Cost}}$

Resource Allocation Game. Game theory is a mathematical method which attempts to solve the exchanges between all players of game to produce certain the best outcomes for themselves. A game comprises of three components, a set of players, all the potential approaches every player will select, and the specific utilities of players linked with the strategy accomplished by every player. At each step, players decide one of their strategies and acquire a utility in return. Each player of a game attempts to maximize its individual utility by choosing the extremely profitable approach as compared to other players' options. In a cloud market, customers are conscious in making the choice, seek to minimize the expenses, and make certain limitations of cost $E = (E_1, E_2, ..., E_N)$ and time $T = (T_1, T_2, ..., T_N)$. Every player can consider a deadline displayed with T^0, and a maximum budget amount, E^0, the optimum objective function of the user is:

$$\min \text{€ s.t. } \sum_{k=1}^{K} e_k^i \leq E^0, \sum_{k=1}^{K} t_k^i \leq T^0$$

Cloud provider is responsible for the execution of tasks and provides resources to the users. A cloud provider develops the requested software/platform/infrastructure services, handles the technical infrastructure required for providing the services, provisions the services at agreed-upon service levels, and protects the security and privacy of the services.

Data center: Data center is used to provide various resources to the users such as virtual machines, storage, hardware, etc. Data center forms the foundation of a broad selection of services presented across the Internet involving Webhosting, ecommerce, and common services such as software as a service (SaaS), platform as a service (PaaS), and grid/cloud computing.

Cloud service management: This service allows service providers to ensure optimal performance, stability, and competence in virtualized on-demand environments.

Provisioning/configuration: Cloud provisioning refers to the processes for the deployment and integration of cloud computing services in the proposed model.

Cloud broker: Cloud broker is an entity that handles the negotiation among providers and consumers. In the proposed architecture, users' request for the resources through brokers.

4 Evaluation

This work uses Cloudsim tool to simulate the cloud-based infrastructures and application service. Cloudsim is created to evaluate the application benchmark study in a controlled environment to reproduce the results. This tool models the datacenter, processors, storage, bandwidth, operating system, virtualization standard and machine location throughout the initialization of the heterogeneous computing resources configuration. Table 1 shows the characteristics of resources used in simulation. Two common distributions, Normal and Pareto, signify preferences about the prices.

4.1 Metrics

Sum of bids for tasks for all users $= m$
Bid for task $= n$
Bid for other competitors $= m - n$
Execution speed $u =$ Length/MIPS
Execution time $= u + u$ (Bid of other competitors/Bid for medium task)

Table 1 Resource characteristics

Characteristics	Parameters
VM monitor	Xen, VMware
Number of PE	2, 4, 8
MIPS rating per PE	100, 200, 300, 400
Memory	512, 1024, 2048
Storage	160G, 320G, 500G
Bandwidth	128M, 256M, 512M

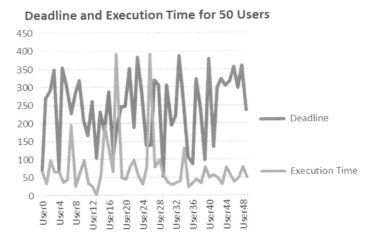

Fig. 2 Deadline and execution time for 50 users

Fig. 3 Payoff for 50 users

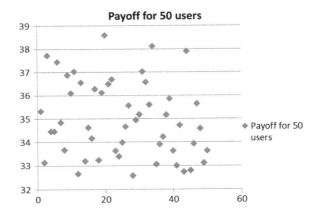

Completion cost = Execution Time * Bid of the user
Minimum deadline: 48.82
Maximum deadline: 390.62

Figure 2 shows deadline and execution time for 50 users. Figure 3 shows payoff for 50 users. Figure 4 shows bids for 50 users.

5 Conclusion

This research work proposes a model to manage the resources in cloud environment and concludes that in the proposed work, the resource allocation problem can be

Fig. 4 Bids for 50 users

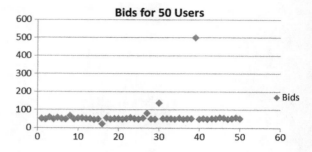

solved by using different metrics such as deadline and execution time and tested on 50 users. The minimum deadline for 50 users is 48.82 and maximum deadline is 390.62 and execution time for user 13 is infinity.

References

1. Alexander, B.: Web 2.0: a new wave of innovation for teaching and learning? Learning **41**(2), 32–44 (2006)
2. Buyya, R., Yeo, C., Venugopal, S.: Market-oriented cloud computing: vision, hype, and reality for delivering it services as computing utilities. In: Proceedings of the 10th IEEE International Conference on High Performance Computing and Communications (HPCC-08, IEEE CS Press, Los Alamitos, CA, USA) (2008)
3. Foster, I., Zhao, Y., Raicu, I., Lu, S.: Cloud computing and grid computing 360-degree compared. In: Grid Computing Environments Workshop, GCE'08, 2008, pp. 1–10 (2008)
4. Lu, S.W., Bharghavan, V.: Adaptive resource management algorithms for indoor mobile computing environments
5. Kailasam, S., Gnanasambandam, N., Janakiram, D., Sharma, N. (2010). Optimizing service level agreements for autonomic cloud bursting schedulers. In: Proceedings of the In ICPP Workshops, pp. 285–294. https://doi.org/10.1109/icppw.2010.54
6. Islam, N., Prodromidis, A.L., Fong, L.L.: Extensible resource management for cluster computing. In: Computer Science Published in Proceedings of 17th International Conference on Distributed Computing Systems (1997) https://doi.org/10.1109/icdcs.1997.603418
7. Czajkowski, K., Foster, I., Karonis, N., Kesselman, C., Martin, S., Smith, W., Tuecke, S.: Resource Management Architecture for Metacomputing Systems (1998)
8. Raman, R., Livny, M., Solomon, M.: Matchmaking: distributed resource management for high throughput computing (1998)
9. Jhawar, R., Piuri, V., Samarati, P.: Supporting security requirements for resource management in cloud computing. 2012 IEEE 15th International Conference on Computational Science and Engineering

Unique Stego Key Generation from Fingerprint Image in Image Steganography

A. Anuradha and Hardik B. Pandit

Abstract The concept of image steganography involves concealing secret message, using stego key and cover image. The stego key is used both at the sender side and the receiver side, for insertion as well as for extraction of secret message. Thus, the stego key and the insertion algorithm are connected closely in image steganography. The process of insertion requires a unique method of embedding data in the original cover image for generating the stego image, such that it is difficult for the hacker to know the presence of secret data. This process in turn depends on the stego key to a large extent. On the other hand, biometrics has been accepted as one of the strongest ways of confirming the identity of any individual. This paper focuses both on biometrics and steganography for the secret transmission of data. Thus, a unique key has been generated from the fingerprint image, as a part of biometrics, where the special features of the image have been extracted. The special features and their corresponding locations have been found. These locations are used for data embedding in the cover image, thus paving the path for a unique way of insertion of secret bits.

1 Introduction

Whenever the concept of image steganography comes into picture, based on the generalized model, three things always come into picture: stego key, cover image, and secret data, resulting in the final stego image. The requirement for the stego key is that it defines the method for hiding the data in the image [1]. The steganographic method can be applied, with the availability of the stego key.

A. Anuradha (✉)
L. J. Institute of Computer Applications, L. J. Group of Institutes, Ahmedabad, India
e-mail: anuradha.acharya@ljinstitutes.edu.in

H. B. Pandit
Department of Computer Science, Sardar Patel University, Vallabh Vidyanagar, India
e-mail: hardik00@gmail.com

Further, the same stego key will be utilized to get the data back from the stego image at the receiver side. It is very much similar to the key being used in cryptography for extracting the hidden data from the stego image [2]. The embedding algorithm must be very strong so that, even if the presence of data is known to the intruder, the extraction of data is nearly impossible. This makes a reference to the concept of open security, which tells that even if the design methodology is known to the public, it should safeguard the data. This idea can be supported with Kerckhoff's principle [3, 4]. The principle says that, the system remains safe, under the circumstances when all the details (including the algorithm) to make public, except the key. So it is the key, whose selection makes everything special [5]. But to ensure the expected sender of the message, some identification can be sent in the coded form. Hence, if a key is generated from the unique identification of the sender, it can support both the demands of authenticity and designing the algorithm for embedding data secretly. Finally for the most efficient and unique identification of any individual, biometrics can be utilized. The biometrics will take care of the identification of the sender and ensure that the correct sender has transferred the secret message. In support of this, only the authenticated persons are authorized to send data. Thus, it should not lead into any security issues, when the system is in the hands of the enemy [6]. Thus, the essential elements which can enhance the robustness of the system are:

- Identification, authentication, and authorization of the sender
- Robustness of the embedding algorithm
- Open security or secret transmission.

2 Review of Literature

The steganography corresponds to "invisible communication," with the objective of securing it from eavesdroppers [7]. It refers to the Greek words "stegos" and "grafia." The meanings are "cover" and "writing," respectively. Thus, it is based on the concept of covered writing. Different steganographic types are there like audio, video and image steganography. Among these types, image steganography has been proved to be the robust one.

The steganographic process begins with the generation of stego key. The key will then be used to embed the required message in a cover image. The key can be generated using any process. Thus, the final stego image will be created, containing the secret message to be transferred in a concealed way to the intended recipient (see Fig. 1).

Different approaches have been used by different authors for image steganography. LSB technique has been used for embedding secret bits in the least significant bits. In [8], based on a random generated key, text is embedded in the least significant bits of the color matrices of RGB image. The concept of pixel value differencing (PVD) and modulo operation (MO) has been implemented in [9], with the intention of increasing the peak signal-to-noise ratio and hiding capacity, by embedding data in non-overloading blocks of image. A combined approach of data

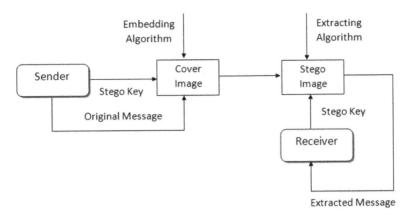

Fig. 1 Model of steganography

compression, data embedding and also neural network for maintaining the quality of the communication channel, for better results [10]. For data embedding, the PVD concept has been applied in this paper. The scheme of least significant bits is used, for embedding in edge positions, in color images [11]. The color image is divided into three components. One component is used as an index to find the edges, while the other two components are used for embedding purpose. So as to come up with a bigger capacity in comparison with the other methods of image steganography, data is embedded by changing the color palette of the image in [12]. This mechanism comes under color palette transformation, where pixels with the same color will be changed into the same color. The method found to be one of the best, according to structural similarity index (SSIM) and peak signal-to-noise ratio (PSNR) values. To increase security and imperceptibility of embedded message, spread spectrum embedding techniques are implemented with chaotic encryption, for spreading the message in the cover image, so that message embedding is possible on the smallest bit of the cover image [13]. Thus, making it difficult for the attacker for guessing the existence of the secret message, it has been implemented the concept of mapping the pixels in an image to 26 alphabets of English alphabets, along with some special characters for maximum use of bits of pixels for data embedding [14]. The idea worked effectively, with different short and long messages leading to less suspicion for the attackers. An improved reversible image steganography method based on pixel value ordering (PVO) is proposed in [15], where every three continuous and neighboring pixels are grouped, in the rows and columns of the image. The pixels with more steganographic difference value which is the difference between maximum and minimum value are used for data hiding, showing higher capacity and acceptable PSNR value. A histogram of oriented gradient (HOG) algorithm is proposed, where blocks of interests find out, based on the gradient magnitude and angle of the cover image [16]. After that PVD algorithm is used to insert data in the prominent edge direction, LSB substitution is used, in the remaining two pixels [16].

3 Limitations of Existing Systems

All the recent methodologies almost tried to follow some unique methods for hiding data. However, none of these methods adopted the combined concept of real-time embedding, dynamic embedding, and biometrics together. The distinctive approach of data embedding always can be defined in the form of a stego key. Consequently, in the current proposal, the stego key has been adopted in the form of biometrics, fingerprint image, and feature extraction, leading to a robust technique of data embedding based on image steganography.

4 Overview of Feature Extraction

"Interest point detection" refers to the identification of interest points in an image, which are actually somehow "special." Actually, an image has many elements like "corners" and "blobs" and "special" features. These features can be extracted using some detectors. Detector is an algorithm that selects the special points in an image satisfying some criterion.

On the other hand, descriptors that describe the image patch around the interest points are a vector. The interest points along with the descriptors are called the "local features." Local features can be referred to some distinct features of an image like points, edges, and small image patches. Normally, local feature detection and extraction are used for applications like image registration, object detection, and classification, tracking and motion estimation [17]. Image registration is the process of aligning images of the same scene, which is used to compare common features in different images. This concept is helpful for discovering whether a river has migrated or to know how an area became flooded [17]. Object detection refers to identifying objects like faces or vehicles in images or videos based on extracted features. In object classification, objects are identified and then they are classified into different categories based on their specific features being extracted. Object tracking refers to estimate the trajectory of an object during its motion [18], whereas motion estimation determines the movement of objects in an image sequence [19].

However, the research paper focuses on the extraction of the unique features of a finger image with the intention of generating a unique stego key, to be used for image steganography. Actually, the location of the unique features of the fingerprint image has been extracted as a part of the stego key.

5 Biometrics and Authentication

Identification, authentication, and authenticity are the three buzzing words in the field of information security, satisfying three important requirements of information, i.e., confidentiality, integrity, and availability. The person with proper identification is considered as the authorized person. Accordingly, the person will be authorized to give a particular role in any defined system. Behalf of identification, different approaches have been used such as password mechanism or tokens. Password refers to knowledge-based authentication mechanism, which requires precise recall [20], and also there are fair chances of its misuse if disclosed somehow. Tokens like bank cards are used for ATM transactions which may be stolen or lost [21]. But these days, IT is dependent on a more trustworthy system known as the biometrics.

Biometric authentication is based on the biological characteristics of any individual, thus verifying the individuality. Biometrics is the measurable, physical or physiological traits of any individual, being used for the verification of the claimed identity [22].

Since biometrics is non-transferable, stable, and enduring in nature, it provides one of the strongest authentication mechanisms. Also, it is a convenient practice of authentication. Fingerprinting, retina, voice recognition, and face recognition are some of the mechanisms of biometric systems, used as a part of authentication. The research has selected fingerprinting, being the most convenient way of identification or authentication. Thus, fingerprinting has been selected not only for getting the stego key to be used for image steganography, but also to identify the valid sender. Hence, the intended recipient can get the message being sent, with the conformity that it is from the authorized sender only.

6 Unique Key Generation

Among different biometric mechanisms, fingerprinting is the best, reliable, and fastest way of personal identification [23]. It can provide the high level of security because of unique identification property [24]. Fingerprints are unique collection of patterns called the ridges (single-curved segments) and valleys (region between two adjacent ridges) [24]. Two important characteristics of fingerprints that make them the best choice for biometric identification are [25]:

- Individuality and
- Persistence.

It is difficult to find two people with the same fingerprints [26]. Most of the research has been done on fingerprints for "fingerprint recognition and matching." Almost all of them have tried to extract the minutiae extraction for recognition and matching. However in the current research, instead of only identifying the ridges or valleys, special features of the fingerprint image have been extracted, where the unique feature includes corners and blob features. These features are called the local features like points, edges, or small image patches.

These features are not for any recognition of matching, but for the unique stego key generation for creating the stego image in the best possible manner. Thus both the objectives of confirming the valid sender as well as concealing of the secret message can be fulfilled simultaneously.

One of the advantages of this approach is that due to the unique key generation, though the length of the message can be known by some steganalysis process, since the research is based on the "Kerckhoff's principle," the intruder cannot extract the secret bits being embedded.

7 The Model

A small portion of the location coordinates, collected from the variable descriptor, has been represented in Table 1.

Fingerprints are used as a part of biometrics vigorously these days for the identification of persons. Thus for unique key generation, the uniqueness of fingerprint image can be used. Thus, the unique features of the fingerprint image and their locations have been extracted.

Figure 2 depicts that, corner features have been detected first, i.e., the FAST features have been extracted using "detectFASTFeatures()" function.

Table 1 Location vector for unique key

	1	2	3
63	316	132	
64	339	134	
65	139	135	
66	282	135	
67	231	136	
68	93	137	
69	222	137	
70	239	137	
71	258	137	
72	331	137	
73	134	138	
74	301	138	

location_valid_points_corners1
763x2 single

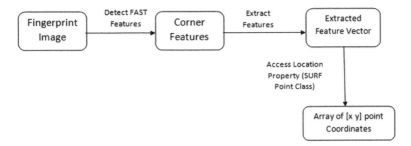

Fig. 2 Model of unique key generation from fingerprint image

Following this, the feature vectors have been extracted using extractFeatures() function.

Along with the descriptors, the model resulted in an array containing the pair of X- and Y-coordinates for the local features.

Though different methods are available for detecting corner and blob features, the research has used "detectFASTFeatures()" function, for detecting corner features, that returns the corner point objects. The minimum accepted quality of corners ("Min-Contrast") has been taken as "0.1", followed by "extractFeatures()" function, for extracting the feature vectors or the descriptors, and their corresponding locations. These locations will be used for embedding the secret bits in the cover image for the purpose of image steganography in the next step of the research.

8 Results

The size of the fingerprint image taken in the research is 470 × 450 (unit 8), being downloaded from Internet [27]. The algorithm could able to extract 883 corner points, in the form of a 883 × 1 cornerPoints, using "detectFASTFeatures()" function. Through "extractFeatures()" function, 763 valid corners have been identified.

Thus, the valid corner locations found is a 763 × 2 matrix (single), containing the X- and Y-coordinates of the locations of the valid corner points (Figs. 3, 4, and 5).

Fig. 3 Original fingerprint image

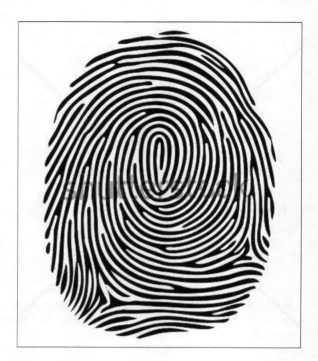

Fig. 4 Detected corner points plotted with insertMarker() function

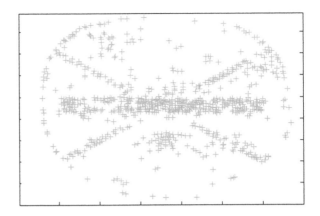

Fig. 5 Plotted corner points for the fingerprint image

References

1. https://www.trustwave.com. Last accessed 10 Mar 2020
2. Bhanupriya Katre, B.: Dynamic key based LSB technique for steganography. Int. J. Comput. Appl. **167**(13), 0975–8887 (2017)
3. http://whatis.techtarget.com. Last accessed 10 Mar 2020
4. https://www.techrepublic.com. Last accessed 10 Mar 2020
5. https://link.springer.com. Last accessed 21 Feb 2019
6. https://en.wikipedia.org. Last accessed 10 Mar 2020
7. Kharrazi, M., Sencar, H.T., Memon, N.: Image steganography: concepts and practice. 22 Apr 2004, 1:49 WSPC/Lecture Notes Series: 9in x 6in MAIN
8. Srilakshmi, P., Himabindu, Ch., Chaitanya, N., Muralidhar, S.V., Sumanth, M.V., Vinay, K.: Text embedding using image steganography in spatial domain. Int. J. Eng. Technol. **7**(3.6) 1–4 (2018)
9. Sahu, A.K., Swain, G.: Digital image steganography using PVD and modulo operation. Internetworking Indonesia J. **10**(2) (2018)
10. Majumder, J., Pradhan, C.: High capacity image steganography using pixel value differencing method with data compression using neural network. Int. J. Innov. Technol. Explor. Eng. (IJITEE). **8**(12) (2019). ISSN: 2278–3075
11. Rashid, RD, Majeed, TF: Edge based image steganography: problems and solution. In: 2019 International Conference on Communications, Signal Processing, and their Applications (ICCSPA), Electronic ISBN: 978-1-7281-2085-0, USB ISBN: 978-1-7281-2084-3, Print on Demand (PoD) ISBN: 978-1-7281-2086-7 (March 2019)
12. Margalikas, E., Ramanauskaitė, S.: Image steganography based on color palette transformation in color space. EURASIP J. Image Video Process. Article number: 82 (2019)
13. Rachmawanto, E.H., Rosal, D., Setiadi, I.M., Sari, C.A., Andono, P.N., Farooq, O., Pradita, N.: Spread embedding technique in LSB image steganography based on chaos theory. In: 2019 International Seminar on Application for Technology of Information and Communication (iSemantic), Electronic ISBN: 978-1-7281-3832-9, USB ISBN: 978-1-7281-3831-2, Print on Demand (PoD) ISBN: 978-1-7281-3833-6 (2019)
14. Al-Husainy, M.A.F.: Image steganography by mapping pixels to letters. J. Comput. Sci. **5**(1):33–38 (2009). ISSN 1549-3636 © 2009 Science Publications
15. Liu, H.-H., Lee, C.-M.: High-capacity reversible image steganography based on pixel value ordering. EURASIP J. Image Video Process. Article number: 54 (2019)
16. Hameed, M.A., Hassaballah, M., Aly, S., Awad, A.I.: An adaptive image steganography method based on histogram of oriented gradientand PVD-LSB techniques. Digital Object Identifier 10.1109/ACCESS.2019.2960254. 17th Dec 2019

17. https://in.mathworks.com. Last accessed 10 Mar 2020
18. https://www.igi-global.com/dictionary/moving-object-detection-and-tracking-based-on-the-contour-extraction-and-centroid-representation/20697. Last accessed 10 Mar 2020
19. https://www.cmlab.csie.ntu.edu.tw/cml/dsp/training/coding/motion/me1.html. Last accessed 10 Mar 2020
20. https://www.usenix.org/legacy/publications/library/proceedings/sec2000/full_papers/dhamija/dhamija_html/node2.html. Last accessed 10 Mar 2020
21. https://thenextweb.com/future-of-communications/2016/02/17/the-ultimate-guide-to-selecting-a-device-authentication/. Last accessed 10 Mar 2020
22. The Irish Council for Bioethics, Biometrics: Enhancing Security or Invading Privacy? www.bioethics.ie
23. Tukur, A.: Fingerprint recognition and matching using matlab. Int. J. Eng. Sci. (IJES) **4**(12), 01–06 (2015), ISSN(e):2319-1813, ISSN (p): 2319-1805
24. Bansal, R., Sehgal, P., Bedi, P.: Minutiae extraction from fingerprint images—a review. IJCSI Int. J. Comput. Sci. **8**(5), No 3, ISSN (Online): 1694–0814. www.IJCSI.org (September 2011)
25. http://biometrics.cse.msu.edu/Publications/Fingerprint/PankantiPrabhakarJain_FpIndividuality_PAMI02.pdf. Last accessed 10 Mar 2020
26. http://kaheel7.com/eng/index.php/gods-creations/580-humans-fingerprints. Last accessed 10 Mar 2020
27. https://openclipart.org/tags/fingerprint. Last accessed 10 Mar 2020
28. Zahed, A., Sakhi, M.R.: A novel technique for enhancing security in biometric based authentication systems. Int. J. Comput. Electr. Eng. **3**(4) (2011)
29. Fridrich, J., Goljan, M.: On estimation of secret message length in LSB steganography in spatial domain. http://www.ws.binghamton.edu/fridrich
30. Zöllner, J., Federrath, H., Klimant, H., Pfitzmann, A., Piotraschke, R., Westfeld, A., Wicke, G., Wolf, G.: Modeling the security of steganographic systems. In: Aucsmith, D. (ed.) Information hiding, LNCS 1525, pp. 344–354. Springer, Berlin (1998)
31. Rana, M.S., Sangwan, B.S., Jangir, J.S.: Art of hiding: an introduction to steganography. Int. J. Eng. Comput. Sci. **1**(1), 11–22 (2012)
32. Kumar, A., Pooja, K.M.: Steganography—a data hiding technique. Int. J. Comput. Appl. **9**(7), 0975–8887 (2010)
33. Sharma, V., Kumar, S.: A new approach to hide text in images using steganography. Int. J. Adv. Res. Comput. Sci. Softw. Eng. **3**(4) (2013). ISSN: 2277-128X
34. Kabay, M.E.: Identification, Authentication and Authorization on the World Wide Web. M. E. Kabay & ICSA (1997)
35. Zaeri, N.: Minutiae-based fingerprint extraction and recognition. INTECH, Open Science/Open Minds
36. Keerthana, M.S.: Fingerprint matching incorporating ridge features using contourlet transforms. Int. J. Innov. Res. Comput. Commun. Eng. ISSN(Online): 2320–9801, ISSN (Print): 2320-9798
37. Parra, P.: Fingerprint Minutiae Extraction and Matching for Identification Procedure. Department of Computer Science and Engineering, University of California, San Diego
38. Barham, Z.S.: Finger print recognition using MATLAB
39. Fridrich, J., Goljan, M., Soukal, D.: Searching for the Stego-Key. http://www.ws.binghamton.edu/fridrich
40. http://blogs.getcertifiedgetahead.com. Last accessed 14 Mar 2019
41. https://securitycommunity.tcs.com. Last accessed 18 Feb 2019
42. http://onin.com. Last accessed 15 Apr 2019

Self-powered IoT-Based Design for Multi-purpose Smart Poultry Farm

Tajim Md. Niamat Ullah Akhund, Shouvik Roy Snigdha, Md. Sumon Reza, Nishat Tasnim Newaz, Mohd. Saifuzzaman, and Masud Rana Rashel

Abstract The purpose of the present work is to make an IoT-based smart poultry farm system. In this work, the power supply is developed using renewable energy mainly with solar energy and nano-hydro. This IoT-based module helps to develop the system's productivity to ensure farm's constant healthy condition. The proposed system collects several types of sensor data from the farm, such as temperature, humidity, toxic gas, water level, and moisture. Then the overall conditions will be controlled and be kept proper level with the developed system automatically. This system stores all the data in the central database for further analysis and getting knowledge from them also gives notifications. Optimal values for the power and health condition are obtained through the stored data. It helps to predict future conditions too. The electrical devices, doors, and dustbins of the farm can be controlled via IoT systems through mobile phones and online platforms from remote locations and from anywhere in the world with Internet connectivity.

1 Introduction

The goal of this project is to produce healthy and sustainable poultry meat and also data analysis for qualitative and quantitative research, where people can get fresh meat with profit. Most importantly, solar energy and hydro energy are used in the proposed system. In Bangladesh, farmers are not conscious about health and automation for growing up chicken so that, most of the time they use the medicine, which is the most injurious for our health. For fresh chicken, farmers have to maintain the medium level

T. Md. N. U. Akhund (✉) · S. R. Snigdha · Md. S. Reza · Mohd. Saifuzzaman · M. R. Rashel
Daffodil International University, Dhaka, Bangladesh
e-mail: tajim.iitju@gmail.com

T. Md. N. U. Akhund · N. T. Newaz · Mohd. Saifuzzaman
Jahangirnagar University, Savar, 1342 Dhaka, Bangladesh

M. R. Rashel
University of Evora, Largo dos Colegiais 2, 7000-645 Évora, Portugal

of automation but due to the reduction of skill. Likewise, the room temperature will be 35 °C when the chicken appears. At present, meat and egg production can adhere only 68–64% of the national demand as the report of Kris Gunnar's, BSc on July 5, 2018 [1]. But if we consider what type of poultry chicken we eat, that is harmful for health. The chickens are given antibiotic injections seven days after birth and when we eat this type of chicken our antibiotic resistance became more weak. Also, the food made for poultry chicken is combined with chromium and arsenic. Chromium is not banished even after cooking because its temperature tolerance is 2900 °C but we are normally cooking at 100–150 °C so that is not possible to destroy chromium. By eating this type of chicken, our kidney becomes affected and it also increases the chance of cancer. Due to arsenic, diabetics and neurological difficulty may attack us. In Bangladesh, there are 64,769 register poultry farms, but that is not enough to maintain public demand said Bangladesh Livestock Research Institute [2]. Including human life, electronic world, voting everything are related with IoT nowadays [3, 4]. A current report shows that there are total 14.2 billion devices connected in 2019 and it will reach 25 billion by 2021. Use of IoT is increasing in robotics and daily life [5, 6]. The embedded devices and sensors create revolution in our world through a new era such as homes, mobiles, supermarket, industry as well as farm instrumentation and voting too [7]. All of these motivated us to develop such an IoT smart farm system with renewable energy for poultry farms.

This work has the following specific objectives:

1. As we consider Bangladesh, we cannot get fresh poultry meat as well as egg. In that case, our fundamental target is to produce healthy and sustainable meat owing to monitoring farming environmental parameter like temperature, humidity, CO, LPG, smoke, and so on.
2. In addition, the total power comes from the solar panel and nano-hydro generator that means we try to introduce the most convenient renewable energy.
3. Another more, by analyzing data we predict the weather and disease of poultry chicken and take the suitable step.

The rest of the paper is organized as follows: related research has been presented in Sect. 2. The system model and methodology have been presented in Sect. 3. Obtained results and related discussion have been presented in Sect. 4, while conclusions and future works have been outlined in Sect. 5.

2 Related Work

Authors of [1] proposed wireless controlling method on their paper where they control the whole system through customer command. By using the application of wireless sensor network, quality of chicken can be improved ultimately that leads to improving human health, and a wearable wireless sensor can detect tainted chickens [2]. Authors of [8] implement Environment Controlled Poultry Management System (ECPMS) using temperature and humidity sensor, gas sensor, and LDR, Raspberry

Pi, and camera monitoring the poultry farm can handle the situations like turn on or off light and fan. Authors of [9] have mentioned IoT-based smart farming systems including GSM in 2019. But GSM systems may not be cost-effective for rural people of Bangladesh. Authors of [10] developed a fire base alarm system using Raspberry Pi and Arduino Uno; if the smoke sensor detected any smoke because of fire, the camera captures the images and provides notification. The main advantages of this system are it will provide the right information to the firefighter and reduce the false report. They have proposed a technique to improve the productivity of a chicken farm by using modern technology of wireless sensor networks. Authors of [11] have proposed an IoT-based electronic voting system, that is automated and controlled with embedded systems. Authors of [12] have proposed an IoT-based system that for food's cultivation and protection, they irritate bats from the fruit trees. Authors of [13] have proposed an IoT-based robot that can be controlled via the gestures of human body parts. In 2018, authors of [14] represent a statistic of Zambia and proposed an IoT-based monitoring system in a poultry house with GSM and GPRS. Here, they just monitored the current temperature but did not develop any controlling system. Moreover, today's smart world looks into smart city [15, 16], security [17, 18], and many more, so toward making the world smarter our research will definitely add great value.

3 System Model and Methodology

The working process of the proposed system is as follows:

1. The total system is consisted of temperature sensor, humidity sensor, gas sensor, water-level sensor, moisture sensor, multiple node MCU, Arduino, OLED display, relay module, HC-05 Bluetooth sensor, LCD display, fan, light, cooler, heater, servomotors, etc.
2. All the sensors collect data from the environment then pass data in Arduino. Arduino sends the data to cloud server through GSM module or Wi-Fi module node MCU and also stores information in SD card.
3. All the electrical devices, doors, dustbins, lights, and fans are automated and those can be controlled via Bluetooth, Wi-Fi, and Internet from anywhere in the world.
4. The comfortable environment for chicken is kept via automatic controlling. User can set the threshold values of temperature, humidity, and water level. If the temperature increased more than the threshold, the fans and cooler will turn on automatically. Heater will turn on if opposite things happen. If the water level of the water pot decreases, the pumps will automatically turn on to supply water.
5. The full system is powered by solar and nano-hydro system. Backup power is stored in the batteries during daytime.

The full developed system concept is illustrated in (Fig. 1). The system flow diagram is mentioned in (Fig. 2).

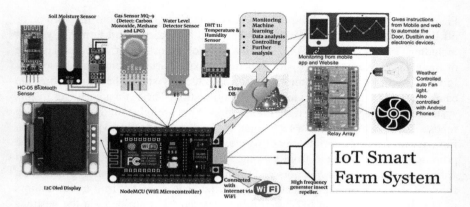

Fig. 1 Conceptual diagram of the system

According to the Bangladesh Livestock Research Institute, the temperature is one of the great facts for growing up poultry chicken naturally. Also, they ensure about 90% of chicken is dead due to lack of monitoring and controlling. This proposed system will control the temperature of the farm as our findings of recommended temperature (Table 1) for particular days of chicken.

4 Result and Discussion

The proposed system results the following features and outputs:

1. Power supply of the system obtained from battery, solar, and portable nano-hydro power generator from water flow of the system.
2. It can measure temperature, humidity, water level, moisture, and toxic gas.
3. Light and fan depend on the temperature level, and when the temperature is below 20 °C, all the heating light will turn on, likewise when the temperature is up to 24 °C, automatically the heating light will be turning off and the cooling fan will be on due to temperature control. This threshold value can be changed by the user.
4. It includes insect repelling system with ultrasonic sound.
5. All the electronic devices can be controlled via Android phone from short distance via Bluetooth and from all over the world via Internet.
6. Humidity between 60 and 80% is sufficient for poultry chicken. The system can maintain that.
7. Gets data in CVS, XML, and JSON format suitable for MATLAB and Python analysis.
8. Monitoring, analyzing, and controlling can be done from anywhere in the world.
9. Auto-water supply sensing water sensor.

Fig. 2 System flowchart

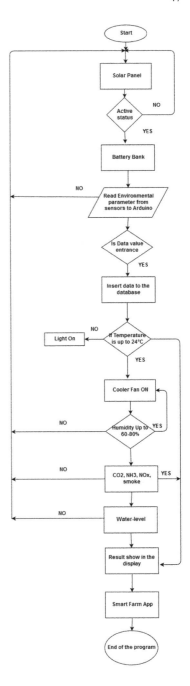

Table 1 Recommended temperature for chicken

Age of chicken	Temperature in degree celsius (°C)
First day	30–32
First week	30
Second week	26
Third week	24
Fourth week	22
Overall Recommend	28 (normal body temperature)

10. Used MQ-135 gas sensor module detects the amount of ammonia (NH_3), nitrogen oxides (NO_x), alcohol, benzene, smoke, and CO_2 in the farm.
11. Shows sensor values, voltage, and current in OLED display.
12. It sends data with Wi-Fi to cloud database.
13. Collected data can be seen from mobile app.
14. Sends data via mobile SMS.
15. Predict and further analysis with machine learning in future.
16. All the data are saved in the cloud database of ThingSpeak. The channel link is: https://thingspeak.com/channels/685524.
17. Through the developed smart farm app, the user can get update and control farm from anywhere. By this app, the user can monitor multiple farm conditions. Also, this app stores all the information about the user to get appropriate information about the farm.

If the system is applied in real-time world, the chicken can live with the natural food rather than artificial or tannery food. In this regard, some statistics are found which prove that natural foods for poultry will increase health and decrease cost (see Table 2).

In overall research for highly productive and hygienic chicken, this kind of feature is really effective for to reach our main goal. The device is already tested in real environment in a poultry farm and huge data are collected. These data may be used to extract valuable information through artificial intelligent algorithms in future. The prototype is shown in (Fig. 3). This system is also installed in real-time environment (Fig. 4).

The collected data can be monitored from developed app and Web site, and the electric devices and doors are controlled via Bluetooth and Internet through Android apps (Fig. 5).

The collected sensor data can be downloaded as csv or Excel format from the cloud server and mobile app (see Fig. 6). This system collected a poultry farms data for 3 months. 117,414 entries have been collected. Among them, 500 values of temperature, humidity, and water level were considered to draw the graph. There were 500 chickens. Among them, only 3 chickens died after applying this system. The collected unanalyzed data are visualized in (Fig. 6).

Table 2 Natural feed cost versus tannery feed for poultry

Materials	Quantity (kg)	Estimated value (TK)
Wheat/crushed corn/crushed rice	800	$(800 \times 15 - 18) =$ 12,000–14,400 tk
Wheat husk	100	$(100 \times 10 - 15) =$ 5000–7500 tk
Rice paddy	500	$(500 \times 10 - 15) =$ 5000–7500 tk
Sesame foul	240	$(240 \times 20 - 22) =$ 4800–5280 tk
Dried fish powder	200	$(200 \times 55 - 60) =$ 11,000–12,000 tk
Oyster powder	140	$(140 \times 20 - 25) =$ 2800–3500 tk
Salt	10	$(10 \times 7.5 - 8.5) =$ 150–170 tk
Total (1990 kg) natural food cost = 40750–50350 tk		
Total cost of 1990 kg with tannery food		
Provita Feed Ltd. (1990 kg)		= 73,630–79,600
Agata Feed Mills Ltd. (1990 kg)		= 70,446–88,555
Bengal feed (1990 kg)		= 71,640–83,580

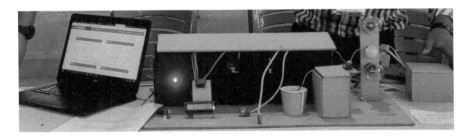

Fig. 3 System prototype

5 Conclusion and Future Work

This century is the era of technology. Things are changing very rapidly. It is important to adopt the technology to give service to all levels of people. IoT-based solution is needed of time to get better result from farming in every sector. This work has developed a device that establishes directional communication between actors. This focuses on poultry industry. Its work is to improve and ensure healthy condition in the poultry farm. This system collects lots of data and stores them in cloud server.

Fig. 4 IoT-based smart poultry farm system in real environment

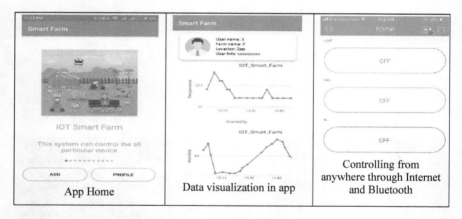

Fig. 5 Real-time data monitoring Android app

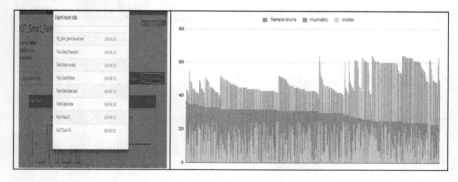

Fig. 6 Data downloading and visualizing (suitable for MATLAB analysis)

Those data can be analyzed and predict about future conditions of the farm. This system's future work will focus on that.

References

1. Mahale, R.B., Sonavane, D.S.: Smart poultry farm: an integrated solution using WSN and GPRS based network. Int. J. Adv. Res. Comput. Eng. Technol. **5**(6) (2016)
2. Ayaz, M., Ammad-uddin, M., Baig, I.: Wireless sensor's civil applications, prototypes, and future integration possibilities: a review. IEEE Sens. J. **18**(1), 4–30 (2017)
3. Akhund, T.M.N.U., Mahi, M.J.N., Tanvir, A.H., Mahmud, M., Kaiser, M.S.: ADEPTNESS: Alzheimer's disease patient management system using pervasive sensors-early prototype and preliminary results. In: International Conference on Brain Informatics, pp. 413–422. Springer, Cham (2018)
4. Sarker, M.M., Akhund, T.M.N.U.: The roadmap to the electronic voting system development: a literature review. Int. J. Adv. Eng. Manage. Sci. **2**(5)
5. Akhund, T.M.N.U.: Study and implementation of multi-purpose IoT nurse-BoT
6. Akhund, T.M.N.U., Sagar, I.A., Sarker, M.M.: Remote temperature sensing line following robot with bluetooth data sending capability
7. Akhund, T.M.N.U.: Remote sensing IoT based android controlled robot. Methodology **9**, 11
8. Islam, M.M., Tonmoy, S.S., Quayum, S., Sarker, A.R., Hani, S.U., Mannan, M.A.: Smart poultry farm incorporating GSM and IoT. In: 2019 International Conference on Robotics, Electrical and Signal Processing Techniques (ICREST), pp. 277–280. IEEE (2019)
9. Mitkari, S., Pingle, A., Sonawane, Y., Walunj, S., Shirsath, A.: IOT based smart poultry farm. System **6**(03) (2019)
10. Paplicki, P.: Simplified reluctance equivalent circuit for hybrid excited ECPMS-machine modelling. In: 2016 21st International Conference on Methods and Models in Automation and Robotics (MMAR), pp. 241–244. IEEE (2016)
11. Sarker, M.M., Shah, M.A.I., Akhund, T.M.N.U., Uddin, M.S.: An approach of automated electronic voting management system for Bangladesh using biometric fingerprint. Int. J. Adv. Eng. Res. Sci. **3**(11) (2016)
12. Akhund, T.M.N.U., Rahman, M.H., Banisher, B.: An approach to create a high frequency ultrasound system to protect agricultural field from bats
13. Akhund, T.M.N.U.: Designing and implementation of a low-cost wireless gesture controlled robot for disable people
14. Phiri, H., Kunda, D., Phiri, J.: An IoT smart broiler farming model for low income farmers. Int. J. Recent Contrib. Eng. Sci. IT (iJES) **6**(3), 95–110 (2018)
15. Saifuzzaman, M., Moon, N.N., Nur, F.N.: IoT based street lighting and traffic management system. In: 2017 IEEE Region 10 Humanitarian Technology Conference (R10-HTC), Dec 21–23 2017
16. Siddique, M.J., Islam, M.A., Nur, F.N., Moon, N.N., Saifuzzaman, M.: BREATHE SAFE: a smart garbage collection system for Dhaka city. In: 2018 10th International Conference on Electrical and Computer Engineering (ICECE), pp. 401–404 (2018)
17. Shetu, S.F., Saifuzzaman, M., Moon, N.N., Nur, F.N.: A survey of Botnet in cyber security. In: 2019 2nd International Conference on Intelligent Communication and Computational Techniques (ICCT), Jaipur, India, 2019, pp. 174–177
18. Saifuzzaman, M., Khan, A.H., Moon, N.N., Nur, F.N.: Smart security for an organization based on IoT. Int. J. Comput. Appl. **165**(10), 33–38 (2017)

Phrase-Based Statistical Machine Translation of Hindi Poetries into English

Rajesh Kumar Chakrawarti, Pratosh Bansal, and Jayshri Bansal

Abstract Statistical machine translation (SMT) is a variant of machine translation where the translations are handled with statistically defined rules. Numerous researchers have attempted to build the framework which can comprehend the different dialects to translate from one source language to another target language. However, the focus on the translation of poetry is less. Reliable and rapid transliteration of the poetry is very mandatory for the execution of the computer to translate the poem from one language to another. The existing approach has several issues such as time consumption, quality of the translation process, and matching of similar words. To overcome these issues, we propose a phrase-based statistical machine translation (PSMT) with special adherence to word sense disambiguation (WSD). Quality of the translation is increased by sensing the ambiguous words with WSD. The Hindi WordNet along with the Lesk algorithm identifies the ambiguous words and senses the exact meaning before the phrase extraction. Finally, the proposed method is compared with machine translation schemes such as rule-based machine translation and transfer-based machine translation. The experimental results suggest that the proposed method performed well with the inclusion of WSD in the PSMT technique.

R. K. Chakrawarti (✉)
Department of Computer Engineering, Institute of Engineering and Technology, Devi Ahilya Vishwavidyalaya, Indore 452020, Madhya Pradesh, India
e-mail: rajesh_kr_chakra@yahoo.com

P. Bansal
Department of Information Technology, Institute of Engineering and Technology, Devi Ahilya Vishwavidyalaya, Indore 452020, Madhya Pradesh, India
e-mail: pratosh@hotmail.com

J. Bansal
Human Resource Development (HRD) Centre, Devi Ahilya Vishwavidyalaya, Indore 452020, Madhya Pradesh, India
e-mail: dr.jayshribansal@gmail.com

© The Editor(s) (if applicable) and The Author(s), under exclusive license to Springer Nature Singapore Pte Ltd. 2021
T. Senjyu et al. (eds.), *Information and Communication Technology for Intelligent Systems*, Smart Innovation, Systems and Technologies 196,
https://doi.org/10.1007/978-981-15-7062-9_6

1 Introduction

India being a multilingual nation has various languages for communication. Yet Indians accentuate on Hindi for communication. Across the country, there is a large population that is not aware of linguistics and semantics of the English language. In this manner, there is a need to build up a machine translation application for the poems that will cross over any barrier between these two languages [1, 2]. Today, most nations acknowledge Indian societies and attempt to get familiar with the Hindi language to learn the Indian culture. Hence, interpretations of Hindi writing and sonnets are extremely basic and significant to understand. The machine translation systems are automated computer programs (software) capable of translating information available in one language (called the source language) into different dialects (called the objective language) [3]. With the support of machine translation frameworks, interpretations of Hindi writing into English are made simple. Existing frameworks on interpretations of Hindi sonnets into English is vital and an inconceivable exercise in machine translation model. In this, lyrics assume a significant role in contrast with the written work of interpretations. Since ballads give sentiments, feelings, expression, and more, the genuine interpretations of the sonnets are extremely significant.

The substantial data for the interpretation remained gathered to interpret the Hindi lyric into English. Since Hindi does not pursue any typical standard but, represented in various ways, a standard-based interpretation has been pursued in which many linguistic principles have been built to actualize in a POS Tagger. Moreover, Hindi words can be written in various ways but it does not pursue a particular spelling design. To address this problem, an accumulation of the corpus in an information base recognizes the right significance of phrase or word concerning its context by the word sense disambiguation (WSD) [4, 5]. WSD has many applications such as document indexing, theme extraction, semantic annotation, genre identification, semantic web search, and information retrieval [6–9]. The machine translation techniques are direct machine translation (DMT), rule-based machine translation (RMT), example-based machine translation (EMT), interlingua machine translation (IMT), knowledge-based machine translation (KMT), statistical-based machine translation (SMT), and hybrid-based machine translation (HMT) [10]. Among various MT systems, the SMT plays a significant role in the translation process. SMT is a technique widely used for translation purposes with the help of statistical analysis in order to formulate rules which are best for the translation of a target sentence [11]. The SMT categorized into four different types, namely word-based, syntax-based, phrase-based and hierarchical phrase-based. The word-based SMT utilizes the words and their neighbor's during the translation process. While the phrase-based methods use phrases instead of words, it also considers the neighboring phrases while translation [12]. The syntax-based machine translation incorporates the syntax representation in order to find the best of the words. The hierarchical phrase-based is a hybrid approach which combines the strength of phrase- and syntax-based methods. It employs the synchronous context-free grammar for the translation purpose [13, 14]. The proposed method employs the phrase-based SMT in order to solve the problems through the statistically devised rules. Some of the problems faced during the translation process are discussed as follows the machine transliteration technique deals with the problem such as retrieval of cross-lingual, probability evaluation of translation, difficult to find the sentence with the prime and highest probability, multiple representations of one word [13]. In the Trans tech system, several words and grammatical rules arise the problem for translating into another language [15]. The example-based machine translation is good for translating the short sentence but it is worst for translating long sentences of the poem [16]. The word substitution in the hybrid approach does not provide desired results as it does not care about the syntactic and semantic constraints of the target language [17, 18]. The machine translation system needed to enhance the performance of the system with complex sentences. The major problem arises in machine translation is due to the unavailability of structural, morphological differences, and word-aligned data during the translation of different languages [19–21].

Contribution

In this paper, we aim to discuss the machine translation system based on the statistical machine translation system (SMT). The study on various SMT and translation models of morphologically rich languages has been carried out. The study gives an insight into how the translation models have been carried out in each of the works. The research finds various challenges associated with the translation of Indian languages into English. One of the challenges is the words sense ambiguity widely found in the Indian language. The sense ambiguity occurs when the certain words sounds and spells similar but the sense of the word differs based on the context of the sentence. Generally, in SMT this problem is not addressed in most of the research. The proposed phrase SMT solves this problem by integrating the WSD with the SMT to improve the translation quality.

Organization

The remaining section of the paper is described as follows: Sect. 1 discusses the background information. Section 2 provides the literature review. Section 3, explains the proposed PSMT method in detail. In Sect. 4, we present the implementation details and results. Finally, Sect. 5 concludes the work.

2 Literature Review

A few sorts of research have been carried out for the poetry translation, accessible in one language to other languages around the globe. Machine translation has rolled out a major improvement in making the Indian language progressively adaptable to learn and comprehend which has been considered by different translation techniques but has to face a few difficulties during translation. The research on the SMT and the details of various translation models carried out are discussed in this section.

Xiong et al. [11] proposed a maximum entropy-based segmentation model for STM. The sentences are spitted into sequences of segments which can be translated. The phrasal extraction is a small module among the collection of modules used in the SMT; after extracting the phrases by maximum entropy, the result is integrated with the SMT. The experiment is conducted with the news domain and is the method that has improved the quality of translation in terms of BELU. Ilknur and Kemal [22] proposed PSMT with the aid of local word reordering. The local word reordering concentrated to obtain the word order of certain English prepositional phrases and verb with respect to the morpheme order of corresponding Turkish verbs and nouns. A morpheme-based language model is used for decoding, and with the aid of word-based model, re-ranking of n best list is handled by the decoder. The decoder output is repaired by correcting the words which have problems like incorrect morphological structure and words which are outside of the vocabulary. Liu et al. [23] proposed a framework where the translation memory is joined with the phrase-based statistical machine translation. The translation memory is integrated with the phrase-based MT through this approach. In the unified framework, different informations are extracted from translator machine (TM) in order to ease the SMT decoding process. The experiment is simulated in a Chinese English TM database. The result shows that there is an improvement in BLEU score. The approach is tested with different models and training data in order to prove the sturdiness of the approach. Pathak et al. [24] proposed a method to translate from Hindi to English with an automatic parallel corpus generation system. The contents of Hindi news and its headline obtained from the media named Navbharat Times are extracted based on the month and year. The unwanted data are removed in the preprocessing step, and with the aid of Google translator API, the crawled contents are translated into English contents. Then align both the news and compare it with the fuzzy string matching algorithm by finding its similarity which was based on Levenshtein distance. Finally, the threshold value was taken from the matching algorithm and if the ratio was more than the threshold value save both the news content as a part of the parallel corpus process. This approach needs to augment the parallel corpus for the translation among two languages. Sharma and Mittal [25] proposed a dictionary-based query translation system that aims to translate the Hindi words into English. In this technique, first, tokenize the queries and before removing the stop words create the multi-word terms by using the n-gram technique for the translation. Next, match the source language terms with the bilingual dictionary if the terms are not translated then move to other cases by calculating the percentage match of the highest common subsequence of the source query. If the query terms are not translated in both the cases then, out-of-vocabulary (OOV) term was used by a rule-based approach would convert them into a roman format. Finally, collect the target language document from the input dataset and the query term was mapped to match the words for the translation. This approach was difficult to translate the named entities into the English language. Subalalitha et al. [26] proposed a template-based information extraction (IE) framework for translating the Tamil poem to English. In this framework, the information can be extracted by template-based information and n-grams-based information. Initially, bilingual mapping was used to translate each word and two features are required to design templates such as term-based and universal networking language knowledge base (UNL KB) which are used to match the words. In the term-based features, there would be an appropriate match between the input and the feature. In some cases, some features were not present in the template, and then UNL KB is used to add semantic constraints for the poem. Finally, the part-of-speech of the words presented in the poem identified by the Tamil morphological analyzer that analyze the noun, verb, adverb, and adjective which extracts the information by n-grams to check its grammar for the translation. This approach has high computational complexity for the translation process. Natu et al. [27] introduced transfer-based machine translation (TBMT) scheme that aims to translate the Hindi text into English. This scheme performed several phases such as preprocessing, tokenizer, part-of-speech (POS) tagging, translation, and grammar check. Initially, the translated sentence should be sophisticated in the preprocessing stage and each word required POS tagging should be segmented into tokens. Then, the Viterbi algorithm used with the hidden Markov model assign sequence of POS tag for the translated words. The segmented words find their translated word from the database for translation. Finally, check the grammar for the words and arranged to determine the final structure of the translated words. This approach needs to enhance the quality of translation gender, number, and tense. Mishra et al. [28] presented the rule-based machine translation (RBMT) scheme to translate Hindi idioms into English. This technique contains two phases such as comparison and translation phases. In the first phase, the system compares the words with the comparison algorithm where the input was searched by the Hindi database. If the input was present in the database, then input belongs to case 1 such as different meanings and different forms in both the language and sends to transfer-based module. Otherwise, it belongs to two cases such as case 2 and case 3 where input idioms with the same meaning and same form, others have the same meaning but different form in both the language in the database, in such a case send them to interlingual-based module. The transfer-based module composed of tokenizer, parser, POS tagger, and declension tagger which translates the idiom by case 2 and case 3. The interlingual-based module composed of input, mapper, database, transfer-based module, and output which translate by case 1. Finally, these modules are used to produce the English translation. This approach needs to enhance the efficiency of the translation of idiom.

3 Proposed Phrase-Based Statistical Machine Translation

The detail of the proposed SMT is discussed in this section. Generally, SMT-based translation methods consist of the collection of small modules which are involved in the translation process.

The PSMT is used to translate the Hindi poetry along with incorporating the WSD approach to sense the disambiguation. The PSMT makes use of the phrases of one or more words in the translation process. The PSMT model first divides the input into phrases. The PSMT is based on the noisy channel model in the information theory.

Fig. 1 Proposed PSMT technique

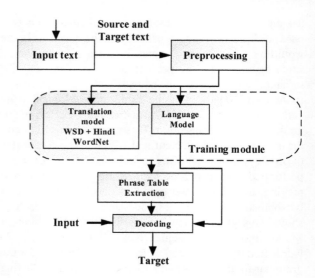

Figure 1 shows the block diagram of the proposed approach. Consider a sentence S of a target language. The sentence S consists of a series of words. The machine translation transfers the source language which is given as $s_1^J = s_1, \ldots s_j \ldots S_J$ into a target language sentence $t_1^I = t_1, \ldots t_i \ldots T_I$. In SMT, the conditional probability is given as $P_r(t_1^I | s_1^J)$. The probability model is used for translation by finding solution for the maximization problem.

$$\hat{t}_1^I = \arg\max_{t_1^I} \{ P_r(t_1^I | s_1^J) \} \tag{1}$$

After applying Bayes' theorem, Eq. 1 is given as

$$= \arg\max_{t_1^I} \{ P_r(t_1^I) . P_r(s_1^J | t_1^I) \} \tag{2}$$

Equation 2 provides the language model $P_r(t_1^I)$ and the target model $P_r(s_1^J | t_1^I)$. The phrase-based translation utilizes the phrase which is the sequence of word. In PBT, the source sentence is segmented into phrase, and each phrase is translated and the target sentence is obtained from the phrase translation.

3.1 Word Alignment Model

Initially, preprocessing is carried with the raw data. In the preprocessing stage, extra spaces and character encoding are resolved by using the AntConc tool from the annotated corpus. With the help of the tokenizer, the Hindi poem is divided into segments by the sequence of words that require POS tagging into units known as tokens. The word alignment is the second step after preprocessing the words. In this proposed work, GIZA++ implementation of the IBM models is employed to perform the word alignment process. The tool runs the algorithm from source to target as well as target to source in both of the directions.

The IBM model evaluates the word-to-word probability of word-to-word alignment for all the source and target word for given sentences. Results of the alignment method create a link between the source and target words. The $\bar{s}$ is a source phrase, $\bar{t}$ is the target phrase, and ($\bar{a}$) is the alignment between source and target which is led by a is valid only if the points do not cross the boundary of the bi-phrase.

3.2 Phrase Table Creation

Once the phrases are extracted with the GIZA++ word alignments, Moses directly use the contents in order to generate phrase table. The sequence of words is stored in phrase table as bi-phrases ($\bar{s}, \bar{t}$) along with the alignment ($\bar{a}$) if the following conditions are satisfied.

The alignment ($\bar{a}$) between the words is led by "a" provided the condition that if there are at least one link and all links from both of their ends ($\bar{a}$) with or none.

The $\bar{S}$ and $\bar{t}$ are word subsequences in the source sentence x and target sentence y, respectively, and both of should not have longer words than K number of words. The phrase extraction finds the same words in both the sentences. For example, consider the two sentences राजू के पास एक महंगा कलम ह, (rahu kae pass kalam hai) and राजू के पास एक महंगा कलम ह(naee nae kalam lo kada). Both of these sentences have the same word kalam. Therefore, there arises a need to exactly sense the meaning of the word. Therefore, the WSD is integrated with the SMT. Once the words are disambiguated, the phrases are extracted.

3.3 Integration of WSD with the SMT

The sense of disambiguation is an important step in the translation process. There are many words which sound similar but their meaning varies based on the sentence. It is found that the word *kalam* has different meaning based on the meaning of the text. The meaning of *kalam* resembles pen as well as trimming based on the context of the sentence. Therefore, it is necessary to distinguish the meaning of the word before reordering of the phrase-based SMT method. A knowledge-based method with the help of Hindi WordNet and Lesk algorithm is employed to sense the disambiguity. The Hindi WordNet is a database with a collection of nouns, adjectives, and adverbs.

3.3.1 Lesk Algorithm

The Lesk algorithm was proposed by Michael E. Lesk in the year 1986. A simplified form of Lesk algorithm is used in the proposed method. The sense of the words is decoded with the principle that the meaning of the word is decided by finding the sense which overlaps the most in WordNet and the given context.

The Lesk algorithm helps to find the correct meaning of words in a context through an individual decision by locating the sense which overlaps the maximum between the dictionary or WordNet definition. An example gives more idea regarding the working of the Lesk algorithm. If the word *kalam* is searched on the WordNet, the senses of the word are found out. The sense which is maximum overlapped will be the output of the algorithm. Consider the example, पास. It means that the person named raju has a costly pen. The words are searched in the Hindi WordNet. In the sentence, the words like महंगा(pass), महंगा(mahanga), कलम(kalam) have different senses based on the context. The search on word net shows various meanings for each of the word. The Lesk algorithm helps us to find the actual words based on the usage in a sentence. According to the Lesk algorithm, the best sense of *kalam* is pen. Similarly, other words are also processed and matched with the correct sense. Once the word disambiguity is identified, the phrase extraction is handled. Figure 2 shows the extracted phrases of the poem.

The features of the phrase table generated with Moses are as follows. The Moses phrase table consists of five features for each bi-phrase. They are phrase translation probability, lexical weighting, phrase inverse translation probability, inverse lexical weighting, phrase penalty. The value of the first four features takes values between 0 and 1. The decoder uses the features directly in order to generate a phrase table. With the aid of the tool Mosses, the phrase pairs are extracted and phrase pair score is calculated according to Eq. 3.

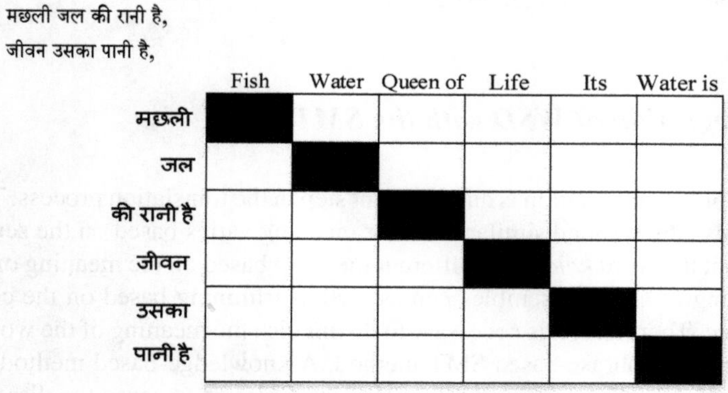

मछली जल की रानी है,
जीवन उसका पानी है,

	Fish	Water	Queen of	Life	Its	Water is
मछली	■					
जल		■				
की रानी है			■			
जीवन				■		
उसका					■	
पानी है						■

Fig. 2 Phrase table

3.4 Reordering Model

The reordering technique is adopted from [29]. The technique utilizes the phrasal dependency tree in order to order the translated words. The dependency relation between the contagious synthetic non-syntactic phrases is used in the model.

3.5 Language Model

The language model employed in this proposed work is obtained from the connecting phrase. The connecting phrases are constructed based on the steps followed in [30]. The probability of a target phrase in SMT is given as

$$P_T(e) = \sum_f P_s(f) * P(e/f) \quad (3)$$

where $P_s(f)$ is the probability of source phrase. The connecting phrases are created at first and then by employing the average probability of the connecting phrases is employed to take a decision such that which of the connecting phrases should be used.

The probability of the connecting phrases is given as

$$p_{\text{connecting}}(w_1^n w_{n+1}^k) = \sum_{n=1}^{k-1} \left(\sum_{\beta} P_{t \arg et}(\beta w_1^n) * \sum_{\gamma} P_{t \arg et}(w_{n+1}^k \gamma) \right) \quad (4)$$

The N-grams are generated based on the steps like splitting, replacing, and renormalizing. A N-gram pruning method based on the phrase table is constructed based on the two conditions that the phrase table already consists of the contents, and the contents are result of the concatenation of two or more phrases in the phrase table. The connecting phrases are created based on the probability as in Eq. 4. With the help of the threshold, each of the probability which is higher than the threshold is retained.

4 Experimental Setup and Result Analysis

(a) *Experimental setup*

The Moses tool is used to evaluate the proposed method. The Moses is an open-source tool widely used for the translation purposes. Adding to Moses python interface in Moses is used. We evaluate the MT output with the python-based evaluation tool since MT evaluation system does not provide interface to evaluate the output. Thus, we calculate the precision and recall through the designed evaluation tool. The Moses possess a BLEU scoring tool named multi-bleu.perl. The Moses also have another one popular tool named NIST mteval script. The text needs to be converted into SGML format. The proposed system employs the NIST mteval script for calculating the BLEU score.

The datasets for the proposed system consist of 3 text files where the parallel dataset consists of 2 text files with 5000 lines where one is for Hindi poetries and second is for translating into English poetries. The third one consists of monolingual dataset containing around 10,000 lines for English Poetries.

(b) *Result analysis*

The details of the performance metrics are provided under this section. The precision method is a general evaluation method in MT. It is calculated from the number of correct words and the output length of the MT. The precision is given as

$$\text{Precision} = \text{Number of correct words/output length of MT} \qquad (5)$$

The recall is obtained from the number of correct words divided by the reference length. The recall is given as

$$\text{Recall} = \text{Number of correct words/reference length of translation} \qquad (6)$$

The BLEU score is used to find the accuracy of the proposed approach for the translation of Hindi poem into English poem, and the evaluation is based on the geometric mean of the modified n-gram precision p, effective corpus length be r, poem translation length be l, N is the number of words.

$$\text{BLEU} = \min\left(1 - \frac{r}{l}\right) + \sum_{n=1}^{N} W_n \log p_n \qquad (7)$$

The proposed PSMT's result is compared with rule-based machine translation (RBMT) and transfer-based machine translation (TBMT) with the same dataset used for the PSMT. In order to study the effect of WSD in the proposed method, the simulation is carried with and without WSD as shown in Table 2.

The proposed PSMT method is compared with TBMT and the RBMT translator as shown in Table 1. The precision metric increased from 83.11 to 94.02%, the recall metric improved from the range of 87.45–93.3%, the BLEU metric increases from

Table 1 Performance metrics comparison

MT	Precision	Recall	BLEU
TBMT	83.11	87.45	0.0695
RBMT	92.05	91.09	0.7663
PSMT	94.02	93.3	0.9024

Table 2 Effect of WSD on the proposed method

Method	Precision	Recall	BLEU
Proposed with WSD	84.05	82.09	0.726
Proposed without WSD	83.02	93.3	0.9024

0.0695 to 80.04%. The proposed system provides more accurate results than the other systems.

Table 2 shows the effect of WSD by comparing the proposed method with and without WSD. There is a change in the performance metrics when the WSD approach is applied. The proposed approach is examined with Hindi sonnet which is translated into English interpretations in Table 3 on three machine translation frameworks.

Table 3 shows that the poems have been converted with a good quality than the existing methods.

Table 3 Proposed approach versus Tbmt and Rbmt for Hindi poetry translation

S. No.	Hindi poetry	TBMT	RBMT	Proposed approach
1.	चांद मामा आओ ना, दूध-बटासा ना हो ना, मीठी लोरी गाओ ना, बिस्तार में है ना ना	Chand Mama Ao Naa Naa, Dhood-Batasa Nana, Sweet Lullaby Sing Naa, Bistar is not.	Come on, visit the moon, don't have to go to batsa, you're not in the sweet, but not in the bihar	The moon may not be able to come, be able to find a sweet, sweet, little gait, it is in the bed is n't it.
2.	बादल राजा, बादल राजा जल्दी से पनि बरसा जा। नन्हे-मुन्ने झूलों में हैं। धरती की तू प्या भुज जा। जल्दी से पनि बरसा जा	Cloud King, Cloud King Water may be showered with water. The little ones are in the swings. Go to the love arm of the earth. water-watering	The king, The King, pour over the clouds with the king of thunder. The little boy is in the swing. You go to Bhuj. Pani Pani Pani.	Rain king, rain the king come quickly to give water. The tiny ones are in dangles. Give you to the earth. rain down with water quickly
3.	हाथी आया, हाथी आया सूद हिलता है हाथ आया चलत तीर्थ हठ अया जुम जुम कर हाथ आया कान हिलता है हाथ आया	Elephant came, elephant came The trunk shakes the elephant came Walking Elephant Arrives Zoomed in, the elephant came Ear Shaking Elephant Arrives	The elephant came and the elephant came sniffing on the nose of the elephant, and the elephant came to him and the elephant came to him and the elephant came to him	The elephant came, elephant came into his hands, shaking his hands, and a mass of inlet-touched his hands, shaking his hands, shaking the hands, shaking his hands.

5 Conclusion

In the present era, machine translation is the important research in natural language processing area. We have introduced phrase-based STM for the interpretation of Hindi poem into the English language. We have developed a translator for Hindi to the English language, which provides the best interpretation with the best precision. The approach has been successful in improving the quality of translation since the WSD model is also included to encounter the word sense ambiguity. The usage of Hindi WordNet and Lesk Algorithm has been able to provide the exact meaning of the words as well as it improved the quality of translation. Future work integrates into a significance examination of the proposed component to deal with the complex Hindi/English sentence, different varieties of the same word and consideration of increasingly linguistic sentence the author/year convention are also acceptable.

References

1. Young, M.: The Technical Writer's Handbook. Mill Valley, CA: University Science (1989). Sinha, R.M.K.: A system for identification of idioms in Hindi. In: 2014 Seventh International Conference on Contemporary Computing (IC3) IEEE, pp. 467–472 (2014)
2. Dutta-Roy, S.: Negotiating between languages and cultures: english studies today. In English Studies in India, pp. 61–72. Springer, Singapore (2019)
3. Alqudsi, A., Omar, N., Shaker, K.: Arabic machine translation: a survey. Artif. Intell. Rev. **42**(4), 549–572 (2014)
4. Daud, A., Khan, W., Che, D.: Urdu language processing: a survey". Artif. Intell. Rev. **47**(3), 279–311 (2016)
5. Bouhriz, N., Benabbou, F., Lahmar, E.B.: Word sense disambiguation approach for Arabic text. Int. J. Adv. Comput. Sci. Appl. **7**(4), 381–385 (2016)
6. Sarmah, J., Sarma, S.K.: Decision tree based supervised word sense disambiguation for Assamese. Int. J. Comput. Appl. **141**(1), 42–48 (2016)
7. Zhou, J., Yang, J., Song, H., Ahmed, S.H., Mehmood, A., Lv, H.: An online marking system conducive to learning. J. Intell. Fuzzy Syst. **31**(5), 2463–2471 (2016)
8. Sreenivasan, D., Vidya, M., Sreenivasan, D., Vidya, M.: A walk through the approaches of word sense disambiguation. Int. J. Innov. Res. Sci. Technol **2**(10), 218–223 (2016)
9. Mittal, K., Jain, A.: Word sense disambiguation method using semantic similarity measures and OWA operator. ICTACT J. Soft Comput. **5**(2) (2015)
10. Pathak, A. K., Acharya, P., Balabantaray, R. C.: A case study of Hindi–English example-based machine translation. In: Innovations in Soft Computing and Information Technology, pp. 7–16. Springer, Singapore (2019)
11. Xiong, D., Zhang, M., Li, H.: A maximum-entropy segmentation model for statistical machine translation. IEEE Trans. Audio Speech Lang. Process. **19**(8), 2494–2505 (2011)
12. Singh, M., Kumar, R., Chana, I.: Neural-based machine translation system outperforming statistical phrase-based machine translation for low-resource languages. In: 2019 Twelfth International Conference on Contemporary Computing (IC3), pp. 1–7 (2019)
13. Singh, M., Kumar, R., Chana, I.: Neural-based machine translation system outperforming statistical phrase-based machine translation for low-resource languages. In: 2019 Twelfth International Conference on Contemporary Computing (IC3), pp. 1–7 (2019)
14. Mohaghegh, M., Sarrafzadeh, A.: A hierarchical phrase-based model for English-Persian statistical machine translation. In 2012 International Conference on Innovations in Information Technology (IIT), pp. 205–208 (2012)

15. Kaur, A., Goyal, V.: Punjabi to English machine transliteration for proper nouns. In: 2018 3rd International Conference on Internet of Things: Smart Innovation and Usages (IoT-SIU), pp 1–7 (2018)
16. Masroor, H., Saeed, M., Feroz, M., Ahsan, K., Islam, K.: Transtech: development of a novel translator for Roman Urdu to English. Heliyon **5**(5), e01780 (2019)
17. Ayu, M.A., Mantoro, T.: An example-based machine translation approach for Bahasa Indonesia to English: an experiment using MOSES. In: 2011 IEEE Symposium on Industrial Electronics and Applications, pp. 570–573 (2011)
18. Kaur, V., Sarao, A.K., Singh, J.: Hybrid approach for Hindi to English transliteration system for proper nouns. Int. J. Comput. Sci. Inf. Technol. (IJCSIT) **5**(5), 6361–6366 (2014)
19. Ye, Z., Jia, Z., Huang, J., Yin, H.: Part-of-speech tagging based on dictionary and statistical machine learning. In 2016 35th Chinese Control Conference (CCC), pp. 6993–6998 (2016)
20. Mall, S., Jaiswal, U. C.: Developing a system for machine translation from Hindi language to English language. In 2013 4th International Conference on Computer and Communication Technology (ICCCT), pp. 79–87 (2013)
21. Chakrawarti, R.K., Mishra, H., Bansal, P.: Review of machine translation techniques for idea of Hindi to English idiom translation. Int. J. Comput. Intell. Res. **13**(5), 1059–1071 (2017)
22. El-Kahlout, I.D., Oflazer, K.: Exploiting morphology and local word reordering in English-to-Turkish phrase-based statistical machine translation. IEEE Trans. Audio Speech Lang. Process. **18**(6), 1313–1322 (2009)
23. Liu, Y., Wang, K., Zong, C., Su, K.Y.: A unified framework and models for integrating translation memory into phrase-based statistical machine translation. Comput. Speech Lang. **54**, 176–206 (2019)
24. Pathak, A.K., Acharya, P., Kaur, D., Balabantaray, R.C.: Automatic parallel corpus creation for Hindi–English news translation task. In 2018 International Conference on Advances in Computing, Communications and Informatics (ICACCI) (pp. 1069–1075). IEEE (2018)
25. Sharma, V.K., Mittal, N.: Cross-lingual information retrieval: a dictionary-based query translation approach. In: Advances in computer and computational sciences (pp. 611–618). Springer, Singapore (2018)
26. Subalalitha, C.N.: Information extraction framework for Kurunthogai. Sādhanā **44**(7), 156 (2019)
27. Natu, I., Iyer, S., Kulkarni, A., Patil, K., Patil, P.: Text translation from Hindi to English. In: International Conference on Advances in Computing and Data Sciences, pp. 481–488. Springer, Singapore (2018)
28. Mishra, H., Chakrawarti, R.K., Bansal, P.: Implementation of Hindi to English idiom translation system. In: International Conference on Advanced Computing Networking and Informatics, pp. 371–380. Springer, Singapore (2019)
29. Farzi, S., Faili, H., Khadivi, S., Maleki, J.: A novel reordering model for statistical machine translation. Res. Comput. Sci. **65**, 51–64 (2013)
30. Schwenk, H., Dchelotte, D., Gauvain, J.L.: Continuous space language models for statistical machine translation. In: Proceedings of the COLING/ACL on Main conference poster sessions, pp. 723–730. Association for Computational Linguistic (2006)

A Smart Education Model for Future Learning and Teaching Using IoT

Swati Jain and Dimple Chawla

Abstract The Internet of things (IoT) is a giant network of heterogeneous things namely people, intelligent objects and devices, information and data all of which collect and share data about each other and the environment which they are a part of. IoT is a revolutionary technology that has already started to upgrade human lives in all aspects: Smart Cities, Smart Healthcare, Home Automation, Safety and Security, and Education. With the advent of the Internet of Things (IoT) in the education sector, the complete education model has changed enormously. It has created a platform for information sharing and communication between the people and the environment in any educational organization. This paper explains the basics of IoT and aims at highlighting all the practical methods for integrating IoT features in education. It classifies the impact of IoT on education into a three-dimensional model with Stakeholders, Applications, and Learning Modes as the dimensions. It classifies the applications into four areas: campus management, class management, disability accommodation, and improving students learning. The author has focused on how these three IoT dimensions will help in making a smarter education model.

1 Introduction

Education sector has always embraced new developments in Information Technology and Communication (IT&C). The term "Education" in today's era is not only restricted to textbooks, but it is more about a digital environment of knowledge and information. The influence of technology is prominent on different aspects of education from increasing student engagement in the learning process through digital content to creating personalized content to improve student's understanding

---

S. Jain (✉) · D. Chawla
Vivekananda Institute of Professional Studies, Delhi, India
e-mail: jainswati3107@gmail.com

D. Chawla
e-mail: chawladel@gmail.com

© The Editor(s) (if applicable) and The Author(s), under exclusive license
to Springer Nature Singapore Pte Ltd. 2021
T. Senjyu et al. (eds.), *Information and Communication Technology
for Intelligent Systems*, Smart Innovation, Systems and Technologies 196,
https://doi.org/10.1007/978-981-15-7062-9_7

and outcomes. One of the most powerful aspects of technology is IoT. In simple terms, Internet of Things (IoT) is made up of any device with an On/Off switch connected to the Internet [1]. It is basically a giant network of interconnected things and devices ranging from mobile phones to fitness bands to jet engines of airplanes. According to a study conducted by Juniper Research [2], the total number of IoT devices and sensors will exceed 50 billion by 2022, up from an estimated 21 billion in 2018 [2]. Since IoT technology is expected to boost remarkably in the upcoming years, the aim of this paper is to explore what are the different aspects of education which are affected by the Internet of Things (IoT) and how they can be used to improve the education sector as a whole along with the improvement of online learning and teaching.

2 Literature Review

According to a report "IoT in Education Market—Global Industry Analysis 2018–2026" [3] IoT in Global education market is expected to grow from 4.8 Billion USD in 2018 to 11.3 Billion USD by 2023, the Compounded Annual Growth Rate of IoT in education would be 18.8% in the forecasted period and education sector had the biggest year on year percentage increase for IoT. The report also predicted that the two major factors responsible for this growth are increased use of connected devices in institutions and fast adoption of E-learning and digital content. Asim Majeed et al. [4] in their paper titled "How Internet of Things (IoT) Making the University Campuses Smart? QA Higher Education (QAHE) Perspective" Proposed a model for smart-university campus using IoT as a service. The use of smart wearable devices, sensors, and actuators were suggested to collect mass data and take meaningful actions. They also explained how IoT can help in making campuses safe through the use of NFC and RFID technology. Narendar Singh et al. [5] and Sri Madhu et al. [6] presented different practical models for automating the attendance process through IoT. They made use of face recognition technology along with IoT PIR sensors to capture images of students present in class and mark their attendance automatically and for faculty attendance, RFID tags were used. Munib ur Rahman et al. [7] in their work on "using IoT and ICT for creating a smart learning environment for students" explained different ways of promoting learning among students such as Virtual Classrooms, Centralized lab servers, home-school connectivity, and gamification of tests and assignments. They suggested implementations of these methods will help those students who are not able to attend classes. Madhu Rao et al. [8] in their work proposed a complete smart campus based on IoT which includes various smart features like smart parking system which can enable students to find an empty parking spot on the campus through a mobile application, smart street light system which works on motion sensors and light intensity automatically reduce when no motion is detected, smart library with features of online book issue requests, status of issued books, availability of books in library. The smart campus will also have other features like smart gardening, smart weather monitoring, and smart office.

3 Three-Dimensional Impact of IoT on Education

The impact of IoT on the education business model can be represented across three dimensions: Applications, Learning Mode, and Stakeholders (Fig. 1).

3.1 Learning Modes

There are different ways in which classes can be conducted using IoT.

3.1.1 Collaborative Learning [9]

It is a concept where students present at different locations of the world can be a part of the same classes through ICT. It takes into consideration the power of intelligence and knowledge of a bigger set of people and improves the learning performance that would be achieved by a single participant alone. This helps to develop critical thinking abilities and problem-solving techniques by facilitating interaction between a diverse set of students.

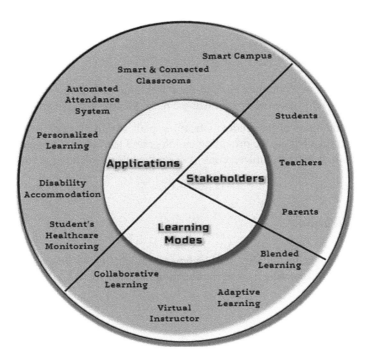

Fig. 1 The three-dimensional impacts of IoT on the education model

3.1.2 Virtual Instructor

In this learning mode, the focus is on making the video lectures available after the classes. Kalimuthu et al. [10] have proposed a system of a virtual classroom with IoT. It suggests the use of Raspberry Pi and webcam to automatically live stream and record the lecture when it starts and uploads on the cloud when it ends. A virtual classroom setup like this will help to improve the learning and understanding of the concepts as the complete subject lectures will be available to students and can be viewed a multiple number of times [10].

3.1.3 Adaptive Learning

This is a data-driven approach that facilitates personalized learning for everyone. It connects the classroom to real-life environmental conditions such as the number of student present, room temperature, and student activity, etc. through the use of various multimedia sensors transmitting data to a central server. It adjusts the path and pace of the teaching based on the data gathered from the student. For example, Ultrasonic sensors can be used to identify the empty seats in the front of the class, light sensors can detect the light intensity of the room and send the value to the cloud for any further action, like turn on the lights when the first person enters the room and turn off when everyone leaves [11].

3.1.4 Blended Learning

This learning mode combines Face-to-Face teaching with E-learning to a certain degree. The benefits associated with using a hybrid learning mode are as follows:
- All the modalities of the student's learning path within the specified course are connected together, hence providing an integrated learning environment [12].
- Students will have complete control over the place, path, time, and pace of learning.

3.2 Applications

Internet of Things technology will enhance the quality of education and will contribute towards creating a smarter education model in the following ways.

3.2.1 Smart Campus

Smart campus is a connected campus with embedded IoT technology making the interaction between environment, people, and objects possible. Embedded IoT sensors and actuators gather data around the campus and transmit it to the relevant applications to further processing and actions [4].

Features of various smart campus around the world:

- Sookmyung Women's University (SWU) in South Korea has connected all its stakeholders together through one mobile app Smart Sookmyung. Through this app, students can mark and monitor their attendance, check availability of seats in the library can and reserve seats if needed.
- Bournville College in the UK controls 400 doors throughout the campus online and monitors foot patterns across campus through access control technology.
- UNC-Wilmington shows the availability of washers and dryers to students through laundry app.
- Boston University uses sustainable trash cans which alerts the authority every time the cans are full. This reduced the average pickup required from 14 to 16 times per week.

3.2.2 Smart and Connected Classrooms

A classroom with an intelligent environment and connected with different kinds of hardware and software modules like Smartboards, projectors, sensors, cameras are considered a smart and connected classroom. These devices monitor various parameters of the class environment such as temperature, humidity, CO_2 levels, etc. and the student's attributes such as attendance, assignments submitted, progress reports, etc. A smart classroom system using these sensors can determine in real-time if the environment is appropriate to boost student's concentration and learning ability. For example, when the temperature rises above a certain level, automatic cooling systems would be turned on to ensure it doesn't affect student's focus on the class. The main purpose of creating a smart and connected classroom like this is to improve the teaching and learning environment [13].

3.2.3 Automated Attendance System

With the aim of eliminating the attendance management and monitoring issues, an automated attendance system can be adopted in Higher education institutions with the help of IoT. The two different models which can be used for automated attendance system are as follows:

- *Radio Frequency Identification (RFID)* technology along with the Raspberry Pi and PIR sensors can create a fully automated attendance system where the RFID tag in students' ID cards will be used to mark attendance. When the student enters the class, PIR 1 will be activated, the RFID reader will read the UID on the students' ID card. The attendance for the student will be marked only when the student enters the class and PIR sensor 2 is activated [6].

- *Face detection and Face recognition* technology where a live stream of students present in class is captured by a camera connected to Raspberry Pi. These images are then compared with stored student images to recognize the students present and mark their attendance [5].

Automated attendance monitoring eliminates the need for manual attendance maintenance and also avoids any mistakes or adoption of unethical ways to mark attendance.

3.2.4 Personalized Learning

The major shortcoming of the traditional education system is that the same course work is designed for everyone. The flexibility to accommodate and acknowledge the learning potential of all the students is not present. With smart classes, each student can be monitored individually and data can be collected for individual's performance in class, assignments, and exams. This repository of data will enable tutors to create a more student-centric learning approach, i.e., weaker students may be given an extra practice assignment to help them cope up with the class.

3.2.5 Disability Accommodation

IoT classrooms can eliminate the need for establishing separate classes and curriculum for students with special needs by integrating the following features in traditional classroom setup:

- Screen Reader: A text-to-speech converter to read out the contents on screen for visually impaired or blind students.
- On-Screen alerts: visual messages can appear on the screen instead of audible notifications for the aid of hearing-impaired students.
- Themes: High-contrast themes for visually impaired students allowing them to set the screen to a more comfortable color theme and text size.
- Speech-to-text: for visually and physically impaired students speech-to-text converter can be embedded for examinations.
- Live streaming: for physically handicapped students facility of live streaming of lectures can be provided to help them learn at the comfort of their home.

3.2.6 Student's Healthcare Monitoring

Student's health plays a vital role in determining his overall academic performance, thus access to health care services is important in any educational institution [13]. Wearable technology is one of the ways to monitor health using IoT. Such healthcare wearable devices demonstrate the use of technology in an efficient manner; few of them are like smart continuous glucose monitoring (CGM), connected inhalers, ingestible sensors, and connected contact lenses.

3.3 Stakeholders

The second dimension of model IoT's impact on education (Fig. 1) depicts that the people who will directly be benefitted from the use of IoT in the education model are Teachers, Students, and Parents. Ways in which each of the stakeholders will be affected are as follows.

3.3.1 Teachers

In a traditional classroom setup, the teacher is responsible from teaching, writing down notes on board, taking attendance, maintaining assignment, and result records to maintaining the classroom environment. From a teacher's perspective, IoT in classes will help manage these responsibilities with ease. For example:

- Automated attendance system where students would be marked present based on their RFID tags.
- Notes will be available automatically on the cloud before every class and all the assignments will be submitted online by the students and all the student records will be managed automatically by the system.
- Different multimedia content such as tutorial videos, infographics, complex formulae, etc. could be easily presented in class with the help of smart-interactive boards instead of using white chalkboards.

3.3.2 Students

From a student's perspective, IoT is going to change the education model as follows:

- With the use of hybrid learning mode, all the subject notes, lectures, assignments, and tests would be available at a common platform and students can see the video lecture at any time and any place.
- IoT classrooms will help in incorporating disability both physical and mental as the students can now learn at their own pace and place without the undue stress of keeping up with the pace of the rest of the class.
- Students will now be free from the need of carrying books and notes to classes. All the resources can be accessed online through laptops, tablets, or smart phones.
- With collaborative learning, students from around the world can be a part of the same class. This helps the students widen their knowledge horizon removing the location barrier.

3.3.3 Parents

Parents as one the stakeholders will be affected by smart campuses and smart classrooms using IoT in the following ways:

- With smart wearable devices, parents can be informed as soon as their child enters the campus, live location of every on-site student can be tracked ensuring complete safety for every student on campus.
- Automated attendance and result systems can keep parents updated on their child's attendance, scorecards, class assessment records, and submission reports.
- Student's healthcare monitoring system on smart campus, student's wearable devices will keep track of student's vital health conditions based on their medical history, for example, blood pressure and heart rate monitoring, which will be available to both students and parents.

4 Conclusion

IoT has become an urgent need for the education industry and their technical and management activities. This paper has shown a great impact of IoT on education by defining three-dimensional models with its applicability areas. Also, this study was motivated by the gaps in teaching methodology and several issues related to practical methods for contributing IoT features in education. However, IoT is bringing new opportunities for schools as well for B-Schools by aiding teachers to secure buildings with connected devices, to embrace more schools from simple classroom systems to smarter education models which will reap students for learning benefits.

References

1. Foote, K.D.: A Brief History of the Internet of Things. https://www.dataversity.net/brief-history-internet-things/ (2016)
2. Juniper Research: IoT—The Internet of Transformation (2018)
3. Markets and Markets Analysis, IoT in Education Market—Global Industry Analysis, Size, Share, Growth, Trends, and Forecast 2018–2026, Report Code: TC 6855. www.marketsandmarkets.com/IoT-Education/Market-Report (2019)
4. https://www.juniperresearch.com/press/press-releases/iot-connections-to-grow-140pc-to-50-billion-2022 (2018)
5. Tew, Y., Tang, T.Y., Lee, Y.K.: A study on enhanced educational platform with adaptive sensing devices using IoT features. In: Proceedings of APSIPA Annual Summit and Conference 2017 (2017)
6. Narendar Singh, D., Kusuma Sri, M., Mounika, K.: IOT based automated attendance with face recognition system. Int. J. Innov. Technol. Expl. Eng. (IJITEE) 8(6S4) (2019). ISSN: 2278-3075

7. Sri Madhu, B.M., Kanagotagi, K., Devansh: IoT based automatic attendance management system. In: International Conference on Current Trends in Computer, Electrical, Electronics and Communication (ICCTCEEC-2017), IEEE, 978-1-5386-3243-7/17 (2016)
8. Madhura Rao, N., Swathi, R., Sneha, M., Kotian, S., Rao, N: An IoT based smart campus system. Int. J. Sci. Eng. Res. **9**(4), 146–151 (2018). ISSN 2229–5518
9. Mahmood Ali, A.M.: How Internet-of-Things (IoT) Making the University Campuses Smart? QA Higher Education (QAHE) Perspective. IEEE. 978-1-5386-4649-6/18 (2018)
10. Satu, M.S., Roy, S., Akhter, F., Whaiduzzaman, Md.: IoLT: an IOT based collaborative blended learning platform in higher education. In: International Conference on Innovation in Engineering and Technology (ICIET). IEEE (2018). 978-1-5386-5229-9
11. el-zakhem, I.: Use of a blended learning approach for an internet of things course. Int. J. Adv. Comput. Eng. Netw. **5**(7) (2017). ISSN: 2320-2106
12. Kalimuthu, M., Prakash, N.K.V., Ponra, A.S.: IoT based virtual classroom. In: National Conference on Science, Engineering and Technology (NCSET) **4**(6), 46–49 (2016). ISSN: 2321-8169
13. Bagheri, M., Movahed, S.H.: The Effect of the Internet of Things (IoT) on Education Business Model. IEEE. https://doi.org/10.1109/sitis.2016.74 (2016)

Internet of Things (IoT)-Based Advanced Voting Machine System Enhanced Using Low-Cost IoT Embedded Device and Cloud Platform

Miral M. Desai, Jignesh J. Patoliya, and Hiren K. Mewada

Abstract The traditional voting system is basically of two major types. One is voting through ballot paper and another one is voting through electronic voting machine (EVM). Voting system through ballot paper requires so much resources as well as security. There will be maximum possibility of malfunctioning in case of ballot paper-based voting machine. Electronic voting machine-based voting system is better than ballot paper-based system, but it is not authenticated. This paper describes the design of smart and secure electronic voting machine based on the IoT platform. The suggested system is more efficient than both traditional systems, as both traditional systems are time consuming and also not authenticated. The proposed system functions into two specific phases. One is user authentication, and another is user voting. Authentication process can be done using fingerprint authentication. Fingerprint database of all the voters is stored in the system initially as prerequisites. If any person wants to vote to any party, the authentication of respective person is to be done by fingerprint matching process. Once the fingerprint matches successfully, the person can vote to any specific party. Data analysis can be done in form statistics of the percentage voting of individual party and is to be uploaded on the web server as well as Google spreadsheet. Due to the fingerprint authentication method, malfunctioning like fake voting and repeat vote can be avoided. As the system is based on the fingerprint authentication, in future it can be linked with the Aadhaar Card of the respective person.

M. M. Desai (✉) · H. K. Mewada
Department of EC Engineering, CSPIT, CHARUSAT, Changa, Gujarat, India
e-mail: miraldesai.ec@charusat.ac.in

H. K. Mewada
e-mail: hmewada@pmu.edu.sa

J. J. Patoliya
Electrical Engineering Department, Prince Mohammad Bin Fahd University, Al Khobar, Kingdom of Saudi Arabia
e-mail: jigneshpatoliya@charusat.ac.in

© The Editor(s) (if applicable) and The Author(s), under exclusive license to Springer Nature Singapore Pte Ltd. 2021
T. Senjyu et al. (eds.), *Information and Communication Technology for Intelligent Systems*, Smart Innovation, Systems and Technologies 196,
https://doi.org/10.1007/978-981-15-7062-9_8

1 Introduction

In democratic ruling system, voting is the important weapon to choose the right candidate as a leader among all eligible candidates who appear in the race of becoming the right leader. An honest "vote" can elect the honest candidate, and ultimately it can build efficient and honest Governance. If the Governance of any country is strong, it can increase standard of living of democracy. People can choose the right candidate—representative of individual party by giving vote to the right party. To make the voting process smooth and transparent, electronic voting machine is used. In earlier days, people were using manual voting system, in which people have to submit the ballot paper of individual party in voting box. At the end of voting process, election committee officers count the ballot paper of individual party. At the end of counting process, committee members declared winner party who has maximum ballot papers. In traditional voting system, there would be chance of malfunctioning like repeat voting, fake voting, mistake in counting ballot papers, etc. Electronic voting system overcomes this problem as counting of ballot papers is to be done automatically by the machine itself. Electronic voting systems are also not up to the marks as design of system firmware is not proper, and there would be a chance of malfunctioning in the design of firmware itself. Sometimes, the authentication process is not properly designed in the electronic voting system [1–3].

To overcome the problems of manual voting system and traditional electronic voting system, it is essential to design smart voting system. The suggested system is the design of smart voting system based on the Internet of Things (IoT) platform. The system is becoming secure as authentication is to be done by scanning the fingerprint of voter. Purpose of doing authentication using fingerprint is to enable the electronic ballot reset for allowing voters to cast their votes and also enables to send the vote details directly to the server. So the authentic person can only vote for the specific party. If user is authenticate, then and then machine will accept the vote. As the system is based on the IoT platform, it will send all the statistics on the web server. The functional block diagram of the system is shown in Fig. 1 [4].

As shown in Fig. 1, when the system is started with the voting of registered voter, the system will check authenticity using biometric authentication process. Once the authenticity is checked, the voting process will be started by allowing voter to give the vote to the specific party. The specific party will be selected by pressing the button for individual party for only once. At the regular interval, the voting data will be uploaded on Google cloud and Google spreadsheet by pressing upload button.

The proposed paper is organized in the given form. Paper starts with the introduction as Sect. 1; the objective of the system is described in Sect. 2. Literature review is described in Sect. 3. Hardware level design is in Sect. 4. Section 5 described experimental setup and software environment. Test results are discussed in Sect. 6. Conclusion is mentioned in Sect. 7. Paper is concluded with the references.

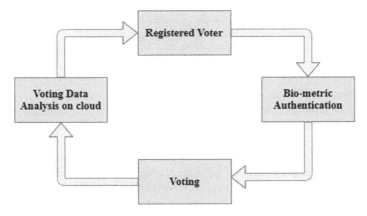

Fig. 1 Functional block diagram of proposed system

2 Objective

The objective of the system is to design smart and efficient electronic voting machine, where biometric authentication can be done as well as statistics of voting can be updated on cloud through IoT media.

3 Literature Review

Motivation of designing smart voting machine based on IoT platform comes from some existing electronic voting machines. Existing electronic voting machines are vulnerable to attacks. Attacks could be physical access or remote access of the system. The attack could be like anyone can change the existing code of the system by malicious code and malicious code can steal votes or changes the votes from one party to another party [5]. One of the existing systems is electronic voting machine using biometric identification. In this existing system, vote can be given by accessing voter's fingerprint. Then it will be matched with the internal database of fingerprints. Fingerprint-based biometric voting machine can be implemented on Arduino using fingerprint sensor module [6]. But the limitation of this type of system is that we cannot keep the record of voting data on to the server. Cryptographic voting scheme is used to provide transparency and enable fast tally. The problem of cryptographic voting machine can be overcome using Bingo Voting [6]. Smart voting machine can also be applicable by linking the Aadhaar Card of the voter with the voting machine [7]. The proposed system can overcome all above-stated issues.

The design of smart voting machine can be done by some hardware component as well as specific software environment. The hardware component consists of

Arduino UNO board, fingerprint sensor module, Wi-Fi module—ESP32 and LCD. The software environment consists of HTTP along with HTML.

Arduino UNO Board is shown in Fig. 2. The Arduino UNO board is ATmega328-based microcontroller which is working on 16 MHz frequency. The board consists of inbuilt ADC and USB connection port [8]. Arduino C is the derived programming language from C/C++ which contains inbuilt hardware-based libraries.

Fingerprint Sensor Module is shown in Fig. 3. Using this fingerprint module, user can store finger print data and can configure it in 1:1 or 1: N mode for identifying the person. The module can interface with any microcontroller using 3V3 or 5 V [9].

ESP-32 Wi-Fi Module as shown in Fig. 4 comes with a built-in Wi-Fi chip as ESP8266, USB connector and rich assortment of general-purpose input–output pins.

Liquid Crystal Display (LCD) is used as the remote notice board display. The pin-out diagram of 16 * 2 size LCD is shown in Fig. 5.

HTTP stands for Hypertext Transfer Protocol which is an application layer protocol. It is used mainly to send and receive information to the webpage and from the webpage. HTTP is the beginning of the modern-day Internet which was invented around 1990s. HTTP is a connectionless protocol, thus the information sent

Fig. 2 Arduino UNO board

Fig. 3 Fingerprint sensor module [9]

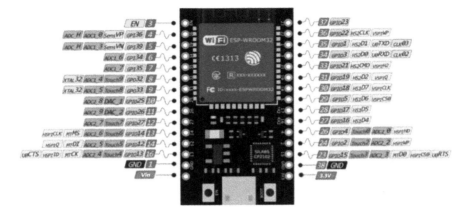

Fig. 4 ESP-32 pin-out diagram

Fig. 5 LCD pin-out diagram

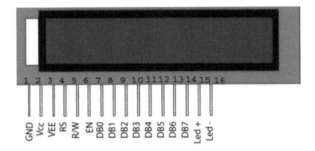

between the client and server does not require a constant connection. The client can make request to any remote server and then the connection is terminated as soon as the request is met. Once the server is ready with the response, connection is re-established and the data is transferred. Another feature of HTTP which adds to the versatility of the protocol is that any type of data can be sent over it. This makes the job of designer much easier as any data can be sent to any server without any external needs being met, thus HTTP is media independent. The server and client are aware of each other only when the data transfer between them is taking place, and the rest of the time they have no information about each other, so HTTP is a stateless protocol. A typical HTTP request will look alike [12].

String ("GET") + /output.php + "HTTP/1.1\r\n" + "www.iotnb.com" + host + "\r\n" + "Connection: close\r\n\r\n"

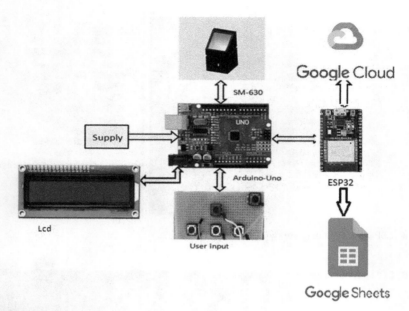

Fig. 6 Hardware level design

4 Hardware Level Design

Hardware level design of the proposed system is shown in Fig. 6. It consists of Arduino UNO as a main system on a chip (SoC) board, ESP32 as a Wi-Fi module, push buttons as user inputs, 16 * 2 LCD as validation display, fingerprint module (SM-630), Google cloud as server and power supply.

5 Experimental Setup and Software Environment

Experimental setup of the suggested system is shown in Fig. 7. It consists of Arduino UNO SoC, ESP32, fingerprint sensor module, LCD and push buttons. The software environment is created using Arduino C and HTML scripting language. The detailed software flow diagram for suggested system is shown in Fig. 8.

6 Test Results and Discussion

Test results of the proposed system are discussed in following figures (Figs. 9, 10, 11, 12, 13 and 14). When system is started, it tries to search the sample of fingerprint.

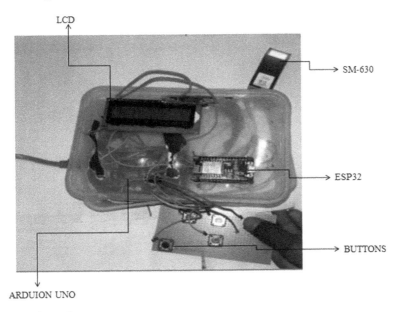

Fig. 7 Experimental setup

When matched sample is identified, it will validate the message on LCD as shown in Fig. 9. According to the voters selectivity, the validate message for individual parties is shown in Fig. 10. If user wants to see the statistics of individual party votes, then he/she has to press upload button. Statistics of individual party's vote on webpage server are shown in Figs. 11, 12 and 13. The same voting statistics are also captured on Google spreadsheet as shown in Fig. 14.

7 Conclusion

IoT-based smart voting system is designed and implemented here. The suggested system is more secure and reliable from any other traditional system, because authentication level is added in the system before voting process. Authentication process is to be done using the method of fingerprint matching. As the IoT facility is added in the system, the statistics of the individual party vote can be uploaded onto the server. Because of the authentication stage, malfunctioning like fake voting, repeated voting can be overcome. Suggested system is limited to do authentication process for 20 users only as fingerprint module used in the system is capable of scanning fingers of 20 users only. The proposed system can be enhanced by linking the Aadhaar card or Voting Card of individual person's with the system database. Authentication process can also be increased in form of the face recognition or iris recognition or palm recognition instead of fingerprint recognition.

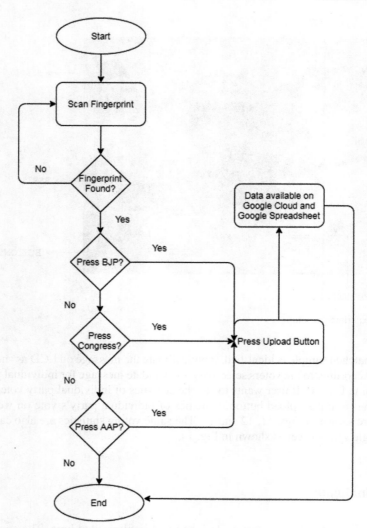

Fig. 8 Software flow diagram

Internet of Things (IoT)-Based Advanced Voting ...

Fig. 9 Fingerprint scanning process

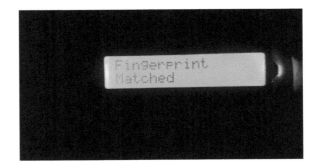

Fig. 10 Validation message after vote for respective party

Fig. 11 Validation message on webpage after every successful voting—BJP

Fig. 12 Validation message on webpage after every successful voting—Congress

Fig. 13 Validation message on webpage after every successful voting—AAP

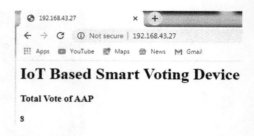

Fig. 14 Statistics of individual voting on Google spreadsheet at an interval of every 1 h

January 21, 2019 at 04:1 Vote_	14	8	8
January 21, 2019 at 05:1 Vote_	12	6	3
January 21, 2019 at 06:1 Vote_	5	6	2

References

1. Srikrishnaswetha, Kone, Kumar, Sandeep, Mahmood, Md: A study on smart electronics voting machine using face recognition and Aadhar verification with IO. Innov. Electr. Commun. Eng. **2019**, 87–95 (2019)
2. Deepika, J., Kalaiselvi, S., Mahalakshmi, S., Agnes Shifani, S.: Smart electronic voting system based on biometrie identification-survey. ICONSTEM-2017
3. Devgirikar, S., Deshmukh, S., Paithankar, V., Bhawarkar, N.: Advanced electronic voting machine using Internet of Things (IOT). NC-RISE 17 **5**(9), 83–85
4. Rezwan, R., Ahmed, H., Biplop, M.: Biometrically secured electronic voting machine. R10-HTC (2017)
5. Lavanya, S.: Trusted secure electronic voting machine. ICONSET 2011 (2011)
6. Bohil, J., Henrich, C., Kempka, C.: Enhancing electronic voting machines on the example of bingo voting. IEEE Trans. Inf. Forensics Secur **4**(4), 745–750 (2009)
7. Bhuvanapriya, R., Rozil Banu, S., Sivapriya, P.: Smart voting. ICCCT 2017 (2017)
8. Homepage on Arduino UNO: Available https://datasheet.octopart.com/A000066-Arduino-datasheet38879526.pdf
9. Homepage on Finger Print Sensor: Available https://www.sunrom.com/p/finger-print-sensor-r305
10. ESP32Datasheet. Available https://cdn.sparkfun.com/datasheets/IoT/esp32_datasheet_en.pdf
11. Homepage on LCD. Available https://www.engineersgarage.com/electronic-components/16x2-lcd-module-datasheet
12. Homepage on HTTP. Available. https://www.tutorialspoint.com/http/http_parameters.html

IoT: Security Issues and Challenges

Er Richa

Abstract Internet of things (IoT) is referred as group of devices and appliances which has capability to sense, accumulate and send data using the Internet without any human intervention that are installed with sensors. Behind smart cities, self-driving cars, smart meters and smart homes, IoT has been the core technology. IoT devices have inherent vulnerabilities also limited resources, lack of explicitly designed IoT standards, with diversified technologies, which have represented a proliferating ground for the growth of specific cyber threats. Since IoT devices are being used increasingly, security is the biggest challenge for the manufacturers. This paper focusses on the security concerns in IoT and the challenges to deal with different types of attacks.

1 Introduction

The Internet of things (IoT) is an effective worldwide network of information which consists of objects that are connected together, such as RFIDs, actuators, sensors, also different instruments and intelligent appliances that are becoming an essential component of the future Internet. A huge number of the IoT solutions have been developed by research institutes, public and private research organizations and many start-ups, small and medium enterprises, large corporations, in the last decade which have made their way into the market. This, in turn, has generated more security and privacy risks for the users of these devices, which is currently one of the biggest challenges of the IoT. This technology is mostly identical to the products which follow the concept of the "smart home," appliances like home security systems, lighting fixtures, thermostats, CCTV cameras and many other home devices which are supported by more than one familiar ecosystems that could be managed with

E. Richa (✉)
IT Department, Netaji Subhash University of Technology, Dwarka, New Delhi, India
e-mail: richa0023@gmail.com

© The Editor(s) (if applicable) and The Author(s), under exclusive license
to Springer Nature Singapore Pte Ltd. 2021
T. Senjyu et al. (eds.), *Information and Communication Technology
for Intelligent Systems*, Smart Innovation, Systems and Technologies 196,
https://doi.org/10.1007/978-981-15-7062-9_9

the devices linked to that particular ecosystem, for example smart speakers and also smartphones.

The phrase "Internet of things" was firstly given in 1999 by Kevin Ashton which focusses on RFID technologies. RFID and sensor technology have become enabler for computers that can perceive, relate and comprehend the world, with no limitations while entering the data manually. Today, computers and, hence, the Internet are too much dependent on humans for information. [1] The foundation of technology Internet of things is the MOSFET (i.e., MOS transistor or abbreviated as metal–oxide–semiconductor field-effect transistor) that was firstly given in 1959 by Dawon Kahngand Mohamed M. Atalla at Bell Labs [2]. MOSFET is defined as the elementary unit of modern electronics, which includes smartphones, tablets, computers and services provided by Internet [3]. It basically scales down the power consumption and paves the way to fabricating device using silicon-on-insulator semiconductor, multicore processor automation which in turn leads to Internet of things.

IoT has stepped in application fields like energy industry, agriculture, transportations and health services [4]. As stated by Gartner, that by 2020, the Internet connected IoT devices will cross 28 billion [5] and approximately the number of the community of human beings would reach 7.8 billion, and hence, with the result, each and every human being will almost possess three devices which can be connected with the Internet.

In my paper, I have discussed the generic architecture of IoT, recent development trends, features of IoT that make it different, potential threats to IoT devices and challenges to secure the devices connected via IoT. Besides the development of many security algorithms, still IoT is encountering many challenges while deployment of measures related to security and standards is also mentioned in this paper. The next part of the paper includes Sect. 2 that shortly explains the architecture of IoT with the evolution of IoT. Section 3 will introduce the security issues in Internet of things. Section 4 explains the recent developments in this technology. Section 5 describes the major challenges in deployment of IoT devices. Section 6 will present the conclusion.

2 IoT Architecture

With the IoT technology, physical objects are enabled to perceive, think and execute jobs with "talk" side by side, to coordinate decisions and for sharing information. IoT also helps in conversion of the physical objects, i.e., it transforms them against their classic nature into sharp while utilizing its elementary technologies like pervasive and ubiquitous computation, sensor networks, communication technologies, embedded devices and Internet applications also protocols [6].

IoT framework is the system consisting of components like sensors, cloud services, protocols, actuators and layers. It basically includes four stages showing various types of components into an advanced and integrated network. Its architecture can be explained with the help of Fig. 1 which includes the different stages:

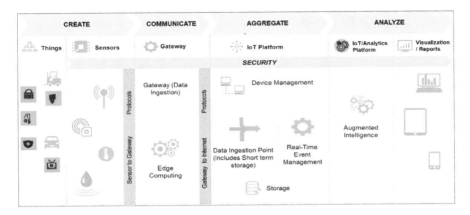

Fig. 1 Framework of Internet of things

sensors and gateways, Internet gateways and data acquisition systems, edge IT and data center and cloud [7].

The different devices categorized in things like smart health, smart farming, postal services, traffic control, infrastructure monitoring sensors and smart homes include electrical appliances, temperature sensors, smart cameras, smart cars where all the devices interact with the help of Internet.

Here these devices consist of sensors which organize the data that is sent downward direction to subsequent layers of the framework, enabling the "things" in IoT. Edge computing helps in processing data and generates insights into local loop and provides analysis and action without human intervention. Gateways are used to communicate data from devices to the cloud and can perform data format and protocol translation. Device management sits on top of gateway and IoT platform to remotely monitor and/or audit devices, manage the lifecycle of devices, diagnose and service connected devices, streamline software updates. Data ingestion point helps with acquiring, importing and managing data for later use or storage in a database. Augmented intelligence can be defined as data in motion (for real-time, perishable insights) and data at rest (for insights that need large volumes of data over time).

Some of the studies have proposed these stages arranged in the five-layer model which consists of basically: the business layer, application and service management, the object abstraction and then perception layer consisting of objects [6]. The IoT devices possess the properties as shown in Table 1.

3 Security Issues in IoT

As IoT means connecting everything to Internet, because of unavailability of general standards and security of the framework in IoT is the biggest challenge. The swapping of facts and details among millions or trillions of objects connected via Internet is

Table 1 Properties of IoT devices

Properties of IoT devices	Description
Identify	Each and every IoT device possesses an identification property like IPv6 or IPv4 address for communicating to different objects
Sensing	Different mechanisms for sensing like mobile applications and smart hubs are employed for collecting the information
Communication	The protocols used for communication for IoT are Bluetooth, ZigBee, LoRa, WiFi, Z-wave, IEEE 802.15.4, SIGFOX and LTE advanced for communicating between objects with the user
Computing	Different processing unit for example microcontrollers and software applications are used for processing of the information that gets attained from the smart objects
Utilities	Various utilities like information aggregation, identity related, smart city and smart homes are given by IoT objects to different users according to the information they get through surroundings
Semantics	In IoT, semantics represents recognizing and analyzing data smartly to provide required services. Various web technologies, like Recourse Description Framework (RDF), are supporting IoT devices for this work

the key functionality of the IoT. It is utmost important to secure swapping of data so as to avoid losing or negotiating privacy. With the increasing speed, IoT devices are proliferating in higher amounts in emerging markets, and so the time will come when the number of humans will be overrun by the number of devices. From homes, offices, automobiles, parking meters, everything is getting smarter because it is connected via IoT devices. So as the speed of Internet is increasing, similarly use of IoT devices is increasing in daily life of people. As stated by Govtech [8], "Everyone will own an average of six to eight IoT devices by 2020." But this is only the amount of the devices which each person has, actual amount of number of the devices every person will use is going to be large. Nevertheless, many users still now lack the privacy protection and management knowledge, while the IoT devices are much smarter and not getting closer to our lives, these can instantly finish number of assignments without any human involvement and suggestions. There are different types of IoT attacks which can be categorized as follows [9].

Lack of Observance on the Part of IoT Device Fabricator This is the main source among many IoT security issues, and the manufacturers do not spend adequate time and good materials on security [10]. For example, an intelligent refrigerator can reveal Gmail login credentials, the fitness trackers which came with Bluetooth are detectable after first pairing, and a smart thermostat can be accessed owing to weak passwords. Some security threats in IoT devices from manufacturers are weak, predictable or encrypted passwords, old and unpatched embedded operating systems and software, hardware concerns, insecure update mechanism, insecure data transfer and storage.

Lack of User Knowledge and Awareness As IoT is a developing technology, and still many people do not know much about it, one of the major IoT security threat is the user's ignorance and lack of awareness of the IoT functionality. In 2010, social engineering was used for Stuxnet (malicious computer worm) attack against a nuclear facility in Iran. The attack targeted supervisory control and data acquisition systems which in turn corrupted 1000 gas centrifuges which made the plant explode. It specifically targeted industrial programmable logic controllers (PLCs), which are mainly a type of IoT device. Although the internal network was inaccessible from the public network to avoid attacks, but Stuxnet spread by plugging a USB stick into one of the internal computers.

Security Problems in Device Update Management For maintaining security in IoT devices, updates in devices are very crucial. Due to lack of built-in secure update mechanism or right after new vulnerabilities are discovered, the devices should be updated. In contrast to computers and smartphones that are automatically updated, there are still some IoT devices that are continually being used without the necessary updates. There is another risk while updating, which is when an IoT device uploads its backup out to the cloud and will suffer a short downtime and if the connection is not encrypted and the update files are vulnerable, a hacker will be able to steal important information.

Lack of Physical Hardening Besides some of the IoT devices can be operated individually without any intrusion from a user, they are required to be physically secured from outside threats. Sometimes, the devices can be physically modified like USB flash drive with malicious software (Malware) when they are located in remote locations for long stretches of time. Recently, this was in news that whenever we put the USB of mobile phones in public places for charging, at the same time data from phones gets stolen from hackers using malware.

Botnet Attacks While an IoT device alone which is corrupted with malware does not cause any real threat, it is a group of them that can collapse anything. In a botnet attack, a hacker uses malware to infect the devices by creating an army of and then to bring down the target commands them for sending thousands of requests per second. The largest distributed denial of service (DDOS) attack occurred in October 2016 for spreading Mirai malware using IoT botnets [11]. Dyn, The Internet service provider was attacked and major services, including Twitter, Netflix, all experienced service disruptions. Hundreds of thousands of IP cameras, NAS and home routers were infected by multiple DDoS attacks.

Industrial Espionage and Eavesdropping Another prominent IoT security issue is privacy invasion. The interactive IoT doll for example Cayla doll comes with a Bluetooth pin, which is used to give access to the toy's speaker and microphone within the 25–30 m radius to anyone. The doll was banned in Germany and was tagged as an espionage device.

High jacking the IoT Devices Ransomware that is designed to block access to a computer system until a sum of money is paid has been named as one of the critical type of malware ever existed. It was about 70% of Washington DC surveillance cameras got infected just before the Trump's inauguration speech with ransomware. and the police were not even able to record for several days.

Data Integrity Risks of IoT Security in Healthcare Data is always circulating with IoT technology advent. Health care is always considered as an easy and soft target. Data is being transmitted, stored and processed. The vulnerabilities found on St. Jude Medical's implantable cardiac have given access to hackers, they altered the pacemaker and shocks, or even worse, and there were issues of ICD battery depletion.

Rogue IoT Devices A rogue device is which replaces an original device or attaches to the network as a member of a group for collecting or altering sensitive information. According to the figures from cybersecurity company Infoblox, fifty percent of the organizations have reported shadow IoT devices on their network, and these devices break the network perimeters and bring the organizations at risk for cyber-attacks.

4 Recent Developments in IoT

The IoT is continuously offering an unlimited reservoir of opportunities, in every field. The Internet of things (IoT) and Internet of everything (IoE) applications are further being diversified through the development of the Internet of nano-things (IoT) [11]. The idea of Internet of nano-things is very much similar to Internet of things except the things in IoNT are very small enough to be termed as nanoscale. Therefore, Internet of things will be utilized in the world that seems to be invisible to human eye.

Edge Computing on the Rise Embryonic

Edge computing is a computation process in which the local points or storage devices are placed closer which helps in gathering data, sorting and filtering it to improve response time and save the bandwidth. Then, this data is passed to the cloud and all of the processes become much faster. Edge computing works on "instant data" that is generated by sensors, and the processes are regularly upgraded, so it is only a matter of time before edge completely takes over cloud computing.

Health care and IoT the Perfect Combination

Patients care and safety are ensured with IoT health care. Real-time information can be collected from patients using sensors and wearable devices which help in making them available to healthcare providers and doctors through mobile devices only. With the help of health monitors, analysis is transferred per minute which are helpful to doctors around the world for analyzing and dealing with the health problems.

IoT Cloud Development

With the cloud services, security features are very crucial, thus lots of efforts are required by technical experts for improvement in the security layers and locks that surrounds the cloud data storage features. Many cloud platforms have been developed for easy and faster connectivity like ThingWorx, Google cloud, Microsoft Azure are few among them. Cloud-based applications that are fabricated on containers are bringing new set of flexibility and performance to formation. Google's open-source platform Kubernetes is a container symmetrical platform, and Docker is the most popular runtime container environment in recent days [12].

The Prominence of Big Data and AI

If we can describe AI as the brain of the system, then IoT is the digital nervous system. It is the artificial intelligence because of which the machines are able to think, analyze and process information similarly as humans do. For implementing AI in an organized manner and on a big scale, the machines should communicate with each other finely. With the help of IoT, this is possible. As stated by Cisco [13], "the Internet of Everything is the intelligent connection of people, process, data and things." The Internet of everything is the connection of physical things to the cyber things together that behaves as one. With this technology, every living, non-living or any virtual object is able to talk to each other via Internet. This virtual object part was not there in IoT which is possible with the integration of IoT with AI only.

Introduction of IoT Operating Systems

With the advent of IoT operating system (OS), the support for newly developed IoT hardware along with the recent standards and techniques is developed for all the communication layers. RIOT is the IoT OS that is specially developed for IoT devices. IoT OS supports latest protocols standards for future intelligent IoTs. Variety of IoT OSs, namely Contiki-OS [13] and RIOT [13] facilitated tremendous growth in this area.

Google Waymo

Google has launched its first self-driving car with this name in 2017. The Waymo is the best example of future IoT. Machine learning, edge analytics and involvement of AI all come together which make this car operate and analyze the physical environment in real time.

5 Challenges in IoT

IoT devices bring effective communication between devices, saves time and cost and have numerous benefits so security is the key challenge in this area. The security in IoT devices is usually not taken care of and treated as a second thought by the IoT device fabricators. Sometimes factors like quick turnaround time required for market demand and lower cost drives the development procedure and device's design. There are very less number of devices that support the protection mechanism, apply some solutions at software level, for example signing firmware that is initiated through the computation of cryptographic hash values. The computed value with the help of private keys is after then signed before attaching the signature to the image that is scanned. So in some devices proper, protection mechanism is followed using different software solutions. Sometimes, if completely focussed on the software-based protection schemes that sometimes leads to vulnerabilities in hardware unexpectedly (e.g., debug interfaces open), which allows for introduction of fresh attacks, for example, the theory in [14] apparently explained that a hardware platform that is not secure will certainly lead to an insecure software stack. Some of the challenges in IoT have been listed below [15–17].

Outdated Hardware, Software and Standards

The hardware and software used in IoT devices should be regularly updated for better performance. Most of the times, the devices provided by manufacturers are not able to operate on their own technologies and services provided. It is very important to standardize IoT for providing better compatibility to all objects and sensor devices.

Information Security and Protection

IoT uses divergent object identification technologies, for example, 2D-barcodes, RFID, etc. Now every kind of objects that are used daily embeds these identification tags which carry the information about object, so it is necessary to prevent unauthorized access and take proper privacy measures. Sensitive data is compromised in IoT devices that should be secured and not to be identified, linked and traced. Certainly, the potential to gain access to IoT devices by exploiting certain vulnerabilities or either by brute forcing their default credentials remains a primary attack vector [18].

Data Encryption and Confidentiality

The sensor devices perform sensing and measurement independently and then pass the data to the information processing unit (IPU) to the transmission system. Good encryption mechanism is required by the sensor devices so as to assure the data integrity at the IPU. The services provided by IoT are used to determine who can see the data, thus, it is required to protect the data from outside intrusion.

Difficulty in Identification of Infected Device

This is one of the most noteworthy challenges, i.e., the designing and application of Internet-scale solutions for addressing the security problem in IoT. It is very important to investigate the factual data, with the help of which IoT maliciousness can be detected at the Internet level. A significant hurdle in this approach is the demand of mechanisms development which can obtain suitable data in a timely manner. With the widespread use of IoT in dissimilar domain prevents clarity of security incidents related to IoT which in turn restricts the appropriate analysis of such data for identifying attribute and mitigate maliciousness.

Spectrum

A dedicated spectrum is required by sensor devices for transmitting data through the wireless medium. With the limitation of spectrum availability, for allowing communication of billions of sensors through the wireless medium, the dynamic cognitive spectrum allocation mechanism should be effective and efficient.

Green IoT

With the advent of IoT, the energy consumption due to network is increased at alarming rate because of rise in data rates and ever increasing number of Internet-enabled services also because of rapidly expanding Internet connected edge devices. So, there will be a significant increase in the consumption of network energy in future IoT. So, green technologies need to be supported for making the devices connected in network efficient in energy.

6 Conclusion

IoT is growing in a very effective manner day by day. It is not only being used in homes but also in each and every place which makes life simpler and easier. We can say that IoT is something that has the world in a hand. With this study tried to provide a review of certain important sides of IoT mainly focussing on the security issues and challenges that are involved with IoT devices. Still there are many issues and challenges related to the security of the IoT being faced. In this area, research focusses are needed so as to address these security issues and challenges in IoT diverse environments with the help of which users can use IoT devices with trust for communication and sharing information safely all over the world.

After analyzing the importance of security in IoT applications, we can say that it is utmost important to have security mechanism in IoT devices and communicating networks. Also to keep the devices safe, it is suggested that default passwords should not be used for the devices and security requirements to be read while using it for the first time. Chances of security attacks can be decreased by disabling the features that are not used. Moreover, study of different security protocols is very important which are used in IoT devices and networks. The IoT deployment requires large research efforts and could be hard to deal with the challenges but remarkable personal, economic and professional benefits can be attained with IoT in the near future.

References

1. Ashton, K., et al.: That internet of things thing. RFID J. **22**(7), 97–114 (2009)
2. Available: 1960: Metal Oxide Semiconductor (MOS) transistor demonstrated. The silicon engine: a timeline of semiconductors in computers. Computer History Museum. Retrieved 31 Aug 2019

3. Available: http://postscapes.com/internet-of-things-history
4. Atzori, L., Iera, A., Morabito, G.: Understanding the internet of things: definition, potentials, and societal role of a fast evolving paradigm. Ad Hoc Netw. **56**, 122–140 (2017). doi:https://doi.org/10.1016/j.adhoc.2016.12.004
5. Rivera, J., van der Meulen, R.: Gartner says the internet of things installed base will grow to 26 billion units by 2020. https://www.gartner.com/newsroom/id/2636073 (2013)
6. Al-Fuqaha, A., Guizani, M., Mohammadi, M., Aledhari, M., Ayyash, M.: Internet of things: a survey on enabling technologies, protocols, and applications. IEEE Commun. Surv. Tutor. **17**(4), 2347–2376 (2015). https://doi.org/10.1109/comst.2015.2444095
7. Guth, J., Breitenbucher, U., Falkenthal, M., Leymann, F., Reinfurt, L.: Comparison of IoT platform architectures: a field study based on a reference architecture. In: 2016 Cloudification of the Internet of Things (CIoT), Paris, 23–25 November 2016
8. Govtech. Future Structure: the new framework for communities (Infographic). Available: http://www.govtech.com/dc/articles/FutureStructure-The-NewFramework-for-Communities.html (2015)
9. WeLiveSecurity.: 10 things to know about the October 21 IoT DDoS attacks. Available: https://www.welivesecurity.com/2016/10/24/10-things-know-October-21-iot-ddos-attacks/ (2016)
10. Available: https://www.intellectsoft.net/blog/biggest-iot-security-issues/
11. Miraz, M., Ali, M., Excell, P., Picking, R.: Internet of nano-things, things and everything: future growth trends. Futur. Internet **10**(8), 68 (2018). https://doi.org/10.3390/fi10080068
12. IoT 2019 in Review: The 10 Most Relevant IoT Developments of the Year. Available: https://iot-analytics.com/iot-2019-in-review/
13. Yang, L.T., Di Martino, B., Zhang, Q.: Internet of everything. Mobile Inf. Syst. (2017)
14. Arias, O., Wurm, J., Hoang, K., Jin, Y.: Privacy and security in internet of things and wearable devices. IEEE Trans. Multi-Scale Comput. Syst. **1**(2), 99–109 (2015)
15. Available: https://readwrite.com/2019/09/05/9-main-security-challenges-for-the-future-of-the-internet-of-things-iot
16. Patel, K.K., Patel, S.M., et al.: Internet of things IOT: definition, characteristics, architecture, enabling technologies, application future challenges. Int. J. Eng. Sci. Comput. **6**(5), 6122–6131 (2016)
17. Gang, G., Zeyong, L., Jun, J.: Internet of things security analysis. In: International Conference on Internet Technology and Applications (iTAP) (August 2011)
18. Nasrallah, A., Thyagaturu, A., Alharbi, Z., Wang, C., Shao, X., Reisslein, M., El Bakoury, H.: Ultra-low latency (ULL) networks: The IEEE TSN and IETF DetNet standards and related 5 g ULL research. IEEE Communications Surveys & Tutorials (2018)

Coronary Artery Disease Prediction Using Neural Network and Random Forest-Based Feature Selection

Aman Shakya, Neerav Adhikari, and Basanta Joshi

Abstract Coronary artery disease (CAD) has become very common nowadays and is the leading cause of death across the world. In this paper, an attempt has been made to develop a model for the classification of CAD by using clinical features, ECG, and features from laboratory tests of patients. Optimum features were selected with the help of domain expert advice and random forest (using mean decreasing accuracy). A neural network-based classification model was developed with the selected features for classification of left anterior descending artery (LAD), left circumflex artery (LCX), and right coronary artery (RCA) in human heart. To get the optimum result from the neural network model, a 10-fold cross-validation method was adopted. The optimized model reached its maximum accuracy, specificity, and sensitivity of 91.397, 90.32, and 95% for LAD; 88.09, 80, and 91.52% for LCX and 90.36, 72.72, and 95.71% for RCA classification, respectively.

1 Introduction

Heart disease, or cardiovascular disease, refers to a range of conditions that affects the heart, including blood vessel diseases like coronary artery disease (CAD). Cardiovascular disease includes conditions that involve narrowed or blocked blood vessels that may lead to heart attack, chest pain, or stroke. Cardiovascular disease is the top killer responsible for 17.7 million deaths per year making 31% of all global deaths [1]. Hospitals and laboratories collect large amounts of data relevant to heart conditions. Increasing access to such huge datasets and corresponding demands to

---

A. Shakya (✉) · N. Adhikari · B. Joshi
Pulchowk Campus, Institute of Engineering, Tribhuvan University, Lalitpur, Nepal
e-mail: aman.shakya@ioe.edu.np

N. Adhikari
e-mail: neerav.adhikari@ntc.net.np

B. Joshi
e-mail: basanta@ioe.edu.np

analyze these data have led to development of new machine learning algorithms for leveraging such data and motivated this work too.

A number of research works have been done in prediction of coronary artery disease using machine learning techniques. However, only limited work has been done for the prediction of blockage for each artery independently. Hence, a contribution and novelty of this work is the diagnosis of blockage of arteries separately---left anterior descending artery (LAD), left circumflex artery (LCX), or right coronary artery (RCA)---by taking into account medical data of patients.

Since the medical data collected has a large number of attributes, feature selection is necessary for choosing relevant features for proper predictive classification of heart disease. In this work, experts (doctors) recommendation along with random forest (mean decreasing accuracy) has been introduced to identify the most relevant features. These features are then used in the classification of LAD, LCX, and RCA with a neural network model using resilient backpropagation algorithm.

2 Related Works

Liu et al. [2] present a hybrid approach that combines the Relief-F algorithm (RA) and random forest for modeling and analysis of complex disease, with improved speed and stability. Paper [3] evaluates classification algorithms, linear discriminant analysis and propose a novel hybrid feature selection methodology for diagnosis of CAD. Different feature selection methods were used, like information gain and gain ratio, Relief-F and chi-squared test. Alizadehsani et al. [4] used information gain for feature selection along with feature creating algorithm to increase the accuracy using Naïve Bayes. In [5], Gini index and information gain were used for feature selection and then C4.5 and bagging algorithms were used to forecast LAD, LCX, and RCA.

A comparison between ANN and SVM for predicting coronary artery disease is presented in [6]. SVM algorithm presented higher accuracy and better performance than the ANN model. Babič and Olejár [7] employed different algorithms like Naïve Bayes, support vector machine, and neural network for classification.

Arabasadi et al. [8] proposed heart disease detection using a hybrid neural network. Different feature selection algorithms (Gini index, weight by SVM, information gain, and principal component analysis) were applied in preprocessing of data. They made use of error backpropagation algorithm with sigmoid exponential function to train their neural network. Samuela et al. [9] proposed an integrated decision support system using ANN and Fuzzy_AHP for prediction of heart failure risk. A fuzzy analytic hierarchy process (Fuzzy_AHP) computes the global weights for attributes using individual contributions. The global weights are then applied to train an ANN classifier for prediction of heart failure risks.

A summary of the related works is given in Table 1. It can be seen that neural networks (NN) give one of the best accuracies for classification of CAD. Random forest has not been used yet in feature selection, although it has shown good results

Table 1 Comparative summary of related works

Paper	Feature selection	Classifier	Accuracy (with dataset)
[5]	Gini index (GI), information gain (IG)	C.45 Bagging	LAD—79.54% (IG and bagging) RCA—68.95% (GI and bagging) LCX—65.09% (No Feature extraction)
[8]	GI, weight by SVM, IG	Neural network (NN)	Switzerland dataset—71.5% Cleveland dataset—89.4% Z-Alizadeh Sani dataset—88.85%
[4]	IG and GI	Bagging, NN, and Naïve Bayes	Bagging—89.44% NN—87.13% Naïve Bayes—47.84%
[2]	Relief-F (RF)	Random forest	CAD—90.28%
[3]	IG, gain ratio (GR), Relief-F	SVM, random forest, Naïve Bayes, bagging	SVM (Z-Alizadehsani)—86.07% RF (Clevland)—83.16% RF (Z-Alizadehsani)—87.12% NB (Clevland)—83.49% NB (Z-Alizadehsani)—80.57% Bagging (Clevland)—83.16% Bagging (Z-Alizadehsani)—87.7%
[7]		SVM and NN	NN (Clevland)—86.32% SVM (Z-Alizadeh Sani)—86.67% SVM (Hungarian)—73.7%
[9]	Fuzzy AHF	NN	81.10%

as a classifier. Hence, in this work, we propose NN model for classification with random forest for feature selection.

3 Methodology

3.1 Overview of the Proposed Approach

An overview of the proposed approach is highlighted in the block diagram in Fig. 1. The Z-Alizadeh Sani dataset was collected from the UCI machine learning repository. Medical domain experts were asked to mark important features out of these.

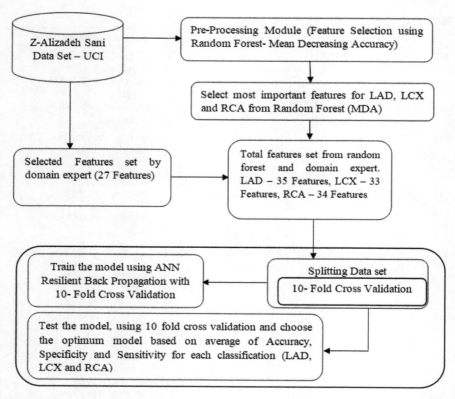

Fig. 1 Overview of proposed approach

Random forest (mean decreasing accuracy) was also adopted for selection of significant features. Then, combined features selected from both the methods were input in a neural network classification model. 10-fold cross-validation was used as limited samples that were available in the dataset.

3.2 Feature Selection

Random Forest-based Feature Selection Random forest provides two straightforward methods for feature selection---based on mean decrease impurity (MDI) and mean decrease accuracy (MDA). MDI signifies the more a feature decreases the impurity, the more important the feature is. The idea of MDA is that if the variable is not important, then rearranging its values should not degrade the prediction accuracy. In this work, important features were selected with MDA from a random forest model generated. Following are the key factors used in experiments for random forest.

No. of variables tried at each split (M_{try}): 7
Number of trees grown in each experiment: 500
Experiment run: 10 times and average of MDA was taken
Lowest Out-of-bag (OBB) error achieved: 16.47 (LAD), 17.34 (LCX), and 17.94 (RCA).

Using these parameters, the important features determined using random forest algorithm are listed in Table 2, for LAD, LCX, and RCA models separately.

Feature Selection Based on Expert Recommendation Machine learning seeks to represent generalizations, that is, not to represent each individual situation, but to group the situations that share important properties. Expert knowledge includes facts

Table 2 List of important features selected from random forest

Features for LAD	Features for LCX	Features for RCA
Typical_Chest_Pain	Age	Typical_Chest_Pain
Age	Typical_Chest_Pain	DM
Region_RWMA	CR	Age
EF_TTE	PLT	Neut
Nonanginal	TG	Poor_R_Progression
Attypical	BP	Attypical
CR	HTN	ESR
Dyspnea	Weight	Lymph
DLP	Sex	FBS
St_Elevation	K	Q_Wave
Lymph	FBS	Nonanginal
Current_Smoker	Attypical	PR
Diastolic_Murmur	Length	Current_Smoker
DM	Current_Smoker	HTN
Q_Wave	DM	Function_Class
Poor_R_Progression	DLP	Diastolic_Murmur
Neut	BUN	Region_RWMA
BP	EF_TTE	Na
FBS	HDL	TG
Function_Class	Q_Wave	WBC
Tinversion	ESR	BMI
ESR	EX_Smoker	
Weight	Tinversion	
Lung_rales	Region_RWMA	
Sex		
HTN		

of related domain and requires specialized information. One distinction of this work is that the feature selection during preprocessing of the dataset was done consulting the domain expert as well.

Dr. Amrit Bogati, Cardiologist, Shahid Gangalal National Heart Centre, Kathmandu, Nepal and Dr. Ravi Sahi, MBBS, MD, DM Cardiology TUTH, Manmohan Cardiovascular and Transplant Centre IOM Maharajgunj, Kathmandu Nepal helped to list down the important features in the total dataset. Among the total listed features in the dataset, following features were marked most important by the doctors---age, weight, length, sex, BMI, DM, HTN, current smoker, ex-smoker, FH, obesity, DLP, typical chest pain, dyspnea, function class, atypical, nonanginal CP, Q-wave, ST elevation, ST depression, T inversion, FBS, CR, TG, LDL, HDL, and EF-TTE.

Input Features for Training Models The input dataset for the neural network training models of each class LAD, LCX, and RCA was made by combining features from both methods mentioned above. The resulting input features selected are as shown in Table 3.

3.3 Neural Network Model for Classification

Separate neural network models were created for LAD, LCX, and RCA, as there were different sets of selected feature inputs for each. For classification of LAD 35 features, for LCX 33 features, and for RCA 34 features, selected were used from preprocessed data. For example, the optimized neural network model with selected features for LAD classification is depicted in Fig. 2. The resilient backpropagation algorithm was used for training the neural network as it is more efficient than conventional back propagation. 10-fold cross-validation resampling procedure was used to evaluate the models as the number of dataset samples was limited.

4 Results and Analysis

The optimized resilient backpropagation neural network model was trained and was tested for LAD, LCX, and RCA classification using 10-fold cross-validation method. To validate the models, different performance metrics were considered like accuracy, specificity, sensitivity, and ROC curve. Maximum accuracy of 91.397% with 90.32% specificity and 95% sensitivity for LAD classification, maximum accuracy of 88.09% with 80% specificity and 91.52% sensitivity for LCX and maximum accuracy of 90.36% with 72.72% specificity and 95.71% sensitivity for RCA were achieved by the proposed model. The confusion matrix of results is shown in Table 4. The ROC curves for the trained LAD, LCX, and RCA models are shown in Fig. 3.

Table 3 List of input features selected using random forest and expert recommendation

SN	Features for LAD	Features for LCX	Features for RCA
1	Age	Age	Age
2	Atypical	Atypical	Atypical
3	BMI	BMI	BMI
4	BP	BP	CR
5	CR	BUN	Current_Smoker
6	Current_Smoker	CR	Diastolic_Murmur
7	Diastolic_Murmur	Current_Smoker	DLP
8	DLP	DLP	DM
9	DM	DM	EF_TTE
10	Dyspnea	Dyspnea	ESR
11	EF_TTE	EF_TTE	EX_Smoker
12	ESR	ESR	FBS
13	EX_Smoker	EX_Smoker	FH
14	FBS	FBS	Function_Class
15	FH	FH	HDL
16	Function_Class	Function_Class	HTN
17	HDL	HDL	LDL
18	HTN	HTN	Length
19	LDL	K	Lymph
20	Length	LDL	Lymph
21	Lung_rales	Length	Neut
22	Lymph	Nonanginal	Nonanginal
23	Neut	Obesity	Obesity
24	Nonanginal	PLT	Poor_R_Progression
25	Obesity	Q_Wave	PR
26	Poor_R_Progression	Region_RWMA	Q_Wave
27	Q_Wave	Sex	Region_RWMA
28	Region_RWMA	St_Depression	St_Depression
29	Sex	St_Elevation	St_Elevation
30	St_Depression	TG	TG
31	St_Elevation	Tinversion	Tinversion
32	TG	Typical_Chest_Pain	Typical_Chest_Pain
33	T inversion	Weight	WBC
34	Typical_Chest_Pain		Weight
35	Weight		

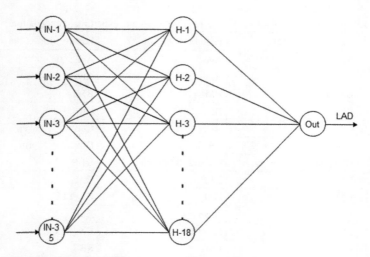

Fig. 2 Optimized neural network for LAD classification

Table 4 Results of trained neural network models for LAD, LCX, and RCA

	LAD		LCX		RCA	
	Predicted	Predicted	Predicted	Predicted	Predicted	Predicted
Actual	Normal	Stenotic	Normal	Stenotic	Normal	Stenotic
Normal	28	5	54	5	67	5
Stenotic	3	57	5	20	3	8

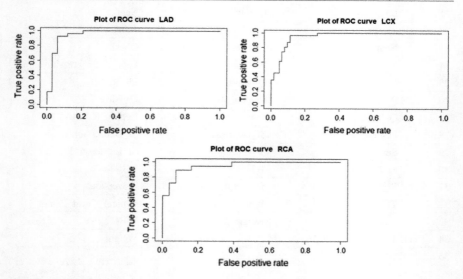

Fig. 3 ROC curve plots for trained LAD, LCX, and RCA models

5 Conclusion

This work presents an approach for classification of coronary artery disease using resilient backpropagation neural network. Feature selection based on random forest (mean decreasing accuracy) and expert recommendation was used to select input features to the classifier effectively. After feature selection, 35 different features were used for LAD, 33 for LCX, and 34 for RCA, out of total 54 features in the dataset. Combining the features selected from both methods, improved results were obtained with classification accuracy of 91.397% for LAD, 88.09% for LCX, and 90.36% for RCA models. The results show distinct improvements compared to accuracy of existing related works due to effective feature selection and training.

In future, more robust classification models may be built by taking large number of data samples from all possible categories of patients and geographical regions along with use of other feature selection techniques like information gain, recursive feature elimination, ant colony, artificial bee colony (ABC), etc.

References

1. Cardiovascular diseases (CVDs) Key facts. https://www.who.int/en/news-room/fact-sheets/detail/cardiovascular-diseases-(cvds). Last accessed 9 Feb 2020
2. Liu, M., Xu, X., Tao, Y., Wang, X.: An improved random forest method base on RELIEFF for medical diagnosis. In: Proceedings of the IEEE International Conference on Computational Science and Engineering (CSE) and IEEE International Conference on Embedded and Ubiquitous Computing (EUC) (2017)
3. Kolukisa, B., Hacilar, H., Goy, G., Bakir-Gungor, B.: Evaluation of classification algorithms, linear discriminant analysis and a new hybrid feature selection methodology for the diagnosis of coronary artery disease. In: Proceedings of IEEE International Conference on Big Data (2018)
4. Alizadehsani, R., Habibi, J., Hosseini, M.J., Sani, Z.A.: A data mining approach for diagnosis of coronary artery disease. Comput. Methods Programs Biomed. 111(1), 52–61 (2016)
5. Sani, Z.A., Alizadehsani, R., Habibi, J.: Diagnosing coronary artery disease via data mining algorithms by considering laboratory and echocardiography features. Res. Cardiovasc. Med. 2(3), 133–139 (2013)
6. Ayatollahi, H., Gholamhosseini, L.: Predicting coronary artery disease: a comparison between two data mining algorithms. BMC Public Health 19, 448 (2019)
7. Babič, F., Olejár, J.: Predictive and descriptive analysis for heart disease diagnosis. In: Proceedings of the Federated Conference on Computer Science and Information Systems (FedCSIS), IEEE (2017)
8. Arabasadi, Z., Alizadehsani, R., Roshanzamir, M.: Computer aided decision making method for heart disease detection using hybrid neural network. Comput. Methods Programs Biomed. 141, 19–26 (2017)
9. Samuela, O.W., Asogbona, G. M., Sangaiahc, A.K.: An integrated decision support system based on ANN and fuzzy_AHP for heart failure risk prediction. Exp. Syst. Appl. 68(C), 163–172 (2016)

A Collaborative Approach to Decision Making in Decentralized IoT Devices

Venus Kaurani and Himajit Aithal

Abstract To make embedded devices (IoT edge devices) decentralized, autonomous, collaborative, and intelligent, we introduce agent technology and multi-agent systems architecture. It targets to overcome the issues faced by centralized IoT systems such as security, single point of failure, latency, and computational complexity. To make these devices decentralized, collaborative, and autonomous, their way of communication and capability to act to the stimuli from the environment need to be enhanced. And for making them intelligent, the devices should be capable enough to make a decision based on the previous actions performed in the ecosystem. In this approach, agent technology has been used as a backbone for making devices autonomous and collaborative. We discuss implementation of the same using automated, IoT enabled digital book trading platform.

1 Introduction

IoT has the potential to optimize and better all the day-to-day activities of human life. But IoT technology, so far, has been centralized, due to which it is facing issues like security, single point of failure, computational complexity at edge, and latency [1]. To overcome all these issues, we introduce a decentralized and collaborative system. By embedding intelligence into the devices and by making them capable of self-learning, we try to reduce the dependence of these devices on centralized platforms for their decision making. The project aims to explore the multi-agent system, to move centralized IoT to a decentralized IoT system. Several agents form a multi-agent system, the multi-agent system when developed as an architectural system

---

V. Kaurani · H. Aithal
Robert Bosch Engineering and Business Solutions Private Limited, Bengaluru, India
e-mail: himajit.aithal@in.bosch.com

V. Kaurani (✉)
Department of Electronics and Communication, Institute of Technology, Nirma University, Ahmedabad, India
e-mail: fixed-term.Kaurani.VenusRajkumar@bcn.bosch.com; 18mece08@nirmauni.ac.in

is called an agent framework [2], which includes all the functionalities to handle the actions of agents, the life cycle of the agents, interaction between agents, and platform services. There are different Agent Frameworks existing, which adhere to the specifications given by the consortium Foundation of Physical Intelligent Agent (FIPA).

FIPA ensures interoperability and a well-defined architectural system of a platform [3]. It does not talk about any intelligence or decision making as an inherent part of the framework. Here in this paper, an attempt has been made to address this gap by embedding decision-making capability to the agent. Further, the agent can achieve intelligence in different ways possible such as based on lookup tables, mathematical modeling, previous experiences, self-learning, and several other strategies. However, a decision taken by a single agent should also consider the overall ecosystem goals. In order to achieve these global ecosystem goals, the agents would need to act collaboratively.

In this paper, we introduce agent technology that enables decision making in autonomous IoT devices and a collaborative approach to achieve system-level goals more efficiently. Section 2 talks about computing in multi-agent systems, and it shows the management of the software agent inside an agent platform. Section 3 gives an introduction to decision making of the agent at an individual level and collaboration between agents to take the decision at the system level. Section 4 describes JADE—a Java-based agent environment, its features and talks about the popularity of JADE in use cases such as market place scenarios and auctions. Section 5 shows implementation and simulation results of book trading example using JADE as an IoT-enabled book trading platform. Section 6 describes a further updated scenario which would be carried out in the future. Finally, Sect. 7 concludes the entire implementation of the market place and summarizes the effective use of the multi-agent system in collaboration and decision making among IoT devices.

2 Multi-agent-Based Computing

Multi-agent system-based computing includes several agents which are capable of performing computations [4]. Each agent is associated with an intermediate source platform, which is called as agent platform, to manage all the agent-related activities. Several such agent platforms constitute an agent framework. Because of its distributed nature, even if an agent platform shuts down, only the agents associated with that platform would be affected. Rest of the operations can be carried out smoothly by the other platforms and agents. The FIPA has specified the specifications of the agent framework about their abstraction model, management model, the action of agents, the life cycle of the agent, agent communication language, communicative acts, and interaction protocols [3].

The brief introduction to some of the important features of the FIPA specification is as follows [5]: The abstraction model includes the specification of services available to agents. It also describes interoperability of the message transport services between agents. It explains the Agent Communication Language (ACL) message representation and its various forms, language content used in the messages, and it supports the services given by the directory (which is a ledger of all agents in the ecosystem) in the agent platform. The agent management model specifies the norms for the framework in which the FIPA compliant agent resides and performs actions. It states that agent management reference model includes logical components such as agent management system (AMS), directory facilitator (DF), message transport service (MTS), and agent platform (AP) which provide the infrastructure to the agents. FIPA-ACL message structure includes some predefined message parameters. The message structure contains sender, receiver, content, and performative information. Performative is the most important parameter of the message. It describes the type of communicative acts. There are 22 different types of communicative acts given by FIPA. Some of them are CFP or Call for Proposal, INFORM, REQUEST, and Accept-Proposal. Using these communicative acts, FIPA has given some predefined interaction protocols such as contract net protocol, brokering protocol, query protocol, and several others.

There are several agent frameworks which are fully FIPA compliant or partially FIPA compliant. Some of the agent frameworks which are FIPA complaint are JADE, JASON, GAMA, and JADEX. JADE—Java-based agent framework has been used to showcase market place scenario of book trading as an example. In this example, each agent acts as an autonomous agent, where every agent residing in an agent platform works cooperatively to reach the goal through negotiations, interaction protocols, and arguments [6].

3 Collaborative Decision Making

For decision making in a multi-agent system, agents need to make decisions individually as well as collaboratively to achieve the system goal. Decision making considers previous experiences and also experiences of other agents in the ecosystem. Decision making for an agent can be classified as quantitative decision making, qualitative decision making, and knowledge-based decision making [7]. A popular model for agent decision making is the belief, desire, intention model (BDI). However, this model looks only from the individual agent's perspective and hence can be thought of as a greedy model. By bringing in collaboration, this paper also tries to enhance existing models for decision making [8].

Agents act based on environmental changes. By introducing intelligent computation capability at the device, we make the process more flexible and capable. Each agent in a multi-agent system interacts with other agents to reach the system-level goal that includes the system-level collaboration and also control over other agents. Considering our example, it includes buyers and sellers which communicate with each other for getting the targeted book.

4 JADE—Java Agent Development Framework

JADE is a Java-based intermediate platform structure for a multi-agent system, and it is fully FIPA compliant [2]. Jade implements the complete agent management specifications of FIPA, which includes basic blocks of agent management system such as agent management systems, directory facilitator, and Message Transport Protocol. Other than the basic functionalities of FIPA, JADE has its own additional features. In JADE, agents register themselves in the main containers with their addresses and services. Agents are the Java threads residing inside the containers, which supports the execution of operations inside the JVM [6]. Briefing of the three main blocks of FIPA specifications is as follows:

Agent Management System (AMS): It is an essential component that manages the directory of Agent Identifier (AIDs) of every agent residing in that Agent Platform. The agent has to register to it to get valid AID. AMS is responsible for managing the operation and life cycle of agents.

Directory Facilitator (DF): DF provides yellow pages service in the platform. Every agent has to register its services under DF, and any agent can subscribe to get information about the services of other agents from the DF. It allows agents to perform certain operations such as search, register, deregister, and modify.

Message Transport Service (MTS): It is responsible for the communication between agents in a platform or even inter-platform. For communication, FIPA has defined Agent Communication Language (ACL), which gives the standard message structure to be followed. Also, there are 22 different communicative act (CA) libraries provided by FIPA, which are used for defining some standard interaction protocols for agent interactions or negotiation [5].

Apart from being FIPA compliant for agent-based development, JADE includes distinctive features such as different predefined behaviors for actions performed by the agent. Also, several kinds of agents are involved in JADE for tracking of the agent's life cycle, actions performed, behaviors, for debugging purpose inside a platform. Some of these agents are sniffer agent, introspector agent, log agent, remote monitoring agent (RMA), and dummy agent. Sniffer agent is used to notify events in a platform and messages between the agents when it subscribed to a particular service. Introspector agent monitors the agent life cycle, queues of the messages of the specific agent to which it is subscribed, agent behaviors, and the reaction of the agent to stimuli. RMA is responsible for monitoring and administrating the hosts and container nodes present remotely or apart from the current container. JADE is also popularly used as a software agent simulation platform for other multi-agent system use cases [9]. In our present work, we include the book trading example as our marketplace scenario and showcase peer-to-peer negotiation on top of existing agent functionality.

5 Implementation of Marketplace Scenario Using JADE—Extended Version of Existing Book Trading Example

For the simulation of IoT-enabled automated book trading example, we envision the books are fitted with smart RFID tags which ensure connectivity. These tags also provide a digital identity to the books which are then used in the market place by JADE. Functionalities of JADE involved in the implementation for the same are specified below [10].

AMS is responsible for the management of platform services, agent activities. DF is responsible for managing the services of agents registered inside the platform. RMA is responsible for monitoring the agents who lie on another platform. The main container includes information about the addresses of agents, AMS, and DF agents residing in a platform [7]. The buyer agent is responsible for buying a book in this example. The seller agent is responsible for all the selling activities inside a platform. Sniffer agents used for sniffing all the messages passed between the agents for interactions and negotiations. Sniffing is a mean to observe the contents of a message passed between two agents. The introspector agent used for debugging operations in the platform, such as observing the behaviors of the agent performing in a defined manner, actions performed by the agent. The agents interact with each other through an ACL message with some specific FIPA communicative acts such as CFP, Accept-Proposal, Reject-Proposal, Request, and Inform.

In this scenario, we have three automated agent types: buyer, seller, and library. All the agents communicate with each other for trading of books. There are two buyer agents, two seller agents, and one library agent. The buyer agent finds which agents have registered themselves as a book-selling agent and sends CFP or Call For Proposal to that seller agent, based on its priority list of the books, wallet value of the buyer, and expected max price of the book. Seller sends back the Propose based on the availability of the book. If it is not available, it sends Refuse. On getting propose from seller, buyer sends Accept-Proposal if the price matches, else it sends a Propose with its expected price to start a negotiation. When a successful purchase of any book occurs, either buyer or seller agent sends INFORM message to end the negotiation. The negotiation ensures that both parties are satisfied with the price of the part being bought/sold. If in case the buyer is not able to procure a book, it checks whether the title exists with the library agent. If it exists, the buyer rents out the book from the library agent, else the buyer goes to the next book in its list. This process is continued till the buyer has no other book to buy.

The buyer agents are modeled to have two behaviors—a Ticker Behavior, which checks after specific timestamps, if there are any changes in the registered list of book-selling agents and library agents and a Cyclic Behavior, which is used for sending our purchase requests to each seller agent. The seller and library agent have only one Cyclic Behavior, which is used to respond to the offers and proposals from the buyer.

There are three scenarios of this book trading example on IoT-based platform: Scenario 1 is the existing example specified in JADE example documents in which there are a buyer agent and multiple seller agents. The buyer sends CFP to sellers, and the seller sends a proposal to the buyer based on the availability of book with the seller along with the price of the book, or else it sends refuse for the book. The buyer on receiving the Refuse or Propose responds with the inform message to indicate the status of transaction and terminates irrespective of the successful buying process or not. The interaction of seller and buyer in this is without any negotiation (Fig. 1).

- The buyer sends CFP (Call for Proposal) for the specified target book.
- Seller sends Propose to the buyer with book price.
- The buyer sends Accept-Proposal or Reject message to the seller with the best price of the book.

Scenario 2 proposes some updates in the available example (i.e., Scenario 1). It includes some advanced features in buyers such as buyer wallet, multiple book list, a priority list for the books, maximum price of a book a buyer is willing to buy for and minimum value of a book a seller is willing to sell for, and seller wallet. Also, a buyer agent can make decision based on the wallet value, a priority list of the book. A library agent is added in the example, along with the buyer and seller agent with the book renting service. The buyer agent sends the CFP to the seller in case it has enough wallet balance, and if it has insufficient wallet balance, it sends CFP to the library agent. Buyer with its max price of a single book starts negotiation, if the seller has the specified book. The buyer agent terminates after buying all the books in its book list. The sequence diagram is the same as in Fig. 2.

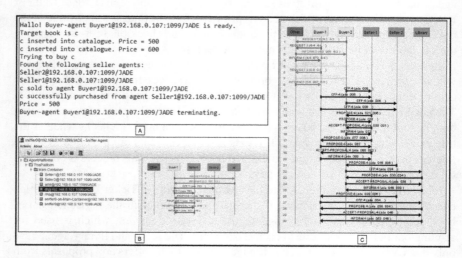

Fig. 1 **a** Operation performed by buyer and seller; **b** Scenario 1; **c** Scenario 2

A Collaborative Approach to Decision Making in Decentralized IoT Devices

- All buyers send Call for Proposal (CFP) for the specified target book with the quantity of book based on its wallet value and book priority list to all the sellers.
- Sellers send Propose to the buyer based on availability.
- Again buyer sends Propose with the desired price to start the negotiation.
- Based on the buyer proposed price seller sends Accept-Proposal or Reject to end the negotiation.
- Buyer sends Inform message to every seller for the successful purchase of the book.
- Buyers send CFP to library agent in case of insufficient balance in the wallet.
- Library agent sends Accept-Proposal or Reject based on the availability of a book and quantity.

Scenario 3 looks at the book trading example with effective interactions of agents, which is beneficial for the entire ecosystem. Buyer agent collaborates with other buyers in the ecosystem to come up with the optimal price it should ask sellers for a

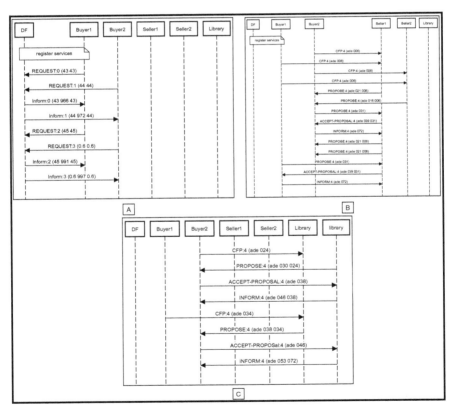

Fig. 2 Scenario 2: **a** checking services with DF; **b** negotiation of buyers and sellers; **c** negotiation of buyers and library

book. However, the contents of the messages change in this scenario, with the buyer agent trying to minimize the cost it has to pay per book. The current simulation uses random pricing strategy; however, the link between strategy and performance need to be investigated.

6 Future Work

For future work, we would bring in collaborative strategies for buyer and seller agents such that they would also have a chance to maximize their profits. We would optimize the strategies for both the agents and showcase the role of governance to maintain ecosystem stability in a decentralized ecosystem.

7 Conclusion

Agent technology proves to be an effective way of making IoT devices decentralized. By adding collaboration among the agents, many drawbacks of centralized systems can be addressed. Collaboration among the agents would not only benefit the individual agents but would help stabilizing the entire decentralized ecosystem. This research and its findings are relevant for many digital economic platforms like energy and manufacturing.

Acknowledgements We want to thank Sri Krishnan V., Mohan B.V., Manojkumar Parmar, Jidnyesha Patil, and Shrey Dabhi from Robert Bosch Engineering and Business Solutions Private Limited, India, for their valuable comments, contributions, and continued support to the project. We are grateful to all the experts for providing us with their valuable insights and informed opinions ensuring completeness of our study.

References

1. Semwal, T., Bhaskaran Nair, S.: Agpi: agents on raspberry pi. Electronics **5**(4) (2016)
2. Park, A.H., Park, S.H., Youn, H.Y.: A flexible and scalable agent platform for multi-agent systems. Int. J. Comput. Inform. Eng. **1**(1), 270–275 (2007)
3. Introduction to FIPA: Accessed on 9 Mar 2020 (2007). http://www.fipa.org/
4. Poslad, S.: Specifying protocols for multi-agent systems interaction. ACM Trans. Auton. Adapt. Syst. **2**(4), 15 (2007)
5. FIPA-Standard Status Specifications: Accessed on 9 Mar 2020 (2007). http://www.fipa.org/repository/standardspecs.html
6. Bellifemine, F.L., Caire, G., Greenwood, D.: Developing Multi-Agent Systems with JADE. Wiley Series in Agent Technology. Wiley, London (2007)
7. Bellifemine, F., Poggi, A., Rimassa, G.: Jade: a fipa2000 compliant agent development environment. In: Proceedings of the Fifth International Conference on Autonomous Agents, pp. 216–217 (2001)

8. Balke, T., Gilbert, N.: How do agents make decisions? A survey. J. Artif. Soc. Social Simul. **17**(4), 13 (2014)
9. Kravari, K., Bassiliades, N.: A survey of agent platforms. J. Artif. Soc. Social Simul. **18**(1), 11 (2015)
10. JADE Examples: Accessed on 9 March 2020. https://jade.tilab.com/documentation/examples/

Analysis of MRI Images for Brain Tumor Detection Using Fuzzy C-Means and LSTM

Sobit Thapa and Sanjeeb Prasad Panday

Abstract The collection and identification of medical images are the most important scientific task. MRI processing is also one of the reliable methods underpinning medical diagnosis. In this research, segmentation is performed by fuzzy C-means that detects the infected region from the images of brain MRI. Gray-level co-occurrence matrix (GLCM) is used to extract the features from the MRI images of the brain which gives the better result for classification than existing system. Long short-term memory (LSTM), a recurrent neural network extension method is used to identify brain MRI images. It gives a more precise and effective outcomes for image of MRI image classification.

1 Introduction

The brain is the biggest and most complex part of human organ. The role of brain is regulating the other organs of the body centrally. Brain tumor is the brain's irregular cell growth that gives the skull pressure and inhibits the human body's normal functioning resulting in inappropriate health behavior. The brain tumor needs to be identified quickly and in short time, MRI imaging plays crucial role in brain tumor diagnosis as it provides a clear resolution of an MRI brain image as shown in Fig. 1, which makes it easy to examine and identify. Many tumors pose a threat to human life. There are basically two types of brain tumors that are labeled into malignant and benign tumors. Tumor with malignant can propagate outside of brain, while

---

S. Thapa
Graduate Student Faculty of Engineering, Nepal College of Information Technology, Pokhara University, Lalitpur, Nepal
e-mail: gahasobit1@gmail.com

S. P. Panday (✉)
Department of Electronics and Computer Engineering, Pulchowk Campus, Institute of Engineering, Tribhuvan University, Lalitpur, Nepal
e-mail: sanjeeb@ioe.edu.np

© The Editor(s) (if applicable) and The Author(s), under exclusive license to Springer Nature Singapore Pte Ltd. 2021
T. Senjyu et al. (eds.), *Information and Communication Technology for Intelligent Systems*, Smart Innovation, Systems and Technologies 196, https://doi.org/10.1007/978-981-15-7062-9_12

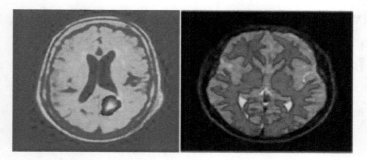

Fig. 1 MRI brain images (*Source*: www.kaggle.com)

tumor with benign are not able to spread. MRI is the most frequently used procedure in medicines that allows doctors to diagnose and control different diseases. Through medical image analysis, segmentation has become an important building block. Nonetheless, conventional segmentation algorithms do not yield better result in MRI image of brain segmentation because different forms of uncertainty include noise, strength, inhomogeneity, and inaccurate levels in gray images [1]. Healthcare professionals use different MR techniques including MRI scanning, computed tomography scanning, etc., to detect the presence of brain tumor. When treating severe brain disorders, they need to pay more attention. The healthcare industry continued to use technique of data mining to identify the presence that type of diseases for decision making better. As a second opinion, the detection system of brain tumor can be used by physicians to correctly diagnose brain tumors [2].

2 Related Work

Astina Minz and Prof. Chandrakant Mahobiya [3] in their work use an effective automatic disease recognition algorithm on brain MRI data that has been prepared using the AdaBoost optimizer. The machine learning approach used in this research was supervised learning for MRI image classification. Their result shows maximum accuracy of 89.90% using this method.

Lavanyadevi et al. worked on "Brain Tumor Classification and Segmentation in MRI Images using PNN" [4]. The recognition of object has been done, and for the reduction of the input data, PCA has been applied. The PCA has capability of decreasing the huge amount of input features into manageable size. The GLCM technique has been used for the extraction of the features.

Sayali et al. worked on "Segmentation of MRI Brain Image Using Fuzzy C-Means for Brain Tumor Diagnosis" [5]. This paper addressed the weakness of the FCM clustering algorithm for MR images segmentation, which can work very quickly and easily, but this algorithm does not guarantee high precision. To improve the accuracy, ant colony algorithm was introduced. Results show that fuzzy c-mean performs well as compared to k-mean algorithm.

Tao Liu et al. worked on "Sign Language Recognition with Long Short-Term Memory" [6]. Authors introduced an algorithm to recognize the Chinese sign language. The algorithm automatically recognizes the time series sign language knowledge through LSTM and is free to develop handicraft applications. The proposed algorithm outperforms existing works in relevant field resulting 63.3%.

In this research, we do the preprocessing and contrast of an image is enhanced, fuzzy c-means for image segmentation and feature is extracted by using gray-level co-occurrence matrix which provides the better classification result. The classification of an image is done by using long short-term memory, latest version of recurrent neural network that can solve the long-term dependencies.

3 Proposed Methodology

This proposed model includes a number of steps, for example, preprocessing and enhancement, segmentation using fuzzy c-means, feature extraction through GLCM, and classification using LSTM as shown in Fig. 2. LSTM is the latest and most popular variation of RNN that has capability to deal with dependencies on long-term time series data. When it comes to human life, a very high accuracy is on demand so as to save human life from disease like brain tumor. So, LSTM provides the high accuracy for recognition of brain tumor.

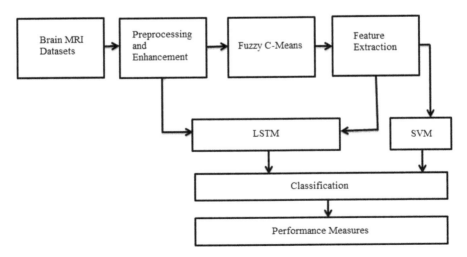

Fig. 2 Proposed research model

3.1 Image Acquisition and Preprocessing

The very first step is acquisition and preprocessing of an image. All MRI images are composed from hospitals, radiologists, and Internet. These images of RGB are converted into the grayscale image. A gray picture is a data matrix, the values of which represent shades of gray, as the grayscale image components have an intensity value from 0 to 255.

3.2 Image Segmentation Using Fuzzy C-Means

Because the structure of brain typically has clear and consistent continuous form and shape, the MR image series has high correlation with its third dimension which may help to segment tumors. The segmentation approach is a process of splitting images into various parts and chunks. The fuzzy c-means method is employed in the segmentation of MRI image. FCM method has been applied to assess the suspect area from the MRI image of brain. This method of clustering approach provides better result in segmentation [7].

3.3 Feature Is Extraction

The GLCM, a statistical approach for texture research, is also known as gray-level spatial dependency matrix which considers the spatial relationships between the pixels. The gray-level co-occurrence matrix works to distinguish the surface of the image by measuring the pixels according to specific values and specifying the spatial relation to be shown in gray-level co-occurrence matrix image types [4]. It is from this matrix we can calculate statistical extraction.

3.4 Classification Is Done by Using Long Short-Term Memory

The LSTM is one of the most latest and popular RNN which has ability to capture the dependencies for a long time. The LSTM was first coined in 1997 by Hochreiter and Schmidhuber and refined by many researchers. They show better performance for a wide range of issues as they are now being more popular [8]. LSTM is designed specially to reduce or even remove the problem of dependences that exists in long-term time series data. All RNNs have the shape of a chain of neural network recurrent sections. LSTMs also have this type of cascading architecture.

The fundamental issue to the LSTMs is considered as the current state of a cell. The popularity LSTM is also because of the capability to insert or delete cell-state details, sincerely controlled by multiple gate structures. As an optional way of passing knowledge, the gates are used. They consist of a neural network layer sigmoid that is computed by performing matrix multiplication.

3.5 Activation Function

An equation in linear form is simple, but its complexity is limited and it has less power to learn complicated functional mappings from data. Arbitrary complicated function which maps the output and gains access to the larger space of hypothesis could be beneficial from the deep representations to know and represent almost everything. Hence, there is a necessity of the non-linear activation function [9]. This makes the network more efficient and allows it the ability to learning something complicated and complex and nonlinear dynamic functional map between outputs and inputs.

3.6 Loss Function

To monitor a neural network's performance, output must be calculated as far as it is from what is predicted. That is the network's loss function, also known as the goal function. The loss function takes network and true goal predictions and calculates a distance ranking, measuring how well the network was doing. Categorical cross-entropy is the best loss function to use in this case. This measures the difference between two distributions of probability: here, in between the network's outputs of probability distribution and the real label distribution.

4 Result and Discussion

Dataset overview

The dataset used in this research is taken from different sources, and real images are taken from local hospital. Image dataset consists of brain images of different persons and are used for performing the experiments and evaluated the value of contrast, entropy, energy, homogeneity standard deviation, and area of the brain tumor after clustering for parametric analysis. The study mainly focuses on the brain image segmentation for the detection of brain tumor. In this research, the system has been tested for both normal and affected brain image.

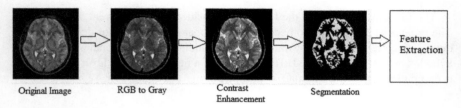

Fig. 3 Preprocessing, segmentation, and feature extraction of an image

Preprocessing, segmentation, and feature extraction

Figure 3 shows the step-by-step procedure for preprocessing, segmentation, and feature extraction. We have used fuzzy c-means for segmentation and gray-level co-occurrence matrix for feature extraction and extract ten different features from the image. After extraction of an image, it is passed to the LSTM model for classification.

Classification

SVM versus LSTM

The accuracy of SVM and LSTM with different number of datasets is depicted in Fig. 4. Both classifiers use the GLCM for feature extraction. SVM with the small number of datasets provides rapid increase in accuracy, and after the datasets above the 2000, it gets slowly increasing and constant at certain point. In case of LSTM, the accuracy slightly increases in few datasets and after increasing the number of datasets, accuracy increases rapidly. It shows that LSTM gives the better result when the numbers of datasets are large. Table 1 shows the final result comparison between SVM and LSTM in our research.

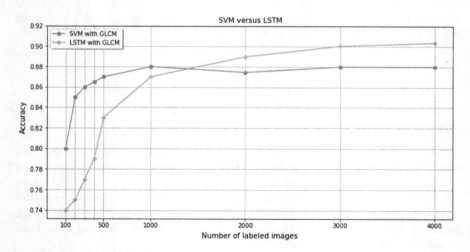

Fig. 4 Analysis of SVM versus LSTM model

Table 1 Comparison of SVM and LSTM result

Measures	Algorithms	
	SVM (%)	LSTM (%)
Accuracy	88	90.35
Sensitivity	90	82.25
Specificity	84	92.75

It has been observed that with LSTM and GLCM above confusion, matrix is generated. This model uses two hidden layers of LSTM model because the performance is similar with higher number of hidden layer which are more time-consuming and more complex. The learning rate is 0.001 because of larger learning rate that decreases the accuracy. The learning rate 0.001 or smaller than that, the accuracy of model will not increase longer. Therefore, in this model, the learning rate use is 0.001.

Loss function

Loss is used to optimize the hypothesis such that we can get the best weight. Typically, loss function minimizes the error in the model. With increasing the number of epoch, log loss is decreasing. The line graph shown in Fig. 5 represents the log loss with number of epoch for training and testing of MRI image. In this research, the loss result for both training and testing is better at the epoch 10–11.

The result comparison from our research and with existing system is tabulated in Table 2.

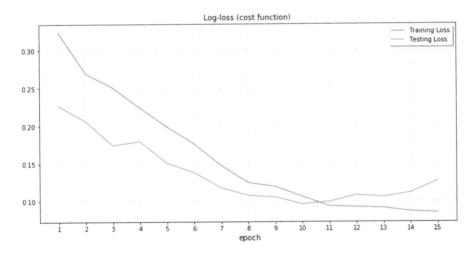

Fig. 5 Log loss with number of epoch

Table 2 Comparison of LSTM with existing work

Method	Technique	Classifier	Datasets	Accuracy (%)
Parveen et al. [7]	ML	SVM (polynomial)	120	87.5
Astina Minz et al. [3]	ML	AdaBoost	50	89.9
Sornam et al. [10]	ML	EML	168	72
Proposed	DL	LSTM	4000	90.35

5 Conclusion

In this proposed system, LSTM-based model has been developed to predict the brain tumor. The MRI of brain image is segmented using fuzzy c-means. For this research, the datasets are collected from different repository as well as real data of images from hospital. The brain tumor is classified using texture-based features. The features were extracted using the co-occurrence matrix on grayscale.

Different activation functions are used for LSTM model. For this model, Adam optimizer algorithm was used to optimize performance and maintain a constant learning rate of 0.001. The proposed model LSTM provides the better result than other algorithms for classification.

From the above table, it is concluded that the LSTM algorithm provides better result than the existing algorithm using GLCM.

6 Future Work

This research has used 10 kinds of features for the classification but other features like information measure of correlation, entropy variance cluster shade, maximum probability, and inverse difference normalized which also may have essential features that can be used for further research. Moreover, different types of optimizer algorithms like Nadam, Adam, SGD, and RMS pro can be used with LSTM. Further, LSTM model can be combined with CNN and other neural networks to be used with model BLSTM.

References

1. Harish, B.S., Aruna Kumar, S.V., Manjunath Aradhya, V.N.: A picture fuzzy clustering approach for brain tumor segmentation
2. Keerthana, T.K., Xavier, S.: An intelligent system for early assessment and classification of brain tumor
3. Miaz, A., Mahobiya, C.: MR image classification using adaboost for brain tumor type. In: IEEE 7th International Advance Computing Conference (IACC) (2017)

4. Lavanyadevi, R., Machakowsalya, M., Nivethitha, J., Niranjil Kumar, A.: Brain tumor classification and segmentation in MRI images using PNN. In: IEEE International Conference on Electrical, Instrumentation and Communication Engineering (ICEICE) (2017)
5. Sayali D. Gahukar et al.: Introduced segmentation of MRI brain image using fuzzy c means for brain tumor diagnosis
6. Liu, T., Zhou, W., Li, H.: Sign language recognition with long short term memory. In: IEEE (2016)
7. Praveen, Singh, A.: Detection of brain tumor in mri images, using combination of fuzzy c-means and SVM. In: 2nd International Conference on Signal Processing and Integrated Networks (SPIN), IEEE (2015)
8. Zhang, Q., Wang, H., Dong, J: Prediction of Sea surface temperature using LSTM. IEEE Geosci. Remote Sens. Lett. **14**(10) (2017, October)
9. Tai, T.-M., Jhang, Y.-J., Liao, Z.-W., Teng, K.C., Hwang, W.-J.: Sensor based continuous hand gesture recognition by LSTM. In: IEEE (2018)
10. Sornam, M., Kavitha, M.S., Shalini, R.: Segmentation and classification of brain tumor using wavelet and zernike based features on MRI. In: IEEE International Conference on Advances in Computer Applications (ICACA) (2016)

A Survey on Recognizing and Significance of Self-portrait Images

Jigar Bhatt and Mangesh Bedekar

Abstract An art of taking picture of oneself through selfie camera of smartphones has become ubiquitous. It has turned to be the effective way for expressing ones emotion nowadays. Researchers are analyzing selfie images for multiple aspects like personality, psychology, behavior. Authentication of such images taken under research is doubtful to be a real selfie image. Thus, researchers attempted identifying images as selfie automatically and segregating such images over the database. There are works that helps in detecting selfies through computer vision and deep learning methods. This research defines positive aspects of such investigations to elucidate on various aspects of analysis and highlights the different points of forethought for examining self portraits along with the future research direction.

Keywords Selfie · CNN · Computer vision

1 Introduction

Social media has become major part of human kind. Social networking is only platform over the internet that has most users, that has 3.48 billion users out of 7.6 billion total population. That means around 1 in 3 people uses social media. Photographs are the prime data that are shared on social media every day, among which most photographs are selfie images.

Selfie was first introduced in 2013 in Oxford Dictionary as word. Selfie are the portrait of oneself taken by themselves through front camera of their smartphones. One can also take selfie with group of people, objects, monuments, or pets. Around 2 billion selfies are taken each day by 14–28 years old demographic. Most selfies are shared on Facebook, Whatsapp, Twitter, Instagram, Snapchat—58, 425, 492

J. Bhatt (✉) · M. Bedekar
School of Computer Engineering & Technology, MIT World Peace University, Pune, India
e-mail: jbhatt462@gmail.com

M. Bedekar
e-mail: mangesh.bedekar@mitwpu.edu.in

© The Editor(s) (if applicable) and The Author(s), under exclusive license to Springer Nature Singapore Pte Ltd. 2021
T. Senjyu et al. (eds.), *Information and Communication Technology for Intelligent Systems*, Smart Innovation, Systems and Technologies 196,
https://doi.org/10.1007/978-981-15-7062-9_13

photographs with the tag of hashtag selfie on instagram each day and more than 235 million selfies till date. 77% and 50% of female and male college students shares selfies on Snapchat, respectively. Overall status represents that total of 70% females and 30% of males take a selfie. As of today, even aged people are deep into selfies. Selfie has become new art of self expression and representation. Selfie is known to be the fifth monument in the world of photography. Generally, 70% of selfie are taken from smartphones, 21% from tablets which are nothing but extended smartphones or can also say minimized laptops, and other 9% are taken from cameras. Selfies are generally arm length photograph from the subject taken by the user, it may even be closer to face of the subject, or at wide angle with the use of selfie stick. Selfie takers usually focus on composition of image, that is what elements or objects should be in the frame while clicking photograph.

Self-portrait images manage to acquiesce phatic communication. The selfie is a distinguished director of embodied social energy because it is kinesthetic power: Image is a product of kinetic bodily movement; and gives aesthetic that is visible form to that movement in images; and it is inscribed in the circulation of kinetic and responsive social energy among users of movement-based digital technologies [1]. Orekh et al. [2] described the culture of sharing selfies on social media in today's world as part of self-acceptance. This may help in building or destroying respect of a person. They also discussed how selfie helps in analyzing character of a person including friends group, hobbies, geographical travel, work place, behavior, and more such things. And how are people so serious with their representation over the virtual world. This studies have made authentication of selfie image critically important and also the segregation of selfie images from other images. People who enthusiastically share selfies critically view those who somehow do not fit in the same league of technological interest and online decorum. This often leads to multiple memberships in technology, networks, spaces, and bodies, the selfie signals, eventually resulting in inherent tensions between materiality, digitality, and identity (Fig. 1).

1.1 Why Selfie Detection?

Aforementioned selfies are the prime data over the Internet, which are analyzed for various aspects. Selfies have raised the phenomenon of narcissism, leadership/authority, grandiose exhibitionism, entitlement/exploitativeness, aloneness, and even deaths. People are intend to click at any moment/occasions of time, like at wedding, office, traveling, death occasions, school, colleges, food, etc. Assemblage of selfies provides insight for different connections between bodies, machines, physical spaces, and networks [3]. Also, the authors provide assemblage of different types of selfies. Researchers define that this different scenario leads to study of emotional and psychological pattern of human being. So it is necessary to evaluate between real and fake selfies, as fake selfies may result into fake assumption pattern. Perspective of users may differ at each aspect and also with use of selfie stick for same user [4].

A Survey on Recognizing and Significance of Self-portrait Images

Fig. 1 Examples of selfie image. *Source* selfie dataset from center of research in computer vision by university of central florida

Personality traits: Several researches have proved that personality of a person can be observed form the belongings and environment. For example, offices of extraverts are decorated and warm, fastidious individuals have well-organized bedrooms, and those who urge to learn and able to accept new challenges have varieties of books, articles, magazines in their bedrooms [5]. Several research have also found patterns about facial expression, gestures, and body posture in pictures [6]. Selfies contain unique patterns than other type of photograps. For example, duck face, a countenance made by pushing lips outward and upward to offer the looks of huge and pouty lips, is usually seen in selfies but not other sorts of portraits. Such cues may reveal new personality expression in photographs [7]. This has also proved the consideration judgment of a person from images. Considerateness involves being kind, agreeable, trustworthy, and willing to help others.

Narcissism: Selfie throws light on narcissism, i.e., love for oneself as psychological perspective of uploader [8]. It leads researchers to convey sentiments analysis on selfie image [9]. Researchers who focuses on usage of social network started paying more attention on narcissism; the presumption is that social media may create or support narcissistic tendencies by serving as channels for self-promotional displays [10]. Survey has proved relationship between narcissism of users of social media and their higher rate of usage [11, 12]. Other studies use natural photographs for experimenters and found personality is based related to smile and cheerfulness [6], while narcissism is related to flashy clothing, makeup, and attractiveness [13]. Social attraction and consideration have become major issue nowadays; people post selfies regularly of their visits, self showoff, and are waiting for others reaction on it.

Deaths: Selfies are casual and more spontaneous, i.e., instantly clicking a picture. Selfies have been proven dangerous by researchers. Many deaths have been registered that have caused due to selfie. For example, a man came ahead of train, fall from school staircase, drowned in river, and many more such cases. Research have been made for studying epidemiological characteristics for selfie that may able to control the cause of death [14]. It is thus important to analyze selfies of death which otherwise become very difficult to differentiate from the sublime experience, making it almost impossible to distinguish between selfie and sublimity [15]. Most of them take selfies while traveling; many tourist destinations have designed special location spots for selfies with high security measures [16]. India is leading in number of deaths caused due to selfie followed by Russia and USA.

The following three aspects define the major research areas of selfies, for which researchers still need explore. One of the major drawbacks for researchers is they need to manually parse through each image of dataset and extract the relevant selfies. Authenticity of an image being a selfie image is very less in the dataset created by scrapping over social media sites. Facial authentication systems have also given raise to fake selfies. So a system is required that extracts only selfie images from huge dataset for the researchers.

Segregation of huge image data over social networking platforms or on cloud file storage systems is critical task. Automation is required to categorize image data into types of images found. One of them is self-portrait images, i.e., selfies. As most of data generated on networking platforms are self-portrait images, it is so required to categorize selfie images from others.

2 Applications

Data segregation is preeminent application for detecting image as selfie. As vast increase in image collection nowadays, it is difficult to segregate images manually in different directories. This can be used for mobile phones, laptops, external drives, cloud storage, or any other storage devices. Even for the database servers, specially used by social media platforms.

Lets take an example of photograph gallery on mobile devices. Figure 2a shows different directories created for camera images and selfie images. But directories are created due to hardware of back camera and front camera. There are events where people take self-portrait image with help of back camera, shown in Fig. 2b. Yet being a selfie image, this image will be technically stored in camera directory instead of

Fig. 2 **a** Screenshot of image gallery of android mobile device, **b** self portrait taken using rear (back) camera

selfie directory. This similar happens when images of objects or other person are taken with front camera.

This makes it important to have proper classification of selfie images from other images for faster access. Today, everyone moves their images to external drive including both hardware and cloud-based drives. Same happens while moving images to drive that user needs to manually create directories for storage.

Social media databases stores image data based on various use cases based on users, locations, album, especially tags (hashtags). Nearly, every image posted on social media websites has various tags with it as caption. Selfie tag is the most used over various platforms like Facebook, Instagram, etc. Instagram has stated that more than 280 million images were posted by 2018 with tag selfie (#Selfie). All this images are stored and even updated by name #Selfie on their servers. But, are all truly selfie images is the issue. There have been cases that posts which are not actually selfie and are images of scenery or objects or even taken images from distance have caption as #Selfie as shown in Fig. 3a, b. As aforementioned researchers collect data from such medium and get some false images, and they need to manually remove them. It is important that system automatically detects actual selfie and stores them independently.

Fig. 3 Use of #Selfie done wrong **a** Image of person taken from a distance. **b** Image of scenery with beautiful sea

3 Selfie Detection Module and Methods

Figure 4 represents architecture of a simple computer vision approach for selfie detection and classification system with following modules: image preprocessing, segmentation, feature extraction, constraints development, and classification. Different modules are mapped with deep learning techniques like CNN. Approach is usually divided into two phases, the training and the testing.

Image Preprocessing: Image preprocessing is that the use of computer algorithms to perform image processing on digital images. It allows a much wider range of algorithms to process the input data. The important goal of image preprocessing is to enhancement of some important image features and/or improve the image data (features) by suppressing unwanted distortions so that our AI-computer vision models can be improved with help of improved data. Most known preprocessing techniques include read image, resizing image, color space conversion, smoothing, denoising, and enhancement. As per the literature, denoising is followed the most as it helps in reducing the distortion of the image and each feature can be captured precisely.

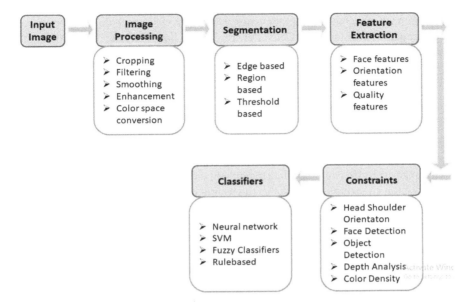

Fig. 4 Selfie detection module

Segmentation: Segmentation helps in dividing the foreground and background of the image. Over here, human face or body is only case to work on as part of foreground of image. Segmentation is processed based on object, region, edges, or threshold. In various literature, segmentation is done based on object that is human body.

Feature Extraction: Image feature is a simple image pattern which supports in describing what we see on the image. The key importance of such image features in computer vision is to transform visual information into the vector space. Image descriptors and neural nets are the ways to extract features from the image. Feature extraction is to process and extract a particular feature on which computer vision model is going to be processed. As per the literature, various features required face, objects, for face, object detection are extracted. Further, this features help in constraining the hypothesis of the research as where hypothesis was angle between face and shoulder object present in foreground, object present in foreground, etc.

Classification: Classifying images into their respective categories like selfie and non-selfie images. Based on similarities of the features extracted from images helps in training the classifier; this trained model helps in testing the new set of images. Classifiers like support vector machine, naive Bayes have been explored by researchers in literature. These classifiers should differentiate between selfie and non-selfie images. Researchers have also accepted the idea of deep learning approaches to classify the images. Convolutional neural network, class of deep neural network have rich area of exploration when it comes computer vision. Computer vision techniques are required to be mapped with of layers of CNN. Few researches have also taken CNN into consideration for classifying selfie images.

4 Literature Survey

Selfie detection is the domain that has not been examined much as of now. The literature defines some of the researches processed for detecting image as selfie.

Priyanka et al. [19] examined data for authenticated selfie, i.e., real or fake based on face detection by using image quality measures into consideration. Face detection was performed using Viola-Jones algorithm; this algorithm uses feature based on Haar feature filters instead of multiplications to spot people's faces, eyes, lips, nose, mouth or upper body. Image quality measures (IQM) and local binary pattern (LBP) were used as cascade feature detectors. The experiment was performed on 200 images of DSO-1 and DSI-1 of digital forensics dataset for face detection of selfie images.

Bhalekar et al. [18] designed a system using object recognition for detecting the given image as a selfie. The process was followed by dividing the image in grids and segmenting foreground and background image. The mathematical model was created based on Gaussian mixture model, exhaustive method, segmentation, selective method, bag of words approach as five inputs to the function. Authors only considered selfie images taken from smartphone's selfie camera. The result focused on a single-faced full selfie.

Annadani et al. [17] proposed selfie detection based on synergy constraint. Synergy constraint defines the orientation of head and shoulder at time of taking selfie; authors states that while taking selfie head is slightly moved toward the hand in which camera is hold. Synergy feature was developed by using canonical correlation analysis (CCA) function on two multidimensional handcrafted features [here, histogram of gradients (HOG) and local binary pattern (LBP)]. This features were mapped to convolutional neural network (CNN), employed with AlexNet as a transfer learning technique. The resulted accuracy was prominent on manually annotated 15,000 images. But, it also resulted into false positive classifying the images that were not selfie as selfie and false negative that the system not detected images that were truly selfie.

5 Discussions and Future Scope

In the twenty-first century, youth clicks pictures or videos on regular basis that too of many kinds, some of them are shown in Fig. 5. There are various ways of segregating multimedia data depending on their key features. Here, selfie images segregation is our keen interest.

A Survey on Recognizing and Significance of Self-portrait Images

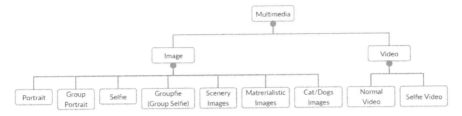

Fig. 5 Taxonomy of multimedia types and its different uses by the youth

Selfie detection is interesting and challenging problem in itself. Only few significant solutions have been made in past decade, there is much of the work is to be done yet. Important question that is needed to be addressed for detecting image as a selfie: "What categorizes selfie?". We believe that a robust selfie detection system should be effective under full variation in:

- Face detection.
- Facial expression detection.
- Head and shoulder orientation. [Head is more facing to the shoulder. Angle between them.]
- Quality of selfie image.
- Color density. [Selfie cam mostly uses aperture f/2.0 that is more darker then lower aperture. So color density defers.]
- Presence of glasses, phone in the image foreground.

Quality: Image quality refers to the weights of different image attributes, like sharpness, color accuracy, noise, lens flare, contrast, tone reproduction, aspect ratio, etc. Selfie images defer in quality from the normal primary camera images and DSLR images. This is because of the aperture used in selfie camera. Most of the selfie camera uses f/2.0 size aperture. Nowadays, selfie camera too have high megapixel resolution, but that does not make much difference as the aperture used only passes certain amount of light (Table 1).

Table 1 Difference in image qualities

Measure	Selfie	Non-selfie
Megapixel	May be similar	May be similar
Aperture	High (f/2.0)	Low (<f/1.5)
Sharpness	Low	High
Noise	High	Low
Contrast	High	Low
Color saturation	Low	High
Lens flare	High	Low

Database Quality: Dataset plays a key role in training the system; any variation in dataset may lead to false output. Today images which are selfie or non-selfie can be taken in any angle, or specifications. There is no perfect way or method to click a image, every image differs from each other in terms of details. This implies that dataset should consider each aspect of images for proper training of model. It has been observed that researchers who tried to examine such system used 90% of selfie images based on some common aspects; it is also required to use more non-selfie images for analysis; this lead into some false results. Also, researchers have to manually annotate each image in dataset, which is stressful. This fact is explicitly evident by use of constrained sizes of image dataset in the studies. Even, training and testing ratio differ a lot from one work to another.

Feature Extraction and Classification Module: Automation in classifying image as selfie image is been probed by some researchers with help of machine learning, deep learning, and computer vision techniques. It is observed classifiers like support vector machine, convolution neural networks have been used for definite result. Proper exertion of deep learning can aid in remodeling the efficiency of the system on large datasets. Classification is completely dependent on features of the image that model need to process on. This implicates feature extraction is the key to define correct selfie. As observed studies for selfie detection have majorly focused on single feature to signify selfie image that resulted in limited scope of detection. More than two features can be taken into count, that is more two computer vision technique working in parallel, for better scope of detection.

6 Conclusion

The script summarizes various studies to automate identification image as selfie image using machine learning, deep learning and computer vision techniques. The review offers well-recognized range of wide computer vision techniques in this field and as such mapping it with deep learning makes it a broad domain of prospective research with proper use of prediction logic. Selfie and groupfies can be separated as part future scope. Even best selfie can be extracted from various similar defining selfies. As observed, selfie detection results approximate output because of multiple exceptional cases. In future, selfies can be more examined with more features and ideas for accurate classification. Moreover, the real-time implementation of the above-discussed studies may be used to practically enhance data segregation methods significantly.

References

1. Frosh, P.: The gestural image: the selfie, photography theory, and kinesthetic sociability. Int. J. Commun. 1607–1628 (1932–8036/2015FEA0002)
2. Orekh, E., Sergeyeva, O., Bogomiagkova, E.: Selfie phenomenon in the visual content of social media. In: International Conference on Information Society, IEEE, pp. 116–119 (2017). https://doi.org/10.1109/i-Society.2016.7854191
3. Hess, A.: The selfie assemblage. Int. J. Commun. 1629–1646 (2015) (1932–8036/2015FEA0002)
4. Bevan, J. L.: Perceptions of selfie takers versus selfie stick users: exploring personality and social attraction differences. In: Computers in Human Behavior, pp. 254–263. Elsevier, Amsterdam (2017). https://doi.org/10.1016/j.chb.2017.05.039
5. Gosling, S.D., Ko, S.J., Mannarelli, T., Morris, M.: E,A room with a cue: Personality judgments based on offices and bedrooms. J. Personal. Social Psychol. 379–398 (2002). https://doi.org/10.1037//0022-3514.82.3.379
6. Borkenau, P., Brecke S., Möttig, C., Paelecke, M.: Extraversion is accurately perceived after a 50-ms exposure to a face. J. Res. Personal. 703–706. https://doi.org/10.1016/j.jrp.2009.03.007
7. Qiu, L., Lu, J., Yang, S., Qub, W., Zhu, T.: What does your selfie say about you? In: Computers in Human Behavior, pp. 443–449. Elsevier, Amsterdam (2015). https://doi.org/10.1016/j.chb.2015.06.032.
8. Weiser, E.B., #Me: Narcissism and its facets as predictors of selfie-posting frequency. In: Personality and Individual Differences, pp. 477–481. Elsevier, Amsterdam (2015). https://doi.org/10.1016/j.paid.2015.07.007
9. Halpern, D., Valenzuela, S., Katz, J.E., "Selfie-ists" or "Narci-selfiers"?: a cross-lagged panel analysis of selfie taking and narcissism. In: Personality and Individual Differences, pp. 98–101. Elsevier, Amsterdam (2016). https://doi.org/10.1016/j.paid.2016.03.019
10. Buffardi, L.E., Campbell, W.K.: Narcissism and social networking web sites. In: Personality and Social Psychology Bulletin, pp. 1303–1314. Sage, London. https://doi.org/10.1177/0146167208320061
11. Panek, E.T., Nardis, Y., Konrath, S.: Mirror ormegaphone? How relationships between narcissism and social networking site use differ on Facebook and Twitter. In: Computers in Human Behavior, pp. 2004–2012. Elsevier, Amsterdam. https://doi.org/10.1016/j.chb.2013.04.012
12. Ryan, T., Xenos, S.: Who uses Facebook? An investigation into the relationship between the Big Five, shyness, narcissism, loneliness, and Facebook usage. In : Computers in Human Behavior, pp. 1658–1664. Elsevier, Amsterdam (2011). https://doi.org/10.1016/j.chb.2011.02.004
13. Vazire, S., Naumann, L.P., Rentfrow, P.J., Gosling, S.D.: Portrait of a narcissist: manifestations of narcissism in physical appearance. J. Res. Personal. 1439–1447 (2008). https://doi.org/10.1016/j.jrp.2008.06.007
14. Jain, M.J., Mavani, K.J.: A comprehensive study of worldwide selfie-related accidental mortality: a growing problem of the modern society. Int. J. Injury Control Safety Promotion 544–549 (2017). https://doi.org/10.1080/17457300.2016.1278240
15. Preez, A.D.: Sublime selfies: to witness death. Eur. J. Cultural Stud. 1–17 (2017). https://doi.org/10.1177/1367549417718210
16. Dinhopl, A., Gretzel, U.: Selfie-taking as touristic looking. Ann. Tourism Res. 126–139 (2016). https://doi.org/10.1016/j.annals.2015.12.015
17. Annadani, Y., Naganoor, V., Jagadish, A. K., Chemmangat, K.: Selfie detection by synergy-constraint based convolutional neural network. In: Conference on Signal Image Technology and Internet-Based Systems, pp. 335–342 IEEE (2016). https://doi.org/10.1109/SITIS.2016.61

18. Bhalekar, M.A., Bedekar, M.V., Aslam, S.: An approach to detect an image as a selfie using object recognition methods. In: Cognitive Informatics and Soft Computing, pp. 111–120. Springer, Berlin (2018). https://doi.org/10.1007/978-981-13-0617-4_12.
19. Priyanka, V.S., Hussain, B., Aneesh, R.P.: Genuine selfie detection algorithm for social media using image quality measures. In: International Conference on Circuits and Systems in Digital Enterprise Technology (ICCSDET), pp. 1–6. IEEE (2019).https://doi.org/10.1109/iccsdet.2018.8821075

Anomaly Detection in Distributed Streams

Rupesh Karn, Suvrat Ram Joshi, Umanga Bista, Basanta Joshi, Daya Sagar Baral, and Aman Shakya

Abstract Anomalies refer to any non-conforming patterns to the expected behavior in the system. The detection of anomaly in real time from logs arriving at very high velocity and are in huge volume requires a distributed framework with high throughput and low latency. In this research, statistical method has been implemented for finding the suspicious associations in Spark Streaming, a highly scalable distributed and streaming framework. The models were deployed in both local mode as well as in cluster mode to perform anomaly detection on server logs.

1 Introduction

Anomaly detection in distributed streams refers to inspection of anomalous patterns in high-velocity streaming data which are also distributed among multiple nodes in a cluster environment. With the increasing number of Internet users and Internet services, there has been exponential growth in network traffic and thereby increasing the number of anomalies in the network system. Normally, most of the production systems deploy some sort of monitoring mechanism that continuously monitors the system behavior for anomalous patterns. But, the traditional and conventional approach of anomaly detection does not perform up to expectations when the volume and velocity of data are significantly high.

Hence, to handle such cases, the framework for anomaly detection should be scalable and perform in real time with high throughput and low latency. Since anomaly detection is a domain-specific topic and the definition of whats normal in itself is vague, it is a conundrum to establish a generic framework that solves all types of anomaly detection problems.

R. Karn · S. R. Joshi · U. Bista
LogPoint, Kathmandu, Nepal

B. Joshi (✉) · D. S. Baral · A. Shakya
Institute of Engineering, Tribhuvan University, Kathmandu, Nepal
e-mail: basanta@ioe.edu.np

Forming a general framework for anomaly detection is a difficult challenge as the definition of normality is typically very domain-specific. One of the difficult tasks for anomaly detection is identifying which statistical technique fits the best for given domain of datasets. The primary factor that determines fitness of test is nature of variable (quantitative or categorical) and correlation between the variables of the dataset. The complexity of volume and velocity while dealing in streaming fashion is addressed using big data technology Apache Spark, which is a distributed processing framework that offers streaming engine integration for real-time analytic of high-velocity data streams. The complexity of message addressing from multiple sources is addressed using Apache Kafka, a distributed, replicated message-log system. Between the log generating/collecting sources and Apache Spark engine, we have implemented Apache Kafka as distributed messaging queue for our real-time anomaly detection initiative.

2 Literature Review

Previously, a lot of research has been done in the field of anomaly detection in real time. Classification technique for anomaly detection involves creating a classification model by learning the patterns from a labeled dataset, then using the model to detect anomalous behavior on unseen data. [1] makes use of Bayesian belief networks for showing statistical dependency among variables. Kernel classifiers and classifier design methods were proposed by Mukkamala et al. [2] to solve anomaly detection problems. Nearest neighbor [3]-based algorithms assign the anomaly score of data instances relative to their neighborhood. The anomaly score can be either set to the average distance of the k-nearest neighbors as proposed in [4] or to the distance to the kth neighbor like the algorithm proposed in [5].

Alnafessah and Casale [6] uses artificial neural network for anomaly detection in real time and also shows that this method outperformed decision tree, nearest neighbor, and SVM but the major drawback is that it employs offline learning.

Clustering method, unsupervised anomaly detection algorithms [7], attempts to detect anomalies on the basis of dissimilarities. The assumption is that measurements that correspond to normal behaviors are similar and thus they lie in the same cluster. Rawat [8] and many more found that in clustering techniques observed data are grouped into clusters, according to a given similarity or measure of distance. Generally, the large clusters correspond to normal data, and the smaller ones or the outliers can be identified as anomalous events [9, 10].

Probabilistic distribution approaches such as information theoretical measures [11, 12] provide insight for anomaly detection on network attacks by using measures such as entropy and information gain/cost. [13] uses combination of two well-known metrics, relative entropy and Pearson's correlation, to detect anomalies over both high-velocity streams and/or large volumes of data at rest.

Several recent works have been observed on anomaly detection for streaming data. In [14–17], chi-square-based statistical anomaly detection technique has been implemented using Spark for real-time anomaly detection. The major drawback of this approach was that it employed a static threshold and failed to address the change in data distribution over time. As the initially built model did not adapt to the changing data over time, a high number of false alarms were triggered for the new data whose behavior is considered normal for that given period.

3 Methodology

A simple explanation about how the system for detection of anomalous behavior has been implemented can be understood from the system block diagram Fig. 1. First of all, the logs from different sources are collected using Kafka which is a distributed messaging system and based on publish–subscribe architecture in which messages are published to "topics" to which subscribers can subscribe. The purpose of retrieving data for visualization and analysis as the whole process occurs in the memory and not in disk level thus preventing the time loss in accessing and retrieving data from the disk.

Similarly, a large number of formatted logs are being processed each second so the work load will be very high if only a single machine is used and there is a very good chance of failure as well as performance degradation. Thus, a spark cluster with multiple machines are used for processing of the formatted logs with load balancing concept so that each machine will contribute to the log processing thus

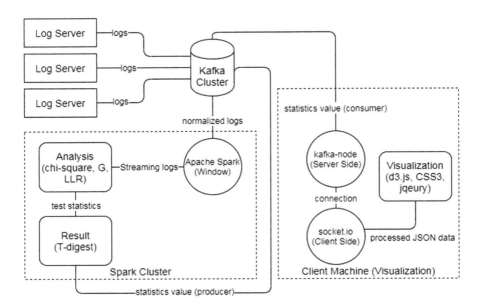

Fig. 1 System block diagram

decreasing failure chances and increasing efficiency as a whole. The results obtained after processing of these logs are again stored back to kafka-cluster and retrieved by another machine which is used for visualization purpose.

Statistical anomaly detection technique has been implemented which deals with statistical operation of the log data to compute test-statistics and thus determining the strength of dependency between particular variables. The drawbacks would be to what threshold the data would be considered non-anomalous and the dependency results also vary corresponding to the interval of window we take thus generating erroneous results. So a probabilistic approach for a certain interval window comparing the statistics results of particular combination to those of other combinations thus computing relative test-statistics.

3.1 System Operation Algorithm

In each window,

1. Receive the JSON logs via streaming through socket or kafka. The logs can be received through stored streaming or live streaming.
2. Parse the JSON logs in form of key-value pair.
3. Choose two fields for analysis. For instance, the two fields could be the combination of (source_ip, destination_port) to detect the port scan.
4. Apply statistical test to calculate log-likelihood ratio values.
5. Using T-digest approach, form a cumulative distribution of the log-likelihood ratio values.
6. Set the threshold at desired percentile (say 99th percentile)
7. Compare the llr values with threshold.

$$if(llr > Threshold) then Anomalous$$

$$else Normal$$

3.2 Anomaly Detection Module

After receiving the information from the logs in required format, first of all, the log-likelihood ratio for every user in each window is calculated. Then, using T-digest, a cumulative distribution of log-likelihood ratio values is created . The high log-likelihood ratio (LLR) values tend to be anomalous. So, we set the threshold at 99.9th percentile of the distribution and flag the extreme 1% as anomalous value. Hence, we have a dynamic threshold according to the distribution and we have solved

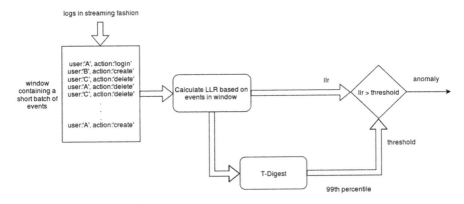

Fig. 2 Using T-digest to set Threshold to desired percentile

the problem of estimating a quantile from distributed and streaming data in a constant memory using T-digest. The whole process is shown in Fig. 2.

As shown Fig. 2, after the logs are processed, the LLR values were calculated. The LLR values are then fed into the T-digest block where a distribution of these log-likelihood values is created. Then we query the required quantile value from the T-digest [18] distribution according to the threshold. In our case, we have set the threshold to 99th percentile of the distribution. Then we compare the log-likelihood values of each observation with the threshold value. If the observed value is greater than the threshold value, then an anomaly is fired.

4 Results and Analysis

An approach that detects the anomalous and suspicious behavior from the streaming network logs, with unlabeled datasets has been implemented. First of all, chi-square test statistic was used to find out dependency between two fields in the logs. From Fig. 3, we can see that we have a very low precision and f1-measure at both the significance level of 95 and 99%. The very low precision and f1-measure reflects that this approach did not suited very well for our streaming case.

In order to overcome the limitations of the static threshold of chi-square statistics, the LLR approach along with T-digest data structure which a probabilistic data structure and fits well to our scenario of anomaly detection has been implemented. The T-digest computes the threshold according to the distribution of data in every window in O(1) complexity. Hence, we have an efficient dynamic threshold in streaming environment. The performance metrics we obtained by using log-likelihood approach with T-digest shows fair results. From the graph in Fig. 4, we can see the variation of precision, recall, and f-measure at different thresholds.

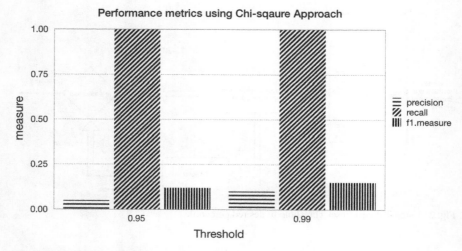

Fig. 3 Variation of performance metrics with significance level

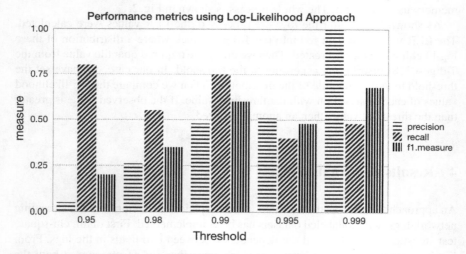

Fig. 4 Variation of performance metrics with threshold

As seen in Fig. 5, the performance in local mode in a machine having specifications as Intel Core I3 2.1 Ghz processor and 4GB RAM is observed with input rate of around 500 events/s. With a window size of 2 s, around 1000 events have to be processed within this 2 s. Around 1000 events are being processed in just 71 ms which means we have capability of processing even more events if we scale our application to a large number of computers.

As seen in Fig. 6, the performance in cluster mode (a spark cluster having two master nodes and four executor nodes, each with a 256MB RAM, i.e., 4 cores and

Anomaly Detection in Distributed Streams

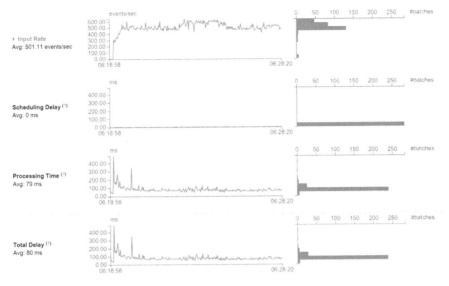

Fig. 5 Performance in local mode

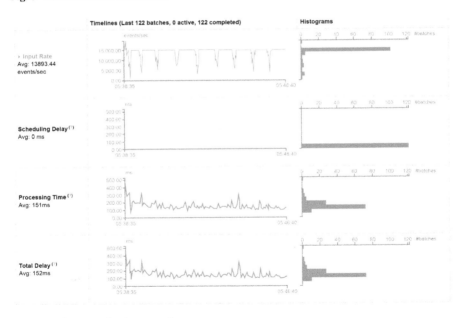

Fig. 6 Performance in cluster mode

2.5GB RAM for driver and executors execution) is observed with the input rate is about 14,000 events/s. Hence, about 28,000 events processed within this time in one machine. With total of two drivers and four executors, we are able to process about 75000 events in 0.1 s which gives us even more capability of processing. Such result in cluster is obtained when we are provided just a small pool for processing as other resources are running in cluster in parallel as well.

5 Conclusions

With the logs being generated at a high velocity, it is necessary to detect and filter out suspicious events from the logs in the real time. The proposed system is a sophisticated real-time framework that detects such anomalous events on the go with the use of Apache Kafka to collect log from different sources in streaming fashion and Apache Spark to provide great processing power for clustering computing. In this paper, statistical approaches for online detection of anomalies using log-likelihood ratio were experimentally evaluated. A possible future work can be to use machine learning with unsupervised neural networks. The application of the stated statistical measure to compute the correlation value and then apply it to the neurons for adjustment of the weight would improve the accuracy of the model to better identify an anomaly and reduce false alarms. The use of self-organizing map (SOM) in the system will be a great enhancement. It is used in reducing the dimension of data visualization.

References

1. Kline, K., Nam, S., Barford, P., Plonka, A., Ron, A.: Traffic anomaly detection at fine time scales with Bayes. In: International Conference on Internet Monitoring and Protection (2008)
2. Mukkamala, S., Sung, A., Ribeiro, B.: Model selection for kernel based intrusion detection systems. In: Proceedings of International Conference on Adaptive and Natural Computing Algorithm (2005)
3. Manocha, S., Girolami, M.: An empirical analysis of the probabilistic k-nearest neighbor classifier. Pattern Recogn. Lett. **28**, 1818–1824 (2007)
4. Angiulli, F., Pizzuti, C.: Fast outlier detection in high dimen-sional spaces. Lecture Notes Comput. Sci. **2431**, 43–78 (2002)
5. Ramaswamy, S., Rastogi, R., Shim, K.: Efficient algorithms for mining outliers from large data sets. In: Proceedings of the 2000 ACM SIGMOD, pp. 427–438 (2000)
6. Alnafessah, A., Casale, G.: Artificial neural networks based techniques for anomaly detection in apache spark (2019)
7. Duda, R.O., Hart, P., Stork, D.: Pattern Classification, 2nd edn (2001)
8. Rawat, S.: Efficient data mining algorithms for intrusion detection. In: Proceedings of the 4th Conference on Engineering of Intelligent Systems (EIS'04) (2005)
9. Patcha, A., Park, J.-M.: An overview of anomaly detection techniques: Existing solutions and latest technological trends. Comput. Networks **51**, 3448–3470 (2007)
10. Chandola, V., Banarjee, A., Kumar, V.: Anomaly detection: a survey. ACM Comput. Survey J. **41** (2009)

11. Gu, Y., McCallum, A., Towsley, D.: Detecting anomalies in network traffic using maximum entropy estimation. In: Proceedings of IMC (2005)
12. Lee, W., Xiang, D.: Information-theoretic measures for anomaly detection. In: Proceedings of IEEE Symposium on Security and Privacy (2001)
13. Rettig, S., Khayati, M.: Online anomaly detection over big data streams (2015)
14. Ahmad, S., Lavin, A., Purdy, S., Agha, Z.: Unsupervised real-time anomaly detection for streaming data. Neurocomputing **262**, 134–147 (2017)
15. Otey, M., Ghoting, A., Parthasarathy, S.: Fast distributed outlier detection in mixed-attribute data sets. Data Mining Knowl. Discovery **12**, 203–228 (2006)
16. Chandola, V., Banerjee, A., Kumar, V.: Anomaly detection: a survey. ACM Comput. Surveys **41**, 1–58 (2009)
17. Solaimani, M., Iftekhar, M., Khan, L., Thuraisingham, B.: Statistical technique for online anomaly detection using spark over heterogeneous data from multi-source vmware performance data. In: IEEE International Conference on Big Data (2014)
18. Dunning, T.: Better anomaly detection with the t-digest (2015)

Envisaging Bugs by Means of Entropy Measures

Anjali Munde

Abstract There are numerous methods for envisaging bugs in software practices. A widespread method for bug estimation is applying entropy of alterations as recommended by Hassan [5]. In the existing literature, investigators have recommended and executed a surplus of bug extrapolation methodologies, which change in relation to precision, difficulty and the input files they need, nonetheless actually rarely any of them has anticipated the amount of bugs in the software established on the entropy. A standard for bug estimation is showed, in the way of a widely accessible data set comprising of numerous software techniques, and a huge divergence of the comprehensive and analytical ability of familiar bug estimation methods, collectively with innovative methods is offered. Based on the outcomes, the working and constancy of the methods with regard to our standard are examined and a number of understandings on bug estimation models are determined. In the present communication, we will acquire a method by taking into concern the amount of bugs logged in numerous announcements of Bugzilla software and compute the distinctive measures of entropy, namely Shannon entropy [18] and Kapur entropy [9] for the variations in several software updates in these time intervals. Eventually, this work has the possibility to shed light on how to utilize simple linear regression to envisage the bugs which are still impending.

1 Introduction

The occurrence of the open-source software (OSS) has confronted numerous models in economics, business strategy, IT management and most significantly the software engineering as it is exposed to one and all with its source code unfilled for alteration and enrichment to all. These source codes undertake numerous variations in order to

A. Munde (✉)
Amity University, Noida Uttar Pradesh,, India
e-mail: anjalidhiman2006@gmail.com

see the novel characteristic outline, feature enhancement and bug repair. An imperative region of interest for OSS is when to announce a latest version. The outsized quantity of operators surpasses the quantity of inventors as the source code is easily accessible and is established in an unrestricted way

The achievement and expansion of the OSS group are obtaining interest incessantly through software trade individuals plus university circles. The trades are becoming concerned through open-source groups and proposing numerous amount of maintenance fluctuating from 'contribution in current plans' near the 'announcement of latest software' certified in open source. While the software develops, this involves variations and the variations in the source code are because of the latest characteristic addition, characteristic improvement and bug corrected. To retain the appreciations of the OSS operators, the group should issue the software in a judicious way. Nevertheless, realistically the situation would not be feasible to issue appropriately, i.e. on frequent root. Here, we have a prerequisite to perform a study that ascertain the appropriate announcement period regarding software.

Present date investigation associated with subsequent announcement obstruction contains the amount regarding conditions circulated in the software. The amount related to conditions is essential in the direction measured for benefit growth and consumed by way of degree in determining the subsequent issue of the software. In closed source situation, announcement period of the software is usually static through prerequisite study stage regarding the plan. After the foremost announcement, consequent announcements happen grounded upon amount of necessities and bug stability. In the open-source situation, it is already established as the adequate extent of alterations in addition to amount of bugs static in the software. A limited open-source schemes issue their software edition on a regular base by means of innovativewise, precise conditionwise, significant areawise, and by means of static time, etc.

In this paper, to ascertain envisaged bugs, linear regression simulations are employed through studying experimental bugs and entropy quantity and to establish forecasted time, linear regression simulation is applied as a result of studying experimental bugs, experimental time and entropy measure. Simple linear regression is employed to envisage bugs using experimental bugs and entropy quantity with Shannon entropy [18] and Kapur entropy [9] with parameter α and β.

The paper is further divided into the following sections. Section 2 provides the review of work available in the literature. Section 3 describes the information theoretic approach. Section 4 describes the data collection process and the methodology adopted. In Sect. 5, the paper is finally concluded with the results.

2 Review of Literature

During the last decades, numerous methods have been considered in the literature for the bug extrapolation to conform the requested requirements of the consumers. Investigators are employed to establish the appropriate period for the announcement of the software. Kapur et al. [10 11] deliberated an optimum resource distribution design to reduce the expenditure of software throughout the analysis stage beneath the dynamic background and employed optimal control theoretic system to review the conduct of dynamic simulations in software analysis stage and operational phase.

Kapur et al. [12] anticipated a realistic technique for ascertaining when to discontinue software examination contemplating breakdown concentration and cost as two elements concurrently. The suggested model is founded on multi-trait efficacy examination which benefits companies to create a judicious result on the ideal technique of the software. Pachauri et al. [16] created an effort to demonstrate the vagueness included in approximated limit of SRGM in inadequate repairing situation employing fuzzy numbers. They also computed the consistency, the ideal announcement period of the software created on cost consistency principles employing fuzzy environment.

Peng et al. [17] recommended an outline to acquire challenging attempt reliant on fault recognition procedure and fault alteration procedure simulations through the consideration of defective restoring and deliberated optimum announcement procedure under several conditions.

Hassan [5] recommended the notion of information theory to determine the quantity of vagueness or entropy of the dissemination in order to measure the difficulty of code changes. He tried to envisage the bugs existing in the software established on earlier imperfections utilizing entropy measures. Ambros and Robbes [1, 2] recommended a standard for estimation of bugs and presented a widespread assessment of acknowledged bug estimation methods.

Singh and Chaturvedi [19] established a method to envisage the bugs of the software founded on recent year entropy employing vector regression and associated with simple linear regression. Singh and Chaturvedi [20] established the complexity of code change and employed it for envisaging bugs. Singh et al. [21] established the hypothetical bugs staying inactive in the software by suggesting three methods specifically software reliability growth models, hypothetical complexity of code changes established models and complexity of code changes constructed models.

Chaturvedi et al. [3] acquired a simulation to envisage the hypothetical alteration complexity in the software and experimentally confirmed the model employing historic code changes facts of open-source software Mozilla's components. Chaturvedi et al. [4] suggested a complexity of code change model to direct the NRP utilizing multiple linear regression model and computed the performance applying numerous residual statistics, goodness-of-fit curve and R2.

Kaur et al. [13] proposed the metrics obtained by means of entropy of variations to evaluate five machine learning systems aimed at envisaging faults.

Kaur and Chopra [14] intends Entropy Churn Metrics grounded on History Complexity Metrics and Churn of Source Code Metric. The investigation evaluated the software subsystems on three restrictions to ascertain which properties of software techniques create HCM or ECM additionally chosen above others.

Kumari et al. [15] suggested fault dependency grounded arithmetical simulations through studying the instantaneous explanation of faults and observations proposed through operators in conditions of the entropy-grounded degrees.

3 Entropy

The information theory associates with evaluating as well as describing the quantity of information. The principle utilizes the info in computing the quantity of ambiguity or entropy of the distribution.

Shannon [18] initiated the notion of entropy in communication theory and originated the topic of information theory. The stochastic technique has a significant property identified as entropy which is extensively tapped in numerous areas.

The Shannon Entropy, H_n given by Eq. (1.1) is stated as

$$H(P) = -\sum_{i=1}^{n} p_k \log p_k \quad (1.1)$$

where $p_k \geq 0$ and $\sum_{k=1}^{n} p_k = 1$

Every component has the equivalent probability of occurrence ($P_k = 1/n$; $\forall k \in 1, 2,\ldots, n$) for a distribution P, we attain greatest entropy. Additionally, a component i has a probability of happening for a distribution P, $P_i = 1$ and $\forall k \neq i$, $P_k = 0$, we attain marginal entropy. Hassan [6–8] has tapped this notion of entropy to calculate the complexity of code change.

Reliable mathematical notions from information theory as Kapur entropy [9] (α–β entropy) have been utilized. These entropies are identified as simplified entropies and because of the existence of restrictions, they are employed for association functions in diverse subjects, for example, coding theory, statistical decision theory, etc. There are numerous measures of entropy however in this analysis, simply Kapur entropy [9] given by Eq. (1.2) has been thought to review outcomes grounded on these entropies. For advanced investigation, additional measures of entropy may be acquired into contemplation and the inquiry and relative review are added subject of study.

Kapur [9] formulated a methodical line to obtain a generality of Shannon entropy [18]. He depicted entropy of order α and degree β explained as follows:

$$H_{\alpha,\beta}(P) = \frac{1}{1-\alpha} \log\left(\sum_{i=1}^{n} p_i^{\alpha+\beta-1} / \sum_{i=1}^{n} p_i^{\beta}\right), \ \alpha \neq 1, \ \alpha > 0, \ \beta > 0, \ \alpha + \beta - 1 \quad (1.2)$$

4 Envisaging Bugs

A project is selected and then subsystems are chosen. The CVS logs are browsed and the bug reports of all subsystems are accumulated. The bugs are obtained through the descriptions and organized on annual base aimed at every subsystem. Additionally, Shannon entropy [18] and Kapur entropy [9] aimed at every interval utilizing these bugs registered in favour of every subsystem that is computed. The bugs are envisaged by applying simple linear regression for upcoming year grounded on the entropy analysed for every existing interval. Regression exploration is a numerical procedure which assists in examining the associations between the variables. Simple linear regression simulation encompasses two variables where one variable is forecasted through another variable. It is the utmost fundamental simulation which is frequently consumed for extrapolation. Simple linear regression system has been employed amid the entropy quantity and experimental bugs existing in every announcement in every time interval to envisage bugs which are impending for every announcement in every time interval.

In this part, the subsequent methods are implemented. Primarily, the information is assembled and pre-administered and then entropy measures are computed.

An open-source software namely Mozilla proposes alternative towards operators and propel novelty next to network. This remains an open software population that fabricates great quantity of schemes as Thunderbird, Firefox mobile, the virus chasing approach Bugzilla. Now, a small number of constituents of Mozilla software are chosen. Bug descriptions of all these constituents from the CVS source are collected. Subsequently obtaining information from this source, yearly amount of variations for every records of these subsystems from the CVS notes is documented.

The succeeding subconstituents of Mozilla open-source software are measured

1. Doctor: A web-built instrument aimed at revising records without operating CVS
2. Elmo: A codename for the latest type of Mozilla localization panel. It is a foundation of web apps, tools and documentation.
3. Graph server: A web-based instrument that deals through performance information gathering and graphing server.
4. Telemetry server: A web service to obtain and deposit telemetry information.
5. AUS: Application Update Service. It is a resourceful web-based instrument to examine the revises of application software accessible on the server through the consumer software.

4.1 Assessment of Entropy Measures

These statistics facts have been employed to compute the probability of every constituent for the 9 time periods from 2011 to August 2019. By means of these probabilities, Shannon entropy [18] is estimated correspondingly for every interval. Likewise, Kapur entropy [9] can be determined. Table 1 exhibited under represents the Shannon entropy:

Table 1 Shannon entropy in all the releases of the Bugzilla OSS

Year	2011	2012	2013	2014	2015	2016	2017	2018	2019
Shannon entropy	0.2441	0.1955	0.3684	0.432	0.4579	0.3727	0.6041	0.5533	0.5717

Table 2 Kapur entropy in all the releases of the Bugzilla OSS

Year	2011	2012	2013	2014	2015	2016	2017	2018	2019
Shannon entropy	0.2947	0.2873	0.4602	0.4675	0.4638	0.4637	0.6875	0.5961	0.5986

The Kapur entropy for quantities of $\alpha = 0.1$ and $\beta = 1.0$ has been reviewed in Table 2 beneath:

As of this investigation, it has been examined that Shannon entropy [18] is highest in the year 2017 and lowest in the year 2012. Kapur entropy [9] is the highest in the year 2017 and lowest in the year 2012.

4.2 Bug Prediction Modelling

SLR model remains as the utmost fundamental paradigm comprising two inconstant around which one inconstant is envisaged through added inconstant. The inconstant in the direction of projection is termed the dependent variable in addition to the regressor which is termed the independent variable. The regression equation given by Eq. (2.1) requires to be extensively tapped to limit the dependent variable Y utilizing independent variable X through subsequent equality:

$$Y = a + bX \qquad (2.1)$$

where a is the y-intercept of the regression line and b is the slope.

The parameters are attained utilizing the SLR method. Subsequently assessing the regression coefficients by means of several amounts of entropy combined with past statistics of detected bugs, a technique remains structured by means of envisaging bugs probable to happen in each interval. Around this analysis, X namely entropy measure, stands to be an regressor and Y, i.e. forecasted bugs probable to happen for every time, is thought to be a regressand. At this point X, namely entropy degree, is diverse around every situation specifically Shannon entropy [18] and Kapur entropy [9].

The regression coefficients remain utilized in envisaging the impending bugs happening into practice. To maintain the bugs for the next year, utilizing various measures of entropy is envisaged. This bug estimation remains advantageous to the software manager in regulating the resource involved within correcting the bugs to sustain the excellency of the software.

The subsequent record given in Table 3 represents the envisaged bugs beside the Shannon entropy measure.

The subsequent record given in Table 4 represents the envisaged bugs beside the identified bugs computed by means of Kapur entropy [9] measure by way of discrete estimates of α and β restriction.

Table 3 Bugs predicted through Shannon entropy in all the releases of the Bugzilla OSS

Year	2011	2012	2013	2014	2015	2016	2017	2018	2019
Y_o	4	12	9	9	5	13	16	22	24
Y	7.0866	5.5639	10.9812	12.9740	13.7855	11.1160	18.3663	16.7746	17.3511

Table 4 Bugs predicted through Kapur entropy in all the releases of the Bugzilla OSS

Year	2011	2012	2013	2014	2015	2016	2017	2018	2019
Y_o	4	12	9	9	5	13	16	22	24
Y	6.0789	5.8157	11.9648	12.2244	12.0929	12.0893	20.0487	16.7981	16.8870

From Tables 3 and 4 computed above, it is determined that the experimental bugs are thoroughly associated by means of experimental bugs through every interval for entropy measures chosen for a fact.

5 Conclusion

A model with the aim of envisaging the bugs which are still to originate in upcoming period by employing simple linear regression is developed. The paper comprised selecting the bugs present in foremost announcements of Bugzilla project with certain data pre-processing. Various measures of entropy specifically, Shannon entropy [18] and Kapur entropy [9] for the alterations in numerous software updates, have been analysed. In the first stage, simple linear regression is related amongst the entropy computed and detected bugs for all announcement through each phase to acquire the regression parameters. Then, the parameters are operated to ascertain the anticipated faults for each announcement in each time period.

References

1. Ambros, M.D., Lanza, M., Robbes, R.: An extensive comparison of bug prediction ap proaches. In: MSR'10: Proceedings of the 7th International Working Conference on Mining Software Repositories, pp. 31–41. (2010)
2. Ambros, M., Lanza, M., Robbes, R.: Evaluating defect prediction approaches: A benchmark and an extensive comparison. Empirical Softw. Eng. **17**(4–5), 571–737 (2012)
3. Chaturvedi, K.K., Bedi, P., Mishra, S., Singh, V.B.: An empirical validation of the complexity of code changes and bugs in predicting the release time of open source software. In: Proceedings of the IEEE 16th International Conference on Computational Science and Engineering, pp. 1201–1206. Sydney, Australia (2013)
4. Chaturvedi, K.K., Kapur, P.K., Anand, S., Singh, V.B.: Predicting the complexity of code changes using entropy based measures. Int. J. Syst. Assur. Eng. Manag. **5**(2), 155–164 (2014)

5. Hassan, A.E.: Predicting faults based on complexity of code change. In: Proceedings of the 31st International Conference on Software Engineering, pp. 78–88. Vancover, Canada (2009)
6. Hassan, A.E., Holt, R.C.: Studying the chaos in code development. In Proceedings of 10th Working Conference on Reverse Engineering (2003)
7. Hassan, A.E., Holt, R.C.: The chaos of software development. In: Proceedings of the 6th IEEE International Workshop on Principles of Software Evolution (2003)
8. Hassan, A.E., Holt, R.C.: The top ten list: dynamic fault prediction. In: Proceedings of ICSM, pp. 263–272. (2005)
9. Kapur, J.N.,: Generalized entropy of order α and β. The Maths Semi, pp. 79–84. (1967)
10. Kapur, P.K., Chanda, U., Kumar, V.: Dynamic allocation of testing effort when testing and debugging are done concurrently. Commun. Depend. Qual. Management **13**(3), 14–28 (2010)
11. Kapur, P.K., Pham, H., Chanda, U., Kumar, V.: Optimal allocation of testing effort during testing and debugging phases: a control theoretic approach. Int. J. Syst. Sci. **44**(9), 1639–1650 (2013)
12. Kapur, P.K., Singh, J.N.P., Sachdeva, N., Kumar, V.: Application of multi attribute utility theory in multiple releases of software. In: International Conference on Reliability, Infocom Technologies and Optimization, pp. 123–132. (2013)
13. Kaur, A., Kaur, K., Chopra, D.: An empirical study of software entropy based bug prediction using machine learning. Int. J. Syst Assur. Eng. Manage. 599–616 (2017)
14. Kaur, A., Chopra, D.: Entropy Churn metrics for fault prediction in software systems. Entropy (2018)
15. Kumari, M., Misra, A., Misra, S., Sanz, L., Damasevicius, R., Singh, V.: Quantity quality evaluation of software products by considering summary and comments entropy of a reported bug. Entropy (2019)
16. Pachauri, B., Kumar, A., Dhar, J.: Modeling optimal release policy under fuzzy paradigm in imperfect debugging environment. Elsevier, (2013). 10.1016/j.infsof.2013.06.001
17. Peng, R., Li, Y.F., Zhang, W.J., Hu, Q.P.: Testing effort dependent software reliability model for imperfect debugging process considering both detection and correction. Reliability Eng. Syst. Saf **126**, 37–43 (2014)
18. Shannon, C.E.: A Mathematical Theory of Communication. Bell Syst. Tech. J. **27**(379–423), 623–656 (1948)
19. Singh, V.B., Chaturvedi, K. K.: Entropy based bug prediction using support vector regression. In: Proceedings of 12th International Conference on Intelligent Systems Design and Applications during 27–29 Nov. 2012 at CUSAT, Kochi (India). IEEE Explore, pp. 746–751. (2012)
20. Singh, V.B., Chaturvedi, K.K.: Improving the quality of software by quantifying the code change metric and predicting the bugs. In: Murgante, B., et al. (eds.) ICCSA 2013, Part II, LNCS 7972, pp. 408–426. Springer, Berlin, Heidelberg (2013)
21. Singh, V.B., Chaturvedi, K.K., Khatri, S.K., Kumar, V.: Bug prediction modelling using complexity of code changes. Int. J. Syst Assur. Eng. Management **6**(1), 44–60 (2014)

TRANSPR—Transforming Public Accountability Through Blockchain Technology

P. R. Sriram, N. J. Subhashruthi, M. Muthu Manikandan, Karthik Gopalan, and Sriram Kaushik

Abstract To ameliorate the deplorable state of public infrastructure construction in growing and underdeveloped countries, and eliminating the rampant corruption involved in them is the primary objective of this research work. We propose to achieve this objective by bringing enhanced transparency in public sector projects. This project aims to deploy a system that would capture all bids for public infrastructure projects on the blockchain. All activities on the project, from the call for tender, procurement of material, choice of contractor, the status of work will be stored on the blockchain. By opening up the blockchain to the public it would be a fully transparent interface. Since the data is stored on the blockchain, it is immutable and every update can be traced to the person who made the update and can be independently verified. By publicly disclosing the status of infrastructure and government projects, a greater degree of accountability can be unlocked. To facilitate this, we propose a methodology called TRANSPR, a concord-based blockchain network. The design of the system is laid out, and the mechanism by which the system maintains state would be elaborated. The coin distribution process will also be detailed, and the formulae would be laid out. As the system is consensus—based, new blocks are established by consensus decisions and not mining. Since there is no mining or documentation

---

P. R. Sriram (✉) · N. J. Subhashruthi · M. Muthu Manikandan · K. Gopalan · S. Kaushik
PricewaterhouseCoopers Services LLP, Chennai, India
e-mail: sriram.pr97@gmail.com

N. J. Subhashruthi
e-mail: subhashruthi.nadadhur@gmail.com

M. Muthu Manikandan
e-mail: muthu.manikandan.m@gmail.com

K. Gopalan
e-mail: karthikmadan@yahoo.co.in

S. Kaushik
e-mail: sriramkaus@gmail.com

© The Editor(s) (if applicable) and The Author(s), under exclusive license to Springer Nature Singapore Pte Ltd. 2021
T. Senjyu et al. (eds.), *Information and Communication Technology for Intelligent Systems*, Smart Innovation, Systems and Technologies 196,
https://doi.org/10.1007/978-981-15-7062-9_16

of execution involved, the energy needs of TRANSPR are minimal, and it is environmentally sustainable. The primary difficulty in the rollout of TRANSPR would be in the adoption phase, as it involves an overhaul of existing processes.

1 Introduction

In todays world, theres little faith in the efficiency of government due to reduced transparency and increased corruption. Large infrastructure investments undertaken by the government have huge costs and take a lot of time to complete. There is little information about the decisions made by the government to choose the contractor, the choice of material, the considerations made before drawing the plan for the project, etc. Putting them all in the public realm will help the public understand the complexities involved in a project. Moreover, it will help other firms undertaking similar projects, by reducing the amount of duplicate work. Sometimes, it looks as if many projects are stalled in a half-finished state. The public has no awareness of the projects progress and the reason behind the stall. Theres no way for the public or the government to know and analyze the track record of a company in finishing projects. Theres no mechanism by which the firms are penalized for delayed delivery of projects. It has now become commonplace for firms to exceed the estimates by several years!.

To overcome the above-mentioned shortcomings, the TRANSparency in Public sectoR—TRANSPR system is proposed. TRANSPR is a blockchain-based system to track everything associated with a project. The blocks on TRANSPR chain contain data about the date of the project start, estimated budget, bids obtained, details of the bids, the contractor's chosen, materials used, current status, slippages in schedules if any, etc. Each project on the blockchain is assigned a unique ID, against which all updates are made. Each project is assigned a set of milestones, and an update has to be made for each milestone. Deadlines are assigned for the milestones and the responsible parties have to justify slippages if any. The parties will be penalized if they miss updating a milestone. The public can connect to the blockchain and view the projects, see their status, and the responsible parties. By openly revealing the status of projects, the public will have increased trust in the government and the system as a whole. The success of TRANSPR lies in increased adoption and usage amongst the government departments.

2 Existing Literature

The paper Bitcoin: A Peer-to-Peer Electronic Cash System by Nakamoto, Satoshi, introduced the concept of blockchain to the world and triggered the blockchain revolution of today. This paper elucidates the fundamental concepts of the blockchain and proposes a solution to the double-spending problem. The paper introduces a timestamp server, which hashes all the data, and circulates it along with the timestamp. For the timestamp server to be implemented, the author introduces a proof of work—finding a sha256 hash with a requisite number of 0 s. Security is guaranteed as long as honest nodes control more CPU than attackers, as the attacker will have to modify all preceding blocks. Only the honest chain will grow the fastest and longest, and as the system by design considers only the longest chain as legitimate, all attacks will fall through. The proof of work difficulty is adjusted by a moving average, ensuring that only a fixed number of blocks are available every hour. This paper also lays the groundwork for incentivizing miners, by allotting them coins, or by charging transaction fees. The proposed system also separates the identities of the individual from the transactions, as all transactions are made only with the private key of the individual, and not the actual identity. This paper is the cornerstone on which the entire technology is built.

Blockchain: Future of financial and cybersecurity by Singh, S., Singh, N talks about the concept of Blockchain and how the transactions are handled using the blockchain mechanism. It explains the characteristics and needs of blockchain and how they work with bitcoin. The paper attempts to highlight the role of blockchain in shaping the future of banking, finance, and the Internet of Things. The working of bitcoin is given in detail along with the advantages of using it. They also give a detailed insight into how bitcoin is used and adopted in various companies.

Decentralizing Privacy: Using Blockchain to Protect Personal Data by Guy Zyskind, Oz Nathan, Alex Sandy Pentland talks about security breaches compromising user's privacy. In this paper, A decentralized personal data management system is described to ensure users own and control their data.They have implemented a protocol that obtains automated access-control manager via blockchain that does not demand trust in a third party. This paper gives an insight into how a new implementation can help users protect their data and be on a private network. Collecting, sharing and storing sensitive data has been simplified on the basis of making legal and regulatory decisions, with the use of a decentralized platform.

Ahmed Kosba, Andrew Miller, Elaine Shi, Zikai Wen, Charalampos Papamanthou have described how all transactions, including the flow of money between pseudonyms and the amount transacted, are exposed on the blockchain in their paper Hawk: The Blockchain Model of Cryptography and Privacy-Preserving Smart Contracts. They have presented a new system called Hawk, which is a decentralized smart contract system that does not store financial transactions in the clear on the blockchain, thus retaining transactional privacy from the publics view. It gives an idea of how security and cryptography can be used in blockchains to make the transactions private.

The paper on A Random Block Length Based Cryptosystem through Multiple Cascaded Permutation Combinations and Chaining of Blocks by Jayanta Kumar Pal, J. K. Mandal, A four-stage character/bit-level encoding approach (RBCMCPCC), where the first three steps take input block of length 128 bit, 192 bit and 256 bit respectively, and yield instant blocks of the equal lengths using identical lengths of keys in each step. After the four steps, the ciphertext is produced. This paper proves that the multistage cipher technique and the use of multiple numbers of keys of non-uniform lengths in various stages of the encoding process along with random session keys have enhanced the security features. The paper by Xiao Wan, Qingfan Hu, Zheming Lu, Manlian Yu on Application of Asset Securitization and Blockchain of Internet Financial Firms talks about how the rapid spread of internet has changed peoples lifestyle. The concept of Accounts Receivable Backed Securitization (ARBS) of JD.com Consumer credit Line and the specific transaction procedure is introduced in this paper. It also establishes the blockchain technology utilized in the asset securitization of JD, primarily contain the basic principles of blockchain technology, the requirement of application in ABS and the application perspective in the future financial industry.

An Overview of Incremental Hash Function Based on Pair Block Chaining by Su Yunling, and Miao Xianghua talks about how the hash function plays a major role in information security. Message authentication, data integrity, digital signature and password protection are extensively made possible via hash function. It introduces an outline of incremental hash function structured on pair block chaining, with research on theory, security, efficiency and application.

"Information Sharing for Supply Chain Management based on Blockchain Technology" by Mitsuaki Nakasumi gives an idea that Major companies and their planning operations are supported by supply chain management systems that provide information analysis and information sharing. The following in based on factual data, as the information between the companies is asymmetric, then resulting to a disturbance of the planning algorithm. It is of utmost importance to ensure reactivity towards market variability by sharing essential information across manufacturers, suppliers and customers.

3 Description and Analysis

3.1 Blockchain

Technological advancements in the field of networking have made it much easier to store and validate millions of transactions. The blockchain is a distributed database that provides an unalterable and semi-public record of a digital transaction. A timestamped batch of transactions are collected by each block which are then included in the ledger or in the blockchain. A cryptographic signature is used to identify each block. These blocks refer to the signature of the previous block in the chain, that is, they are back-linked, and that chain can be traced back to the very first block created. As such, the blockchain contains a record of all the transactions made which are uneditable. Whenever a transaction needs to be added, a network of computers carries out a series of computations defined by algorithms, identifies the parties involved in the transaction and validates the transaction. The transactions properties, the validity of the sender and receiver, verification hat the sender contains enough coins to make the transaction, verification that theres no other pending overlapping transaction etc. are verified. It is then added to a digital ledger and forwarded to an irreversible chain of transactional blocks. The valid and verified transactions are recorded and are traceable. The original source code for Bitcoin is where block chain was defined first.

Public key cryptography is used to validate all transactions on the blockchain digitally, using asymmetric cryptography. The private key remains inconspicuous. It is held by one of the parties. The other key is the public key. Both the keys together are called the key pair and are used together in encryption and decryption operations. Each key can only be used in conjunction with the other key in the pair, which shows that their relationship is reciprocal. The keys in the pair are exclusively tied to one another: a public key and its corresponding private key are paired together and are related to no other keys. The public keys and private keys have a special mathematical relationship between the algorithms which make the pairing possible. To undo an operation which one of the keys made, the keys must be used together. Each individual key cannot be used in isolation. Thus, the private key cannot be distinguished from the public key. As opposed to symmetric keys, the mathematical relationship that makes key pairs possible contributes to one disadvantage of key pairs. The algorithms used for the key pairs must be strong enough to make it impossible for people to decrypt the information that has been encrypted, using their public key, through brute force. A public key uses its one-way nature and its complexity to prevent people from breaking information encoded with it. The recipient uses the public key to sign and encrypt the message that is to be sent and only the designated recipient can decrypt that transaction with their private key. Public key cryptography is not only used to encrypt messages but also to authenticate an identity. It is also used to verify that the message (transaction on the blockchain) has not been altered

The dispersible aspect of the blockchain makes it possible to transfer data about all transactions to all the nodes in the network. This enables the blockchain to maintain one ledger as opposed to many conflicting ledgers. To update a blockchain, all distributed copies must be reconciled so that all of them contain the same version of data. A maximum number of the nodes in the blockchain must concur which happens via a consensus process. Each new transaction and block is verified computationally, making the process emergent and not a task that occurs at a stipulated time or interval. Each block present in the blockchain is made up of a list of transactions and also contains a block header. The header contains at least three sets of metadata: (1) Structured data about the transactions in the block (2) Timestamp and data about the proof of work algorithm (3) Reference to the parent block through a hash. The hash of its header is used to recognise each block in the blockchain. New blocks are created through a process called mining. It validates new transactions and adjoins them to the chain. In order to mine new blocks, miners present on the network compete to solve a unique, and difficult math puzzle. The proof of work of the solution to the puzzle is included in the block header to verify the block. The difficulty of this problem has increased exponentially since the time bitcoin was created, thereby increasing the computational power. The Bitcoin becomes even more secure because of the increasing complexity of the proof of work algorithm

4 Achieving Consensus Without a Proof of Work

A major issue with bitcoin and other blockchain solutions is the increasing costs of mining new blocks. This alienates miners with a run of the mill hardware, as they do not have enough power to participate in the network. Only miners with enormously powerful machines will be able to mine new blocks. In order to overcome this problem and reduce the time for confirmation, many alternative algorithms have been proposed. One of them is the Ripple Consensus protocol, which forms the basis of the Ripple network. This relies on consensus between nodes over a Proof Of Work (PoW) to maintain state.

The Ripple network uses a commonly shared ledger, which is a distributed database to store information about all Ripple accounts. It is used for Real-Time Gross Settlement (RTGS) across central banks. It is based on the Ripple Protocol, which confirms the transactions through a network of servers. These servers are managed independently and communicate with each other on the network. Each server can propose a change to the shared ledger and is considered as a transaction. A new entry is created (opened) every few seconds, and the last closed entry in the ledger is the record of all Ripple accounts, as determined by the network. The servers come to a consensus about a set of transactions, creating a new last closed entry.

The goal of the distributed consensus process is for each server to apply the same set of transactions to the current entry, thereby arriving at the same last closed entry. The servers will receive transactions from other servers on the network continuously, and it determines what transaction to apply, based on where it originated from. The servers maintain a unique node list or UNL to determine this. The transactions that are agreed to, by a supermajority of peers are considered valid and are added to the entry. If theres no consensus, this implies that the transaction volume is high or the network latency is too great. The servers again try to reach consensus in such cases, reducing disagreement until supermajority is reached. The outcome is that disputed transactions are rejected and transactions that are agreed to by most nodes are added.

5 TRANSPR

The proposed TRANSPR system uses the RIPPLE consensus algorithm to achieve consensus amongst the nodes. There can be a number of participants in the TRANSPR chain. The major portion of the servers will be run by the government, and the government will have to control more than 75 Percentage of the servers on the network to protect against attacks. The firms that undertake government projects will have to join the TRANSPR network, and attach a server for each project in progress. The server will have to be removed from the network once the project is complete.

The servers on the network use a coin called TRANSPR coin or TRC to transact. The coins are strictly controlled and are generated beforehand. Like XRP, TRC is

$$N = (\sum_{i=0}^{n} (c_i + d_i) + D)$$

Fig. 1 Where, **N** is the number of coins in the possession of the firm, **c** is the total duration of a project, **d** is the agreed deviation of a project, **D** is the total leftover coins, from other projects, **n** is the total projects in progress

limited to 100billion in total. The coins are strictly controlled by the government and are allotted to the firms during the inception of the project. The number of coins allotted to the firm can be calculated according to the following formula. $N = R + D$ Where, N is the number of coins, R is the estimated duration of the project (Weeks) D is the accepted delay (Weeks)

The coins are allotted on a per-project basis. Each firm has to use its servers to update the status of the projects each week. They need to spend one TRC for each update, thereby ensuring that the firm runs out of coins after the accepted delay period ends. Once the firm runs out of coins, it will not be able to send updates to the network. If an update is missed, the server is marked Dead and the government will demand an explanation and a revised estimate from the firm. The government will then release more coins, according to the revised estimate, after the firm pays a penalty for slippages in the schedule. In case the firm completes the projects before the schedule, they will be allowed to keep the coins as a reward. If the firm wants to use the coins to bolster the slipping schedules of another project, they will be allowed to do so. A firm will have to sell the coins to the government if it has no projects scheduled to begin in the next two years. The total number of coins that a firm can hold at any point in time can be defined as (Fig. 1),

The public can connect to the network over the internet and view the status of the projects. They will not be able to make any changes to the data. The officials will have elevated access to the network and will be able to flag suspicious updates, that warrant additional scrutiny. An officer will have to manually approve major milestone reports sent by the firm, for records. As the chain records each and every update made, it will be easier to audit and find the exact person responsible for the change, thereby eliminating corruption. The firms will be sending updates to the projects by spending TRC every week. The update is broadcasted across the network, and until each node agrees to the content of the update, it will not be added to the blockchain. As the government will control more than 75 Percentage of the chain, firms can't do a hostile takeover by presenting false updates.

6 Discussion

The openness of the blockchain is a double-edged sword. By openly displaying the amount spent, the work put in, the source of raw materials etc. could lead to a few issues. The firms might discover that the suppliers charge a premium for them, whilst selling it cheap to the others. Such issues could be avoided, by implementing the ideas outlined in the Hawk System [3] and [4]. Moreover, there is the question of security of the blockchain itself, which can be easily achieved using the ideas presented in [5], [6] and [7]. By ensuring that only the users who are authenticated, and have the proper privileges can access the blockchain, a semblance of security can be achieved. The ideas around this are enumerated in [8].

At its crux, the TRANSPR network is similar to a supply chain management system. The advantages and disadvantages of running a supply chain management system on a blockchain are well outlined in [9]. The major criticisms of blockchains are about its limited scalability and the increasing cost of running the network. TRANSPR can easily overcome the scalability issue, as the state is maintained through consensus and not through a Proof of Work. However, the costs will still be high, as the government and companies have to maintain the servers that participate in the network. Moreover, per TRANSPRs design, as the number of contractors increase, the government will have to keep adding servers, to main the 75 Percentage majority. For TRANSPR to be successful, there needs to be increased adoption.

Firms and Governments will need to be incentivized to adopt TRANSPR. For firms, joining the network will help them reduce their costs, as they will not be paying bribes anymore. Moreover, it gives them an incentive to finish projects quickly, so that they can get incentives from the government by selling the TRANSPR Coins. This will lead to more investment in RD, to develop better techniques. As a result, more jobs will be created, and increased hiring will be seen in these sectors. The government will be able to better plan for the future, as there will be increased trust that the firms will deliver on or ahead of time. There can be regulatory frameworks around the system to ensure that the firms do not cheat by overestimating the timelines.

The TRANSPR network relies on consensus as opposed to proof of work as followed by Bitcoin and other alternative coins (Altcoins). As a result, the energy needed to sustain the network is very less, as opposed to bitcoin, which uses almost as much energy as a small town to validate a transaction. The whole objective of the TRANSPR network is to improve the living conditions of the people, by improving governance, one can say that it does so, without damaging the environment. The impact of TRANSPR on the society cannot be overstated. It will usher in an era of increased trust in the government, by enabling greater transparency. TRANSPR can help reduce instances of corruption by publicly displaying the progress of projects, parties involved, the amount invested, the materials used, cost of acquisition of materials, cost of labor, environmental impacts etc. As corruption is eliminated, more companies and contractors will bid for infrastructure projects, confident of getting their due. As a result, the overall quality of infrastructure improves. As corruption is eliminated, the government will be able to put its funds to better use, as the projects will be completed at cheaper prices. Only a minuscule portion of the savings that the government stands to make will be spent in running the TRANSPR network. Since the government can spend more, lots of jobs will be created, thereby leading to a lush and growing economy. A true democracy will be achieved in which the public can trust the government's initiatives and question if anything is amiss.

This will be of great use in developing and underdeveloped countries, where there is rampant corruption and mismanagement of funds. Introducing TRANSPR in such countries will be a paradigm shift, as the quality of governance will increase manifold. Moreover, TRANSPR can be easily adapted to incorporate a variety of use cases—such as health campaigns, criminal records storage, travel records, tax records, etc. As a system with a potentially unlimited number of use cases, TRANSPR can transform the way governments operate.

7 Conclusion and Future Work

From an analysis of the existing systems used to manage public sector projects in developing countries, it is seen that theres little to no accountability on the software front. The news is littered with scams emerging due to politicians and companies swindling government funds. As a result, the confidence in the government is less. Introducing blockchain into the mix will break the nexus between the corrupt politicians and the officials in swindling, as transactions become immutable. Moreover, as everything is public, theres no place to hide, and corrupt people can be easily identified. The TRANSPR system is incorruptible by design and hence will be a major factor for change. As theres practically no way for corrupt people to edit the records, and get away without any traces, a majority of corruption will be reduced. As a result, the faith of the people in government and governance will increase, thereby leading to economic prosperity. The challenges faced by TRANSPR will be in adoption, as an inherently corrupt government will not be inclined towards adopting a tool which eliminates their source of income. Moreover, the technology behind blockchains are still evolving and will take time before they are fully stable. As a result, there might be a few hiccups during the initial phases. As a part of future work, the TRANSPR system can be enhanced in a way that it is incorporated directly into all the block chains so that adoption of a new tool will not come into place. Also, more enhancements for TRANSPR can be made to increase the usability.

References

1. Nakamoto, S. Bitcoin: A Peer-to-Peer Electronic Cash System. 2008
2. Singh, S., Singh, N.: Quot;Blockchain: Future of financial and cyber security,quot; 2016 2nd Interna- tional Conference on Contemporary Computing and Informatics (IC3I), Noida, 2016,pp. 463-467
3. Kosba, A., Miller, A., Shi, E., Wen, Z., Papamanthou, C.: Hawk: the blockchain model of cryptography and privacy-preserving smart contracts. In: 2016 IEEE Symposium on Security and Privacy (SP), pp. 839–858. San Jose, CA (2016)
4. Zyskind, G., Nathan, O., Pentland, A.: Decentralizing privacy: using blockchain to protect personal data. In: 2015 IEEE Security and Privacy Workshops, pp. 180–184. San Jose, CA (2015)

5. Pal, J.K., Mandal, J.K.: A random block length based cryptosystem through multiple cascaded permutation-combinations and chaining of blocks. In: 2009 International Conference on Industrial and Information Systems (ICIIS), pp. 26–31. Sri Lanka, (2009)
6. Wan, X., Lu, Z., Hu, Q., Yu, M.: Application of asset securitization and block chain of internet financial firms: take Jingdong as an example. In: 2017 International Conference on Service Systems and Service Management, pp. 1–6. Dalian (2017)
7. Yunling, S., Xianghua, M.: An overview of incremental hash function based on pair block chaining. In: 2010 International Forum on Information Technology and Applications, pp. 332–335. Kunming (2010)
8. Yu, R. et al.: Authentication with block-chain algorithm and text encryption protocol in calculation of social network. In: IEEE Access, vol. 99, pp. 1–1
9. Nakasumi, M.: Information sharing for supply chain management based on block chain technology. In: 2017 IEEE 19th Conference on Business Informatics (CBI), pp. 140–149. Thessaloniki (2017)
10. Shi, J., Zhu, Q., Jing, N.: Research on life cycle of power financial products based on block chain factor. In: 2017 4th International Conference on Industrial Economics System and Industrial Security Engineering (IEIS), pp. 1–5. Kyoto, Japan (2017)
11. Judmayer, A., Stifter, N., Krombholz, K., Weippl, E., Bertino, E., Sandhu, R.: Blocks and chains:introduction to bitcoin, cryptocurrencies, and their consensus mechanisms. In: Blocks and Chains:Introduction to Bitcoin, Cryptocurrencies, and Their Consensus Mechanisms, vol. 1, pp. 123. Morgan Claypool (2017)
12. Halpin, H., Piekarska, M.: Introduction to security and privacy on the blockchain. In: 2017 IEEE European Symposium on Security and Privacy Workshops (EuroSamp;PW), pp. 1–3. Paris (2017)
13. Natoli, C., Gramoli, V.: The blockchain anomaly. In: 2016 IEEE 15th International Symposium on Network Computing and Applications (NCA), pp. 310–317. Cambridge, MA (2016)
14. Wright, A., De Filippi, P.: Decentralized blockchain technology and the rise of lex cryptographia. SSRN Electron. J. (2015)
15. Atzori, M..: Blockchain technology and decentralized governance: is the state still necessary?. SSRN Electron. J. (2015)
16. Conoscenti, M., Vetro, A., De Martin, J.C.: Blockchain for the internet of things: a systematic literature review. In: 2016 IEEE/ACS 13th International Conference of Computer Systems and Applications (AICCSA), pp. 1–6. Agadir (MAR) (2016)
17. Yli-Huumo, J., Ko, D., Choi, S., Park, S., Smolander, K.: Where is current research on blockchain technology?a systematic review. PLOS ONE 11(10), (2016)

The Scope of Tidal Energy in UAE

Abaan Ahamad and A. R. Abdul Rajak

Abstract This paper focuses on the scope of tidal power in today's era. It describes tidal electricity and the various techniques of using tidal strength to generate energy. In short, it discusses every approach and presents various techniques of calculating tidal strength and electricity most effectively. The paper additionally also discusses a literature survey where a comparison between different hindrances and factors contribute to the generation of tidal energy in different parts of the world, mainly focusing on USA and South Korea. In the later sections, the scope and advancements of UAE in the generation of tidal electricity have been discussed.

1 Introduction

Tidal strength, additionally referred to as tidal energy, converts the energy possessed by high tides into usable electricity. The Rance Tidal Power Station was the first ever wide-spread tidal power plant which started its operation in 1966.

Tidal energy has suffered initially due to the high cost of installation of equipment and non-availability of information about its execution on the internet. As a result, tidal energy is not yet exploited to its best potential, but tidal power carries great potential with itself for the future generations. Tidal energy is more reliable than any other form of energy such as wind or solar, as it can be easily predicted. Advancements in the layout and turbine technology have boosted its total availability and also managed to cut down environmental costs.

A. Ahamad · A. R. A. Rajak (✉)
BITS Pilani Dubai Campus, Dubai International Academic City, Dubai, United Arab Emirates
e-mail: abdulrazak@dubai.bits-pilani.ac.in

A. Ahamad
e-mail: f20160181@dubai.bits-pilani.ac.in

© The Editor(s) (if applicable) and The Author(s), under exclusive license to Springer Nature Singapore Pte Ltd. 2021
T. Senjyu et al. (eds.), *Information and Communication Technology for Intelligent Systems*, Smart Innovation, Systems and Technologies 196, https://doi.org/10.1007/978-981-15-7062-9_17

1.1 Generation of Tidal Energy

Tidal energy is derived from tides which is in turn formed by the combination of Earth's rotation and the forces acted upon by Sun and the Moon. Low tides and high tides are caused because of the gravitational pull of the Moon and the Sun. In total, movement of Sun and Moon relative to Earth, Earth's rotation and the coastal area's geography, are the factors that determine the tide at a particular location [1].

Tidal power generation can be accomplished in three ways:

- Tidal stream generator
- Tidal barrage
- Dynamic tidal

Tidal Stream Generator

Tidal stream generator, also known as tidal turbines, utilizes the movement in the water body (tides) to generate energy. Their functioning is very similar to that of wind turbines. This is the most cost-efficient and environment-friendly way of generating tidal energy. Different turbines with different efficiencies give distinct power output [2]. The output power can be calculated using the following formula:

$$P = \frac{\xi \rho A V^3}{2} \quad (1)$$

where

P The power generated in watts
ξ Turbine efficiency
ρ Density of water (seawater is 1025 kg/m^2)
A The sweep area of the turbine (in m^2)
V^3 Velocity of the flow

The maximum possible energy extraction from a strait connecting two large basins accurate to 90%, is given by:

$$P = 0.22 \, \rho g \Delta H \text{max} \, \Delta Q \text{max} \quad (2)$$

where

ρ Density of water (seawater is 1025 kg/m^2)
g Gravitational acceleration (9.81 m/s^2)
ΔHmax   Maximum differential water surface elevation across the channel
ΔQmax   Maximum volumetric flow rate through channel

Tidal Barrage

Tidal barrage is a dam-like structure that extracts energy from tidal forces generated by the movement of water across rivers. First, a barrage is built across the basin that

The Scope of Tidal Energy in UAE

Table 1 Various tidal barrage techniques

1	Ebb generation	Sluice gates are opened after a sufficient head is created due to the fall of sea water level. Turbines generate power while the gates are closed. Once gate is opened, basin is filled again
2	Flood generation	Through the turbines, the basin is filled. Turbines generate only during tide flood. This is less efficient as the amount of overhead (water level) is very less compare to that of ebb generation
3	Pumping	Turbines can be also be powered in reverse to increase the water level at high tide, in the basin. Due to the aforementioned correlation, this energy is helpful in returning the energy during generation
4	Two-basin schemes	One basin is vacated during low tide and the other is filled during the high tide. Turbines present in the middle of both the basins are responsible for generation that is continuous in this case

supports movemnet of water. The most imprtant job of the turbines is for generating power as water flows in and out of the bay. They produce power when the water level outside relative to the level inside changes [1]. The energy generated from a barrage turbine is calculated by the following formula (Table 1):

Dynamic tidal power

Most recent and not completely tested method of tidal every generation is dynamic tidal power, or DTP. It proposes a T-shaped structure. Dam-like stretch extending from coast to few kilometers inside the ocean. The end of it would be a perpendicular barrier. This barrier intercepts huge and powerful coast parallel tidal waves, in turn generating electricity. The length ranges from 30 to 60 km. The long dam-like structure also blocks the horizontal acceleration and it is long enough to generate a water level difference on either side (of the dam). Another way of generating power is by using low-headed turbines.

2 Designing a Beneficial Power Plant for Effective Tidal Power Generation in the United Arab Emirates

As mentioned earlier, the energy obtained from the tidal barrage is given by

$$\mathbf{E = 0.5 A \rho g h} \tag{3}$$

where

ρ Density of water (seawater is 1025 kg/m^2)
A   Is the horizontal area of the basin
H   Is the vertical tidal ranger

Assuming the tidal height of a tide on a particular day in the UAE (based on average approach) is 3 m.

The surface area of tidal energy harnessing plant is 10 km² (3.2 × 3.2 km)
This is almost equal to 10×10^6 m².
The specific density of water is 1025.18 kg/m².
Mass of water = Volume of water × Specific gravity

$$= \text{area} \times \text{tidal range} \times \text{density}$$

$$= 10 \times 10^6 \text{m}^2 \times 3 \times 1025.18 \text{ kg/m}^2.$$

$$= 30.7 \times \times 10^9$$

Potential Energy of water in basin = 0.5 × density

$$\times \text{ gravitational acceleration} \times \text{tidal range}^2$$

$$= 0.5 \times 10 \times 10^6 \times 1025 \times 9.81 \times 3^2$$

$$= 4.52 \times 10^{11} \text{ J Approx}$$

Since we have two tides occurring naturally everyday
$= 4.52 \times 10^{11}$ J × 2
$= 9 \times 10^{11}$ J Approx

Now calculating the mean power generation potential = Energy Potential/Time in 1 day
$= 9 \times 10^{11} / 86400$ s
$= 1.047 \times 10^7$
$= 10.47$ MW

Let us assume we can achieve a minimum of 40% efficiency.

Hence, the daily power generated would be 10.47 MW × 40% which is almost 4.2 MW of power generated on a daily basis with only a minimal tidal height of 3 m.

3 Literature Survey

.After an intensive research on tidal energy as a whole, this paper not only visualizes the power of tidal energy in the United Arab Emirates, but also focusses on a comparative study of how and why other countries were successful or not in utilizing tidal energy efficiently. The path to utilizing tidal energy productively has always been a challenge for engineers, but from the beginning of twentieth century, engineers have come up with innovative methods to harness electricity in places where tidal waves are minimal. Even after years of technological development, it is really sad to see that tidal energy generation is still in its infancy. There are very few areas in the world where commercial sized power plants are being optimized effectively. The first ever tidal power plant was located in La Rance (France). The largest power plant is located in the heart of South Korea, the Sihwa Lake, which took almost a decade to construct.

England, China, Russia and Canada are few other countries who have a great potential to utilize the power of tidal energy based on their geographical location, technological advancement, feasibility and many other factors. Since the United Arab

The Scope of Tidal Energy in UAE

Emirates resembles the USA in a number of factors, this paper also gives a glimpse of the few tidal power plants in USA. As said earlier, the engineers are working on innovative ways to build a cost effective plan, which is not only ecofriendly but will also give back returns, benefit customers and make good profit. The next section of the paper will focus on the advantages and disadvantages of opting for tidal power plant in accordance with power plants in USA and the largest power plant in the world, the Sihwa Lake Power Plant.

3.1 Based on Land and Pollution

If there is one stand out feature about Tidal power plants, it is the fact that tidal power is a green power source giving out practically zero house gasses. Not only is it environmental friendly, but also takes very little space unlike other power plants. The Sihwa Lake Tidal Power Plant has an installed capacity of 250 MW.

The USA relies highly on wind energy to for their power generation. Comparing tidal plants to some of the wind energy farms such as the Roscoe wind farm in the states (TEXAS), they take 400 km^2 of area. We can clearly see the massive advantage tidal power plants hold against wind farms in terms of area.

Wandering into solar farms, such as the Tengger solar park in China which covers 43km^2 of area and Bhadla solar park in Rajhastan covering up to 45Km^2 of area, we can easily arrive at the conclusion that be it Solar parks or wind farms, they have no comparison to wind power plants.

After analyzing the different countries and their ways of generating power, it is suffice to say that even small countries can compete with land-rich countries provided they have a sufficiently long coast line [3].

3.2 Based on Facility/Cost

One of the most important factors is the humungous feasibility and cost when it comes to tidal plants. Analyzing the Sihwa Lake once again, after 4 decades of coming up with the initial idea, the concept of tidal energy generation took a step as a method to tackle the problem of increasing oil prices and significantly lower the concentration of greenhouse gasses. When there is a high tide, water is made to flow from the West Sea to Sihwa Lake via the **ten turbines**, generating electricity. Each of the ten turbine generators has a capacity of almost **25 MW** allowing the plant to produce **552GWh** of electricity every year supporting a population of half a million (Table 2) [4].

It is very important to note that all these countries went for tidal energy power plants including South Korea because they did not have oil or other resources. UAE has oil and that adds to its advantage.

Table 2 The following table summarizes the functionality of the Sihwa Lake Power Plant

Geographical location	Sihwa embankment, Ansan City, Gyeonggi-do, Korea
Capacity:Power(MW)	254 MW (10 × 25.4 MW turbines)
Number of gates	8 sluice gates (15.3 m × 12 m, Culvert type)
Yearly generation	552.7 GWh
Project construction period	2003–11 (commercial operation began in 2011)
Approximate Project cost	USD 560 million

3.3 Replenishing Aquatic Life

Once again focusing on the Sihwa Lake, around 160 million tons of water flow through the floodgate and water wheel. Note that this is about half the total quantity of water present at Sihwa Lake. It goes without saying, the circulation of water between the outer sea and the lake has improved the water quality by multiple folds [5]. The following are some of the many advantages the tidal power plant has brought in the environment:

- The Chemical Oxygen Level in the Sihwa Lake saw an exponential drop from 17 to 2 ppm two decades back, which has improved the aquatic condition.
- Not only for animals, the 12.5 km long embankment of the Sihwa lake is a popular site for leisure activities attracting over 1.4 Million tourists annually [5].

The major reason for the high cost associated with tidal power plants is the advanced engineering required to build large barrages, construct power plants, install them and connect them to the power grid. Over the past few decades, many new technologies have emerged but there is a lack of one supply technology across the globe which increases the price of every component. The United Kingdom is a striking example of the above. The UK government has invested almost 51 million pounds in the initial stages of the project this project is expected to produce over 400 Megawatts by 2020. However, during the construction process, the wildlife was adversely affected and would take almost 7--8 years to construct.

4 The Scope of Marine Energy in United Arab Emirates

Over the past few decades, tidal energy has been rejected as a source of renewable energy in the United Arab Emirates. Solar and wind energy solutions have dominated this region for as long as time can recall. If the solar panels are well maintained, solar energy is an unparalleled source of energy in the United Arab Emirates.

All this being said, there is a possibility that we are undermining the potential of marine energy in the Gulf. Many projects have tried to focus on bringing home the concept of tidal energy home; however, this project takes a different perspective. To begin with, the United Arab Emirates is surrounded by water from almost all the sides, having a coastline of 1318 km. There are a number of places within UAE where the difference in water levels is sufficient to generate electricity. Although the concepts and technologies are already prevalent, the process is not commercially viable. Marine energy in the UAE can be divided into.

4.1 Tidal Energy

Majority of the paper focusses on countries opting for tidal energy. Tidal energy focuses majorly on two aspects, the vertical tidal range and the total area of the tidal basin. While writing this paper, I observed the tidal range in parts of Abu Dhabi. The tidal coefficients range from 60 to 70, whereas the tidal range vary from 1.1 to 1.75 m approximately. This is a characteristic of very weak tidal currents. To capture tidal currents, you need a tidal basin. As of now, there are a number of tidal basins in Abu Dhabi and a few near palm Jumeirah. In Dubai, the palm islands can act as natural creeks and lagoons [3].

4.2 Wave Energy

The one major distinguishing factor between tidal and wave energy is the fact that harvesting power though wave energy does not require any specific landscape, geography or tidal basins. This allows us to utilize the infinite power of wave energy over long Coastlines. This is the biggest advantage the United Arab Emirates possesses, its vast coastline.

This type energy can be achieved by systems depending on the change in water height between the trough wave and crest. It can also be gained by oscillations of the wave. If you compare the waves in UAE, specifically during the north westerly winds, they tend to have periods of 7--8 s and reaching a height of 1--2 m. The biggest drawback with any marine energy system is that during unnatural conditions such as a storm or tsunami, the mechanical and electrical components often tend to get damaged. However, the moderate climate here in the United Arab Emirates is a favorable condition for the same.

Despite investments in marine energy by various sponsors, the technologies have been developing very slowly.

"Oceanlinx and Wavebob went out of business, Wavegen was folded back into parent company Voith, AWS Ocean Energy scaled back its activities, and Ocean Power Technologies cancelled two of its main projects. Siemens sold its tidal power unit in 2015 after losing faith in the Tidal Energy Sector." The Australian Business Review" [6].

It goes without saying that marine energy is not the first option, keeping in mind the effectively low energy prices in UAE, but this sector of energy generation cannot be neglected as it holds a number of advantages over solar energy, both in the environmental and economic zones.

5 A Comparative Study: It All Adds up

Without a doubt, tidal energy is one of the most budgeted forms sources of renewable energy costing more than $4 million for its implementation. Despite this, looking at the long run, tidal power plants once again prove to be economical. Taking the example of the Sihwa plant, it has 240 MW of production producing over 495GW hours of electricity annually. This means that the plant produces 2.08GW hours of energy for every 1 MW at the cost of 2.4 annually generated kW hour. Almost every plant runs for more than 4 decades, and roughly assuming a plant running for 40 years would only cost $0.06 per kW h. It is important to note that the cost of the Sihwa tidal power plant was recovered within two decades. These statistical facts easily direct us to the fact that the route to tidal power generation is a fruitful one in the long run.

6 Conclusion

The paper presents conclusions and recommendations on the role of tidal power in a number of countries, primarily UK, South Korea and the United Arab Emirates. The project aimed at being as transparent as possible on how the data were accumulated and portrayed. The entire paper is an analysis that has been drawn after a series of conclusions. Over the course of writing the project, I interacted with DEWA professionals working in the energy sectors gaining insights into UAE's future plans. A lot of the material in the paper is inspired by talks I attended on energy conferences held in UAE (Snaggs) where top notch energy industrialists pitched their ideas on utilizing energy effectively.

The paper lays out clear set of recommendations and challenges to the government to ensure the exploitation of marine power in an effective way. The paper indirectly hints at many aspects the government has to keep in mind with the increasing environmental degradation [7]. DEWA along with the UAE government has to take a wider action on climate change as this is an environmental opportunity. Without a doubt, it will be impossible to implement the concept of marine energy alone, the indulgence of private sectors and investors is a must. This coast line-rich country has every potential of becoming the leading energy generating sector in the world. The vision of His Highness Sheikh Mohammed bin Rashid Al Maktoum, Vice-President and Prime Minister of the UAE and Ruler of Dubai, is the implementation of 10X, taking Dubai ten years into the future in the next two years. I believe that the harnessing of marine energy in UAE can be a great asset to this initiative [8].

References

1. Tousif, R., Taslim, B.: An effective method of generating power, Int. J. Sci. Eng. Res. **2**(5) (2011, May)
2. Mohammed, L.: Using tidal energy as a clean energy source to generate electricity, Young Scientist (2018)
3. Electricity grid interconnection in the middle East. https://www.energydubai.com. Accessed on 29 Nov 2019
4. Sihwa Lake tidal power station. https://www.hydropower.org/blog/technology-case-study-sihwa-lake-tidal-power-station. Accessed on 03 Oct 2019
5. K Water: Water quality improvement and increase in biodiversity, MEIS national Marine environmental watering system (2005)
6. Tidal, wave to remain'trifling' to 2020, the Australian Business Review. https://www.theaustralian.com.au/,https://www.emirates247.com/news/emirates. Accessed on 15 Oct 2019
7. Knight, O., Hill, K.: Turning the tide, tidal power in the UK, Sustainable development commission (2007)
8. Interview with Dewa official. Abu Fowaz Abdul Rauf Asst Mgr.- Tech. Coord, DEWA (DUBAI Electricity and Water Authority)

Knowledge Engineering in Higher Education

Shankar M. Patil, Vijaykumar N. Patil, Sonali J. Mane, and Shilpa M. Satre

Abstract Data analysis shows significant part in assessment provision regardless of the kind of industries like industrial or an education institutes. EDM anxieties with emerging techniques for determining knowledge from didactic domain data. EDM techniques are applied to analyze undergraduate learners' assessment. The performance focuses on problematic of domain-wise and subject-wise strong/weak graduates in their respective domain knowledge. In this research work, extract valuable knowledge of data gathered from Bharati Vidyapeeth (BV) College of Engineering. The data includes complete one batch of semester examinations during 2015--2019. HDFS and Map-Reduce framework is used because data is non-relational and unstructured from various departments. Using Map-Reduce classification students are assessed on their performance of knowledge domains as a strong and weak graduate in the given batch.

1 Introduction

Nowadays, huge amounts of data are collected in education institutes. The educational organizations in India are presently facing some problems such as recognizing undergraduates need, personalize training and forecasting the excellence of student

---

S. M. Patil (✉) · V. N. Patil · S. J. Mane · S. M. Satre
Department of Information Technology, Bharati Vidyapeeth College of Engineering, Navi Mumbai, India
e-mail: smpatil2k@gmail.com

V. N. Patil
e-mail: vijay.patil.karad@gmail.com

S. J. Mane
e-mail: sonalijmane9@gmail.com

S. M. Satre
e-mail: shilpa.m.shelar@gmail.com

© The Editor(s) (if applicable) and The Author(s), under exclusive license to Springer Nature Singapore Pte Ltd. 2021
T. Senjyu et al. (eds.), *Information and Communication Technology for Intelligent Systems*, Smart Innovation, Systems and Technologies 196,
https://doi.org/10.1007/978-981-15-7062-9_18

communications. Data excavating is the procedure of determining motivating knowledge from big quantity of data deposited in database or other information accountability [1]. There are several areas in which data quarrying methods play a significant role. Educational data mining (EDM) delivers a group of procedures to benefit institutions to overwhelm these problems, in directive to progress erudition practice of learners as well as growth of institute's revenues [2]. Physical data examination makes congestion for big data investigation. The changeover would not arise mechanically; data mining is essential. Data digging development permits employer to scrutinize data from dissimilar sizes classify it and summarize the association, identify throughout mining process [3]. EDM is an encouraging area of innovation. This study aims to analyze student's performance using data mining techniques.

The issue of clarification and data examination of pedagogy performance is extensively investigated. Data quarrying methods are the favorable practices to excerpt valuable information in this impartial. The data gathered through dissimilar applications need an appropriate technique of removing information from huge sources for well judgment creating. Knowledge discovery in databases (KDD), frequently named data mining, objectives are finding the valuable knowledge from outsized groups of data. In this viewpoint, data mining can evaluate pertinent information outcomes and produce diverse views to comprehend extra learner's performance.

The key impartial of higher teaching institutions is to deliver excellence learning to their learners and to develop the superiority of administrative judgments. One approach is viewing to attain a high level of excellence in the higher edification organization by finding and examining pertinent information outcomes. To evaluate student performance, different perspectives need to be evaluated such as internal examination scores, external examination scores, and practical scores.

In this work, we have collected information like internal and external examination scores from the college desktop application system. The collected information is used to calculate the performance investigation at the semester end. This paper will benefit to the learners and the educators to improve the strengths of the scholar by knowing their domain understanding. This paper work also classifies strong/weaker learner and takes exceptional care to decrease failure ratio and taking suitable action for the forthcoming assessments. Caring of live checking performance and countermeasures before the end semester evaluation absolutely benefits to improve learners' performance.

The remaining paper is prepared as follows: Section 2 represents survey work in educational mining data, Section 3 defines planned framework of student assessment checking. Section 4 describes real work and experiments; Section 5 shows results of student performance. Lastly, we accomplish conclusion of this paper in Section 6.

2 Related Work

Researchers have suggested systems to monitor student performance in terms of result [4–6]. BI techniques are stated by the researchers to the education field [1,

5], customer experience [7], and big data analytics software [8]. Most commonly used techniques for student analysis used are students' learning experience using automated data analysis techniques [9], process mining-based tool [10], association rule mining algorithm [2, 11].

Doctor F. and Iqbal has suggested an intellectual structure for observing learner performance using summarization methods. The fuzzy linguistic summarization (LS) method is used to excerpt weighted uncertain instructions from the grouped labeled data. This structure is used to recognize weaknesses, to deliver additional custom-made response on student's improvement [4].

Ryan Baker has proposed mining method to use an intellectual method. The obtainability of huge group sets in functional set-ups occurred future in edification than additional fields. However, storing educational data is not just important, but processing that with data mining techniques will give good patterns to decide improvements in student study and decision in syllabus or teaching aid usage [5]. Shargabi and Nusari offered mining methods to determine dynamic results and compute the contribution of academic performance to benefit in judgment making [1].

Zhu et al. have done the investigation on course score of semester check based on varied weighted association rules mining algorithm. Using this method directly compared to Apriori algorithm, obtain added valuable relationships among the chapters, and scores, institutions and chapters at the similar threshold standards [11].

Morais et al. suggested systematic method to contract through e-learning data for observing learners performance using data bunching, analytical demonstrating and regression approach, K-means Algorithm and multivariate analysis method [2]. Pieedade and Santos proposed conceptual and technological infrastructure and integrated into the System. It is essential to identify and analyze the learners' unsuccessful and to suggest actions against the issues. Unique measures often pointed out to rise the achievement raise is related with the learners carefully monitoring and with the guesstimate of the tutor to the learners' day-by-day semester events [3]. Ahmadvand A et al. presented a data mining method for assessing the risk of undergraduate student retention. The author has presented students' retention as a binary decision problem [12].

In higher education, researchers have not focused on domain-wise performance evaluation. The proposed method analyzed student performance according to batch and domain for the higher education institute.

3 Proposed Framework

3.1 Students' Academic Data

The aim of this research is to assess the performance of students domain-wise and subject-wise. The complete proposed framework for subject-wise and domain-wise strong-weak, performance evaluation is as shown in Fig. 1.

Student academic data includes internal assessments and semester end examinations. The internal assessment (IA) is assess centered on the presentation of learner in learning activities such as internals examination (IE), laboratory work (LW), assignments, and attendance. The internal valuation is measured by a teacher based on student's performance in IA. The university semester examination (USE) is a final examination at the end of the semester. To pass subject, student has to acquire minimum marks in both the IA and final semester examination. Entire data used in this study has collected from various departments in the form of excel sheets and word documents. The data used in this study is acquired from BV's engineering college of B.E.I.T students of session 2015--2019. The size of the data is 480 *4 = 1920 student's records that include 47 subjects each branch IA and USE record files. Data is stored in different files like USE examination and IA: Term work, practical, and internal examination on various servers in departments.

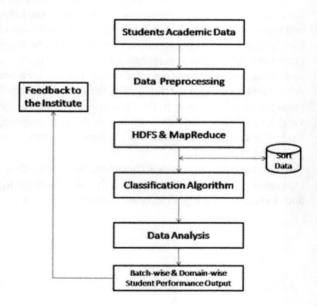

Fig. 1 The framework of the proposed student's performance system

3.2 Data Processing

Data preprocessing is the main and leading stage in the data examination procedure, where few preprocessing tasks and variable transformation are carried out. Initially, data is shuffled from term work, practical and internal examination data sets from raw examination records. After that, data is segregated into branches, semesters, and subjects. Finally, data is normalized for further processing in the Map-Reduce algorithm. The constant variables are transformed into discrete variables, which deliver abundant understandable opinion of the statistics.

3.3 HDFS and Map-Reduce

The most frequently used strategy to approach BDA are distributed file systems (DFS) and Map-Reduce engine. A noticeable proponent of such a system is Hadoop [13]. Hadoop is an open-source technology composed of diverse engines such as a Map-Reduce engine and DFS engine. The data to be examined is kept in the distributed file system and then processed using the Map-Reduce engine. The results are again stored in the file and directly streamed to educational applications.

Map-Reduce is powerful framework and programmers can code components of effort as a map and reduce functions.

- "*In-mapper joining*," the combiner moved the work into the mapper. Instead of producing middle result for each input key-value pair, the mapper totals incomplete outputs of many input records and produce single middle key-value pairs after certain quantity of local combination is completed.
- The interrelated arrangements of **pairs** and **stripes** for keeping path of combined evens from a big amount of remarks.
- The *pair's* tactic saves track of every combined occasion distinctly; however, the *stripes* method keeps track of entire actions that co-occur with the identical result. Though the *stripes* method is knowingly additional effective, it needs memory on the demand of the size of the event space.
- *Order inversion:* the key idea is to change the ordering of calculations into a categorization problem. Through cautious orchestration, one can send reducer the outcome of a calculation (e.g., an aggregate statistic) before it encounters the necessary data deliver that calculation.
- *Value-to-key change:* provides a climbable answer for subordinate sorting. Through moving chunk of the value into the key, one can exploit the Map-Reduce implementation framework itself for categorization. Based on Map-Reduce notion, an extensional Map-Reduce classification algorithm for performance measure is suggested. It permits the user to stipulate data mining procedure's to given restrictions and attempts to make the data mining process to be completed before the target.

3.4 Classification Using Map-Reduce

The workflow of Map-Reduce is shown in Fig. 2. IA and USE data are split into four files. The split datasets are loaded into HDFS. The uploaded data files of HDFS send to Map-Reduce job. The map phase processes data in parallel completely and shuffle an outcome groups to the sort phase. Sorted data set is input toward reduce stage.

The reduce phase processes the < *key, value* > pairs formed in the sort phase. The output generated from reducing phase is input to a merge process to produce results of < *key, value* > pairs. An extensional Map-Reduce classification algorithm is used for the student's performance measure. It permits operator to stipulate a classification procedure in mining data to assess the individual, subject-wise, and batch-wise performance. After that data is visualized for well understanding. This algorithm categorize the student data into numerous stages like students individual performance, strong/weak subject-wise performance, and domain-wise performance. This algorithm firstly add student data as stated by the input constraint of data range correspondingly.

The parameters are carefully chosen using the selection feature techniques called, correlation-based feature selection (CFS). The CFS approximations and grades are the subsection of CFS features than individual features. It selects the group of characteristics that are extremely related with the class, in addition to those attributes that are in low inter-correlation.

The Map-Reduce classification algorithm takings the learner data segment and group manifoldness into interpretation. Data locality is the important issue that disturbs the effectiveness of Map-Reduce sorting procedure. The data locality means, the classification's operation code and the classification's input data are on the same computing machine or on the same frame. When the code and data are on the same node, the efficiency is higher than on the same rack. If the data and code are on the same node, it would avoid the data broadcast in the net and significantly decrease the

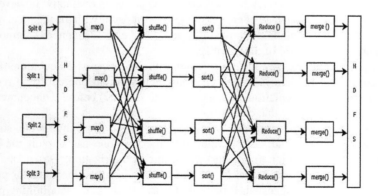

Fig. 2 Map-Reduce working process

delay. Therefore, in the huge scale data handling applications, transferring the code would be "inexpensive" than moving data.

4 Implementation

4.1 Individual Performance

Bharati Vidyapeeth College of Engineering is one of the leading engineering college in India, was established in 1990. BV College of Engineering is well known for excellence in edification leading new technical fields. Through an enlightened mission of "Social Transformation through Dynamic Education," the institute is constantly motivated for the all-inclusive growth of the learners. The institute has dedicated to train and nurture budding technocrat. Two thousand students and 140 staff stepping ahead together to meet academic and specialized superiority. Quality of education is given by six department, i.e., Computer Engineering, Chemical, Electronics and Telecommunication, Information Technology, Instrumentation, and Mechanical Engineering.

In the early phase of the course, an individual evaluation method was used for the assessment of the student, where scholars asses their personal understanding of subjects at the end of semester examination. The individual student performance measure is shown in Fig. 2. Student's individual performance is measured using Eq. 1. The students will know the obtained aggregate score in internal and external examinations. The intention is to know the progress of internal and external assessment of the individual student and to report whichever area of weakness or development.

$$\text{Total Performance} = \sum_{i=1}^{m} \sum_{j=1}^{n} (IA_{ij} + USE_{ij}) \qquad (1)$$

where
n = number of subjects performance
m = number of semester

4.2 Subject-Wise Strong/Weak Performance

Experiments were performed using Map-Reduce for modeling classification of students in relation to their performance on *IA* and *USE*. The Map-Reduce model was used to evaluate subject-wise strong performing students by applying threshold of 60% (threS), i.e., student above threS, and weak performing students by applying threshold of less than 40% (threW), i.e., student below threW as measured in Eqs. (2), (3) and (4).

The Map-Reduce methodology was used to separate the range of the input data values into three different levels, low ($\leq$40), medium (>40 $\leq$60), and high (>60) marks, respectively. The result was represented using two crisp interval sets for each of the percentage rank limitations mirrored in the data exactly, these were: less than 40% marks and greater than 60% marks.

$$\text{Strong Subject} = \begin{cases} 1, \text{ if Subject} > 60 \\ 0, \quad \text{Otherwise} \end{cases} \quad (2)$$

$$\text{Weak Subject} = \begin{cases} 1, \text{ if ubject} \leq 0 \\ 0, \quad \text{Otherwise} \end{cases} \quad (3)$$

$$\text{Subject Total} = \sum_{i=1}^{n}(IA_{ij} + USE_{ij}) \quad (4)$$

4.3 Domain-Wise Performance

In our work, we have created the eight groups different domains of subjects in all semesters. For example, we have shown in Table 1, all the subjects and their codes for the information technology branch. Subject domains are decided based on the interrelation of subjects in all semester subjects, in Table 2, eight domains are shown. These domains are used for the calculation of the domain-wise performance of the students using Eq. (5). Similarly, subject domains are calculated for other disciplines.

$$\text{DSP} = \frac{\sum_{i=1}^{D}(IA_i + USE_i)}{(D * D_{\max})} \quad (5)$$

where,

D is subjects under a domain.

$D_{\max}$ maximum marks for given subjects in that domain.

5 Results

We have performed the experiment on open-source technology, i.e., Ubuntu 16.04 operating system with laptop configuration 4 GB RAM, 1.2GHZ processor and 500 GB HDD. The achieved results from the same configuration are described as follows:

Table 1 The eight semester's subjects and code of information technology

First year		Second year		Third year		Final year	
Subject code	Subject name	Subject code	Subject name	Subject code	Subject name	Subject code	Subject name
101	Applied mathematics-I	301	Applied mathematics-III	501	Computer graphics and virtual reality	701	Software project management
102	Applied physics-I	302	Data structures and algorithm analysis	502	Operating systems	702	Cloud computing
103	Applied chemistry-I	303	Object oriented programming methodology	503	Microcontroller and embedded systems	703	Intelligent system
104	Engineering mechanics	304	Analog and digital circuits	504	Advanced database management systems	704	Wireless technology
105	Basic electrical and electronics engineering	305	Database management systems	505	Open-source technologies	705	Elective-1
106	Environmental studies	306	Principles of analog and digital communication	506	Business communication and ethics	706	Project 1
201	Applied mathematics-II	401	Applied mathematics-IV	601	Software engineering	801	Storage network management and retrieval
202	Applied physics-II	402	Computer networks	602	Distributed systems	802	Big data analytics
203	Applied chemistry-II	403	Computer organization and architecture	603	System and web security	803	Computer simulation and modeling
204	Engineering drawing	404	Automata theory	604	Data mining and business intelligence	804	Elective-2

(continued)

Table 1 (continued)

First year		Second year		Third year		Final year	
Subject code	Subject name	Subject code	Subject name	Subject code	Subject name	Subject code	Subject name
205	sStructured programming approach	405	Web programming	605	Advance internet technology	805	Project 2
206	Communication skills	406	Information theory and coding				
207	Workshop						

Table 2 Subjects Domain Groups

Domain name	Subject codes	Domain name	Subject codes
Operating system	403, 502, 602, 505.	Project management	206, 506, 601, 701,705,706,805.
Programming	205, 302, 303,405,501, 605.	Electronics	105,102,202,304, 306,503.
Database	305, 504, 604,802.	Logical ability	101,201,301,401, 404, 406,703, 803.
Network	402,603,702,704,801, 804.	Non-IT	103,104,106,203,204,207.

Here, we have analyzed the individual student performance with Map-Reduce-based classification. First, we have to select the unique id of the student, and then individual performance is displayed from semester one to semester eight of first-year entry-level student's and third to eight semesters of second-year entry-level students. Sample result of unique id ITH15001 is shown in Fig. 3. In this result, semester six is shown the poor performance of this candidate out of eight semesters as compare to his remaining all semesters. The subject-wise strong performance graph is shown in Fig. 4 and the subject-wise weak performance graph is shown in Fig. 5. Data on 41 students of the first-year and second-year entry-level, consisting of 77 students, engaged in information technology branch was used.

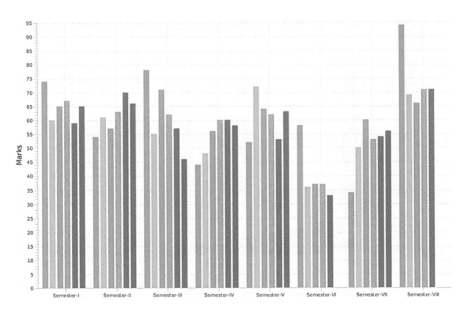

Fig. 3 The student performance of semester I to VIII

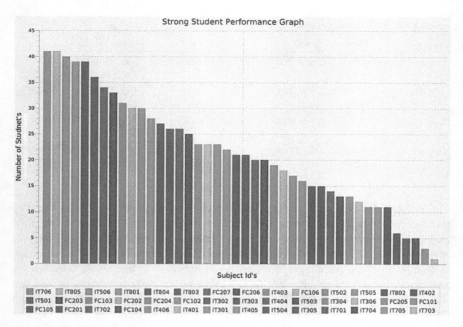

Fig. 4 The subject-wise strong student performance

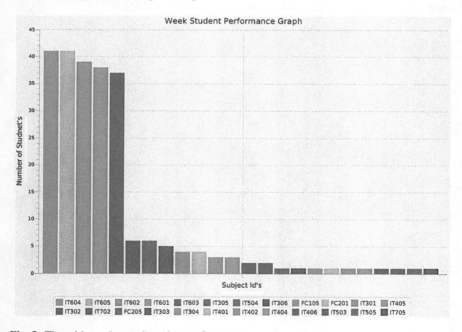

Fig. 5 The subject-wise weak student performance

The data contained of four contributions, such as internal examination, term work, practical, and theory examination. The particular output of the data stated subject-wise strong/weak performance, how many numbers of students achieved more than first grade in the subjects and less than 40% marks.

Figure 4. The graph shows the subject-wise strong performance of the batch. Here, we have to select the batch year to measure the performance, like IT15 this is for 2015--2019 batch. We have used greater than equal to 60 thresholds to measure strong performance. In this graph, the result shows that in two subjects all students get more than 60% marks. But in one subject code IT703 only one student gets more than 60% mark.

Figure 5. The graph also shows the subject-wise weak performance of the same batch. The same procedure is used to measure the weak performance of the batch; here, we have used less than 40% threshold for measuring the performance. The graph shows that in five subjects (for example IT604, IT605, IT602, IT601, IT603) where students got less than 40% marks because these subjects are difficult to them hence, the student result is poor in grade-wise.

Figure 6 shows the result of the domain-wise performance of the 2015--19 Batch. The result shows that students have got very good marks in their respective subject domain. Therefore, performance in programming, network, electronic, management, and logical ability domains for this batch is good. However, database and operating system domains have less performance as compared to other domains. In non-IT domainn all other subjects are included. This domain-wise performance of the IT branch result shows that the students are very good in programming, networking and management domains.

The Map-Reduce classification modeling accuracy was tested on the novel data where the model achieved 90% correctness. The results shows, student data classification Map-Reduce algorithm improves the learner data investigation method of student performance. The proposed architecture using Map-Reduce technique gives required accuracy and efficacy in the results and is extensible for any other academic environment.

6 Conclusion

In this research, we have used Map-Reduce classification algorithm to measure the achieved performance of individual student, subject-wise and domain-wise strong/weak student's performance. Our proposed framework measures the performance of students for each semester or yearly. This framework demonstrates how the Map-Reduce classification approach work meritoriously to observe learners' improvement and performance. An organization needs to have an imprecise earlier knowledge of registered students to analyze their performance in upcoming academics specific domains. We have analyzed the student's academic performances by knowing the student's achievements. Although the data is unstructured and non-relational from the examination department, HDFS and Map-Reduce techniques

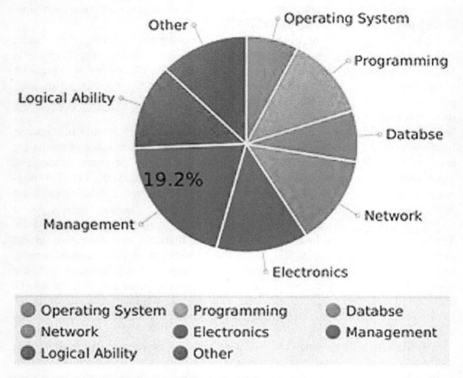

Fig. 6 Domain-wise students performance in the batch

proved to be useful in analyzing student academic performances. The classification result indicates institutional domain expertise and student understanding in subject domains. It is clear from the result that programming and network domain is strong in our research work. As future work, these result will be enhanced with increasing scopes and capacity of data such as attendance, student transfer data, and residence data.

References

1. Al-shargabi, A., Nusari, A., Discovering vital patterns from UST student's data by applying data mining techniques computer and automation engineering (ICCAE). In: 2010 The 2nd International Conference, vol. 2, pp. 547–551. (2010)
2. De Morais, A., Araujo, J., Costa, E.: Monitoring student performance using data clustering and predictive modelling. In: Frontiers in Education Conference (FIE), 2014 IEEE, pp. 1–8. (2014)

3. Piedade, M., Santos, M., Business intelligence in higher education: enhancing the teaching-learning process with an SRM system information systems and technologies, (CISTI), 2010. In: 5th Iberian Conference, pp. 1–5. (2010)
4. Doctor, F., Iqbal, R.: An intelligent framework for monitoring student performance using fuzzy rule-based linguistic summarization fuzzy systems (FUZZ- IEEE).In: 2012 IEEE International Conference. pp. 1–8. (2012)
5. Baker, R.: Educational data mining: an advance for intelligent systems in education intelligent systems. IEEE **29**, 78–82 (2014)
6. Lord, S.M., Layton, R.A., Ohland, M.W.: Multi—institution study of student demographics and outcomes in electrical and computer engineering in the USA, IEEE Trans. Educ. **58**(3), (2015, August)
7. Hua Zhu, X., Qiong Deng, Y., Ling Zeng, Q.: The analysis on course grade of college-wide examination based on mixed weighted association rules mining algorithm. In: International Conference on Computer Application and System Modelling (ICCASM), vol. 14, pp. V14–530–V14–533. (2010)
8. Un, L., Zhihua, C., Web data mining based learners personalized information modelling e-Education, e-Business, e-Management, and e-Learning. In: 2010. IC4E '10. International Conference, pp. 454–459. (2010)
9. Spiess, J., T'Joens, Y., Dragnea, R., Spencer, P., Philippart, L.: Using big data to improve customer experience and businessPerformance. Bell Labs Techn. J. **18**, 3–17 (2014)
10. Chen, X., Vorvoreanu, M., Madhavan, K.: Mining social media data for understanding students learning experiences learning technologies. IEEE Trans. 246–259 (2014)
11. Zhu, Y., Zhang Wang, F., Shan, X., Lv, X.: K-mediods clustering based on map-reduce and optimal search of mediods. IN: 9th International Conference on Computer Science and Education (ICCSE), pp. 573–577. (2014)
12. Ahmadvand, A., Bidgoli, B., Akhondzadeh, E.: A hybrid data mining model for effective citizen relationship management: a case study on tehran municipality E-education, E-business, E-management and E-learning. In: IC4E '10. International Conference on 2010, pp. 277–281. (2010)
13. Hang, Y., Chen, S., Wang, Q., Yu, G.: i2Map-reduce: incremental map-reduce for mining evolving big data knowledge and data engineering. IEEE Trans. Learn. Technol. 1–1, (2015)

Building a Graph Database for Storing Heterogeneous Healthcare Data

Goutam Kundu, Nandini Mukherjee, and Safikureshi Mondal

Abstract Healthdata is unstructured and heterogeneous. Storing and managing healthdata therefore require NoSQL data models, so that they can be efficiently accessed. In this paper, we propose a graph-based data model for presenting patient and doctor relationships when a treatment episode starts. On top of this data model, a doctor recommender system and an online healthcare delivery system have been built which can be used for treating patients remotely.

1 Introduction

In a healthcare domain, the stakeholders maintain complex relationships among themselves and with other related entities belonging to this domain. A patient visits doctors having different specialization depending on the current symptoms. Whenever a patient visits a doctor, a treatment episode starts. Doctors diagnose the diseases and prescribe medicines to each patient or advise to go through few clinical tests. A patient may visit a doctor multiple times during one particular treatment episode. On the other hand, when a treatment episode ends between a patient and a doctor, a new treatment episode may again be started between the same pair of patient and doctor for a new set of symptoms. Other stakeholders like health assistants, nurses, persons involved with administration also participate in these processes, and therefore, the data and relationships relating to them are also important. Ontological descriptions of

---

G. Kundu (✉) · N. Mukherjee
Department of Computer Science and Engineering,
Jadavpur University, Kolkata, West Bengal 700032, India
e-mail: goutam97kundu@gmail.com

N. Mukherjee
e-mail: nmukherjee@cse.jdvu.ac.in

S. Mondal
Narula Institute of technology, Kolkata, West Bengal 700032, India
e-mail: msafi.cse@gmail.com

© The Editor(s) (if applicable) and The Author(s), under exclusive license to Springer Nature Singapore Pte Ltd. 2021
T. Senjyu et al. (eds.), *Information and Communication Technology for Intelligent Systems*, Smart Innovation, Systems and Technologies 196,
https://doi.org/10.1007/978-981-15-7062-9_19

healthcare domain [1] characterize the relationships among the entities and highlight their properties effectively.

However, while translating the ontological description into relational schema, it is found that relational databases are unable to capture the complex relationships between the stakeholders and various entities in the healthcare domain. The tables in relational databases store the elements in a particular set as columns in the table and thus effectively express the relationships among the entities of the same type (such as a set of patients in a patient table). However, relational databases cannot effectively capture the relationships that exist among different types of entities, such as a patient and a doctor. In other words, relational databases cannot efficiently express the relationships among individual data elements. On the contrary, graph databases can nicely express the relationships among individual data elements by creating a node for each data element and connecting the nodes through edges in order to establish relationships among them. In a graph database, data can be modeled using a high number of data relationships. Such models are flexible enough to incorporate new data or data relationships. Moreover, querying data relationships in real-time is also possible.

In this paper, we present a preliminary model to represent doctor-patient relationship using graph databases. The model keeps track of visits made by the patients to the doctors and based on the number of visits, it associates a trust factor to the doctors, and finally, we build an online doctor recommender system on top of this model. Although, an online system has been described here, the graph database model can be used for any healthcare application storing patient-doctor relationship.

The paper is organised as follows. Section 2 gives an overview of graph databases. Section 3 presents a literature review discussing various applications built on top of graph data models. Section 4 describes the features of Neo4j, one of the popular graph database. Section 5 presents a graph data model which stores the patient doctor relationships, and Sect. 6 presents an online doctor recommender system. Section 6 concludes the paper.

2 Graph Databases

A graph database is a collection of nodes and edges, where nodes represent objects and edges represent associations or relationships among the objects. Both, objects and relationships have properties which describe their specific characteristics or attributes. Real-world objects are generally related to each other through complex relationships which can be well understood using human intuition. Usual databases, including relational databases, column-oriented databases, document-oriented databases, etc., are unable to effectively store and manage these relationships, whereas a graph database can conveniently capture the major characteristics of these relationships and use them suitably for intelligent applications. Graph databases are schema less and therefore suitable for storing semi-structured information. Social networks, semantic web, geographic applications, and bioinformatics are significant class of applications in which data have a natural representation in terms of graphs [2]. Virgilio et. al. introduce strategies for modeling data using graph databases [2].

Graphs are traversed in order to respond to user queries. While accessing related objects, graph traversal performs much better in comparison with queries involving join operations in RDBMS. Emil Eifrem conducted an experiment to observe the performance difference between a relational database and a graph database for a certain type of problem called arbitrary path query. Graphs are traversed in order to respond to user queries. While accessing related objects, graph traversal performs much better in comparison with queries involving join operations in RDBMS. Emil Eifrem conducted an experiment to observe the performance difference between a relational database and a graph database for a certain type of problem called "arbitrary path query" [3]. The experiment demonstrates that for certain type of data, graph databases perform much better compared to relational databases. Therefore, an application, in which data may not have a natural graph representation, but relationship among the data is at the core of the application, can also be built on top of graph databases. Performance of RDBMS also deteriorates with the increase in data complexity as observed in [4]. Performance can be further improved by building indexes where indexes use properties either to map to nodes or to relationships.

Neo4j is one of the most popular graph databases. Other popular graph databases are AllegroGraph, InfiniteGraph, and OrientDB. ACID transactions with rollback support are available with these databases. Moreover, many programming languages can be integrated with these databases for building applications. Queries are satisfied through graph traversal. Often traversal is necessary only for a subgraph and not the entire graph, thereby reducing query response time. In Neo4j APIs for both, breadth-first-search and depth-first-search are available which can be used for graph traversal. This paper uses Neo4j for implementing the data model developed by the present research. The experiment demonstrates that for certain type of data, graph databases perform much better compared to relational databases. Therefore, an application, in which data may not have a natural graph representation, but relationship among the data is at the core of the application, can also be built on top of graph databases. Performance of RDBMS also deteriorates with the increase in data complexity as observed in [4]. Performance can be further improved by building indexes where indexes use properties either to map to nodes or to relationships. Neo4j is one of the most popular graph databases. Other popular graph databases are AllegroGraph, InfiniteGraph and OrientDB. ACID transactions with rollback support are available with these databases. Moreover, many programming languages can be integrated with these databases for building applications. Queries are satisfied through graph traversal. Often traversal is necessary only for a subgraph and not the entire graph, thereby reducing query response time. In Neo4j APIs for both, breadth-first-search and depth-first-search are available which can be used for graph traversal.

This paper uses Neo4j for implementing the data model developed by the present research.

3 Related Work

Scott Carey describes seven enterprise applications using graph databases [5]. These applications include network and IT operations, identity and access management, recommendation engines, fraud detection, etc. William McKnight describes use of graph databases in Master Data Management [6].

Park et al [7] proposes "3NF Equivalent Graph" (3EG) transformation to automatically construct healthcare graphs from normalized relational databases. They have used a set of real-world graph queries, such as finding self-referrals, and evaluated their performance over a relational database and its 3EG transformed graph. The authors have demonstrated that using graph representations, the user accessibility to data can be enhanced.

4 Neo4j Graph Databases

A graph database model can be successful for efficiently managing real-time heterogeneous data, if it

- supports real-time updates with fresh data streams
- supports an expressive and user-friendly declarative query language to give full control to data scientists
- supports deep-link traversal (3 hops) in real-time, just like human neurons sending information over a neural network, and
- can scale out and scale up to manage big graph.

Neo4j is an open-source, NoSQL graph database that provides an ACID-compliant transactional backend for any application. Its source code is written in Java and Scala. It supports for flexible property graph schema that can adapt over time, with addition of new relationships to speed up the access to domain data when the application needs changes. It also supports constant time traversals in big graphs along the depth, as well as breadth due to efficient representation of nodes and relationships. It is agile because anytime a new node or a relationship can be added. Even the properties of a node or a relationship can be added or updated anytime.

5 Modeling Patient-Doctor Relationship

In this paper, we model the doctor-patient relationship using a simple graph-based model. Thus, whenever a doctor registers with the system, a node in created in Neo4j database. Details of the doctors including his or her email id, specialization, location (in the form of latitude/longitude of the clinic, health center, etc.), and other required fields are stored as the properties of these nodes. Neo4j stores node properties as

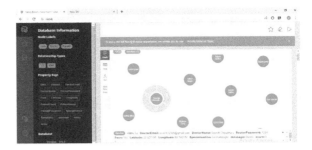

Fig. 1 A set of registered doctors

key-value pairs. In our implementation, doctor's email id is the primary key value to a doctor's node. When a patient registers with the system, another node is created with all details of the patient. For a patient, his or her demographic data is stored as the node property.

When a patient selects a doctor and confirms visit to the doctor, an edge is created from the patient to the doctor with list of symptoms, visit date, and predicted disease as the edge attributes. Thus, the nodes in the graph model represent doctors and patients and their information is stored as properties of the nodes, whereas patients' visits to the doctors' are represented as edges in the graph model with properties related to the visits. Figure 1 depicts a screenshot of Neo4j visual tool displaying nodes related to the registered doctors.

5.1 Creating a Visit Edge

An edge connects a patient node with a doctor node when a patient decides to make an appointment with a doctor. However, in order to book an appointment, a patient needs to login to the system. When a patient logs in, he/she will be required to provide the symptoms to illness they are currently facing. At this stage, the goal is to automatically predict the possible disease that the patient may have. For prediction of the possible disease, a disease-symptom graph [8] has been used (Fig. 2).

Disease-Symptom Graph Database A Disease-Symptom Graph [8] is similar to a Decision Tree Classifier in which symptom nodes form the root nodes and disease nodes become the leaf nodes. There are intermediate nodes which help to classify the initial symptoms. Decision Tree Classifier has been used for classification through application of straightforward question-answering technique. It poses a series of carefully crafted questions about the attributes of a test record. Each time it receives an answer, a follow-up question is asked until a conclusion about the class label of the record is reached. In this research work, the model makes use of a training set with several symptoms and their associated diseases and is trained with it. Once the decision tree is constructed, classifying a test record (here, a patient with symptoms) is straightforward. Starting from the root node, we apply the test condition (here,

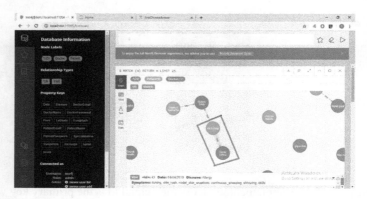

Fig. 2 A visit edge

symptoms) to the record and follow the appropriate branch based on the outcome of the test. It then leads either to another internal node, for which a new test condition is applied, or to a leaf node. When we reach the leaf node, the class label (here, disease) associated with the leaf node is then assigned to the record.

After deciding the probable disease, the patient is provided with a list of doctors. The list of doctors is prepared by the system either according to the visit and revisit scores of the doctors or according to the patient's current location or on the basis of both. These criteria are explained in the next section. From the list of doctors, the patient selects a doctor and confirms the visit. An edge is created from the patient to the selected doctor with the list of symptoms, visit date, and disease as the edge attributes, and a notification goes to the selected doctor. Figure 3 shows a visit edge connecting a doctor to the patient. On top of our graph-based patient-doctor relationship model, we implement a doctor recommender system to aid the patients to find a doctor and book an online appointment. The recommendation is provided by the system on the basis of the probable disease and a system-decided trust factor associated with each doctor.

Trust factor: The primary selection criterion is based on a doctor's score. This score is set to 0 by default when the doctor registers for the first time. Whenever any patient visits to that doctor, the score is incremented by 3. Now if that patient visits to the same doctor within 15 days of the patient's first visit to that doctor, the score is incremented by 1 indicating that the patient trusts the doctor and having benefit from the doctor's treatment. But if within that 15 days of visit, the patient again logs in to the system and visits another doctor, then the new doctor's score is incremented by 3 as usual for the first visit, but the previous doctor's score is decremented by 1. This is done with the assumption that the patient is losing trust to the previous doctor and is not getting benefited from the treatment. In this way, the score of a doctor will be calculated.

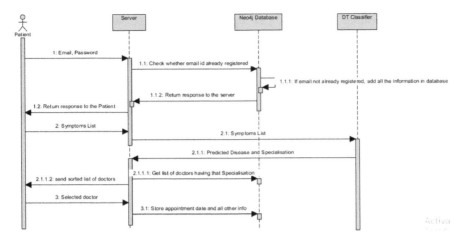

Fig. 3 Sequence diagram showing recommendation by the system

In addition to the above-mentioned score, the distance between the patient and the doctor is also taken into account. Whenever the doctor registers, the doctor's current location is accepted and is always used as the doctor's location. When any patient logs in with his/her login information, his/her current location is taken and is used as the patient's location. The location is collected in the form of latitude and longitude, and the distance between these two latitudes and longitudes are calculated to find the distance.

After the calculation of both parameters, the trust is calculated by assigning a weight to both of them. Here, for experiment purpose, a weight of 0.7 is assigned to the Doctor's score as it is an important measure and a weight of 0.3 is assigned to the distance between the patient and the doctor's current location. Thus, Trust = 0.7 * Score + 0.3 * Distance. A sequence diagram showing the events after patient login to the system is depicted in Fig. 3. Thus, when a patient log in, he/she is provided with a list of doctors sorted based on the trust factor as mentioned above. The patient then selects a doctor and books an appointment with the doctor based on the doctor's scheduled time. The actual visit to the doctor can be a physical visit or may be done within a mobile-based remote healthcare delivery system [9]. In our implemented system, the doctor treats patients remotely using mobile devices. Thus, when the doctor logs in, he/she selects the patients having appointments during that time. The doctor checks the symptoms, may have an audio or video chat with the patient, and prescribes required medicines and medical tests for them which are forwarded to the patients.

Figure 4 shows a screenshot from our mobile app. The screen displays a list of doctors prepared by the system based on the patient's symptoms and the trust factor.

Fig. 4 List of doctors prepared by the system and presented to an allergic patient

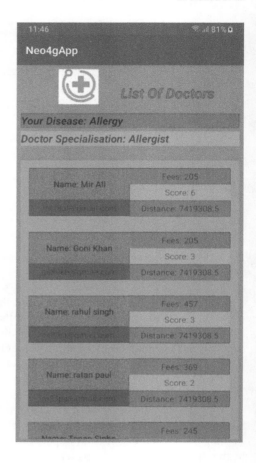

6 Conclusion

Relational databases are unable to capture the complexities in the relationships between different entities in healthcare systems. NoSQL databases are required to store the healthcare data which are voluminous, heterogeneous, and unstructured. This paper presents a graph-based data model to store patient–doctor relationships based on the treatment episodes. Based on the data model, we develop a mobile-based healthcare service delivery scheme.

The data model and the healthcare delivery system are still at their preliminary stages. In future, we shall further modify the model to support more complex applications.

References

1. Sen, P.S., Banerjee, S., Mukherjee, N.: Ontology-driven approach to health data management for remote healthcare delivery. In: Proceedings of the 7th ACM Workshop on ACM Mobile Health 2017, p. 2. ACM (2017)
2. De Virgilio, R., Maccioni, A., Torlone, R.: Model-driven design of graph databases. In: Conceptual Modeling, pp. 172–185. Springer International Publishing, Berlin (2014)
3. Adell, J.: Performance of graph vs. relational databases. https://dzone.com/articles/performance-graph-vs, Feb (2013)
4. An introduction to nosql, graph databases and neo4j. https://www.slideshare.net/featured/category/technology, April 2012
5. Carey, S.: Seven enterprises using graph databases: popular graph database use cases, from recommendation engines to fraud detection and search. https://www.computerworlduk.com/galleries/data/7-most-popular-graph-database-use-cases-3658900/, May 2017
6. William McKnight, May 2018
7. Park, Y., Shankar, M., Park, B., Ghosh, J.: Graph databases for large-scale healthcare systems: a framework for efficient data management and data services. In: 2014 IEEE 30th International Conference on Data Engineering Workshops, pp. 12–19, Mar 2014
8. Mondal, S., Mukherjee, N.: A bfs-based pruning algorithm for disease-symptom knowledge graph database. In: Information and Communication Technology for Intelligent Systems: Proceedings of ICTIS, vol. 2(417) (2018)
9. Das, R., Mondal, S., Mukherjee, N.: More-care: mobile-assisted remote healthcare service delivery. In: 2018 10th International Conference on Communication Systems & Networks (COMSNETS), pp. 677–681. IEEE (2018)

Analysis and Optimization of Cash Withdrawal Process Through ATM in India from HCI and Customization

Abhishek Jain, Shiva Subhedar, and Naveen KumarGupta

Abstract ATMs have become a prominent part of banking and finance system in India. ATM is widely deployed to cater banking needs of the customer. One of the objectives of ATM is to reduce the hustle and bustle which are involved in a transaction when done through the teller. Nowaday's, function of ATM is not limited to cash withdrawal, balance enquiry, and so forth, but ATM now can be used for fund transfer, passbook printing, etc. Still cash withdrawal is the most frequent operation that the customer performs on ATM. This research aims to identify opportunities for change in the current ATM cash withdrawal process and to suggest an efficient cash withdrawal process flow. Study is carried out on an ATM simulator, designed specifically for the research.

1 Introduction

The Indian Financial System is experiencing a major shift and it is evident in the way financial service is being delivered to end customers. One such example is ATM which is the most commonly used distribution platform for financial services. ATM is a cash rending teller machine. The ATM network provides banks with the opportunity to pursue professional and cost-effective models. Now, banks can serve customers outside the banking halls with the introduction of ATM. Different banks are spreading their capabilities and customer segment by improving its ATM services

---

A. Jain (✉)
ServiceNow Software Development India Pvt. Ltd., Hyderabad, India
e-mail: abhishek.jain520@gmail.com

S. Subhedar
Cognizant Technology Solutions, Pune, India
e-mail: shivasubhedar05@gmail.com

N. KumarGupta
Mastercard, Pune, India
e-mail: naveengupta1489@gmail.com

© The Editor(s) (if applicable) and The Author(s), under exclusive license to Springer Nature Singapore Pte Ltd. 2021
T. Senjyu et al. (eds.), *Information and Communication Technology for Intelligent Systems*, Smart Innovation, Systems and Technologies 196,
https://doi.org/10.1007/978-981-15-7062-9_20

Table 1 Top 10 banks with respect to ATM counts in India by Sep 2019

Top 10 banks	Onsite	Offsite	Total ATMs
State Bank of India	25,588	32,979	58,567
Axis Bank Ltd.	5250	12,065	17,315
ICICI Bank Ltd.	5304	9855	15,159
Hdfc Bank Ltd.	6131	7383	13,514
Bank of Baroda	9332	3821	13,153
Punjab National Bank	5328	3657	8985
Canara Bank	4731	4070	8801
Union Bank of India	3949	2849	6798
Bank of India	2470	3355	5825
Syndicate Bank	4146	407	4553

and reach across the country. ATM service has been one of the main characteristics for enhancing customer experience in the banking industry. Customer experience via ATM is affected in many ways, such as finding nearby ATM counters, facilities, 24 * 7 service availability, fast transactions, and multi-function ATM counters.

As indicated by World Bank, ATMs per 100,000 adults have increased to approximately 40.55 in 2015 from approximately 18.9 in 2005. This development in India (International Monetary Fund, 2015) is 19.70 in 2015 from 2.29 in 2005 [1].

Between 1996 and 2001, the number of ATMs in USA increased from 152,000 to 345,000. By comparison, RBR forecasts that the Indian ATM market will increase from 110,000 in 2012 to 400,000 by 2017 [2]. In last five years (from 2014 to 2019), approx. 8.8204% of ATM counters have increased in India [3] which recommends the worth and reliance of ATM counters in Indian market. The following is the statistics about top 10 banks in India regarding 'ATM counters' is presented [3]. In Table 1, rundown of top 10 banks is given with absolute number of ATM (Onsite +Offsite).

As of late, there have been numerous innovative advances in banking and finance space, and ATM is certainly one of them. Kevin Curran and David King considered the various banks of UK as for their ATM navigation menu and recommended the OptiATM menu plan [4]. According to Maenpaa, Kale, Kuusela, and Mesiranta, security can stand for the reliability of an innovation and an overall belief on the part of the user that banking transactions can be completed confidentially and safely [5]. Chinedu N. Ogbuji, Chima B. Onuoha, Uyo, Nigeria, Emeka E. Izogo, Abakaliki, Nigeria has discussed the negative impacts of the automated teller machine as a channel for conveying banking services in Nigeria [6]. Deepti Aggarwal, Himanshu Zade, and Anind K. Dey have recommended changes in hardware foundation of an ATM machine to improve usability and security [7]. Sri Shimal Das and Smt. JhunuDebbarma have discussed improving security at ATM with the inclusion of biometric technique (fingerprint) [8]. Hiroko Akatsu and Hiroyuki Miki have talked about variables influencing to an old individual while utilizing ATM machines from ease of use and accessibility point of view [9]. However, the dominant part of the study focuses on non-Indian environment, which makes this review more prominent than ever. Today, individuals do not need an ATM card or recollect any ATM stick with the aid of biometric and other advancement headways, yet they can process exercises at ATM counters. Today, individuals do not have to insert an ATM card at the system with the assistance of NFC, only tap can wrap up. New ATM machines and the enhancement of the general biological system of ATM will take a certain period of time, study and human activities to enter India. However, seeking openings is another way to enhance the overall environment in the new ATM network.

The primary objective of this study is to identify areas for improvement in the current money withdrawal process via ATM in Indian setting, and to suggest skilled streams that may decrease overall processing time for withdrawal of money.

2 Method

A total of 40 subjects, volunteered to take part in this (Age group from 18 to 55 years, $M = 32.25$; $SD = 7.90$). The criterion used for selecting subject is (a) subjects have exposure to banking services and (b) interacting with ATM machines and bank branches for cash withdrawal. There is an additional sampling done in selecting and dividing subjects based upon their education (literate, low literate, etc.), profession (service, business, students, non-working, housewife, people with low income, etc.), and gender. We have varied the software interface of ATM machines for cash withdrawal process using a simulator and recorded the processing time for cash withdrawal.

We have distributed overall study into two different categories, a. analysis of ATM machine's cash withdrawal process and b. understanding human conduct with ATM machines at ATM counter. For the first category, we analyzed ATM machine software flows of selected banks (top 10 banks w.r.t. ATM counts in India). For second category, we talked with customers at genuine ATM counters as for their ATM cooperation experience, torment focuses, and issues in the withdrawal procedure. Quantitative data was gathered using an online survey with 97 respondents. We designed two ATM system simulators and introduced them to users for performing a task (cash withdrawal) in order to preserve proper security and confidentiality of an individual banking details. Total time taken by each user in performing task is recorded and analyzed.

3 Findings

3.1 Current Withdrawal Process in Majority of ATMs

In the analysis of different top banks ATM cash withdrawal process, it has observed that the cash withdrawal process comprises arrangement of steps differing from 7 to

Fig. 1 Generalize flow of cash withdrawal from ATM in India

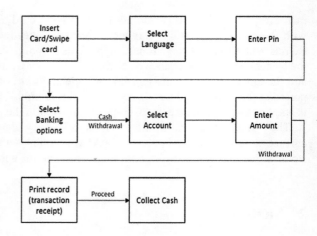

11 roughly. A few stages are reliably noted in all ATMs procedure despite the fact that the sequencing of steps is not steady in all machines. Based upon this analysis, a conventional ATM machine money withdrawal stream is exhibited beneath (Fig. 1). In this flow, constant steps are included.

There is one more special cash withdrawal process found in some ATM machines of different banks, i.e., 'Fast Cash' or 'Quick Cash' (ICICI, n.d.) [10]. In this money withdrawal process, customer can pull back some fixed measure of cash from ATM in less steps. This procedure is making 'amount selection' step faster, yet the remainder of the steps remains similar.

3.2 Common Patterns While Selecting Choices

As per user interviews and survey report, we can see that there are several trends across various user groups in the cash withdrawal process. The pattern is related to choosing similar choices in the process of cash withdrawal. Approximately for cash withdrawal, 80% of respondents pick same account form (Savings/Current). 95% of respondents use one specific language either native or English for processing. Approx. 72% of respondents have some specific amount which they use frequently to withdraw money. The majority of users said that they need a transaction receipt only when they withdraw big amounts [11]. While in user interviews phase, some respondents mentioned that 'Why ATM machines don't know which language I use?' and 'ATM machine should remember which Accounts I select.'

3.3 Navigation

A money withdrawal process in UI (programming) side fluctuates from bank to bank. Indeed, even steps are different, and distribution of action buttons is diverse in nature. Customers need to push physical buttons on the screen/console in some ATM machines, while some devices have a touch screen or both.

3.4 Learnability

Due to different navigation pattern of cash withdrawal process and uncertain nature of action buttons, customer face problem in learning the ATM machine interaction. At the point when customer goes to any new ATM machine (of a similar bank or other), learning glitch comes which expands exchange time and likelihood of mistakes.

3.5 Status of ATM Machine

There is no prompt system feedback to say about the state of the ATM equipment, such as ATM working state, cash availability status, and currency availability status.

3.6 ATM Machine Hardware

In India, there are a variety of ATM machines present, ranging from basic (oldest) to advanced (with biometrics and other technology based).

3.7 Accessibility

Less numbers of ATM machines and counters are accessibility complaint. It makes difficult for physically disabled people to execute and perform activities on machines.

4 Redesign: Optimization of ATM Cash Withdrawal Process

On the basis of the research findings, regular pattern and similar behavior responses are observed while transacting cash withdrawal at ATM by customer. With the aid of these observations, a new approach is conceptualized for the method of 'cash withdrawal' from an ATM. User personalization, configuration capability, default power, and intelligent human behavior learning are main concepts taken in terms of process conceptualization. Banking bodies can analyze transaction data and draw some key insights related to user transaction style such as which language one user prefers to use ATM machines, what is his/her most selected account type for cash withdrawal, what is his/her regular withdrawal number, and what is the scenario when he/she prefers to collect transaction receipts in printed format. With the aid of these insights, users can be provided with intelligent and customized flow that fits their actions and transaction pattern. In the current scenario, banking clients are linked to banks with multiple point of service (POS) such as branch, Internet banking, and mobile device. Banking customers should be able to define their preferences for ATM cash withdrawal processes such as preferred language, form of account, fast amount and instructions for printing, and de-nominations. Below are some of the opportunistic customization to enhance the user experience of ATM machines.

4.1 Language

The system can select customer's preferred language by analyzing previous transactions or language specified by customers. This way selection of language can be done by the system itself rather than consumer selecting.

4.2 Account Type

Customers can set the account type as preferred account for cash withdrawal, i.e., savings or current. By this, customer will not have to select account type for cash withdrawal every time.

4.3 Withdrawal Amount

In an ATM preferences survey analysis (Fig. 2), it has encountered that some customers have a specific amount to withdraw from the ATM which makes their withdrawal process unique. This insight can be used to provide preferred amount of cash withdrawal on the enter amount screen, thereby allowing customers to choose the frequent/preferred amount instead of entering amount each time. Customers can provide some preferred amount (e.g., 2500 INR), or by analyzing, customers ATM usage data log system can suggest preferred or regular amounts.

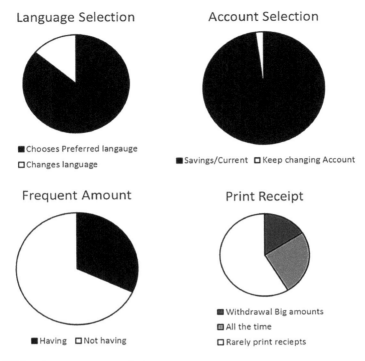

Fig. 2 ATM preferences survey analysis

4.4 Print Summary

Customers can set their preference of 'printing summary statement,' i.e., print every time, print when withdrawing amounts greater than x (e.g., 1000), never print, etc. By making this as preselected, customers may not have to select this every time.

New suggested personalized process flow of cash withdrawal from an ATM machine is presented (Fig. 3). Some steps from generalized flow (Fig. 1) are customized and skipped here, which makes this flow short, simple, and fast.

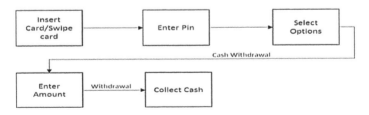

Fig. 3 New personalized flow of cash withdrawal process in ATM

Table 2 Table captions should be placed above the tables

Traditional withdrawal process		Redesigned withdrawal process	
Average time (s)	Standard deviation	Average time (s)	Standard deviation
48.16	10.41	28.27	6.64

In general ATM cash withdrawal process, every steps & sub-task play an important role. In redesigned approach of cash withdrawal process, some steps are eliminated and made pre-filled or defaulted from user's interaction side only. Selection of values from those steps is happening with the help of technology side.

5 Results

Average processing time and standard deviation of current and news suggested flow is presented in Table 2.

6 Discussions

The present research was guided by one major goal, i.e., reducing overall processing time of cash withdrawal process in ATM. Task flow of cash withdrawal processes was varied for this reason, and we found that subject matter took less time to process cash withdrawal via new flow. Approximately, 41.3% of output is improved by upgrading existing flow to new, resulting in less waiting time for an person in the queue of ATM counters and improved transactions/activities from an ATM every day. ATM machines will be so smart in the near future that customers just need to pick the sum and collect cash, rest machines will do so.

7 Conclusion

Owing to digital inclusion banking systems, engagement with ATM machines in India is growing on a regular basis. There is a need for powerful ATM machines in this digital India. Not only should current ATM machines be replaced with more modern, reliable, and secure ATM machines, it is also important to upgrade existing infrastructure. This paper was an attempt to analyze current ATM machines with regard to intelligence, efficiency, and their interaction. We have defined few areas for implementing customization; some processes can be personalized to enhance performance and user experience. In this study, we also introduced a new customized cash withdrawal system that showed an improvement in efficiency of 41.3%.

References

1. International Monetary Fund, F.A.: Automated Teller Machines (ATMs) (per 100,000 Adults)—India. Retrieved from The World Bank. http://data.worldbank.org/indicator/FB.ATM.TOTL.P5?end=2015&locations=IN&start=2005&view=chart (2015)
2. Hirsch, D.: White label ATMs will drive massive expansion in Indian market. Retrieved from ATM Marketplace. https://www.atmmarketplace.com/blogs/white-label-atms-will-drive-massive-expansion-in-indian-market/ (2013)
3. RBI.: BANKWISE ATM/POS/CARD STATISTICS. Retrieved from Reserve Bank of India. https://rbi.org.in/Scripts/ATMView.aspx?atmid=61 (2017)
4. King, K.C.: Investigating the human computer interaction problems with automated. Comput. Inf. Sci. **1**(2) (2008)
5. Maenpaa, K.K.: Consumer perceptions of Internet banking in Finland: the moderating role of familiarity. J. Retail. Consum. Serv. **15**(4), 266–276 (2008)
6. Chinedu, N., Ogbuji, C.B.: Analysis of the negative effects of the automated teller machine (ATM) as a channel for delivering banking services in Nigeria. Int. J. Bus. Manag. **7**(7) (2012)
7. Deepti Aggarwal, H.Z.: Demography based automated teller machine. India HCI (2012)
8. Sri Shimal Das, S.J.: Designing a biometric strategy (fingerprint) measure for enhancing ATM security in Indian e-banking system. Int. J. Inf. Commun. Technol. Res. **1**(5) (2011)
9. Miki, H.A.: Usability Research for the elderly people. Oki Tech Rev **71**(3, 199) (2004)
10. ICICI. (n.d.): ATM banking—cash transaction. Retrieved from ICICI Bank. https://www.icicibank.com/Personal-Banking/insta-banking/atm-banking/atm-banking.page
11. ATM Preference: Retrieved from SurveyMonkey. https://www.surveymonkey.com/results/SM-XNK8PLSB/ (2017)

Survey of Mininet Challenges, Opportunities, and Application in Software-Defined Network (SDN)

Dhruvkumar Dholakiya, Tanmay Kshirsagar, and Amit Nayak

Abstract The electronic devices like laptop, smart phone, television, Webcams, tablet, smart watch, desktop PC, servers, and electronic security are connected to any network directly or indirectly. The concept of computer networks is very old, and nowadays, all electronic equipments are connected to Internet. The Internet provide communication all over the World still end users facing problems sometime due to scarcity of resources and increasing of population day by day needs implementations of new hardware devices like routers and switches are very expensive and very hard working to manage them. The research team of computer network is searching solution for networks with more programming and less need of hardware. Mininet is a network emulator, and in that it has feature of virtual network with switches, hosts, routers, etc. On a single Linux kernel with Mininet, you can learn software-defined network and test software-defined network and software-defined applications. Mininet is used to deploy big network on a virtual host with limited resources. It is an open-source software and is in under active development.

1 Introduction of Mininet

Mininet is created at Stanford University by team of professors to do research in networking area and to run different network topologies [1]. Mininet is developed

---

D. Dholakiya (✉) · T. Kshirsagar
Department of Information Technology, B.Tech Information Technology Engineering,
Chandubhai S. Patel Institute of Technology, Charusat University, Anand, India
e-mail: d18it137@charusat.edu.in

T. Kshirsagar
e-mail: d18it136@charusat.edu.in

A. Nayak
Department of Information Technology, Devang Patel Institute of Advance Technology and Research, Charusat University, Anand, India
e-mail: amitnayak.it@charusat.ac.in

© The Editor(s) (if applicable) and The Author(s), under exclusive license to Springer Nature Singapore Pte Ltd. 2021
T. Senjyu et al. (eds.), *Information and Communication Technology for Intelligent Systems*, Smart Innovation, Systems and Technologies 196,
https://doi.org/10.1007/978-981-15-7062-9_21

Fig. 1 Mininet architecture

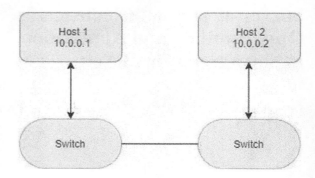

to create the software-defined network with OpenFlow controller, It has Ethernet facility, switches, and multiple hosts. It has inbuilt support for different types of controllers and switches [2]. Mininet has its own Python API.

Mininet is a network emulator that runs only in Linux operating system. It can run collections of hosts, switches, routers, and communication links. It gives us facility like lightweight virtualization for creating a Mininet network looks Like real computer network. In Mininet, the behavior of host is like a real computer machine. You can perform any task in the form of Python scripts like you doing in real computer machine (but that resource or package is needed to install in Linux operating systems). The Python scripts that runs in Mininet are able to send packets among the programmed hosts like a working of real Ethernet interface with its link speed and propagation delays. The packet gets processed like in real computer network that the switch does process of packets (see Fig. 1).

The Mininet network's hosts, switches, links, and controllers behaves like actual hardware of network equipment [2]. They are just one kind of software rather than hardware. In Mininet, it is possible to deploy your created Mininet network in real life, but you have to add some changes in it.

2 Features of Mininet

Mininet provides facilities to build custom topologies with switches, routers, and also you can create data center. In Mininet, the switches are programmable using the OpenFlow protocol, and the software design network that runs in Mininet can be easily implemented on hardware [1]. Mininet is an open-source software, in which you can edit it, fix bugs, and add any functionalities in it using github. At present, Mininet is in under development.

2.1 Starting with Mininet Emulator

Mininet is Linux-based tool which gets you control over the networking to install the Mininet in Linux system that uses the command SUDO which is the command from which you can run the program with security privileges of another users [2]. To download the packages of Mininet (see Fig. 2).

Mininet provides you the default topology: Single Topology, Linear Topology, Tree Topology, and Custom Topology [1]. After installing Mininet, the $sudo mn command is necessary to begin with Mininet, after you trigger this command, the output is default single topology, and in that there are 2 host and 1 switch (see Fig. 3).

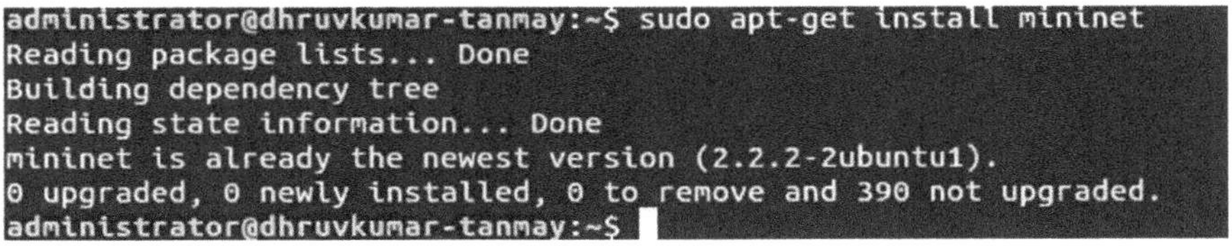

```
administrator@dhruvkumar-tanmay:~$ sudo apt-get install mininet
Reading package lists... Done
Building dependency tree
Reading state information... Done
mininet is already the newest version (2.2.2-2ubuntu1).
0 upgraded, 0 newly installed, 0 to remove and 390 not upgraded.
administrator@dhruvkumar-tanmay:~$
```

Fig. 2 Command $sudo apt-get installed Mininet in terminal

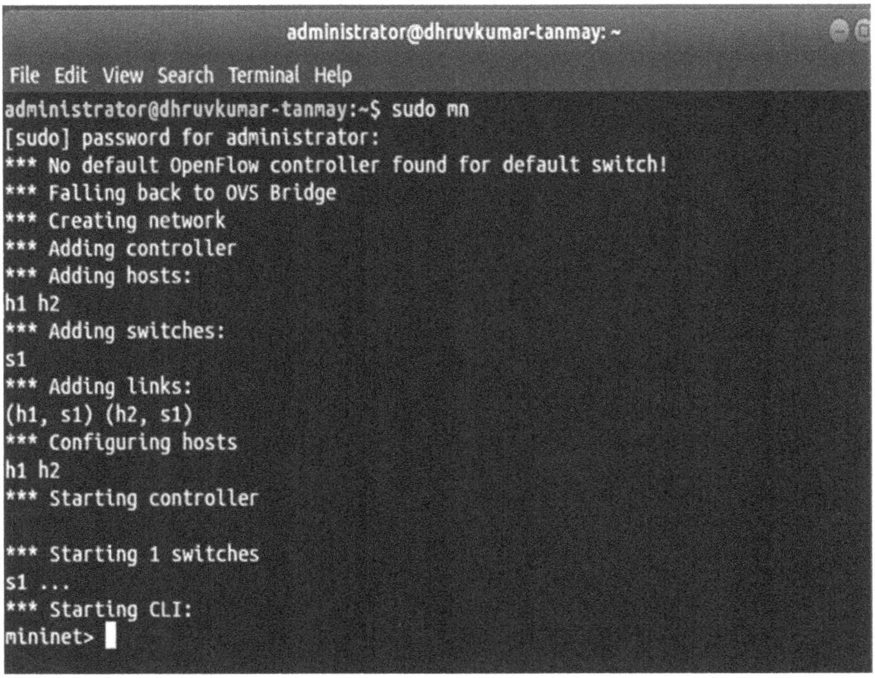

```
                    administrator@dhruvkumar-tanmay: ~
File Edit View Search Terminal Help
administrator@dhruvkumar-tanmay:~$ sudo mn
[sudo] password for administrator:
*** No default OpenFlow controller found for default switch!
*** Falling back to OVS Bridge
*** Creating network
*** Adding controller
*** Adding hosts:
h1 h2
*** Adding switches:
s1
*** Adding links:
(h1, s1) (h2, s1)
*** Configuring hosts
h1 h2
*** Starting controller

*** Starting 1 switches
s1 ...
*** Starting CLI:
mininet>
```

Fig. 3 Command $sudo mn in terminal

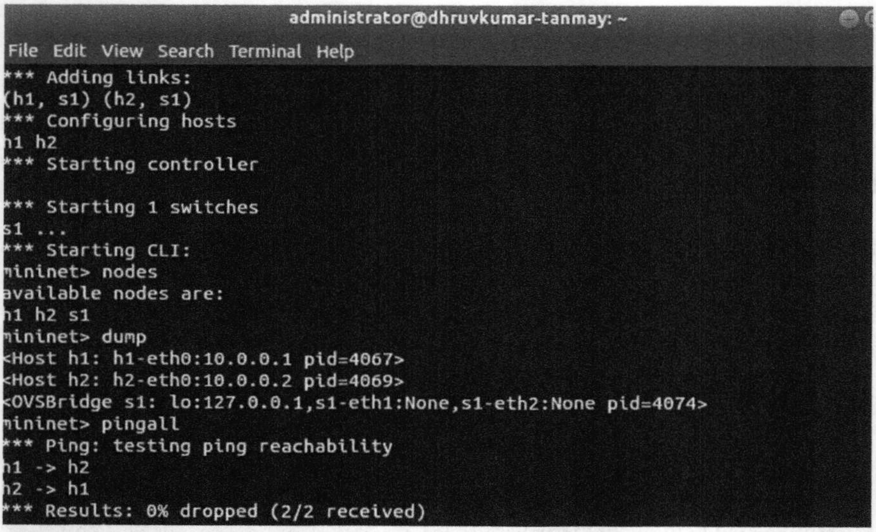

Fig. 4 Commands nodes, dump, and pingall in terminal

Some basic commands of Mininet are list nodes, dumps, and pingall in terminal (see Fig. 4).

Ping command is used to know whether the host can communicate with other host on network, and it sends some data packet on network to that computer and that computer replies back (see Fig. 5).

If the topology not work properly, then delete all objects and nodes (see Fig. 6).

3 Software-Defined Networking

Software-defined networking (SDN) is a future technology in the next-generation network that allows network management to centrally manage the network in the traditional network that the system has control plane and data plane. We cannot add extra feature to the network operating system [3]. As it is product of company which is embedded, we are not allowed to add feature. This is limitation of traditional network, software-defined network overcomes this limitation, and in that there is separation of control plane and data plane [4]. Control plane kept centralized and data plane itself in hardware. This is implemented using OpenFlow protocol and API [5, 6]. From this, a switch acts as router, and router acts as switch using the Mininet command, so control plane is decoupled from the hardware.

Execute or run software for general purpose control plane software. Decouple from software unique to the networking. Usage of merchandizing services meaning that all switch, router, and other network device to be managed by software-defined network

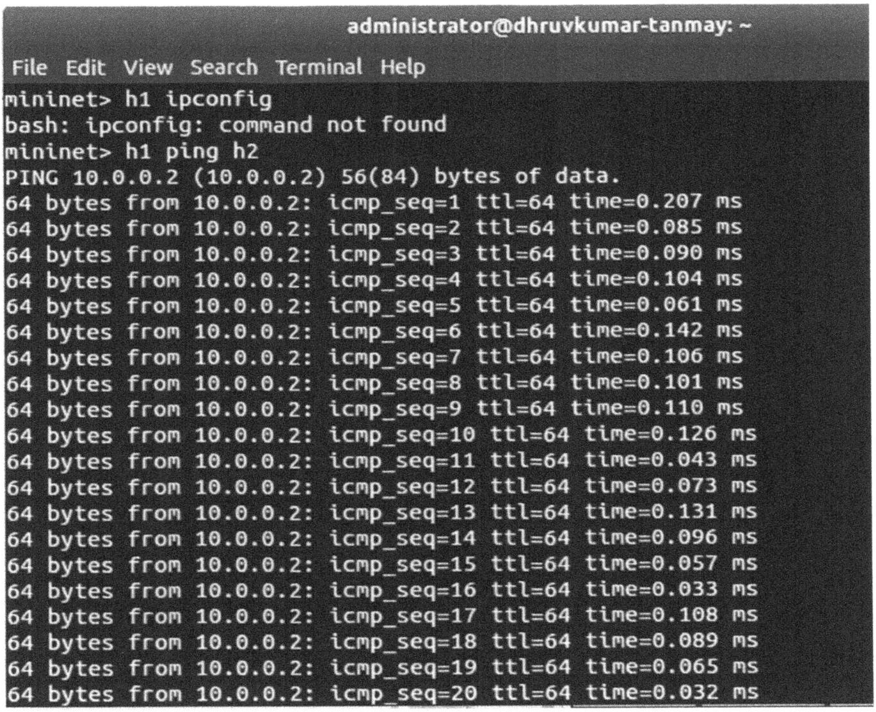

```
administrator@dhruvkumar-tanmay: ~
File  Edit  View  Search  Terminal  Help
mininet> h1 ipconfig
bash: ipconfig: command not found
mininet> h1 ping h2
PING 10.0.0.2 (10.0.0.2) 56(84) bytes of data.
64 bytes from 10.0.0.2: icmp_seq=1 ttl=64 time=0.207 ms
64 bytes from 10.0.0.2: icmp_seq=2 ttl=64 time=0.085 ms
64 bytes from 10.0.0.2: icmp_seq=3 ttl=64 time=0.090 ms
64 bytes from 10.0.0.2: icmp_seq=4 ttl=64 time=0.104 ms
64 bytes from 10.0.0.2: icmp_seq=5 ttl=64 time=0.061 ms
64 bytes from 10.0.0.2: icmp_seq=6 ttl=64 time=0.142 ms
64 bytes from 10.0.0.2: icmp_seq=7 ttl=64 time=0.106 ms
64 bytes from 10.0.0.2: icmp_seq=8 ttl=64 time=0.101 ms
64 bytes from 10.0.0.2: icmp_seq=9 ttl=64 time=0.110 ms
64 bytes from 10.0.0.2: icmp_seq=10 ttl=64 time=0.126 ms
64 bytes from 10.0.0.2: icmp_seq=11 ttl=64 time=0.043 ms
64 bytes from 10.0.0.2: icmp_seq=12 ttl=64 time=0.073 ms
64 bytes from 10.0.0.2: icmp_seq=13 ttl=64 time=0.131 ms
64 bytes from 10.0.0.2: icmp_seq=14 ttl=64 time=0.096 ms
64 bytes from 10.0.0.2: icmp_seq=15 ttl=64 time=0.057 ms
64 bytes from 10.0.0.2: icmp_seq=16 ttl=64 time=0.033 ms
64 bytes from 10.0.0.2: icmp_seq=17 ttl=64 time=0.108 ms
64 bytes from 10.0.0.2: icmp_seq=18 ttl=64 time=0.089 ms
64 bytes from 10.0.0.2: icmp_seq=19 ttl=64 time=0.065 ms
64 bytes from 10.0.0.2: icmp_seq=20 ttl=64 time=0.032 ms
```

Fig. 5 Command h1 ping h2 in terminal

[7]. IDC expands the concept of SDN by stating: "Datacenter SDN architectures include software-defined overlays or controllers that are abstracted from the underlying network equipment, providing intent- or policy-based management of the entire network. This results in a network of datacenters that is more matched with client workload requirements by automatic (thereby faster) provisioning, programmatic network management, ubiquitous task-oriented control, and direct alignment with cloud orchestration systems where appropriate. Software-defined network controller have a common logical architecture (see Fig. 7).

3.1 Benefits of SDN

Centralized networking system control assists in networking systems development and provides support for end users. The software-defined networks provide flexibility to use automated SDN programs to configure, protect, manage, and maximize network resources. APIs provided by computer network systems promote the implementation of common network services such as routing, security, access control,

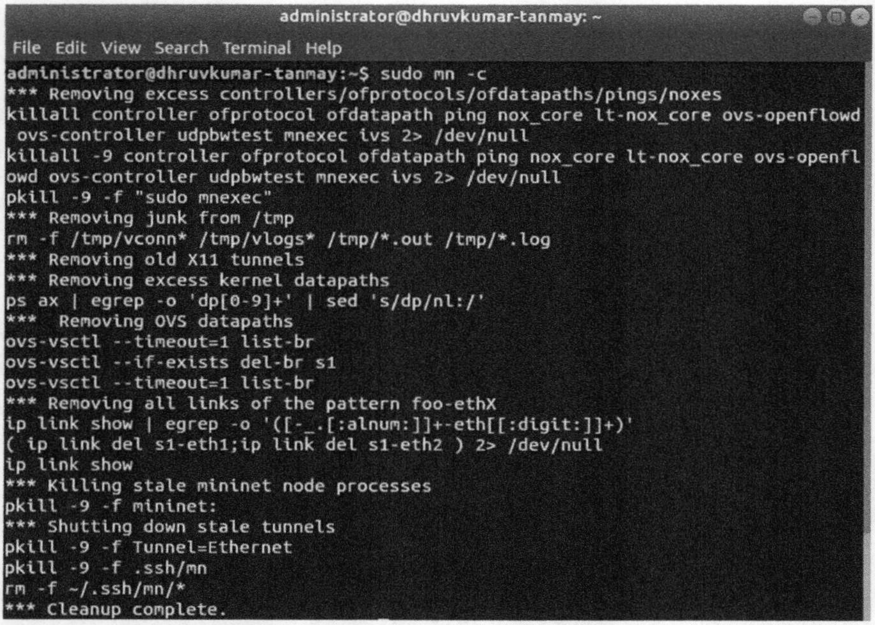

Fig. 6 Command $sudo mn-c in terminal

multicast, bandwidth management, optimization of storage, use of energy, and enable policy management. SDN offers plenty of versatility that allows network programming and management at a scale that conventional networks have struggled to offer. Through decoupling control and data planes, SDN results in benefits such as high efficiency, programmability, and a consolidated view of the network. SDN is seen as a tool for providing a versatile and scalable architecture which is the way to make future networks substantially smarter, scalable, and versatile. It provides network architects an opportunity to provide a genuinely creative, versatile, and scalable network solution with the aid of SDN, network virtualization, and network orchestration, which will be more efficient and cost effective.

3.2 Challenges of SDN

3.2.1 Reliability

To choose the supervisor for the whole network, the latency ratings are exchanged among the controllers through a newly designed east/west bound interface.

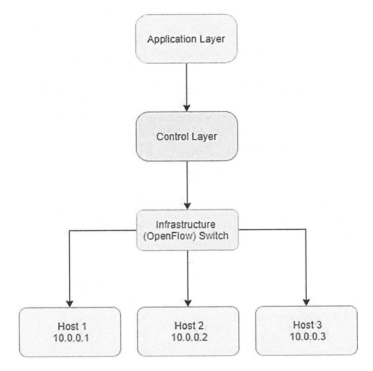

Fig. 7 Software-defined networking architecture

3.2.2 Scalability

Network scalability is a challenge with ever-growing storage demands and slow processor speeds. But this size can be extended and improved with software-defined networking (SDN) to transmit vast amounts of data at a faster rate.

3.2.3 Security

SDN network security has to be in a software-defined network (SDN) anywhere. SDN security needs to be embedded into the infrastructure and provided as a service to protect the quality, credibility, and privacy of all related resources and information.

3.2.4 Interoperability

On the lower layer, interoperability of connectivity is how an SDN network interacts with other networks. This is where the use of existing standard protocols makes the greatest sense.

4 Challenges of Mininet

Using Mininet, you can deploy large network on the limited recourse of single host. Mininet is built to do research in SDN and OpenFlow [1, 6]. It provides control convenience and reality at very low cost. Mininet hardware test beds are fast but very expensive. You can use other simulator NS2 and NS3. It is very cheap, but it only gives realization of real network sometime of its slow and code its complex. To deploy the real network which you have built in Mininet you need more modifications [3, 8].

4.1 Application of Mininet

4.1.1 Software-Defined Network

Software-defined networking technology is a network management technique that allows for dynamic, programmatically optimized network configuration to enhance network performance and monitoring, making it more like cloud computing than conventional network management.

4.1.2 Network Functions Virtualization

Network functions virtualization is a network architecture idea that utilizes the virtualization technology to imagine the whole network node groups as building blocks that can use a connection or chain to construct networking networks together [8].

4.1.3 Visual Network Description

The virtual network description (VND) is a GUI tool that helps to manually build this network through simulation. With this method, the necessary components can be manually set up in SDN, and the properties of various links, switches, and controllers attached to it can be specified to simulate this network using Mininet [7].

5 Conclusion

It can run code on small networks as well as big network. It is a tool to run SDN codes on emulator. Real life system is very difficult to reconfigure. In Virtual machine you can easily change but issues are scalability. Emulator are good alternative but same code cannot be deploying on real time hardware. There are performance issues. The core principle of SDN has fed into a number of networking patterns. Distributing processing resources to distant facilities, pushing data center operations to the edge, embracing cloud infrastructure, and enabling technologies on the Internet of Things—any of these activities can be rendered simpler and more cost-effective with a correctly designed SDN. SDN calls for a variety of health advantages. A consumer may split a network link between an end-user and the data center and provide specific protection configurations for the various network traffic categories. A network may have one low-security, public-faced network that reaches no classified details. The role of SDN in pushing toward private cloud adoption and the hybrid cloud seems a normal one. Big SDN players including Cisco, Juniper, and VMware have also made attempts to bring the data center and cloud environments of businesses together. Intent-based networking (IBN) has a number of elements, but essentially it is about allowing network managers the freedom to identify what they want the network to do, and making an integrated network management framework establish the desired state and execute policy to guarantee that what the company needs occurs. "We are now at a stage where SDN is best known, where most datacenter network customers are acquainted with its usage cases and benefit propositions, and where an increasing range of businesses feel that SDN solutions provide realistic advantages," Casemore said. "With SDN acceleration and the transition toward software-based network integration, the network is regaining lost ground and heading into greater harmony with a surge of emerging task workloads that produce substantial business performance."

References

1. Mininet Homepage. http://mininet.org/, last accessed 2020/02/14
2. Introduction to Mininet. https://github.com/mininet/mininet.wiki.git/, last accessed 2020/02/16
3. Kreutz, D., Ramos, F.M.V., Verissimo, P.E.P., Rothenberg, C.E., Azodolmolky, S., Uhlig, S.: Software-defined networking: a comprehensive survey. Proc. IEEE **103**(1) (2015)
4. Sharma, K.K., Sood, M.: Mininet as a container based emulator for software defined networks. Int. J. Adv. Res. Comput. Sci. Softw. Eng. **4**(12) (2014)
5. HP OpenFlow and SDN Technical Overview. Technical Solution Guide. White paper. Version: 1. Sep 2013
6. Software-Defined Networking: The New Norm for Networks. ONF White Paper, 13 April 2012
7. Nunes, B.A., Mendonca, M., Nguyen, X., Obraczka, K., Turletti, T.: A survey of SDN-past, present and future of programmable network. Commun. Surv. Tutorials IEEE **16**(3), 1617–1634 (2014)
8. Han, B., Gopalakrishnan, V., Ji, L., Lee, S.: Network function virtualization: challenges and opportunities for innovations. Commun. Mag. IEEE **53**(2), 90–97 (2015)

Some Novelties in Map Reducing Techniques to Retrieve and Analyze Big Data for Effective Processing

Prashant Bhat and Prajna Hegde

Abstract As we are living in the era of digital data, it has become necessary to make use of it in intelligent ways. The data generation is not only increasing, but the rate with which data generates is also increasing. Huge data obtained from many sources can be processed and analyzed so as to get useful information. But, the problem is with volume of data, velocity with which data increases, and also different variety and complex structure of data. Storing such large amount of data and the process of retrieving huge data when required are time consuming. One of the solutions for effective processing of Big Data is parallel processing. The software solutions like Hadoop provides way to store and also to implement parallel processing of Big Data. In most of the situations, Big Data cannot be stored in a single system. Distributed File System that can run on different clusters can be used to process Big Data. By using MapReduce model, large dataset can be computed on commodity hardware clusters. This paper presents a novel work of implementing MapReduce technique to analyze and retrieve data. An attempt is made to retrieve data by adopting MapReduce technique. A task is divided into number of sub-tasks, and these sub-tasks can be processed simultaneously by different processors or number of commodity hardware. A novel and effective way of implementing MapReduce is represented in this paper. In other words, this work examines the method and the outcome of MapReduce technique, which is a solution to the problem of processing huge amount of data.

1 Introduction

Big Data is related to complex, large amount of data growing with high speed from many autonomous sources. This data can be analyzed using advanced tools. Structuring huge amount of data is called Big Data Analytics. Since data is large and complex, it becomes challenging to store, maintain, and process data. To solve the

---

P. Bhat · P. Hegde (✉)
School of Computational Sciences and IT, Garden City University, Bangalore, India
e-mail: prajna_rh@yahoo.co.in

© The Editor(s) (if applicable) and The Author(s), under exclusive license to Springer Nature Singapore Pte Ltd. 2021
T. Senjyu et al. (eds.), *Information and Communication Technology for Intelligent Systems*, Smart Innovation, Systems and Technologies 196,
https://doi.org/10.1007/978-981-15-7062-9_22

problem of Big Data, parallel data processing can be applied. Parallel processing can be implemented using advanced tools like Apache Hadoop [1]. By using the techniques like MapReduce, processing of Big Data is made easy [2]. Big Data can impact core business and can give competitive advantage. Big Data Analytics [3] can add value to the business organizations.

Large amount of data generated from day-to-day transactions are called operational Big Data. Online transactional data of an organization, online shopping data, data generated from social media, etc., are the examples for operational Big Data.

Data which is more complex than operational Big Data [4], and are more advanced than operational Big Data are called analytical Big Data. Real-time decisions can be taken with the help of analytical Big Data. Weather forecast data that can be analyzed to protect people from calamities is an example for analytical Big Data.

This paper deals with Big Data and the way in which one can effectively analyze it. Analyzing Big Data results in the information retrieval. This extracted information can be utilized in organizations for increasing their efficiency. The contribution of this paper is using MapReduce tool to analyze the data and extract useful information out of it.

2 Related Works

Author Beakta [5] says that Big Data is emerging with lot of opportunities as well as challenges to deal with large amount of data. Big Data is giving opportunities for the new researcher to extract useful information from massive amount of data. Specialization of data analysis is needed to analyze unstructured complex data in the form of text, image, videos, etc. Here, author tried to explain overview, scope, and advantages of Big Data. Also, this paper puts light on opportunities of Big Data in health care, technology, etc. Along with this, Hadoop and components are also well explained.

Authors Vivekanath and Baplest [6] explained Big Data as structured, unstructured, or semi-structured data in large quantity. Data mining techniques can be used to get information out of it. Big Data is characterized by 3 Vs, volume, velocity, and variety. By storing Big Data in platforms like Hadoop and processing it by MapReduce, Big Data can be analyzed. This paper concentrates on text analysis, audio/video analytics, social media analytics, and also predictive analytics. This paper gives applications of Big Data Analytics. Along with this, author proposed methods of improving techniques of Big Data Analytics.

Author Ramadan [7] says that there is a need for non-traditional method to analyze data in social media, industry applications, and science. Big Data strategies and techniques can be used to organize, store, and process large amount of data. These large datasets cannot be stored and processed by traditional methods and techniques, or even using single computer. This led to the development of different analytical methods to deal with large, heterogeneous data. Big Data has characteristics like veracity which means data comes from different sources. Other characteristics are volume, velocity, and variety. These characteristics make Big Data difficult to store, process, and manage. Volume indicates data with large quantity. Velocity indicates data streaming time and latency. This paper deals with different data analysis tools. This paper reviews large amount of commercial or free tools. Author focused on characteristics of some tolls that helps scientists to select best and suitable tool for the particular data which they have to deal with.

Authors Bhosale and Gadekar [7] says Big Data defines technologies to collect, store,, manage, and analyze large sized datasets with high data streaming time and complex structure. The complex structure, i.e., structured, unstructured, or semi-structured, makes conventional data management methods incapable to manage Big Data. Big Data arrival rate varies, and also data arrives from different sources. Parallel processing can be used to process Big Data using inexpensive, efficient way. Big Data requires new architecture, techniques, and algorithms to analyze and extract useful patterns. Hadoop is a platform for analysis of Big Data. Hadoop is an open-source software that has distributed processing of Big Data using commodity software. It is designed in such a way that it can have single server to thousands of machines which makes the system highly fault tolerant.

Author Komal [8] says that Big Data has increased the need of research in data field. Digital information is increasing with high speed mainly in unstructured form collected from Facebook posts, likes, tweets, blogs, news, articles, YouTube videos, Web sites clicks, etc. Through mobile phones and laptops, billions of people grasp, upload, and store information. Traditional data handling tools are insufficient as data is enormous, heterogeneous, high velocity. Digital marketing and E-commerce are gaining more important in business industry which depends on online transactions. Big Data Analytics is a gift for business industry. Big Data Analytics helps to extract meaningful patterns, unknown correlations in consumer market, preferences of clients, buying attributes, etc. This paper provides in-depth overview of comparative assessment of latest tools and frameworks which are used for Big Data Analytics.

Authors Shobha Rani and Rama [9] say that there is a rapid growth of data along with its analysis mainly because of Web-based applications and mobile computer technology. People are facing challenge with increasing data which helps in decision making. Conventional relational DBMS's cannot handle Big Data. Traditional tools and techniques of data mining are not suitable to handle Big Data. Powerful algorithms are required to process Big Data. The companies like Google, Yahoo, etc., adopted parallel processing algorithms like MapReduce. In the world of Big Data, MapReduce is playing essential part in meeting the raising demands on computing system which needs to manage huge amount of data. Big Data Analysis in parallel and distributed system can be achieved by most popular programming model, i.e., MapReduce. It has high scalability. Hadoop is distributed open-source framework which can store and process Big Data. This paper highlights MapReduce along with Hadoop for processing and analysis of Big Data.

Authors Ghazi and Gangodkar [10] says that applications running on Hadoop clusters are increasing. The organizations have found model that works efficiently in distributed systems. The complex nature of Big Data can be hidden by using

Hadoop Distributed File System and MapReduce. As Hadoop is getting famous, it is necessary to study its technical details. This paper concentrates on components of Hadoop, MapReduce. MapReduce makes use of Job Tracker and Task Tracker which monitors and executes job. HDFS has Name Node, Data Node, and Secondary Name Node which maintains distributed storage efficiently. In-depth study given in this paper is useful for developing large-scale distributed applications. Also, the details given helps in computing intensive applications.

Authors Maitrey and Jha [11] presented a paper which says that there is a huge growth in data with the development of computer technology. This has become challenging task for the data scientists who needs to process large amount of data. Various fields are failing to make use of Big Data which can help in taking useful decisions. New patterns can be discovered from large datasets using data mining techniques. Classical data mining techniques cannot be applied on Big Data. The algorithms that are efficient concurrently working are required to reach scalability and performance requirement to handle large datasets. Threads, MPL MapReduce, and mash-up are developed which are parallel processing algorithms. Rigorous problems can be computed using MPL MapReduce. But, it is difficult to implement MPL MapReduce algorithm in reality. Many MapReduce architectures are developed to manage Big Data. Google is one of them. Another architecture is open-source MapReduce software, that is, Hadoop. This paper specifies Hadoop, implementation of MapReduce for analytical processing of data.

Authors Sarkar et al. [12] presented a paper which says as the importance of Big Data is growing, the search for technique that can effectively process Big Data is also increasing. MapReduce programming technique is one of such techniques. This paper presents detailed discussion of MapReduce and its overview. Along with these, authors explained working of MapReduce with few examples. This paper lists advantages of using MapReduce, disadvantages, and challenges in implementing MapReduce. Authors also specified various solutions for the challenges of Big Data problem. This paper puts light on scope of current, future research in the field of Big Data, and MapReduce. Also, authors mentioned the applications of Big Data.

3 Proposed Methodology

Datasets with millions of records can be processed efficiently by breaking down the tasks into parallel processes. MapReduce is a part of Hadoop ecosystem. It is a programming model that helps to describe the solution using parallel tasks [13]. Then, parallel tasks can be combined to form final desired output. MapReduce is an effect method for handling Big Data. It contains Map and Reduce [14] functions. Map function writes the result on temporary storage. Only one copy of redundant data is processed. Based on the output key, data are redistributed in such a way that data related to one key are placed in one node. Reduce function processes the group of output data and key parallelly. Hadoop MapReduce [15] provides storage with high fault tolerance and reliable data processing as it clones data pieces into different nodes.

The logic behind MapReduce [16] is that workload is divided among machines in the clusters. Degree of parallelism depends in size of input data.

4 Components of MapReduce Framework

4.1 Input Data

Generally, the input for the mapper is stored in Hadoop Distributed File System (HDFS) [17]. It is then passed to the mapper one line at a time.

4.2 Mapping Phase

Mapper takes the input in raw format and converts into key-value pair as shown in Fig. 1. Key is nothing but the factor based on which data is aggregated. Mapper assigns value for each key.

4.3 Reducer Phase

After this, MapReduce will shuffle and sort the key-value pair automatically. Values having same key are then combined.

From this set of shuffled values, Reducer processes and then produces a new set of values that are desired. And this output is written into Hadoop File System.

5 Results and Discussions

This section focusses on in-depth study of Big Data processing. As discussed earlier, one of the easiest ways to process Big Data is by using MapReduce programming approach. Here, an attempt is made to explain the way in which MapReduce works. This can be explained by taking an example dataset as shown in Fig. 2. Dataset contains following attributes.

- Code_module: It indicates the module for which student has been taken admission.
- Code_presentation: It indicates code for which student has taken admission. Letter 'B' indicates starting of presentation from February and 'J' represents starting of presentation from October.

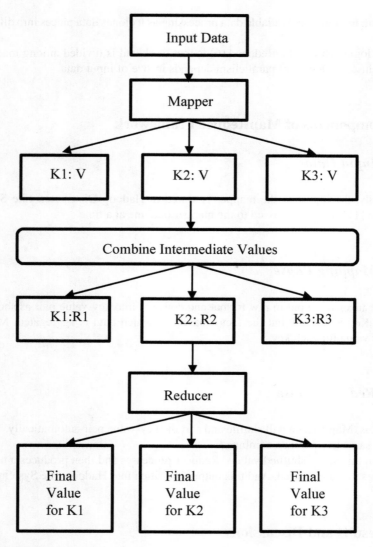

Fig. 1 MapReduce framework

Fig. 2 Dataset

- id_student: It indicates a unique identification number for the student.
- Gender: It indicates gender of student.
- Region: It indicates geographic area to which the student is belonged to while taking admission.
- Highest_education: It represents highest level of education qualification to enter into the module presentation.
- imd_band: It represents index of multiple depravation band of the place where student used to live during module presentation.
- age_band: It indicates age of student.
- num_of_prev_attempts: It represents number of attempts made by the student.
- studied_credits: It represents module credit points of student.
- Disability: It indicates whether student has disability or not.
- Final_Result: It represents final result of student.

To analyze this dataset using MapReduce technique [12], steps are as follows.

5.1 Step 1

- Write one program to implement Map function stdMappaer.java. Specify name of the package as "student." This is followed by importing library packages.
- Then, create stdMapper class extends MapReduceBase implements Mapper <LongWritable, Text, Text, IntWritable>
- Map () method is the main part of Mapper class. It takes four arguments. A key-value is passed each time when map () method is called.
- Map () function receives input text and splits it into words using tokenizer.
- String valueString = value.toString(); String[] onestddata = valueString.split(",");
- 11th index (final_result) is used to form the pair, value 1.

5.2 Step 2

- Create a program called stdReducer.java to implement Reduce function. It also starts with name of the package that is student.
- Create public class stdReducer extends MapReduceBase implements Reducer <Text, IntWritable, Text, IntWritable> {
  Text and IntWritable are data types of key-value pair that are passed to reducer.
- Reduce () function copies key-value and also assigns value 0 to frequency count.
- While loop is used to loop through the set of values related to key and to calculate final frequency value by adding all the values.

  while (values.hasNext()) {
  IntWritable value = (IntWritable)

    values.next();
    frequencyOutput += value.get();

- Final count is obtained by output.collect(key, new IntWritable(frequencyOutput)).

5.3 Step 3

- Create a file called stdDriver.java. Here, driver class is defined.
- It is the responsibility of driver class to set MapReduce job to run Hadoop system. In this class, job name, data types of input/output values, and also names of mapper and reducer classes are specified.
- Following lines of code starts executing MapReduce job.
- try {
- JobClient.runJob(job_conf);
- } catch (Exception e) {
- e.printStackTrace();
- }

5.4 Step 4

- Once these programs are created, next step is exporting class path
- Export CLASSPATH = "$HADOOP_HOME/share/hadoop/mapreduce/hadoop-mapreduce-client-core"
  2.2.0.jar: $HADOOP_HOME/share/hadoop/mapreduce/hadoop-mapreduce-client-common-2.2.0.jar: $HADOOP_HOME/share/hadoop/common/hadoop-common-2.2.0.jar: ~/MapReduceTutorial/student/*:$HADOOP_HOME/lib/*

  Then, all three java files are compiled and put in package directory.

- Create a text file and type the name of the main class in the file. Then, create jar file called result java
- Then, start Hadoop
- Copy studentInfo dataset to be processed into inputdata folder in Hadoop Distributed File System (HDFS).
- After that, run MapReduce job $HADOOP_HOME/bin/hadoop jar result.jar/inputdata/outputresult
- Result can be seen by opening "localhost:50070" port. Select file system. And choose outputresult. And then choose part-00000
- Result can also be seen using command line by typing the command $HADOOP_HOME/bin/hdfs dfs-cat/outputresult/part-00000 as shown in Fig. 3.

The novel work presented in this paper shows how a large dataset can be effectively analyzed by two steps Map and Reduce. An attempt is made in this paper to analyze

Block Pool ID: BP-60715368-127.0.1.1-1573386604022

Generation Stamp: 1051

Size: 79

Availability:

- prajna-VirtualBox

File contents

```
"Distinction"   3024
"Fail"          7052
"Pass"          12361
"Withdrawn"     10156
"final_result"  1
```

Fig. 3 Output shown in port "localhost:50070"

a dataset and useful information is extracted as one can know about the total number of students passed, failed, and the students passed with distinction. The algorithm shown in this paper gives an idea about how MapReduce technique [18] can be implemented to analyze Big Data.

6 Conclusion

Big Data can be efficiently processed and analyzed using one of the components of Hadoop, that is MapReduce. The main strategy of MapReduce is dividing a task into set of independent tasks [9]. This improves the speed of processing and reliability of the system.

This paper gives insight into MapReduce technique and the way with which MapReduce works. This paper helps to get an idea how information can be extracted from Big Data [19]. The idea behind MapReduce is moving the process to the system where the data resides, not moving the data where the processor resides.

References

1. Rajput, N., Ganage, N., Thakur, J.B.: Review paper on Hadoop and map reduce. Int. J. Res. Eng. Technol. **06**(09) (2017)
2. Deshai, N., Venkataramana, S., Saradhi Varma, G.P.: Big data and Hadoop: a review paper. Int. J. Comput. Sci. Inf **2**(2) (2015)
3. Vivekanath, P., Baptist, L.J.: Research paper on big data Hadoop MapReduce job scheduling. Int. J. Innov. Res. Comput. Commun. Eng. **6**(1) (2018)
4. Tardío, R., Maté, A., Trujillo, J.: An iterative methodology for big data management, analysis and visualization, pp. 545–550. https://doi.org/10.1109/bigdata.2015.7363798 (2015)
5. Beakta, R.: Big data and Hadoop: a review paper. Int. J. Comput. Sci. Inf. 2(2) (2015)
6. Vivekananth, P., Leo John Baptist, F.: An analysis of big data analytics techniques. Int. J. Eng. Manag. Res. **5**(5), 17–19 (2015)
7. Ramadan, R.: Big data tools-an overview. Int. J. Comput. Sci. Softw. Eng. **2**. https://doi.org/10.15344/2456-4451/2017/125 (2017)
8. Komal, M.: A review paper on big data analytics tools. Int. J. Tech. Innov. Modern Eng. Sci. IJTIMES **4**(5) (2018)
9. Shobha Rani, C., Rama, B.: MapReduce with Hadoop for Simplified analysis of big data. Int. J. Adv. Res. Comput. Sci. **8**(5) (2017)
10. Ghazi, M., Gangodkar, D.: Hadoop, MapReduce and HDFS: a developer's perspective. Procedia Comput. Sci. **48**, 45–50 (2015). https://doi.org/10.1016/j.procs.2015.04.108
11. Maitrey, S., Jha, C.K.: MapReduce: simplified data analysis of big data. Procedia Comput. Sci. **57**, 563–571. ISSN 1877-0509 (2015)
12. Sarkar, A., Ghosh, A., Nath, A.: MapReduce: a comprehensive study on applications, scope and challenges. Int. J. Adv. Res. Comput. Sci. Manag. **3**, 256–272 (2015)
13. Sudha, P., Gunavathi, R.: A Survey paper on map reduce in big data. Int. J. Sci. Res. IJSR **5**(9) (2016)
14. Dhavapriya, M., Yasodha, N.: Big data analytics: challenges and solutions using Hadoop, map reduce and big table. Int. J. Comput. Sci. Trends Technol. IJCST **4**(1) (2016)
15. Pol, U.R.: Big data analysis using Hadoop MapReduce. Am. J. Eng. Res. AJER **5**, 146–151 (2016)
16. Khezr, S., Navimipour, N.: MapReduce and its applications, challenges, and architecture: a comprehensive review and directions for future research. J. Grid Comput. **15**, 1–27 (2017). https://doi.org/10.1007/s10723-017-9408-0
17. Shvachko, K., Kuang, Sanjay Radia, Robert Chansler, F.: The Hadoop distributed file system. In: 010 IEEE 26th Symposium on Mass Storage Systems and Technologies (MSST), Incline Village, pp. 1–10.https://doi.org/10.1109/msst.2010.5496972 (2010)
18. Malik, L., Sangwan, S.: MapReduce algorithms optimizes the potential of big data. Int. J. Comput. Sci. Mobile Comput. **4**(6), 663–674 (2015)
19. Kaur, I., Kaur, N., Ummat, A., Kaur, J., Kaur, N.: Research paper on big data and Hadoop. Int. J. Comput. Sci. Technol. **4**(10) (2016)
20. Harshawardhan S. Bhosale, Devendra P. Gadekar.: A Review paper on big data and Hadoop. Int. J. Sci Res. Publ. **4**(10) (2014) ISSN 2250–3153

A Study Based on Advancements in Smart Mirror Technology

Aniket Dongare, Indrajeet Devale, Aditya Dabadge, Shubham Bachute, and Sukhada Bhingarkar

Abstract Smart devices are exciting and upcoming advancements in the field of Internet of things (IOT). It is an application of Raspberry Pi, from furniture in the residential and commercial sector to everyday devices like watches and headphones, every object is becoming smarter and hence, making human lives much easier and efficient. Users can actively interact with these devices using voice commands. These smart devices can also assist a user in their personal daily activities as well as in medical use. Currently, various variations of the smart mirror based on software and hardware compatibility as well as based on applications and uses are available in the market. The goal of this paper is to understand the working, various applications, and purposes of the smart mirror according to the varying features in different versions. This paper critically analyses the different versions of the smart mirror and its different features and applications.

1 Introduction

A Smart Mirror is a display device which assists the user in handling daily functions in an easier and productive manner. Commercially, there are many different types of

---

A. Dongare (✉) · I. Devale · A. Dabadge · S. Bachute · S. Bhingarkar
Maharashtra Institute of Technology College of Engineering, Pune, Maharashtra 411038, India
e-mail: aniketd423@gmail.com

I. Devale
e-mail: indrajeet.devale@gmail.com

A. Dabadge
e-mail: adityadabadge@gmail.com

S. Bachute
e-mail: bacshub007@gmail.com

S. Bhingarkar
e-mail: sukhada.bhingarkar@gmail.com

© The Editor(s) (if applicable) and The Author(s), under exclusive license to Springer Nature Singapore Pte Ltd. 2021
T. Senjyu et al. (eds.), *Information and Communication Technology for Intelligent Systems*, Smart Innovation, Systems and Technologies 196,
https://doi.org/10.1007/978-981-15-7062-9_23

Smart Mirrors available in the market having varying features and functionalities. Smart Mirrors usually consist of a Raspberry Pi board as the processor and consist of a two-way mirror to allow the user to view the display of the device by projecting it on a screen. The common functionalities consist of weather updates, clock and alarm settings, and events and reminders. Functionalities are integrated in the devices using various application program interfaces (API) available for public access [1]. Some of the products also consist of dynamic features like facial and voice recognition. A persuasive effort is being made to integrate various other useful functionalities in the Smart Mirror using various tools and software through research. In most of the devices, user input is usually accepted through voice as well as touch [2].

According to [3], the maximum usage of smart mirrors is for personal and day-to-day functions. Productivity and ease of life are of utmost importance in this age. The Smart Mirror assists the user by providing a variety of basic functionalities in the form of weather updates, news updates, clock and alarm settings, all available by integrating APIs into the system. The user can access the Smart Mirror by either voice using the mic or touch using a touchpad [1, 4, 5]. Settings like alarms and events can be periodically set as per the user's convenience. Although these functionalities are similar to that of a smartphone, this device can be accessed hands-free and can be termed as "smart furniture" in a residence.

Smart Mirrors can be used along with Augmented Rendering (AR) in the fashion industry to virtually try on new clothes or cosmetics. South Korean 3D fashion design software firm CLO Virtual Fashion has already unveiled their mirror to try on clothes virtually [6]. AR of cosmetics, makeup features, and even accessories is now possible without the user trying on any products. Since the Smart Mirror is a smart device, it can also recommend other similar products according to the user's choices and preferences.

Since a Smart Mirror is a part of smart devices found in residences, it can also act as a device which assists in monitoring the health of a user and regularly update and send reports of patients to respective authorities in the medical field. According to [7], a Health Monitoring System (HMS) can be used to achieve the above tasks of constant updating and sending of user reports to medical authorities. This can be used for people with health issues like Dementia, Parkinson's disease, and Alzheimer's disease. Video cameras used on Smart Mirrors monitor the behavior of a user so that in an emergency, the Smart Mirror can detect and identify distress and update the authorities.

2 Related Work

In [1], the authors have come up with an idea for an "open platform for discrete display development." Their main aim is to design a mirror which can be accessed by an average user and also an advanced developer. In the paper, they have stated to create such a device by making the User Interface easy to understand but also adding various advanced features that an advanced developer can make use of. In addition to this, in [2], the authors have put forward the idea of a Smart Mirror with an infrared frame network supported by an infrared frame. The objective of the infrared frame is to save energy when not in use by waking up the module automatically only when needed. In [3], the authors have recommended a mirror which can recognize the user's emotion and can play music accordingly. The main objective of the mirror is to multitask. The mirror studies the facial expressions of the user to understand the user's emotion by classifying it using a pre-trained classifier. In [4], the authors suggest a biometric system which will improve the security of the mirror system. It makes use of facial recognition and voice recognition for improved security. The main objective is to defend the system against traditional threats like hacking and man-in-the-middle threats.

In [5], the authors have suggested a Smart Mirror enabling which provides life-logging function for the user that stores the portrait of the user's face over a period of time allowing him/her to notice changes in their face over the duration of that period. The main objective behind this proposed idea is to create a life-logging device that does not require the user to give instructions to the mirror for capturing the portrait every time. In [7], the authors have discussed the use of modern technology in the health sector. They propose the use of Remote Health Monitoring Systems (RHMS) which can keep track of various health parameters of people to improve the health sector. RHMS can identify a person's deteriorating health by studying the user's health parameters and can instantly inform the medical services. They state that their main objective is to save time and health costs by constantly informing the users with their health parameters. In [8], the authors discuss detecting schizophrenia in users by their facial expressions. By comparing the facial expression of the user with the facial expressions of the schizophrenic patients, the users can be alerted to get treatment. The main objective is to save the user's time by helping him/her with proper diagnosis without constantly having to travel to medical centers for medical examination. In [9], the authors have proposed a Smart Mirror that incorporates 3D face reconstruction, IR-based face tracking, and OpenGL material extensive rendering approach to deliver the augmented made-up face. The mirror can tell the user how a certain type of makeup will look on him/her without the user actually having to try the makeup. The aim of the authors is to save money by not wasting any makeup products and also save user's time.

3 Smart Mirror Design

Most designs of the Smart Mirror have been similar to that of a regular mirror as shown in Fig. 1, rectangular in shape mounted along with a display and a decorative frame for external good looks. The hardware design does not consume much time for it's construction as regular mirror frames can also be used and it requires minimal expertise in carpentry.

Fig. 1 Standard two-way mirror to see the reflection which has a display mounted behind it

3.1 Hardware Design

Majorly, the Smart Mirrors are designed by using a Raspberry Pi board, which is connected to a monitor, and the monitor is mounted behind the mirror. Different types of hardware systems are used in constructing the mirror, use of infrared sensors with an infrared sensor box along with camera, microphone and relay are seen in modelling the smart mirror. Smart Mirrors also have a touch control module [2], which has an automatic wake-up module due to the infrared sensor, where when the user approaches the mirror, the smart features are available. We have found a varying use of different monitors of different sizes like a high-brightness 4 K LCD built-in behind the half-mirror surrounded by the wooden frame, as shown in Fig. 2, also has a Kinect v2 RGBD camera on top of frame [5]. A HUD Mirror has also been designed where a two-way mirror is used to allow the LEDs mounted behind to illuminate the given information.

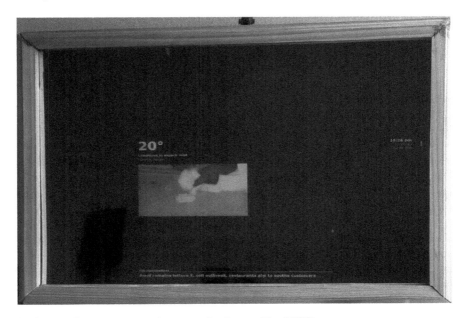

Fig. 2 4K LCD screen mounted on a wooden frame with a RGBD camera on top

3.2 Software Design

As discussed in the related work section, the proposed system includes discrete functionalities. The software design mainly focuses on the working structure of the following functionalities:

- Portrait log
- Gesture and speech UI
- Life rhythm visualization
- Touch control module
- Automatic wake-up Module
- User registration and authentication
- Emotion Detection

The portrait logging function records the user's face to compare and notice changes. The recorded images can be accessed by the user using hand gestures. By raising the left hand, the mirror will display older images or the right hand for newer images. Face detection based on OpenCV library is performed to display faces. The mirror also displays a historic representation of the portraits in a sequential manner. Body composition sensor is integrated with the smart mirror for visualizing various health-related indices such as body weight or body fat percentage.

Touch control module conforms infrared frame with Raspberry Pi to create an infrared network inside the infrared frame for basic touch interaction with the Smart Mirror. This method does not require high mirror configuration.

Automatic wake-up module---Infrared sensors incorporated with the smart mirror reduces the power consumption. The smart mirror features are active when the user approaches the mirror.

User registration (face detection and prediction)---This module captures images of the user. These images are processed using OpenCV cascade classifier to identify the user's face. A cascading classifier is a pre-trained model with positive and negative examples. If a face is identified in an image frame, the frame is saved. This step is repeated until 20 pictures of the user are saved and assigned a unique integer and image count. All the images are loaded in an array and passed into a recognizer for training.

User authentication (detection and prediction)---It starts with a recognizer object. The live images are passed to the recognizer and the prediction method is called. This method analyzes the image with the model file (face database) and returns a unique id of the user whose facial features are closest to the current image. The prediction model also returns an integer value which represents how close the image is to the identified user [4].

Emotion detection---Emotions expressed by the user can be classified into six classes: neutral, happy, sad, angry, surprise, and disgust. A pre-trained Haar classifier is used in order to classify the emotions expressed by the user. There is an exact ratio between the position of different aspects of the person's face when they express a certain type of emotion. A haar cascade is a collection of files which contain such ratios for a vast number of faces [3]. A study was carried out to show how certain types of music can elevate the person's emotion. The following Fig. 3, shows music and emotion correlation.

The above graph represents the correlation between the emotion of the people and the music that they are listening to the most at that time. The *x*-axis describes the number of people feeling a particular emotion, while the *y*-axis plots the comparison

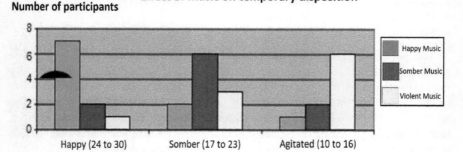

Fig. 3 Effect of music on emotions

among the type of music that they enjoyed the most at that given moment. A database is used to keep a list of songs to be played. Haar classifier detects the emotion of the person and accordingly, the most suitable song is selected from the database which is played using Pafy library.

4 Discussions

Global Home Automation Market size was esteemed at $39,607 million in the year 2016, and is anticipated to reach at $81,645 million by 2023, developing at a compound annual growth rate (CAGR) of 11.2% from 2017 to 2023, and with this evolving rate of home automation industry, more and more people have been inclined toward developing this technology. Smart Mirrors can be seen as a new optimistic lead in this area, where many researchers have tried to implement different software functionalities where one can life log and can be assured of the security of the technology with multi-factor authentication which is essential when one tries to log personal information. Smart mirrors have been implemented with gesture, speech recognition, and facial recognition for ease of access for a user. Various applications of the Smart Mirror are possible due to the ease of incorporating many unique modules.

5 Conclusion

The idea of Smart Mirror has been a great addition to home automation. The market around these technologies can now be expanded with the simplest of ideas and turning them into masterpieces. Surveying different Smart Mirrors throughout the market has made the process of design of hardware very easy. Software design, on the other hand, can be manipulated with skills of your own with regards to the idea one carries to implement Smart Mirror. Smart Mirrors have been implemented in various sectors of society and now can be expanded to many more. This technology can help Do It Yourself (DIY) developers to accomplish innovative ideas, as they are cost friendly and require a certain amount of skill set.

Smart Mirrors have evolved from setting out calendar events and wake up screens to life logging and 3D make-ups, and this might just be the beginning to a whole new front of home automation.

References

1. Chen, J., Koken, M.: SmartMirror: A Glance into the Future. Santa Clara University, Computer Engineering Senior Theses (2017)
2. Jin, K., Deng, X., Huang, Z., Chen, S.: Design of the smart mirror based on raspberry Pi. In: 2018 2nd IEEE Advanced Information Management, Communicates, Electronic and Automation Control Conference (IMCEC 2018)
3. Iyer, S.R., Basu, S., Yadav, S., Vijayanand, V.M., Badrinath, K.: Reasonably intelligent mirror. In 3rd IEEE International Conference on Computational Systems and Information Technology for Sustainable Solutions 2018
4. Njaka, A.C., Li, N., Li, L.: Voice Controlled Smart Mirror with MultiFactor Authentication. In: 2018 IEEE
5. Tani, M., Umezu, N.: Development of a smartmirror with life log functions and its evaluation. In: 2018 IEEE 7th Global Conference on Consumer Electronics (GCCE 2018)
6. Sungjin, P., CLO Virtual Fashion Inc.: 10 Minute 3D Clothing: CLO Virtual Fashion Inc. Exhibits a Speed Fashion Technology at SIGGRAPH 2012. BusinessWire.com (03 Aug 2012)
7. Baig, M.M., Gholamhosseini, H.: Smart Health Monitoring Systems: An Overview of Design and Modeling. Springer Science, Business Media New York (2013)
8. Kohler, C.G., Bilker, W., Hagendoorn, M., Gur, R.E., Gur, R.C.: Emotion Recognition Deficit in Schizophrenia: Emotion Recognition Deficit in Schizophrenia: Association with Symptomatology and Cognition. 2000 Society of Biological Psychiatry
9. Mahfujur Rahman, A.S.M., Tran, T.T., Hossain, S.A., El Saddik, A.: Augmented rendering of makeup features in a smart interactive mirror system for decision support in cosmetic products selection. In: 2010 14th IEEE/ACM Symposium on Distributed Simulation and Real-Time Applications

Unidirectional Ensemble Recognition and Translation of Phrasal Sign Language from ASL to ISL

Anish Sujanani, Shashidhar Pai, Aniket Udaykumar, Vivith Bharath, and V. R. Badri Prasad

Abstract Several countries around the world enable their hearing and speech impaired citizens to streamline communication among themselves and with the able through standardization of sign language. This standardization is typically present in two dialects; alphabetical and phrasal. Alphabetical sign recognition has been extensively researched, however, the phrasal variant, lesser so. We propose a novel approach to recognizing phrasal American Sign Language sequences through an ensemble of the SIFT feature recognition framework with a custom probabilistic prediction measurement and a convolutional neural network, preprocessed through optical flow. We further work on generating the intermediate English language representation and its translation to corresponding sequences in the Indian Sign Language.

1 Introduction

Individuals who are hearing-impaired and/or speech-impaired communicate with others, often through sign language, be it alphabetical or phrasal. A number of countries around the world have standardized sign languages for their citizens in terms of vocabulary, grammar, and the actions themselves. Conversations in sign language, like any other, require all participants to understand the semantics and implications of the verbiage for coherence. This poses an obvious challenge to groups among which not all members know the language, and within which members might belong to different geographic regions. Individuals located in the United States of America who are deaf, hard of hearing and/or children of deaf parents (CODA) primarily use

---

A. Sujanani (✉) · A. Udaykumar · V. Bharath · V. R. Badri Prasad
PES University, Bengaluru, Karnataka 560085, India
e-mail: ansujanani@gmail.com

V. R. Badri Prasad
e-mail: badriprasad@pes.edu

S. Pai
PES Institute of Technology (now PES University), Bengaluru, Karnataka 560085, India
e-mail: shashidharpai95@gmail.com

© The Editor(s) (if applicable) and The Author(s), under exclusive license
to Springer Nature Singapore Pte Ltd. 2021
T. Senjyu et al. (eds.), *Information and Communication Technology
for Intelligent Systems*, Smart Innovation, Systems and Technologies 196,
https://doi.org/10.1007/978-981-15-7062-9_24

standardized American Sign Language (ASL), estimated at about 500,000 speakers [1]. Similarly, the Indian subcontinent consists of between 1.8 million and 7 million native Indian Sign Language (ISL) speakers [2]. This divide between region-specific standardization coupled with the fact that sign language, as any language, takes time to learn and converse in, implies a large communication gap across social, educational and professional forums among many others. Most sign languages are split into two dialects, alphabetical and phrasal.

Prior approaches have involved applying skin detectors, greyscaling, and using edge-oriented histogram modelling for alphabetical sign recognition [3]. Irish and American Sign Language recognition has been attempted through principal component analysis coupled with K-nearest neighbor models and convolutional neural networks [4, 5]. Ethiopian Sign Language recognition has been proposed through implementing Gabor filters and PCA for feature extraction coupled with artificial neural networks (ANN) for recognition [6]. Hardware-based approaches have been made that involves wearable technology fitted with flex, inertial, contact sensors, etc. and having software-calculated flex and relaxation metrics for each sign [7, 8]. Though numerous approaches have been made at recognizing alphabetical signs, relatively fewer have been made in recognizing phrasal actions, including the use of fused 3D CNN and recurrent neural networks (RNNs) [9] and CNN-LSTM fusion networks for Chinese Sign Language Recognition [10]. While most of the above approaches propose, with quite high accuracy, a recognition system based on static images of the alphabetical dialect, we propose an approach combining concepts of traditional computer vision and machine learning to recognize phrasal video sequences of ASL gestures, provide an intermediate representation in English, and further translate it to its corresponding semantics in ISL. We approached the problem by training, testing, and optimizing multiple models on the American Sign Language Lexicon Video Dataset (ASLLVD), Corpus [11–13], and custom-captured videos.

2 Feature Detection and Feature Matching

See Fig. 1.

2.1 Data Augmentation

Due to the fact that languages cannot have complete one to one correspondence in their vocabulary, we selected 15 commonly used English words in their ASL and ISL representations. Each video has a frame rate of 60FPS and is split in a 1:8 frame ratio; one frame is extracted for every eight frames. This set is manually iterated, and 'critical frames' are extracted. Critical frames refer to those frames of the video that contain the actions that are representative of the sign in the video and where boundaries especially those of the arms are not blurred or ambiguous. These

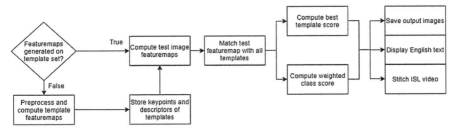

Fig. 1 Process flow for the feature detection and matching model

are the images that we intend to extract characteristic features from. We crop each critical frame manually to retain a minimal representation of the hands. We now apply a sharpening filter, (sigma: 0.5, radius: 1) to each critical frame while taking care to keep the image natural and not overly sharpened. Since the dataset is quite limited in terms of true source and even more so in terms of common vocabulary, we extend the dataset by capturing images of individuals executing the same signs in various environments with different lighting, camera distance, clothing, object presence, and sign nuance variation. The number of images corresponding to a class is proportional to their statistical frequency in the English language. This implies that relatively common words are more likely to be predicted over those that are less common. The same processes of cropping and sharpening are performed on the extended set. A concern that might arise is the difference in parameters of these images. After preprocessing, the controlled set templates contain variable resolution images, while the extended set contain 1920 × 1080 resolution images; however, this turns out to be beneficial to us, as it makes our model more versatile. All images are then subjected to contrast shifts and mirror imaging to generate more templates to be matched with. The dataset spanned a total of over 3850 frames that were split in a train-test ratio of 4:1 (Fig. 2).

Fig. 2 Processed images for FDFM models. Left to right: blue, airplane, house, again

2.2 Feature Detection

Scale-invariant feature transform [14] is a highly successful approach to detecting localized features in images. This algorithm unaffected by scale variance is its biggest advantage. Objects whose features have been detected in a specific configuration will be recognized in future instances irrespective of their translation, scaling, or rotation due to the nature of feature description. We recognize features through localized contrast shifts, as defined by edges, corners, sharp color gradients, and shadows. The same features may not be detected at a different scale of the image when enlarged or compressed, and therefore, we first build a blob detector through Laplacian of Gaussian operations on this image with different σ values, which acts as a scaling factor. For efficient computation, we approximate the Laplacian of Gaussian computation by taking the difference of Gaussian blurring in two images, having different σ values. For every difference of Gaussian map, we search for local extrema over scale and space through an eight-neighbor space of each pixel, across all spaces generated. A pixel p that turns out to be the local extrema of a section s in a scale space S is said to be a potential keypoint. To make these keypoints rotation independent, gradient magnitude and direction are calculated, and the same keypoint is generated in different directions. Descriptors are now created by taking a 16×16 neighborhood around the keypoint and calculating its feature vector.

2.3 Unweighted and Weighted Feature Matching

Accuracy of this model is measured in two ways. We first consider the best matched template to be the prediction and increment a best-score based on correctness. We also consider the sum of the distances produced by the top eight feature matches between the test image and all classes and divide each class-sum by the number of templates within that class. This provides us with a weighted scoring system in which more common words should have more instances, giving them a higher score, making their prediction more probable. In the case of video inputs, we repeat the above procedure while incrementing global class counters depending upon the values of the current-frame class score counters.

3 Convolutional Neural Network

This class of neural networks deals with applying convolution through filters on input recognition followed by pooling—extraction of the detected features by reducing dimensionality. The output of the last pooling layer passes as input into a flattening layer, converting the 2D representation of an image into a 1D vector. This vector then serves as input into a multi-layer perceptron network, which uses forward propagation and activation functions to arrive at the output layer. Error during training is calculated based on the difference between current and the expected output, followed by backpropagating the error through the network. Stochastic Gradient Descent is used to find the direction of local minima on the error curve while the magnitude of change is a function of this direction along with the learning rate and input value to the neuron. Error is propagated through the convolutional block by changing the values of each convolutional mask toward the direction of the local minima. The filters are initally set to random values and 'learn' the appropriate coefficients to recognize features for the data it has seen so far.

3.1 Data Preprocessing Through Optical Flow

We approach video input through compression rather than frame extraction. Motion vectors of objects throughout the video sequence are captured into a single image

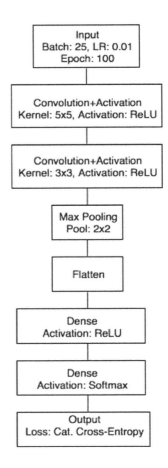

Fig. 3 Network architecture

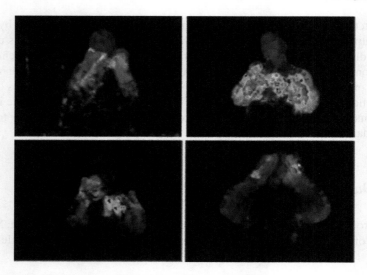

Fig. 4 Processed images for convolutional neural network (CNN)

through the dense optical flow algorithm [15]. This method is advantageous as it enables the model to process only the required information about the gestures being made in the video, irrespective of camera and subject parameters. We calculate the magnitude and direction of vectors formed by moving pixels, tracked by their intensity values between every consecutive pair of frames. HSV space color coding is performed for each vector, proportional to direction and magnitude of that vector. All videos from the controlled set and the extended set go through this process, resulting in over 7500 images. The set thus generated is split into training, validation, and test data in the ratio 14:3:3 (Figs. 3 and 4).

3.2 Network Architecture and Hyperparameters

4 ISL Video Compilation

After intermediate English language representation is obtained, either through the FDFM framework, or the CNN or their hybrid result, a series of corresponding ISL video sequences is compiled together and played.

5 Experimental Results

Accuracy for the SIFT based feature detection and matching approach is measured primarily in two ways. The first being the predicted label versus expected label as dictated by the single best matched template. The second metric is based on the top three classes predicted by the weighted framework. Hybrid template accuracy refers to the union of the best template and weighted class predictions. Ensemble accuracy refers to the percentage of test samples that at least two of the single best FDFM, weighted FDFM, and CNN models converged on, predicting the same result (Fig. 5; Tables 1, 2 and 3).

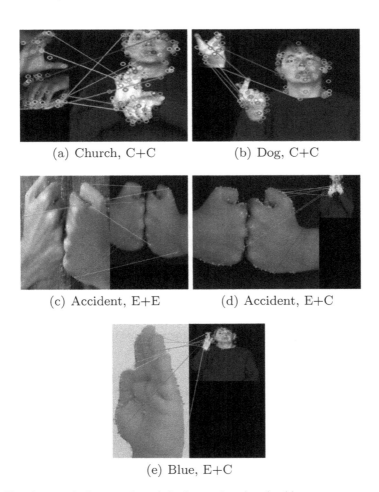

(a) Church, C+C   (b) Dog, C+C

(c) Accident, E+E   (d) Accident, E+C

(e) Blue, E+C

Fig. 5 Template-matched outputs through the feature detection algorithm

Table 1 Feature detection and feature matching (SIFT) accuracy

	Single best template (%)	Hybrid templates (%)
Controlled dataset	92.10	74.28
Extended dataset	80.60	61.54

Table 2 Convolutional neural network (CNN) accuracy

	Train	Validation	Test
Controlled dataset	71.14%	67.46%	72.62%

Table 3 Ensemble accuracy

	Majority (Single best, weighted template, CNN)
Controlled dataset	81.14%

6 Conclusion

Through the above system, we propose a novel method for recognition of ASL by aggregating responses from traditional SIFT based image processing feature extraction and optical flow-based CNN predictions. A final result may be interpreted through the combined confidence in coincidental responses from both systems. We hope to further augment the system, be it in terms of the dataset or the models in an attempt to further bridge the gap that currently exists between the able and the impaired. Future work will include exploration of Natural Language Processing (NLP) treatment of the output from the stated systems in order to analyze the intermediate English representation seen so far and provide more accurate current-frame recognition. In terms of data preprocessing, optical flow of the feature detection and matching preprocessing can be explored and might provide better recognition by both systems. Approaches using LSTMs and 3D convolutional networks have shown immense promise in recognition of features due to their capability of spatial and temporal information processing.

Acknowledgements A special extension of gratitude to all the individuals who contributed to the extended dataset during this study. No copyrighted images have been used in this paper.

References

1. Mitchell, R., Young, T., Bachleda, B., Karchmer, M.: How many people use asl in the United States? Why estimates need updating. Sign Lang. Stud. **6**, 03 (2006)
2. Jaiswal, N.: With a deaf community of millions, hearing india is only just beginning to sign. 01 2017. https://www.pri.org/stories/2017-01-04/deaf-community-millions-hearing-india-only-just-beginning-sign

3. Pansare, J.R., Ingle, M.: Vision-based approach for american sign language recognition using edge orientation histogram. In: 2016 International Conference on Image, Vision and Computing (ICIVC), pp. 86–90, Aug 2016
4. Oliveira, M., Chatbri, H., Little, S., Ferstl, Y., O'Connor, N.E., Sutherland, A: Irish sign language recognition using principal component analysis and convolutional neural networks. In: 2017 International Conference on Digital Image Computing: Techniques and Applications (DICTA), pp. 1–8, Nov 2017
5. Jiang, D., Li, G., Sun, Y., Kong, J., Tao, B: Gesture recognition based on skeletonization algorithm and CNN with ASL database. Multimedia Tools Appl. 10 (2018)
6. Admasu, Y., Raimond, K.: Ethiopian sign language recognition using artificial neural network, pp. 995–1000 (2011)
7. Elmahgiubi, M., Ennajar, M., Drawil, N., Elbuni, M.S.: Sign language translator and gesture recognition. In: 2015 Global Summit on Computer Information Technology (GSCIT), pp. 1–6, June 2015
8. Ahmed, S.S., Gokul, H., Suresh, P., Vijayaraghavan, V.: Low-cost wearable gesture recognition system with minimal user calibration for ASL. In: 2019 International Conference on Internet of Things (iThings) and IEEE Green Computing and Communications (GreenCom) and IEEE Cyber, Physical and Social Computing (CPSCom) and IEEE Smart Data (SmartData), pp. 1080–1087, July 2019
9. Ye, Y., Tian, Y., Huenerfauth, M., Liu, J.: Recognizing American sign language gestures from within continuous videos. In: 2018 IEEE/CVF Conference on Computer Vision and Pattern Recognition Workshops (CVPRW), p. 2145, June 2018
10. Yang, S., Zhu, Q.: Continuous Chinese sign language recognition with CNN-LSTM, p 104200F (2017)
11. Neidle, C., Thangali, A., Sclaroff, S.: Challenges in the development of the American sign language lexicon video dataset (ASLLVD) corpus (2012)
12. Neidle, C., Vogler, C.: A new web interface to facilitate access to corpora: development of the ASLLRP data access interface (DAI) (2012)
13. American Sign Language Lexicon Video Dataset (ASLLVD). http://secrets.rutgers.edu/dai/queryPages/search/search.php
14. Lowe, D.: Distinctive image features from scale-invariant keypoints. Int. J. Comput. Vis. **60**:91 (2004)
15. Farnebäck, G.: Two-frame motion estimation based on polynomial expansion. In: Proceedings of the 13th Scandinavian Conference on Image Analysis. LNCS 2749, Gothenburg, Sweden, pp. 363–370, June–July 2003

Unravelling SAT: Discussion on the Suitability and Implementation of Graph Convolutional Networks for Solving SAT

Hemali Angne, Aditya Atkari, Nishant Dhargalkar, and Dilip Kale

Abstract NP-complete class of problems have a wide range of applications in the real world. With the purpose of finding new ways to solve these problems, this study looks to use SAT as a representative of the NP-complete class. It looks to assimilate SAT with the intention of solving it by way of modeling its instances using graph neural networks, particularly a graph convolutional network. Modeling SAT formulae as graph representations is motivated by the fact that properties of the SAT equations can be aptly depicted by visualizing clauses, literals, and the corresponding relationships between them as a graph.

1 Introduction

This study is an endeavor to understand the issues and various factors that have to be considered to use graph neural network (GNN) architectures to solve the satisfiability problem (SAT). SAT is a part of the NP-complete class of problems that have real-life manifestations in a wide range of realms. Exact solutions to these problems are extremely expensive to compute. Hence, approximate algorithms that use heuristics have been employed to solve them. The heuristic approach demands the study of the properties of these problems along with careful consideration of new approaches to solving them. The NP-complete class of problems are all polynomial-time reducible to each other which implies that solving one of them by an efficient method results

---

H. Angne (✉) · A. Atkari · N. Dhargalkar · D. Kale
MCT's Rajiv Gandhi Institute of Technology affiliated to Mumbai University, Andheri West, Mumbai, Maharashtra 400053, India
e-mail: angne.hemali@gmail.com

A. Atkari
e-mail: adityaatkari@gmail.com

N. Dhargalkar
e-mail: nsdhargalkar@gmail.com

D. Kale
e-mail: dilip.kale@mctrgit.ac.in

© The Editor(s) (if applicable) and The Author(s), under exclusive license to Springer Nature Singapore Pte Ltd. 2021
T. Senjyu et al. (eds.), *Information and Communication Technology for Intelligent Systems*, Smart Innovation, Systems and Technologies 196, https://doi.org/10.1007/978-981-15-7062-9_25

in generating an efficient solution for all problems in the class. Hence, in this study, just as in Atkari et al. [1], we focus on solving SAT, with an implication of solving the entire NP-complete class.

Graph neural network models are a type of artificial neural network that work on graphical inputs and can be interpreted to have a neural network on each of the edges of its input graph. It has been found that the graph representations produced from SAT instances exhibit properties of the instances within their graphical structures aptly (Nudelman et al. [2]), owing to which an approach of employing graph neural networks to model SAT instances is appropriate. Furthermore, Bunz et al. [3] and especially Selsam et al. [4] talk about how not only graph neural networks are instrumental in uncovering the theoretical properties of SAT and solving it, but modeling these instances can also give insights into explaining how neural networks make decisions.

This study encompasses getting familiar with the different graph representations of SAT, making use of Uniform Random 3-SAT (Hoos et al. [5]) datasets, and looks to then figure out an approach to store these graphs, a way to implement a graph convolutional network, eventually train this model on said graphical instances and analyze the results.

2 Literature Review

Being an extrapolation of the findings of Atkari et al. [1], this study seeks to improve on the underwhelming f1-scores obtained using statistical machine learning models in the former. Furthermore, with the intention of using graph neural network models, this study will directly make use of graph representations discussed in Atkari et al. [1] which have been used to derive numerical features as a part of feature engineering done for the machine learning models. Delvin et al. [6] likewise makes use of statistical machine learning models although models divergent datasets. The SATLIB Uniform Random 3-SAT datasets will be used in this study as described in Hoos et al. [5].

Among other properties of SAT, Nudelman et al. [2] discuss three graph representations that each captures a set of characteristics of SAT instances effectively, namely the variable graph, the clause graph, and the variable-clause bipartite graph. The degree measure of the nodes encompassing these graph representations have been widely used to derive a subset of numerical features that are then put to use for modeling SAT instances, notably in Xu et al. [7]; a SAT solver chooses from a portfolio of solvers that are each dominant on different instance types making use of empirical hardness.

Bunz et al. [3] make the direct use of the variable graph representation from Nudelman et al. [2], referred to as variable-variable graph representation, to model instances using GNN. It further discusses SAT properties and how rather simple graphs are in fact representative of said properties. It points out notably how not only

GNNs will give a new insight on the theoretical properties of SAT but also how SAT in turn could help explain the properties of neural learners.

Selsam et al. [4] make use of a graph representation that performs a round of message passing in two stages. Literals and clauses are visualized as nodes, and in the first stage, literals pass messages to the clauses they participate in. In the second stage, clauses pass messages to the literals that participate in each of them; at the same time, each literal passes a message to its conjugate. This trained NeuroSAT model not only is able to classify instances as being SAT or UNSAT but also finds a satisfying assignment to the SAT instance. Message passing is used in a GNN to consolidate a node's neighborhood (that is, its adjacent nodes, the nodes adjacent to said nodes, and so on) information within of the node.

Selsam et al. [4] give insight into the implications of NeuroSAT being able to encode a satisfying assignment of variables for a satisfiable instance. It does this by reasoning as to how the network went about getting to its final configuration. This assignment of variables is found by unsupervised clustering of nodes in their final configuration into two clusters. NeuroSAT is speculated to reason in the following way: the best way that it can tell an instance is SAT is if it is able to find a satisfying assignment for it in the first place.

The architecture of the GNN, the message passing algorithm, and the various GNN implementations have been discussed in Zhou et al. [8]. In this study, the graph convolutional network has been employed, which, as Zhou et al. [8] states, does not support inductive learning. Owing to this, we had to batch the graphs of each individual instance together before feeding them to the graph convolutional network using the deep graph library (Wang et al. [9]).

Defferrard et al. [10] and Kipf et al. [11] discuss the architecture and working of graph convolutional networks, which this study implements to model the SATLIB instances.

3 Graph Representations of SAT

Graphs capture the characteristics of SAT instances to effectively help compute how probable it is for a given instance to be satisfiable. The following are some of the representations identified.

3.1 Graph Representation

Variable Graph: The variable graph consists of variables, one variable for a pair of positive and negative literals, within an instance considered as graph nodes. In this representation, an edge occurs between variable nodes if they occur in the same clause together in the instance at least once.

Variable-Clause Bipartite Graph: The variable-clause bipartite graph considers all variables and clauses of an instance as nodes, where variable and clause nodes are distinctly identified. No variable nodes are connected to each other, likewise for the clause nodes. Edges only occur between a variable-clause pair, and hence, it is bipartite in nature. A variable node is connected to a clause node if the literal corresponding to the variable occurs in that clause within an instance.

Clause Graph: The clause graph considers clauses within an instance as nodes. If a clause contains a literal and another clause contains the conjugate of said literal, then these two clauses are connected to each other by an edge.

NeuroSAT Representation: As discussed above, NeuroSAT considers the literals as well as clauses from an instance as nodes. It takes a two-stage approach of message passing, wherein the first set of directed edges are from literals to the clauses they participate in, and the second set of directed edges are from the clauses to the literals that participate in them as well as from a literal to its conjugate literal.

4 Dataset and Processing

The SATLIB Uniform Random 3-SAT datasets have been used to train and test the graph convolutional network. These have been described in Hoos et al. [5], having a varying number of variables and clauses described as cnf files that are preprocessed to make lists in Python. Each list for an instance contains a list for each clause within which are included the literals that participate in the clause represented as integers. A negative literal is represented as a negative integer.

The study as of now makes use of two graph representations, namely the variable graph and the variable-clause bipartite graph. Moving forward the NeuroSAT representation and the clause graph will also be put to use. Initially, two-dimensional lists were used as the data structures to store instance graphs which eventually changed to implementing the graphs using the NetworkX library from Hagberg et al. [12]. This library has tools to represent, carry out operations on, and visualize complex networks from various different realms.

Figure 1 shows a variable graph with nodes labelled as integers, and Fig. 2 shows a variable-clause bipartite graph with variable nodes labelled starting with a 'v' and clause nodes with a 'c' followed by integers. Figure 3 shows a section of the variable-clause bipartite graph of the same SAT instance. Both these graphs are built for a single instance of a boolean expression with 20 variables and 91 clauses using NetworkX functions.

As discussed previously, graph convolutional networks do not support transfer or inductive learning, owing to which the individual graphs of each of the instances are first combined together using the deep graph library's batch method. This then allows the entire stack of instance graphs to be given to the graph neural network to be trained and tested on.

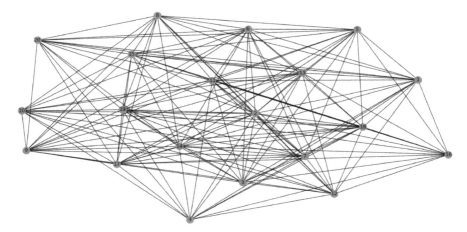

Fig. 1 Variable graph of a SAT instance

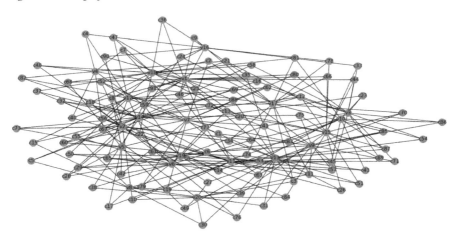

Fig. 2 Variable-clause bipartite graph of a SAT instance

5 Graph Neural Network Implementation

Various graph neural networks (GNNs) are built using graph convolutional networks (GCNs). GCNs are a specific implementation of graph neural networks. A graph neural network performs convolution operation on a graph, instead of on an image composed of pixels. Just like CNN which aims to extract the most important information from the image to classify the image, a GCN passes a filter over the graph, looking for essential vertices and edges that can help classify nodes within the graph.

After multiple efforts with the tf-gnn models, we came across many issues and resorted to moving to PyTorch implementations instead of the former TensorFlow

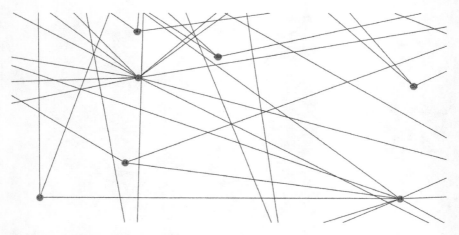

Fig. 3 Zoomed-in variable-clause bipartite graph of the same SAT instance

implementations. As of now, PyTorch provided some of the best graph convolutional network models that were used by us. Primarily, the deep graph library built on top of PyTorch was used by us to model the various graph representations and for implementing the GCN models.

In practice, we can use DGL implementation that makes efficient use of GPU for various calculations. In the DGL batch process, graphs of variable sizes can be an input to the network. In this method, a bunch of graphs is processed at once during the training phase for classification. The initial node features of the graphs are the node degrees. After a certain number of rounds of graph convolution, a graph readout is performed by averaging overall features for each graph in the batch.

For the various graph representations, we used NetworkX [12], a Python language package for exploration and analysis of networks and network algorithms. Although we faced problems in integrating the NetworkX graphs with the DGL models, we modelled DGL representations of the graphs using the graph convolutional network. This was also owing to the fact that the DGL graphs had the functionality of being batched together to perform the training operation as discussed in the previous sections.

6 Conclusions

With evidence from the studies reviewed, it is certain that instances of SAT represented as graphs do exhibit characteristics of SAT and that graph neural network architectures can be used to model them.

In this study, we managed to discuss four different graph representations with respect to the Uniform Random 3-SAT datasets from the SATLIB repository. From these, the variable graph and the variable-clause bipartite graph representations have been visualized using the NetworkX library. The GCN was implemented using the deep graph library which is built on top of PyTorch.

The data structure obtained using the NetworkX library was not used as an input to the GCN model because all the instance graphs could not be batched together. The DGL batch process was successfully used for this purpose instead. Such a batching is required owing to the fact that the GCN model does not support inductive learning. This implemented GCN model has further been trained on the batched data.

7 Future Scope

The framework which was described for classification of SAT instances will be deployed for analysis. The efficiency of the model will be measured on various metrics like the f1 score, precision, and recall.

The implications of the graph convolution network correctly classifying the SAT instances are tremendous as the value of each node in the network can be used to predict the satisfiability of each clause within the instance. A comparative study will be performed to compare this approach with other machine learning, deep learning models, and existing approximate algorithms. Currently, we hypothesize these comparisons to be based on the time and space complexities of these algorithms. This will also be subject to the understanding of such complexities with respect to machine learning and deep learning models.

This study will look to uncover some theoretical properties of satisfiable and unsatisfiable instances that owe themselves to the satisfiability problem.

The process of using graph convolutional network, alone, to classify the SAT instances is described in this paper. Likewise, other different forms of graph neural networks can be employed similarly for classification.

References

1. Atkari, A., Dhargalkar, N., Angne, H.: Employing machine learning models to solve uniform random 3-SAT. In: Jain, L., Tsihrintzis, G., Balas, V., Sharma, D. (eds) Data Communication and Networks. Advances in Intelligent Systems and Computing, vol. 1049. Springer, Singapore (2020)
2. Eugene, N., Devkar, A., Shoham, Y., Leyton-Brown, K.: Understanding random SAT: beyond the clauses-to-variables ratio. In: IEEE Transactions on 27th International Conference on Advanced Information Networking and Applications Workshops (2013)
3. Bunz, B., Lamm, M., Merlo, A.: Graph neural networks and Boolean satisfiability. In: Proceedings of the 28th Annual ACM Symposium on Applied Computing, pp. 1852–1858. ACM New York, NY, USA, March 2013
4. Daniel, S., Lamm, M., Bunz, B., Liang, P. Dill, D.L., de Moura, L.: Learning a SAT solver from single-bit supervision. In: Proceeding of the International Conference on BioMedical Computing (2012)

5. Hoos, H.H., Stützle, T., Gent, I.P., van Maaren, H.: Toby Walsh. An Online Resource for Research on SAT. Kluwer Academic Publishers, SATLIB (2000)
6. Devlin, D., O'Sullivan, B.: Satisfiability as a classification problem. Proc. J. Global Res. Comput. Sci. **4**(4), (2013)
7. Xu, L., Hutter, F., Hoos, H.H., Leyton-Brown, K.: SATzilla-07: the design and analysis of an algorithm portfolio for SAT. In: Bessière, C. (eds.) Principles and Practice of Constraint Programming—CP: CP 2007. Lecture Notes in Computer Science, vol. 4741. Springer, Berlin (2007)
8. Zhou, J., Cui, G., Zhang, Z., Yang, C., Liu, Z., Sun, M.: Graph neural networks: a review of methods and applications. Computer Science. ArXiv, Mathematics (2018)
9. Wang, M., Yu, L., Zheng, D., Gan, Q., Gai, Y., Ye, Z., Li, M., Zhou, J., Huang, Q., Ma, C., Huang, Z., Guo, Q., Zhang, H., Lin, H., Zhao, J., Li, J., Smola, A.J., Zhang, Z.: Deep graph library: towards efficient and scalable deep learning on graphs. In: ICLR Workshop on Representation Learning on Graphs and Manifolds (2019)
10. Defferrard, M., Bresson, X., Vandergheynst, P.: Convolutional neural networks on graphs with fast localized spectral filtering. In: Proceedings of the 30th International Conference on Neural Information Processing Systems (NIPS'16), pp. 3844–3852. Curran Associates Inc., Red Hook, NY, USA (2016)
11. Kipf, T., Welling, M.: Semi-supervised classification with graph convolutional networks. ICLR, 2017 (2016)
12. Hagberg, A.A., Schult, D.A., Swart, P.J.: Exploring network structure, dynamics, and function using NetworkX. In: Varoquaux, G., Vaught, T., Millman, J. (eds.) Proceedings of the 7th Python in Science Conference (SciPy2008), Pasadena, CA USA, pp. 11–15, Aug 2008

CES: Design and Implementation of College Exam System

Punya Mathew, Rasesh Tongia, Kavish Mehta, and Vaibhav Jain

Abstract The caliber of a student is reflected by the results he obtains in his institution's examinations. Properly conducting exams as well as maintaining results via a system thus is mandatory. The performance of a result system is measured by the efficiency it works with. Owing to the large magnitude and complexities of the tasks involved, an automated and easy to use result system is required in every educational establishment. A manual method of handling all the associated tasks won't suffice because of the high chances of errors, complexities and computations involved. The proposed solution is a web application developed via JSP Servlets and MySQL database that handles all the result related activities right from the enrollment of the student to his result generation. As per the need of the institution, the grades and the semester grade point averages (SGPA's) are displayed in a class-wise manner. The application also provides interfaces to maintain and modify schema, allocate faculties and submitting marks to the respective users.

1 Introduction

Result System is an integral part of any educational institute be it a school or a university. It requires maintaining parameters like confidentiality, accuracy, transparency and efficiency while generating results. Any error or delay in its processing may

P. Mathew (✉)
Accenture, Pune, India
e-mail: punyamathew@gmail.com

R. Tongia
Indian Institute of Technology Bombay, Mumbai, India
e-mail: vrsnpdr@gmail.com

K. Mehta
Quantiphi Analytics Solution Private Limited, Bengaluru, India
e-mail: mehtakavish1407@gmail.com

V. Jain
Institute of Engineering and Technology, DAVV, Indore, India
e-mail: vjain@ietdavv.edu.in

© The Editor(s) (if applicable) and The Author(s), under exclusive license to Springer Nature Singapore Pte Ltd. 2021
T. Senjyu et al. (eds.), *Information and Communication Technology for Intelligent Systems*, Smart Innovation, Systems and Technologies 196,
https://doi.org/10.1007/978-981-15-7062-9_26

cause troubles in the smooth functioning of students' individual careers. The manual methodologies used in colleges makes it nearly impossible to achieve these parameters to a satisfiable extent, especially when the associated data is large—around thousands of students enrolled in various courses.

Any college which depends on manual paper and excel-sheet based method for various tasks like marks management, attendance management, etc. suffers from various drawbacks like: Human errors, Management Difficulty, Higher chances of records getting lost or damaged, Lack of transparency between faculty and students, Not environment friendly and highly time consuming.

These can be eliminated with help of a software based solution; a point interface which will allow faculties to perform all the above tasks on their PC's via separate logins. The Credit Based Choice System (CBCS) followed in most colleges allows students to select subjects of their choice and thus further emphasizes the need for an automated digital system. It can be broadly divided into (Fig. 1).

The system provides a means to enroll students in different courses like BE/ME/M.Sc. in accordance with the CBCS (Credit Based Choice System) Schema. The class wise lists of students containing their Roll Number, Enrollment Number and Name are uploaded in CSV format and acts as one of the inputs for the system. Faculties can also easily be assigned to their respective classes and subjects, as per their areas of expertise. The attendance and marks of students are recorded to be further used for computing SGPAs (Semester Grade Point Averages) and thereby generating result sheets and grade cards. To meet the desired requirements, we have developed a web application using HTML, CSS, JavaScript, Java Servlets, JSP's and MySQL.

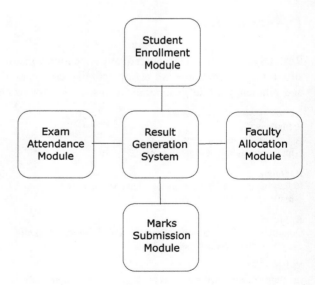

Fig. 1 System architecture of proposed result system

2 Literature Review

Many researches on automated result processing have been made and some of them are reviewed. Beka [1] developed a computerized result processing system that will help them in faster computing and immediate result generation once the students graduate. But this system actually provides means to provide student marks/data individually via searching and not class-wise.

Emmanuel [2] proposed a solution by developing a software Application to ease the processing of the results using PHP (Hypertext processor) scripting language and MYSQL Relational Database Management System in designing the database. Again, this system failed to handle some of the most basic errors (division by zero errors by the operators) and same as [1], didn't have a option of handling students class-wise.

Ukem [3] used Adobe Dreamweaver, PHP and MySQL to develop an application for processing of students results but focused it majorly on reviewing and updating results of individual students throughout their stay at the university. Although this was suitable for result generation but no mechanism existed for allocating lecturers and examiners.

Rajmane [4] proposed a digitized student information management system for the academic activities of an institute. But it handles one student at a time and thus is inefficient in providing (submitting) multiple students records.

Akiwate [5] again proposes a web based student information system. It provides the students details and summarized result in a nice way but doesn't get into the marks entry and examination attendance part. And similar to [3] doesn't handle faculty allocation.

A lot of systems have been designed using various languages and technologies but none of them satisfies the need of present day colleges. Most of the systems work via providing marks and related data student wise, while present day organizations opt class-wise, i.e. a large group of students are considered to be classified as same and their data is handled accordingly. Students are registered separately in most of these systems which is quite a tedious task for any operator. Our system takes advantage of the class-wise distribution of students and does a lot of work by itself. For instance, we will just require the whole class list as input once and for all and the system handles the rest. This not only improves the ease of use but also saves a lot of time and work. A lot of checks and validations could be applied while entering grades and marks which could save a lot of additional work to be done correcting them.

3 Proposed System

As shown in the proposed system workflow in Fig. 2, to tackle the problems arising from manual ways of work, a digitized solution is needed which would fulfill our objectives. For example, allow faculties to edit marks in case of errors, on-line record storage and easy viewing, editing and backup for future reference. As a whole it would

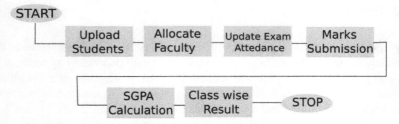

Fig. 2 Proposed system workflow

avoid unnecessary paper work and errors. The complete system is based on client server architecture in which there would be 5 kinds of end users: Faculty, EDP team member, Exam team member, CV team member and Panel Team Member.

- **Upload Students for Classes**: The EDP (Electronic Data Processing) user uploads multiple files on the click of a single button and the students are uploaded according to the schema.
- **Allocate Faculty/Subject**: The Panel Team member will allocate the faculties to their respective subjects, class-wise. The panel team member is responsible for allocating faculties for the subjects of his/her department. Firstly, a list of classes having atleast one subject of the panel member's department is shown, wherein the user can select a particular class and then allocate the faculties to their respective subjects of the class using searchable drop downs.
- **Update Exam Attendance**: The Exam team member can mark the absentees, detained students and UFM (Unfair Means) cases of exams held on a particular date. Exam Attendance Summary can also be viewed here.
- **Marks Submission**: Faculty Allocation and Exam attendance updation is a prerequisite for Marks Submission as only the marks of present students should be allowed to be entered by the respective faculty. The faculty user is provided an interface where he firsts has to choose between the Regular and Ex student option. Then a drop down list of the classes where he teaches is shown from where he can select a particular class for marks to be entered. JavaScript validations (like marks should range between 0 and 20 etc.) are also applied for error detections. Once the faculty is fully satisfied, he can Generate the CSV, which will be further used for Result Generation. Once the Generate CSV button is clicked, faculty cannot make any changes until he is re-granted permission by the CV (Central Evaluation) user.
- **Class Wise Result Generation**: EDP user then generates a class wise result which is then used to generate the report cards.
- **SGPA Calculation**: Once the CSV is generated by the faculties, the system calculates the SGPA for each student as:

$$\text{SGPA} = \frac{\sum C_i * G_i}{\sum C_i}$$

where C_i: Credits Of ith subject, G_i: Grade equivalent score of ith subject.

4 Implementation

4.1 Database Design

The major relations on which the whole design depends are described below:

- **Class Table**: This consists of mapping of individual classes to their respective class ids, department (id's) and course (via the department). The students of the same class thus have the same associated classid. The classes also have an associated year depending upon the course. As multiple sections can exist for a class in a department, a field exists for the same.
- **Schema Table**: It consists of the associations of subjects (subject codes) to the classes (class ids) along with the time of year when they are taught i.e. spring/autumn. As a subject can be taught as an only practical, only theory or both depending on the class; a field for subject type also exists.
- **Allocation Table**: This allocates the faculties to their respective subjects and classes. It also maintains the status of whether the faculty has submitted the grades or not.
- **Result Table**: This is used to store the mid semester tests and end semester exam marks of students for each subject. The fields theory grade, best 2 test marks are computed automatically using triggers.
- **Faculty Table**: The profiles of faculties are maintained here. Also, the faculties are assigned an id which is used throughout the system and in the other relations.Every faculty is assigned an unique id (Fig. 3).

4.2 Tools and Technologies

We have used JSP-Servlets for server side programming for the website as it is simple, open source, interpreted, faster and has a wide support. HTML, CSS, Javascript was used as front end language for website, while Bootstrap was used as front end framework because it is easy to use and is lightweight. The relations have been implemented using MySQL database and LDAP server was used for login authentication to enhance security.

4.3 Results and Discussions

As shown in Fig. 4, the EDP User can view the results of student class-wise. The grades of students in various subjects along with their SGPA's are displayed. As shown in Fig. 5, the Panel User can allocate internal examiners to for the subjects of his department.

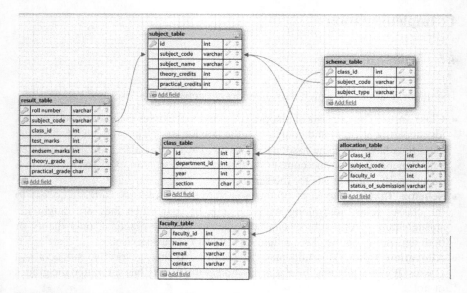

Fig. 3 Database schema of proposed system

Institute of Engineering And Technology,DAVV

4 year Computer Engineering 'B'

Sr No.	Roll Number	Enrollment Number	Student Name	Distributed Computing	Cloud Computing		Artificial Intelligence		Machine Learning		Project Phase - I	Comprehensive Viva - VII	Result	SGPA
				CER7C1	CER7C2		CER7C3		CER7E1		CER7P1	CER7V7		
				4	4	1	4	1	4	1	7	4		
				Th	Th	Pr	Th	Pr	Th	Pr	Pr	Pr		
1	15C7069	DE15255	ABHISHEK CHOUHAN	C	B	B+	F	B+	C	A	A+	B	Fail	5.77
2	15C7101	DE15065	AARUSHI GUPTA	B+	A	B+	B+	A	B+	A+	A+	A+	Pass	7.97
3	15C7103	DE15069	ABHISHEK PATEL	A	A+	B	B	A	B+	A	A+	B	Pass	7.63
4	15C7104	DE15070	ADITI PATHAK	A+	B	B	B	A+	B+	A	A	A+	Pass	7.57
5	15C7105	DE15072	ADITYA BAGDARE	B	A	C	C	B+	C	A	A+	B+	Pass	6.90
6	15C7106	DE15125	MRIDUL AGRAWAL	B	B+	B	P	B+	C	A+	A	A	Pass	6.60
7	15C7107	DE15073	AISHWARYA PATEL	A+	O	A	B+	A	A	B+	O	A	Pass	8.70
8	15C7108	DE15075	AKSHAT SHRIVASTAVA	B+	A	B	C	A	B+	A	A+	B+	Pass	7.37
9	15C7109	DE15077	AMAN BAGHEL	B+	B	C	P	A	P	A	A	B	Pass	6.17
10	15C7110	DE15079	ANKIT CHAUDHARY	A	A+	C	C	A	B	B+	A	B+	Pass	7.20
11	15C7111	DE15083	ANUJ KOLI	A+	A+	A	O	A+	A+	A	A+	A	Pass	8.93
12	15C7113	DE15088	AYUSH GUPTA	A+	A+	A	A+	A+	A	B+	O	A+	Pass	9.00
13	15C7114	DE15095	CHIRAG LATHI	A	A	B	B+	A	B+	A	A+	B+	Pass	7.77
14	15C7116	DE15100	DEEPANSH SRIVASTAVA	A	B	B+	B	A	B+	B	A	B	Pass	6.97
15	15C7117	DE15101	DEVYANI LAMBHATE	A+	O	B+	A	A	A	A	A	A+	Pass	8.50
16	15C7118	DE15102	DIVYANSHI CHATURVEDI	A+	A+	B+	A+	A	A+	A+	A	A	Pass	8.53
17	15C7119	DE15103	GARVIT GUPTA	A	B+	A	C	A	B	A	B+	B+	Pass	6.83
18	15C7120	DE15104	GLADIOLUS VASUNIYA	A+	A+	B	B+	A+	B+	B+	A	B	Pass	7.67

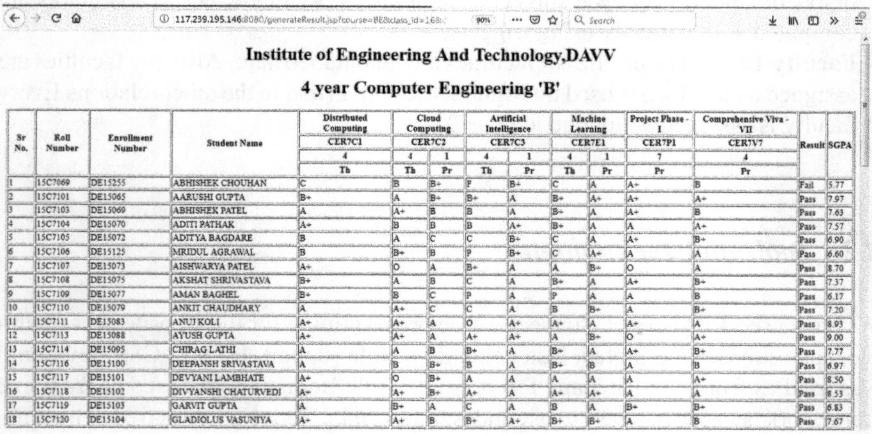

Fig. 4 Classwise result as displayed by proposed system

CES: Design and Implementation of College Exam System

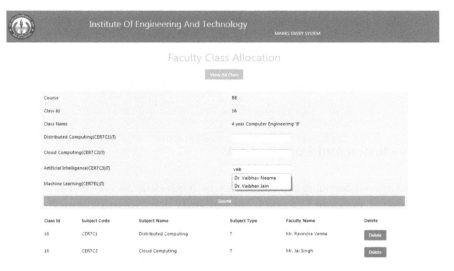

Fig. 5 Interface for allocating internal examiners in proposed system

This developed software was unanimously accepted and is still used by the Exam Department of IET-DAVV College and all the stakeholders have accepted it as well.

4.4 Performance Enhancements

JavaScript Validations present on the interfaces helps preventing erroneous entries from faculties to be submitted. Triggers are used as well to support faster updates and insertions.

5 Challenges Addressed

In our system, we tried to address the following challenges.

- **Acceptability**: For the success of any Software, acceptance from the users is a must. In our case, we were fortunate to have full cooperation from the administration as well as the faculties.
- **Scalability**: Another challenge we faced was handling large amount of data associated with the result generation process. The tools which we have used, efficiently manages the present amount of data being produced and would be able to handle further increase in the amount of data too.

- **Security**: Safeguarding against attacks is an important factor in any system. When it comes to Result Generation, the data associated is highly confidential. The LDAP server provides secured access to the users with login authentication.
- **Digital Compatibility**: The users of the system may not necessarily have the required expertise of using software. Thus it was important to provide user friendly interfaces.

6 Conclusion and Future Work

Maintaining confidentiality and consistency in Result Generation is of at most importance. Manual ways can be helpful, but scalability becomes an issue as the size of the data grows. Online Result Generation systems, such as the one presented in this paper, have a number of positives over manual ways, including scalability, the reduced chances of human error, the ability to generate result in a timely manner, and the ability to back up data for future use.

At present, our system is provides an easy to use interface for entering marks and generating practical exams attendance and grade sheets. It also is able to compute best 2 out of 3 test marks and subsequently total marks and theory grades based on test and exam marks automatically. Allocating internal and external faculties for subjects of both regular and ex students as well as sending emails to the faculties about allotment is done seamlessly. Additionally, the system facilitates the editing of schema for all types of courses offered and generating a yellow statement (A summarized exam attendance). In future, we plan to extend our System to support handling of Alternate Schema which can be used to handle marks of students who go for internships in the final semester. We also aspire to provide an interface for the users, wherein they can provide their system feedback directly to the developers and we wish to incorporate a framework in our system. We are currently working on the prospect of adding a 'smart suggestion' feature for allocating faculties based on their available areas of expertise. We also look to make the software more generalized enough to be adapted by any engineering college in future.

Acknowledgements We would like to thank IET-DAVV Exam department and faculty members for providing valuable inputs and suggestions for developing the proposed system.

References

1. Beka, A.P., Beka, F.T.: Automated result processing system: a case study of Nigerian University. Int. J. Res. Emerg. Sci. Technol. **2** (2015)
2. Emmanuel, B., Choji, D.N.: A software application for colleges of education students results processing. J. Inf. Eng. Appl. **2**(11), 12–18 (2012)
3. Ukem, E.O., Onoyom-Ita, E.O.: A software application for the processing of students results. Global J. Pure Appl. Sci. **17**(4), 487–502 (2011)

4. Rajmane, S.S., Mathpati, S.R., Dawle, J.K.: Digitalization of management system for college and student information. Res. J. Sci. Technol. **8**(4), 179–184 (2016)
5. Akiwate, B., et al.: Web based student information management system using MEAN Stack. Int. J. Adv. Res. Comput. Sci. Software Eng. **6**(5), 357–362 (2016)

ANN-Based Multi-class Malware Detection Scheme for IoT Environment

Vaibhav Nauriyal, Kushagra Mittal, Sumit Pundir, Mohammad Wazid, and D. P. Singh

Abstract Internet of things (IoT) is one of the fastest-growing technologies. With the deployment of massive mobile and faster networks, almost every daily-use item is worked upon to be connected to a network. Such a massive network will exchange gargantuan data every second and give way to security threats. In this paper, we aim to classify the different categories of attack and non-attack data packets which could be used as a standard to apply the redemption techniques and enhance the security features to ensure the data integrity and a foolproof IoT network and propose an ANN-based multi-class malware detection scheme to classify different types of attacks.

1 Introduction

With the omnipresence of Internet of things (IoT) network and the increasing number of devices connected to the network, it becomes vulnerable to the attacks; therefore, it is important to take measures to secure the network and the devices. However, it is difficult to take measures until we know the type of attack packet. The introduction of

---

V. Nauriyal (✉) · K. Mittal · S. Pundir · M. Wazid · D. P. Singh
Department of Computer Science and Engineering, Graphic Era Deemed to be University, Dehradun, Uttrakhand 248002, India
e-mail: vaibhavnaudiyal92@gmail.com

K. Mittal
e-mail: mittalkushagra@hotmail.com

S. Pundir
e-mail: sumitpundir1983@gmail.com

M. Wazid
e-mail: wazidkec2005@gmail.com

D. P. Singh
e-mail: devesh.geu@gmail.com

© The Editor(s) (if applicable) and The Author(s), under exclusive license to Springer Nature Singapore Pte Ltd. 2021
T. Senjyu et al. (eds.), *Information and Communication Technology for Intelligent Systems*, Smart Innovation, Systems and Technologies 196,
https://doi.org/10.1007/978-981-15-7062-9_27

cyber physical systems(CPS) in the Internet of things (IoT) and the monitoring and actions taken based on physical changes encompass the assets of critical importance and have further increased the vulnerability. The deployment of such an enormous system in the smart upgradation projects such as smart cities, smart railway stations, smart toilets and smart roads brings threats along with it. To deal with such a threat in the Internet of things (IoT) system, a number of schemes have been proposed to deal with these attacks and secure the network [1]. We have proposed an ANN-based multi-class malware detection scheme to classify different types of attacks.

1.1 Overview of IoT and the Problem

With the rapid increase in mobile devices and increased use of wireless sensor networks (WSNs), the main focus shifted from providing the solution to a problem to build a robust autonomous self-learning systems. The demand for network connectivity and better security has there upon increased considerably. Internet of things focuses on four major components—first being the sensing, second comes heterogeneous access, followed by information processing and application [2]. Apart from this, security and privacy play the most important role in data transmission and receiving. In an Internet of things (IoT) network, huge amount of data is generated, transmitted and received every second. In order to make the network foolproof, it is very important to protect the network so that it can withstand any situation under attack. To make it viable, it is important to identity the attack packets and train the system to neutralize it without compromising the network and the system.

Since, millions of interconnected devices will be operating under the smart projects, their constant exposure to the open network will highlight their weakness, flaws and might reveal additional network security breaches which could be exploited by the hackers [3]. Breach of just one device could affect and compromise all other devices in the network without having direct communication with them. In order to make the system reliable or better termed self-reliable, it is necessary to train the system in such a way that it can identify the attacks and take measures before compromising the devices and the network along with the complete system.

1.2 Motivation

Since, the governments across the globe have come up with massive projects of developing urban and sub-urban cities into smart cities in order to increase the ease of living and promoting technology, automation and innovation, it becomes obvious that a gargantuan IoT system will be build up to upgrade sufficiently everything on the network. Be it, smart parking, smart divider, smart traffic signals, smart electric and water metres, smart traffic plans or anything else in the network, everything will work on the IoT sensors and devices that will participate in the flow of data and information between the devices [2, 3]. Considering the size of the network and the life cycle of the system being set up, authentication, confidentiality, data integrity and security of the network that these devices will be connected to will be the most challenging, yet achievable job. While working in the direction to secure the IoT network, we need to primarily identify the type of attack before taking the remedial approach. Hence, to secure the network and the system from various kinds of attacks and to make it robust and foolproof, we have proposed machine learning-based solution to classify multi-class intrusions in an IoT network.

1.3 Contribution

The organization of this paper is as follows:

Section 2 describes the work done in securing an IoT network. Section 3 discusses the detailed machine learning methodology and algorithm used in threat detection. Section 4 covers the comparative study of the different schemes applied in intrusion detection. Section 5 covers the detailed outcome of the work and upcoming research opportunities in IoT security and concludes the paper.

2 Related Work

Chaabouni et al. [2] have worked on identification of the security threats and reporting them to the system administrator. The author has also provided machine learning and data learning solutions to detect and report the intrusions along with the non-ML techniques, suggesting that ML techniques may or may not always be applied to the data that is why he has separately used and worked upon different sets of data, i.e., ML and non-ML-based. Since, the gargantuan data generated needs to be transferred from end devices to server for further processing; it is limited by the battery life of the sensors and devices, thus constraining the overall ML application.

Hassija et al. [3] have provided solutions using block-chain techniques to deal with the security threats in the IoT-based networks using distributed, centralized and shared ledger where each entry is tightly coupled with the previous entry using hash key. Apart from block-chain solution, the author has also provided machine learning, fog-computing and edge-based solutions to overcome security threats.

Wang et al. [4] introduced two models to detect the malicious nodes on the basis of type of sensors. The models were classified as single and multiple node detection models. He used the following parameters—the distance between the source and destination of the malicious nodes, malicious node detection probability and the mean distance between source and destination of the malicious node. The malicious node detection was carried out using a number of sensor nodes, the distance between source and destination nodes and various other parameters.

Wang et al. [5] introduced three different types IDS models combined together—(1) sink level also known as intelligent hybrid intrusion detection system, (2) cluster head, (3) node level. It used a self-learning technique where it isolates a new abnormal packet and passes it through the detection module to confirm if it is a threat packet. It used rules to detect a malware and reduce time.

Salehi et al. [6] introduced two-phase-based sinkhole intruder detection. First, it isolates suspicious nodes by authenticating the data in all nodes, and then, in the second phase, it identifies the intruders from a group of suspicious nodes by checking traffic information flow.

Wazid et al. [7] proposed hierarchical WSN and sinkhole intrusions related to it. Firstly, it detects the malicious nodes (sinkhole) on the various network performance measuring parameters like identification, trace path from source to destination, battery depletion rate, etc. As soon as the node comes under suspected category, the second phase confirms the attacker node under the following categories—message delay, message medication and message dropping in sinkholes.

Alaparthy Morgena et al. [8] proposed a multi-level intrusion detection system based on wireless sensor network environment principled on human body immune system. It used battery span, data size and data transfer rate as the parameters to detect the intrusion. In this, a few nodes were placed as immune nodes with some extra processing capabilities and then they form a network amongst themselves to perform pathogen-associated molecular pattern (PAMP) analysis to find the attack.

3 Proposed Methodology

The dataset used is the standard IEEE Internet of things (IoT) environment dataset created exclusively for academic purpose [9]. A simulated Internet of things (IoT) network consisting of various wireless devices was connected together. All packets except Mirai botnet category were captured on simulation.

Following are the four categories of attacks simulated to generate the dataset:

Man in the middle (MITM)
Denial of service (DoS)
Mirai botnet
Scanning.

Dataset has total four files: file-1 (65.718 MB) contains 1,046,270 row and 18 columns, file-2 (63.946 MB) contains 989,400 rows and 18 columns, file-3 (61.298) contains 1,036,612 rows and 18 columns and file-4 (31.513 MB) contains 550,980 rows and 18 columns. We merged the categorical data into one and used 80% for and 20% for testing.

The proposed machine learning methodology is to classify the attack and the non-attack packets. Machine learning is an application of artificial intelligence (AI) that provides systems the ability to automatically learn and improve from experience without being externally programmed, i.e., learn from examples. In the year 1959,

Arthur Samuel, the father of machine learning (ML), defined ML as "the field of study that gives computers the ability to learn without being explicitly programmed". The primary aim of machine learning is to allow the computer system to learn on its own without any human intervention. Machine learning is very popular in the predictive analysis.

Based on the type of data available, machine learning is broadly classified into two categories: supervised and unsupervised learning.

- Supervised learning is based on learning from labelled training data which means that training data includes both the input and the desired results.
- Unsupervised learning is based on learning from non-labelled and non-preclassified dataset. Unsupervised learning studies how systems can infer a function to describe a hidden structure from unlabelled data.

Since, the dataset available to us is unstructured; we have used Wireshark, a packet analysing tool to study the nature of packets. We studied the packets in detail, classifying them under each category of attack and selected 17 attributes that the learning is based upon. While studying the data packets, we manually compared the attack and non-attack packets under each category to identify at least few features (independent) to distinguish between the attack and non-attack packets. Such a feature was necessarily added to the dataset and labelled as a primitive entity. Some other features that depended upon the primitive entities were labelled as secondary.

As a pre-processing operation, we categorically encoded non-integer attributes in order to achieve better model performance. In order to identity the type of non-attack packet, an attribute with encoded numeral is manually added with each category to differentiate amongst different categories. Besides this, all non-attack packets are treated the same.

Since, the proposed scheme used supervised learning techniques named ANN artificial neural networks (ANN) to train and test our data.

Algorithm Used: Artificial neural networks (ANN*)*—The inventor of the neural network defined it as—"A computing system made up of a number of simple, highly interconnected processing elements, which process information by their dynamic state response to external inputs". ANNs are composed of multiple nodes, which imitate biological neurons of human brain. The neurons connected to links and they interact with each other. There is a hidden layer between input and output. The result of these operations is passed to other neurons. The output of each node is called its activation [1, 3, 8].

Table 1 shows the confusion matrix of the prediction on the trained dataset.

4 Comparative Study

Here, we have compared our results with the previously proposed schemes used for intrusion detection for IoT and WSN environment by other researchers. Pundir [1] discusses various intrusion techniques in wireless sensor networks (WSNs) and

Table 1 Confusion matrix of the propose work

Type of attack	Non-attack	DDOS	M.ACK	Mirai host BF	MITM	M.UDP	M.HTTP	Scan HP	Scan ports
Non-attack	476,913	10	81	0	16	0	1041	4	7
DDOS	1	12,844	0	0	0	0	0	0	0
M.ACK	4	0	15,269	0	0	0	0	0	0
Host BF	719	0	0	0	3	0	0	0	0
MITM	1474	0	0	0	608	0	0	0	0
M.UDP	0	0	0	0	0	190,261	0	0	0
M.HTTP	3288	0	0	0	1	0	17,090	1	1
Scan HP	506	0	0	0	1	0	0	1509	439
Scan ports	122	0	0	0	0	0	234	577	1620

Table 2 Comparison of different schemes with the proposed scheme

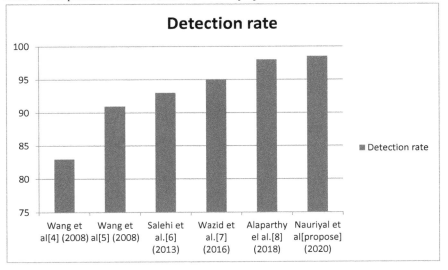

Internet of things (IoT), Wang et al. [4] introduced a single-sensing and multiple-sensing detection model with adetection rate of 83.00%. Wang et.al. [5] introduced an integrated intrusion detection system (IIDS) with a detection rate of 90.96%. Salehi et.al. [6] introduced an intrusion detection base station with adetection rate of 93.00%. Wazid et. al. [7] introduced intrusion detection by cluster head with adetection rate of 95.00%. Alaparthyet.al. [8] introduced an immune theory-based multi-level intrusion detection with a detection rate of 98.00% which proves our work is better than others' techniques which are compared in Table 1.

Table 2 gives an overview of the comparison of the different schemes with the proposed scheme.

Table 3 gives the detailed comparative study of different schemes with the propose scheme.

5 Conclusion

It is evident from the comparative study of different approaches that artificial neural networks (ANNs) give an improvement in result by 0.5%. This concludes that amongst all other techniques mentioned, machine learning model using ANNs gives better result.

We could further apply different machine learning algorithms to compare with the current results and compare them; this could fetch us better results. Since, besides the massive data flowing through the network, huge amount of data is also stored,

Table 3 Detailed comparative study of different schemes with the proposed scheme

Scheme, year	Technique used	Detection rate (DR) %	False positive rate (FPR) %
Wang et al. [4]	Single-sensing and multiple-sensing techniques	83	N/A
Wang et al. [5]	Integrated intrusion detection system	90.96	2.06
Salehi et al. [6]	Intrusion detection by base solution	93	10
Wazid et al. [7]	Intrusion detection by cluster head	95	1.25
Alaparthy el al. [8]	Immune theory-based multi-level intrusion detection	98	N/A
Nauriyal et al. [propose work] (2020)	Multi-class intrusion detection	98.5	1.5

even it becomes vulnerable to attacks; hence, security and privacy of data remain two big challenges in this field.

There is not one, but many different types of IoT devices connected together; even if a single device in the IoT network breaches security, it can compromise the complete network and hence could be worked upon.

References

1. Pundir, S., Wazid, M., Singh, D.P., Das, A.K., Rodrigues, J.J.P.C., Park, Y.: Intrusion detection protocols in wireless sensor networks integrated to internet of things deployment: survey and future challenges. IEEE Access **8**, 3343–3363 (2020)
2. Chaabouni, N., Mosbah, M.: Network intrusion detection for IoT security based on learning techniques. IEEE Commun. Surv. Tutorials (2018)
3. Hassija, V., Chamola, V., Saxena, V., Jain, D., Goyal, P., Sikdar, B.: A survey on IoT security: application areas, security threats, and solution architectures. IEEE Access **7**, 82721–82743 (2019)
4. Wang, Y., Wang, X., Xie, B., Wang, D., Agrawal, D.: Intrusion detectioninhomogeneousand-heterogeneouswirelesssensornetworks. IEEE Trans. Mobile Comput. **7**(6), 698–711 (2008)
5. Wang, S.-S., Yan, K.-Q., Wang, S.-C., Liu, C.-W.: An integrated intrusion detection system for cluster-based wireless sensor networks. Expert Syst. Appl. **38**(12), 15234–15243 (2011)
6. Salehi, S.A., Razzaque, M.A., Naraei, P., Farrokhtala, A.: Detection of sinkhole attack in wireless sensor networks. In: Proc. IEEE International Conference on Space Science Communications (IconSpace), Malacca, Malaysia, Jul 2013, pp. 361–365
7. Wazid, M., Das, A.K., Kumari, S., Khan, M.K.: Design of sinkhole node detection mechanism for hierarchical wireless sensor networks. Secur. Commun. Netw. **9**(17), 4596–4614 (2016)

8. Alaparthy, V.T., Morgera, S.D.: A multi-level intrusion detection system for wireless sensor networks based on immune theory. IEEE Access **6**, 47364–47373 (2018)
9. Kang, H., Ahn, D.H., Lee, G.M., Yoo, J.D., Park, K.H., Kim, H.K.: IoT network intrusion dataset. IEEE Dataport (2019)

Secure Web Browsing Using Trusted Platform Module (TPM)

Harshad S. Wadkar and Arun Mishra

Abstract According to the Internet usage statistics published on June 2019, around 59% of the world's human population uses the Internet. The web browser is one of the most used applications to access the Internet. How many of these 59% population trust the browser and the environment they are using? Thus, there is a necessity to configure the browser for secure browsing as well as the configuration and the browsing system needs to be protected from malicious intent and process. The paper proposes an architecture based on a trusted platform module (TPM) to configure the browser for secure browsing and to provide trust in the browsing environment. The proposed system will help general web users and system administrators to maintain and control a secure browsing environment in the enterprise.

1 Introduction

With the ease of access and good speed, the popularity of the Internet is growing. Online banking, gaming, digital media deliveries have become increasingly popular over the last few years. The web browser provides the interface to access these services.

The web browser is one of the most used interfaces to access web services. With the enhancement in usability features, availability on multiple platforms (desktop, handheld devices), high-speed networks, the web browser has become an integral part of any Internet users' software usage list. The web browser is one of the most used software applications nowadays.

The web browser is designed to send a request on behalf of the user to the server, receive the response from the server, execute (interpret) the received contents on the user's machine, and render the data in a human-readable format.

H. S. Wadkar (✉) · A. Mishra
Defence Institute of Advanced Technology, Pune, India
e-mail: wadkarharshads@rediffmail.com

A. Mishra
e-mail: arunmishra@diat.ac.in

© The Editor(s) (if applicable) and The Author(s), under exclusive license to Springer Nature Singapore Pte Ltd. 2021
T. Senjyu et al. (eds.), *Information and Communication Technology for Intelligent Systems*, Smart Innovation, Systems and Technologies 196,
https://doi.org/10.1007/978-981-15-7062-9_28

During the browsing process, a lot of data is generated in the form of cookies, logs, and records. This data is stored in the user's system as well as some part of it crosses the client's system boundary, knowingly or unknowingly to the user.

For non-hacker or non-security experts, this information stored on the machine or in transit seems to be completely irrelevant. However, this information not only helps the hacker to understand the user's browsing patterns, but also helps to gain access to the user's machine and execute malicious scripts on the user's (also called as a victim) machine or even other systems using the victim's machine. The data generated, stored on the user's system and in transit can be captured, read by an attacker or malicious website or web analytical engine. The information such as CPU, operating system, installed softwares, user-specific data can be used by an attacker to gain control of users (victims) system and even in turn control other systems using the already compromised clients (victim) system. Also, the user's information can be used to analyze a user's browsing pattern so that targeted attacks can be carried out by malicious websites or advertisements can be sent to the user by analytical engines.

Figure 1 shows a user's system in which a browsing process is running on the given operating system. The browsing process uses browser configuration and data files for application loading and throughout the user's browsing activity.

- User is accessing legitimate website 2.
- User is accessing legitimate website 3, but the analytical service running on the website is trying to generate the user's behavioral pattern.

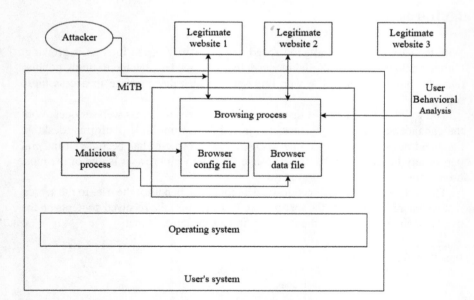

Fig. 1 User system—browsing process, behavior analysis, MiTB attack

- User is accessing legitimate website 1, the communication between the user and the website is known to the attacker and the Man-in-the-Browser (MiTB) attack is taking place.
- Malicious process running under the control of an attacker can read, write, modify browser configuration, and data files.

The constant improvements and addition of new features in the browsers, these browsers have opened the flood gates of attacks to the attackers to take control of the victim's system using the browser as the primary medium. Cross-site scripting (XSS) attack, security misconfiguration, and sensitive data exposure are listed as top 10 vulnerabilities as per the Open Web Application Security Project (OWASP) report [1].

Browser security misconfiguration leads to an information leak, cross-site scripting, cross-site request forgery, and insecure data transfer. We suggest that by configuring web browsers for secure web communication and controlling access to browser configurations, browser-based attacks can be mitigated or reduced.

To minimize or reduce the impact of these security threats, web browser developers may require redesigning and/or redeveloping the browsers. However, this redesign and implementation tasks will be tedious and complex. At present, the browser developers are developing security patches to overcome the vulnerabilities in the browser software itself but these efforts are not sufficient.

To prevent any website from tracking the user, tracking protection should be enabled. To prevent data leakage during the browsing process, the communication should be done using a strong and available transport layer protocol suite (e.g., Transport Layer Security 1.2 (TLS 1.2)). A framework has been proposed to detect browser misconfiguration and correct it (if needed) [2]. The prototype classifies the browser state into a very bad, bad, medium, good, or very good state. The prototype also helps the user to configure his/her browser to the recommended configuration (configuration proposed by us), thereby mitigating the browser-based attacks.

Even if the browsing software is regularly updated and configured for secure browsing, but how the user can trust the integrity of the browser and its configuration.

For example, if a malicious program is able to bypass the anti-malware system running on the PC and then compromise the operating system (OS) and then in-turn modifies the browser's security.

The present work is an attempt to provide a trustworthy environment to the user while using the browser. The use of security hardware (trusted platform module—TPM) running inside the PC provides a means of a trustworthy environment.

The remaining paper is organized as follows. In Sect. 2, information use of trusted computing related to secure storage is presented. The referred research articles are given in Sect. 3. In Sect. 4, we present our approach to secure browser configuration using a trusted platform. We discuss the effectiveness of the proposed system in Sect. 5. We conclude in Sect. 6.

2 Background

In this section, we briefly introduce a trusted platform module and browser (Firefox) configuration files.

2.1 Trusted System

In this section, we present information related to trusted system—trusted platform module, root of trust—measurement and storage. The reader is requested to refer trusted computing books [3–8] for more detailed information.

1. Trusted Platform Module (TPM)
   Figure 2 shows the different building blocks (components) of trusted platform module [9].

   (a) I/O Block: It controls the information flow between different components of the TPM using the bus.
   (b) Non-Volatile Random Access Memory (NVRAM): The NVRAM preserves the values stored in the TPM even if there is no electric power given to the system.
   (c) Attestation Identity Key (AIK): It is an RSA key, resides inside the NVRAM. The AIK is used for attestation.
   (d) Platform Configuration Register (PCR): A PCR is a 20-byte register used by the TPM to store application measurement (integrity) value. When the system is started (or rebooted), all the PCRs are initialized (Registers 0–16 & 23 to 0, 17–22 to -1).
   The PCR value can be modified by executing the PCRExtend function.
   PCR = HASH (PCR + newValue)
   (e) Program code: The code contains TPM's initialization sequence.
   (f) Random Number Generator (RNG): The RNG is used to produce random numbers (nonce) required for cryptographic operations.

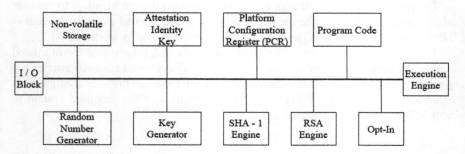

Fig. 2 TPM building blocks

(g) Key Generator: The key generator produces RSA keys.
(h) Secure Hash Algorithm 1 (SHA-1) Engine: The engine is an SHA-1 implementation used to generate a hash of data.
(i) Rivest–Shamir–Adleman (RSA) Engine: The engine is involved in cryptographic operations—encryption, decryption as well as data signing operations.
(j) Opt-in: The opt-in helps the user of the TPM to take ownership of the TPM as well as to configure it.

2. Root of trust measurement

   Figure 3 shows the root of trust measurement process.

   (a) When the user starts or reboots the PC, the Core Root of Trust for Measurement (CRTM) runs. The CRTM measures the integrity (hash value) of BIOS. If the measured hash is correct (unchanged), then the CRTM updates (extends) the PCR value. It also saves the details of the component (BIOS) measured and its corresponding hash value (integrity value) in the Stored Measurement Log (SML) and then passes the execution to the BIOS.
   (b) The BIOS checks the integrity of bootloader and if found correct, then loads the bootloader and gives the control to it.

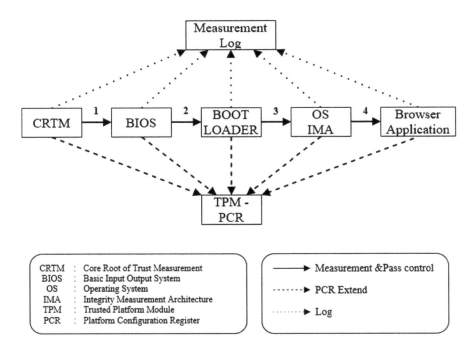

Fig. 3 Root of trust measurement

(c) The bootloader is responsible to check the correctness of the OS kernel and loading the OS.
(d) The OS loads the browser if the browsing application executable and configuration found to be unmodified.

3. Root of trust storage

Figure 4 shows the root of trust storage process.

The TPM can be used to store data in an encrypted format which is bound to a specific TPM, an asymmetric key (storage key—SK), and software configuration measured in PCR. This is called a TPM sealing operation. The key (SK) used for encryption is secured by the Storage Root Key (SRK). The SRK is stored in NVRAM.

To decrypt the sealed (encrypted) data, we should have the same TPM, key (SK) and PCR value used during the sealing process.

The composition of data sealed using TPM is given below [10, 11]. :

```
struct TPM_SEALED_DATA
{
    TPM_PAYLOAD_TYPE  payLoad ;
    TPM_SECRET        authData ;   // authorization data
    TPM_NONCE         tpmProof ;
    TPM_DIGEST        storedDigest ;
    UNINT32           dataSize ;
    BYTE              *data ;

} ;
```

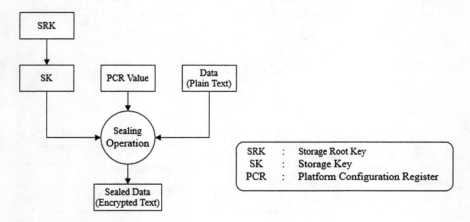

Fig. 4 Root of trust storage—sealing data

2.2 Browser Configuration and Its Loading

1. Browser configuration files
   The browser configuration parameters and their values are stored in the browser configuration files.
   For example, Consider Mozilla Firefox browser, it uses different configuration files, namely mozilla.cfg, prefs.js and user.js [12].

   (a) mozilla.cfg
   The mozilla.cfg is a configuration file Firefox uses to store the settings which are locked. The locked preference does not allow the user to modify the value as it is shown as disabled in the user interface (about:config).
   The lockPref (prefName, curValue) function assigns the value curValue to the preference prefName.
   Following are the example entries in mozilla.cfg :
   lockPref ("network.cookie.cookieBehavior", 1) ;
   lockPref ("security.tls.version.max", 4) ;
   lockPref ("browser.formfill.enable", false) ;
   lockPref ("webgl.disabled", true) ;
   (b) prefs.js
   The prefs.js file does not contain all the preferences but contains only those preferences which are modified by the user.
   (c) user.js
   The user.js file contains the preferences modified by the user. Each time Firefox is loaded the user.js is loaded. When the browser application is closed, the preferences set in the user.js file are stored in pref.js.
   Following are the example entries in user.js :
   user_pref ("network.cookie.cookieBehavior", 1) ;
   user_pref ("security.tls.version.max", 4) ;
   user_pref ("browser.formfill.enable", false) ;
   user_pref ("webgl.disabled", true) ;

2. Browser configuration loading
   When the browser application is started, it loads the browser configuration files in following order [12]:

   (a) Load mozilla.cfg
   (b) Load prefs.js
   (c) Load user.js

   Preferences added last will take effect on the working of the browser application instance.
   For example, if initially mozilla.cfg sets the preference X to 'a', prefs.js sets X to 'b' and finally user.js sets X to 'c' then the Firefox runs with the last edited entry for X that is X set to 'c'.

3 Related Work

In this section, we present recently published information on browser-based attacks like stealing browser history, cross-site scripting, and SSL-based attacks.

To give better page load performance, the web browser caches web pages, images, and urls visited by the user. This browsing history is stored in plain text format on user's machine. The attacker can find information of user's age, gender and location using the stored history. The attacker can then prepare phishing web pages to cheat the user , making the user accessing attacker's web page instead of genuine web page. Using the user's browsing history, history sniffers or analytical engines can perform the user's behavioral analysis.

Lukasz Olejnik et al. performed an experiment to study the uniqueness of web browsing histories [13]. They found that the browsing histories were unique for about 69% of users and 38% of the stable. As the browser histories were unique and stable, the authors claimed to map users' personal interests as well using the collected data.

In cross-site scripting (XSS) attack, the attacker initially injects malicious code into a genuine-looking website. When any user visits the website, then the malicious code injected by the attacker gets executed on the user's browser. This results into sensitive data leaks, cookie theft. The stolen cookies then allow the attacker to start new session with the user visited website and access the website or login to the website.

Philipp Vogt et al. developed a mechanism, to allow or block the sensitive data going out from the browser [14]. In this, JavaScript code was allowed to execute first on the user's browser. If the JavaScript code tried to send any sensitive data to the cross-domain server, a warning message was shown to the user. The data was allowed to cross the browser's boundary if the user agreed to do so.

Secure Socket Layer (SSL) and Transport Layer Security (TLS) protocols suffer attacks due to flaws in their design and/or implementation.

Abeer E. W. Eldewahi et al. presented information about SSL-based attacks like the POODLE, Lucky13 [15]. Use of TLS 1.2 protocol for communication, stopping of TLS compression were few of the countermeasures suggested by the authors to prevent SSL-based attacks.

4 Proposed System

To prevent any malicious process or unauthorized user to modify the secure configuration of the browser configuration file, we suggest the use of a trusted platform module.

The proposed system has three modules.

1. Module 1

    Module 1 is used by the system administrator to configure the browser for secure browsing. In this module, the following steps are performed:

(a) The browser configuration controller (BCC) assesses the browser configuration for secure browsing.
(b) The BCC suggests the recommended secure browsing configuration (if not found correct in step a) to the system administrator.
(c) Using the configuration setting function of the BCC, the system administrator configures the browser configuration parameters to the recommended values.
(d) The browser configuration with the recommended values [2] is stored in the mozilla.cfg file, so that these values will not available to the user for any modification using the user interface.

2. Module 2
Module 2 is used by the system administrator to seal the browser configuration file to the TPM PCR configuration.
This sealing operation is carried out in two phases:

(a) Phase 1:
Phase 1 makes use of the root of trust measurement.
This measures BIOS (B), bootloader (BL), OS (O), and Firefox executable (F).
$PCR_{23} := 0$
$PCR_{23} := HASH (PCR_{23} + B)$
$PCR_{23} := HASH (PCR_{23} + BL)$
$PCR_{23} := HASH (PCR_{23} + O)$
$PCR_{23} := HASH (PCR_{23} + F)$
(b) Phase 2 :
The phase 2 encrypts the mozilla.cfg file along with the configuration value of TPM PCR 23 using the command:
$ tpm_sealdata - -infile mozilla.cfg - -outfile mozilla_enc.cfg - -pcr 23

3. Module 3
Module 3 is used by the system administrator when s/he feels the need to modify browser configuration or to update the browser application executable.
In this scenario, the system administrator performs the following actions in the given order:

(a) Terminate the browser instance
(b) Perform unseal operation to decrypt mozilla_enc.cfg file
$ tpm2_unseal -c mozilla_enc.cfg–outfile mozilla.cfg - -pcr 23
(c) Make necessary browser configuration and write the modified configuration in mozilla.cfg file
(d) Update the browser executable (if a new patch is available)
(e) Seal (encrypt) the mozilla.cfg as per module 2.

5 Security Analysis of Proposed System

5.1 Security Against User Behavior Analysis

User's behavior analysis can be conducted by the analytics engine using one or more of the following :

1. The session recordings in the form of cookies stored on the user's machine by the website visited by the user.
2. The web pages visited, actions like cut-copy-paste and mouse clicks performed by the user.
3. The browser configuration captured by the user's visiting website. This information in-turn leaks the user's machine information as well.

To overcome this user behavior analysis, the browser needs to be configured such that only first-party cookies are allowed to store on the user's machine as well the cookies will be deleted after the browser session is closed.

The browser configuration controller (module 1 of the proposed architecture) helps the user to assess browser configuration, modify it to the recommended settings. This will minimize the user or his (her) browser information leaked to the analytical engine.

For example,
lockPref ("network.cookie.cookieBehavior", 1) ;
lockPref ("browser.formfill.enable", false) ;

5.2 Security Against MiTB

If the data in transit between the user's browser and the user visited website is in plain text and the attacker is trying to sniff the data, then the attacker will be able to read, understand and modify data. To overcome this MiTB attack, the communication should be in SSL/TLS. For this, the browser needs to be configured to allow TLS1.2.

The browser configuration controller (module 1 of the proposed architecture) helps the user to modify the browser configuration to allow only TLS1.2. This will minimize the SSL/TLS-based attacks like SLOTH, DROWN, FREAK, and in-turn the MiTB attack.

For example,
lockPref ("security.tls.version.max", 4) ;

5.3 Security Against Malicious Process Trying to Modify a Browser Configuration File

Our proposed architecture (module 2 of the proposed architecture) performs two operations before the browser is instantiated—chain root of trust and seal and unseal browser configuration file.

1. The chain root of trust ensures that no malware like rootkit gets executed before the operating system kernel is loaded into RAM. This prevents kernel-mode rootkit to intercept system calls. This also prevents user-mode rootkit to start any key-logger and begin storing user-made keystrokes and send the keystrokes to the attacker.
2. The browser configuration file (mozilla.cfg) is sealed using the platform's configuration and unsealed only if the same platform configuration is available at the time of unsealing operation.
   Unless the mozilla.cfg is unsealed and configuration is used for browser instantiation, the browser will not be loaded in the memory and the user cannot use it. This will ensure that the browser will run in a secure mode.
3. The browser configuration (mozilla.cfg) file contains configuration listings using lockPref function. This ensures that the configuration parameters set and locked in mozilla.cfg and will not be available to the user to modify thereby the browser configuration remains unchanged once the browser is instantiated (it is loaded in memory).

6 Conclusion

Our approach of using trusted platform module (TPM) surely not of the first of its kind in providing a trusting environment, but yes—probably it is for the first time TPM is used in providing a secure browsing environment.

We have presented an architecture to safeguard the browser configuration files which contain browser parameter settings used for secure web browsing. With the help of a trusted platform module which is tamper-resistant, the architecture also provides a dependable and secure environment for the user to use the browser.

Future work includes preparing a working prototype of the proposed architecture. We would like to study the impact on the performance and usability of the browser based on our prototype.

The use of the Internet using handheld devices (smartphones and tablets) is growing. To configure the browser running on these devices and provide a trusted environment for the browsing process, one needs to make use of a Mobile Trusted Module (TPM-Mobile). We would like to develop a prototype of our model to be used on handheld devices.

Acknowledgements We thank Deepti Vidyarthi, Anita Thite, and Makarand Velankar for their helpful suggestions and feedback. We would also like to thank the anonymous reviewers for their constructive inputs.

References

1. OWASP: Top 10-2017 Top 10. https://www.owasp.org/index.php/Top_10-2017_Top_10
2. Wadkar, H., Mishra, A., Dixit, A.: Framework to Secure Browser Using Configuration Analysis. Int. J. Inf. Secur. Privacy. IGI Global. https://doi.org/10.4018/IJISP.2017040105
3. Proudler, G., Chen, L., Dalton, C.: Trusted Computing Platforms TPM2.0 in Context. Springer, Berlin (2014)
4. Mitchell, C.: Trusted Computing. IET Professional Applications of Computing Series **6** (2005)
5. Smith, S.: Trusted Computing Platforms: Design and Applications. Springer, Berlin (2005)
6. Balacheff, B., Chen, L., Pearson, S., Plaquin, D., Proudler, G.: Trusted Computing Platforms: TCPA Technology in Context. Prentice Hall PTR, Upper Saddle River (2002)
7. Parno, B., McCune, J., Perrig, A.: Bootstrapping Trust in Modern Computers. Springer, Berlin (2011)
8. Arthur, W., Challener, D., Goldman, K.: A Practical Guide to TPM 2.0: Using the Trusted Platform Module in the New Age of Security. Apress (2015)
9. Tomlinson, A.: Introduction to the TPM. In: Smart Cards, Tokens, Security and Applications, pp. 155–172. Springer, Berlin (2008). https://doi.org/10.1007/978-0-387-72198-9_7
10. Challener, D., Yoder, K., Catherman, R., Safford, D., Van Doorn, L.: A Practical Guide to Trusted Computing. Pearson Education, London (2007)
11. TPM Main Specification Part 2, TPM Structures, Specification version 1.2, Revision 62, October 2, 2003. Trusted Computing Group. https://trustedcomputinggroup.org/wp-content/uploads/tpmwg-mainrev62_Part2_TPM_Structures.pdf
12. A brief guide to Mozilla preferences. https://developer.mozilla.org/en-US/docs/Mozilla/Preferences/A_brief_guide_to_Mozilla_preferences
13. Olejnik, L., Castelluccia, C., Janc, A.: On the uniqueness of web browsing history patterns. Ann. Telecommun.-annales des télécommunications, pp. 63–74. Springer, Berlin (2014). https://doi.org/10.1007/s12243-013-0392-5
14. Vogt, P., Nentwich, F., Jovanovic, N., Kirda, E., Kruegel, C., Vigna, G.: Cross Site Scripting Prevention with Dynamic Data Tainting and Static Analysis, p. 12. NDSS (2007)
15. Eldewahi, A., Sharfi, T., Mansor, A., Mohamed, N., Alwahbani, S.: Cross site scripting prevention with dynamic data tainting and static analysis. In: 2015 International Conference on Computing, Control, Networking, Electronics and Embedded Systems Engineering (ICCNEEE), pp. 203–208. IEEE, New York (2015)

CP-ABE with Hidden Access Policy and Outsourced Decryption for Cloud-Based EHR Applications

Kasturi Routray, Kamalakanta Sethi, Bharati Mishra, Padmalochan Bera, and Debasish Jena

Abstract Electronic health record (EHR) stores not only the patient's health-related data but also the sensitive individual information. Consequently, data security and access privacy are the greatest concern of EHR applications while sharing the data through the cloud. Ciphertext policy attribute-based encryption (CP-ABE) is one of the popular one-to-many encryption schemes, which helps to attain fine-grained access control in cloud domain. In CP-ABE, since the access policy is attached to the ciphertext, anyone with the ciphertext can see the data owner's policy that may cause privacy leakage. In this paper, we proposed a CP-ABE cryptosystem that supports access policy obfuscation and outsourced decryption enabling efficient and secure operations in cloud environment. Linear secret sharing (LSS) scheme is used for supporting any monotonic access structures, thereby improving the access policy's expressiveness. We have incorporated prime-order bilinear group and matrix-based LSS scheme that increases the computational efficiency of the cryptosystem. Finally, we have evaluated the performance of our CP-ABE cryptosystem using practical implementation utilizing Charm framework.

K. Routray (✉) · B. Mishra · D. Jena
International Institute of Information Technology, Bhubaneswar, India
e-mail: a118003@iiit-bh.ac.in

B. Mishra
e-mail: Bharati@iiit-bh.ac.in

D. Jena
e-mail: debasish@iiit-bh.ac.in

K. Sethi · P. Bera
Indian Institute of Technology, Bhubaneswar, India
e-mail: ks23@iitbbs.ac.in

P. Bera
e-mail: plb@iitbbs.ac.in

© The Editor(s) (if applicable) and The Author(s), under exclusive license to Springer Nature Singapore Pte Ltd. 2021
T. Senjyu et al. (eds.), *Information and Communication Technology for Intelligent Systems*, Smart Innovation, Systems and Technologies 196,
https://doi.org/10.1007/978-981-15-7062-9_29

1 Introduction

Cloud computing technology offers computing platform and services accessibility anytime and anywhere on pay as per use model. It is potential to transform operations with ease at affordable cost that attracts many industries and organizations like automotive, health care, banking, retail, education, etc. Healthcare industry is continuously improving in the last few decades with the advancement in technology. Digitization of patients' health records facilitates faster transmission, collaboration, and coordination among different health service providers for better diagnosis and treatment. Because of availability of up-to-date health-related information and faster sharing of crucial medical data at right time, patients are provided with better medical services and care. Cloud computing could help the healthcare industry deliver more services faster and at low cost. Healthcare systems that work on information-centric models can be benefited with cost-effective infrastructure, flexible and scalable computing requirements of cloud platform. Electronic health record (EHR) systems may use cloud services to effectively store, share, maintain, and protect data confidentiality of patient's sensitive information.

Although cloud computing provides various benefits, one of the limitations that hinders its wide scale adoption is the violation of information security and trust. The patient's confidential health record is usually stored in an open and shared third-party server. This makes the user's data susceptible to theft, loss of confidentiality, unauthorized access, and violation of data integrity. So, the EHR application providers utilize cloud platforms with caution because of the various security risks involved. Therefore, there is need of effective mechanism to enforce data security and privacy in EHR systems. In addition, faster data accessibility and lightweight computation for low-power devices is desirable by the EHR system users.

CP-ABE has gained popularity for cloud data management because of its efficiency, expressibility, and fine-grained data access controls. In CP-ABE, ciphertext embeds the access policy and only users can decrypt the ciphertext whose attributes match the owner defined access policy. But in CP-ABE scheme, anyone having the ciphertext can see the access policy that may cause violation of privacy. This paper proposes a solution to hide the user's access policy while performing encryption and decryption of EHRs using a modified CP-ABE scheme. Also, for enhancing the expressibility of access policy and computational efficiency for faster sharing of critical EHR data, we have used LSS scheme [1] for formal representation of access structure. For faster computation, we incorporated prime-order bilinear groups [2] as they are demonstrated to outperform composite-order bilinear groups. Our scheme also supports outsourcing decryption where the computationally intensive tasks are delegated to the cloud. Finally, we prototyped the proposed scheme using Charm platform and the experimental results show its practical usability in EHR application.

Organization of the paper is as follows: Some of the previous work on ABE is discussed in Sect. 2. In Sect. 3, the details of proposed scheme and mathematical construction are presented. Section 4 examines the practical implementation of the proposed scheme in Charm tool and performance analysis. Finally, in Sect. 5, we conclude the paper.

2 Related Work

ABE concept was first proposed by Sahai and Waters [3]. Eventually, two variants of ABE schemes were formulated by Goyal: key-policy-based encryption [4] and ciphertext policy-based encryption (CP-ABE) [5]. In KP-ABE, ciphertexts are associated with set of attributes and user's private key is embedded in the access policy whereas in CP-ABE attributes are associated with private key and access policy are located on ciphertext.

Related to EHRs applications, Ibraimi [6] proposed a new variant of CP-ABE scheme which employs multiple authorities for securing EHRs across different security domains. Another CP-ABE scheme was proposed by Narayan [7] where the broadcast variant of CP-ABE was used to encrypt the EHR and allows users revocation. Li [8] proposed a CP- ABE scheme with major focus on security but lacks in guaranteeing access control and encryption mechanism. Qian [9] suggested a privacy-preserving multi-authority CP-ABE scheme with user and attribute revocation. Recently, Joshi [10] designed a centralized, attribute-based authorization mechanism that uses CP-ABE and permits delegated secure access to EHRs. However, most of the discussed ABE schemes have shortcomings that they are less expressive, computationally less efficient, and prone to access policy privacy issues.

3 Our Proposed CP-ABE Cryptosystem

This section describes the architecture of our proposed cryptosystem along with formal mathematical construction of the scheme.

3.1 Scheme Overview and System Architecture

The general architecture of a collaborative EHR management is shown in Fig. 1. The main components of EHR systems are patient, cloud system, and EHR system clients.

Here, patients are the data owners whose health-related data are collected and stored in the cloud using various platforms like mobile devices, desktops, laptops, and sensor-enabled devices. The cloud systems provide a powerful decentralized computing environment for storing and processing patient's data. The EHR system clients are end users such as doctors, hospitals, and pharmacy who want to use the data to make reasoned decisions.

Figure 2 depicts the architecture of our cloud-based EHR scheme. It comprises of main entities of the system as trusted authority, key generation center (KGC), cloud storage center (CSC), data owner, and EHR user.

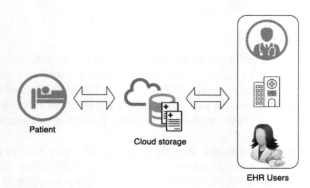

Fig. 1 General architecture of EHR application

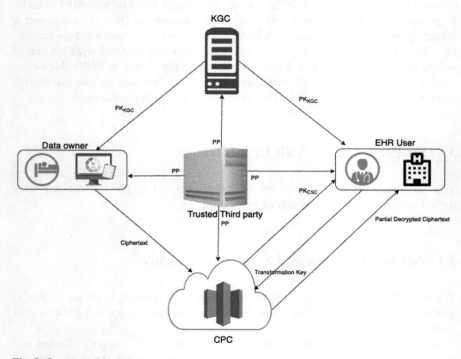

Fig. 2 System architecture

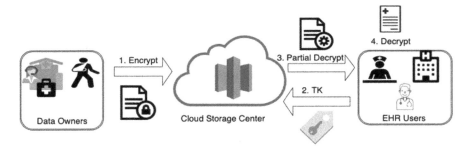

Fig. 3 Data flow in proposed scheme

Figure 3 demonstrates the data flow scenario of our proposed scheme. In this scenario, using the access policy data owner encrypts the EHR data and transfers it to the CSC. When the EHR user wants to retrieve the data, he/she can generate the transformation key (TK) and forward it to the CSC. Thereafter, the users get the partially decrypted ciphertext if the decryption key match the access policy defined on the encrypted EHR. Finally, the user successfully decrypts the partially decrypted ciphertext using his/her private key.

3.2 Mathematical Construction

Our scheme comprises of four phases namely system initialization, key generation, data encryption, and outsourced data decryption (similar to cryptosystem defined in [11]).

3.2.1 System Initialization

System initialization setup defines three setup procedure—GlobalSetup, KGCSetup, and CSCSetup. First the trusted authority executes the GlobalSetup() procedure to generate the system public parameters (PP). Then using PP, KGC and CSC run KGC-Setup() and CSCSetup() procedures and generate their respective private and public keys.

$GlobalSetup(\lambda) \rightarrow GP$: In this algorithm, trusted authority with λ as system parameter chooses two multiplicative cyclic group G_0 and G_1 with same prime order p. The parameter g is a generator of G_0. The bilinear map e on G_0 is defined as $e : G_0 \times G_0 \rightarrow G_1$. Besides, let H_0 and H_1 be two random oracle hash functions. The hash function $H_0:\{0, 1\}^* \rightarrow G_0$ maps an element in $\{0, 1\}^*$ to an element in G_0 while another hash function $H_1 : G_0 \rightarrow \{0, 1\}^{\log p}$ helps to obscure the attributes of access policy. The published public parameters are as follows:

$$PP = \{G_0, G_1, p, e, g, H_0, H_1\}$$

The generated global parameters PP are accessible to all the system's entities.
$KGCSetup(PP) \to (SK_{KGC}, PK_{KGC})$: The KGC invokes procedure by randomly choosing two exponents $\alpha, \beta \in \mathbb{Z}_p^*$ and using the published public parameter PP computes its public key PK_{KGC} and private key SK_{KGC} which are defined as:

$$PK_{KGC} = \{e(g,g)^\alpha, h = g^\beta\}$$
$$SK_{KGC} = \{\beta, g^\alpha\}$$

$CSCSetup(PP, ID_S) \to (SK_S, PK_S)$: The CSC with identity ID_{CSC} randomly chooses an exponent $\gamma \in \mathbb{Z}_p^*$. Using γ, public parameter PP and ID_{CSC} as inputs CSC calls CSCSetup procedure to compute its public key PK_{CSC} and private key SK_{CSC} which are defined as follows.

$$PK_{CSC} = \{g^\gamma\}$$
$$SK_{CSC} = \{H(ID_{CSC})^\gamma\}$$

3.2.2 Key Generation

In this phase, KGC executes KeyGen() procedure using GP, set of user's attribute A, and secret key SK_{KGC} to generate user's personalized secret key.
$KeyGen(A, SK_{KGC}, PP) \to (SK_{U_t})$: The KGC randomly chooses r_t for each user U_t and r_i for each attribute in attribute set A that is $i \in A$. It generates secret key components for each attribute $r_i \in A$ of user U_t which are calculated as:

$$SK_{U_t} = \{D = g^{\frac{\alpha+r_t}{\beta}}, \forall x \in A, D_{1,i} = g^{r_t} \cdot H_0(i)^{r_i},$$
$$D_{2,i} = g^{r_i}, D_{3,i} = H_0(i)^\beta\}$$

The user then combines the generated components to compute the private key as:

$$SK_{U_t} = \{D, \{D_{1,i}, D_{2,i}, D_{3,i}\}_{i \in A}\}$$

3.2.3 Data Encryption

The data owner executes Encrypt() algorithm to convert the EHR into encrypted form before sending it to cloud storage.
$Encrypt(PP, PK_{KGC}, PK_{CSC}, M, \tau) \to (CT, ID_o, g^a)$: Data owner with identity (DO) U_a calls Encrypt() to encrypt EHR data M using access policy τ, public

key of KGC PK_{KGC} and CSC PK_{CSC}. It outputs the ciphertext with obfuscated access policy.

For access policy obfuscation, this algorithm randomly chooses two vectors $\mathbf{v} = (s, v_2, ..., v_n)$ and $\mathbf{v}' = (0, v'_2, ..., v'_n) \in \mathbb{Z}_p^n$. Then, it calculates $\lambda_x = A_x \mathbf{v}^\top$ as shares of s for each row x of τ. In similar manner, the shares of 0 are calculated as $w_x = A_x \mathbf{v}'^\top$ for each row x of τ. After this, $a \in \mathbb{Z}_p^*$ is randomly selected and $s_x = e((g^\beta)^a, H(\rho(x))$ is computed for each row i of τ. The access policy attributes are hidden by substituting the original attributes $\rho(x)$ with $H_1(s_x)$.

Encryption is done by computing the value of $K_S = e((g^\gamma)^a, H(ID_{CSC}))$, where $H(ID_{CSC})$ is identity of the CSC. Finally, algorithm computes the components of ciphertext and ciphertext CT as:

$$C_0 = h^s, \quad C_1 = M \cdot K_S \cdot e(g,g)^{\alpha s}$$

$$C_{2,x} = g^{\lambda_x}, \quad C_{3,x} = H(x)^{\lambda_x + \omega_x}$$

$$CT = \{\tau, C_0, C_1, \{C_{2,x}, C_{3,x}\}_{x \in \{1,2,...,l\}}\}$$

After computing the ciphertext CT, the data owner forwards (CT, ID_a, g^a) to the CSC. Here, ID_a is the identity of data owner.

3.2.4 Outsourced Data Decryption

Outsourced data decryption phase decrypts the ciphertext with the help of CSC. The CSC partially decrypts the ciphertext to significantly reduce the overhead of intensive computational task for the EHR user. This phase constitutes three algorithms $GenTK_{out}$, $PDecrypt_{out}$, and $Decrypt_{out}$.

$GenTK_{out}$ $(PP, SK_{U_t}, A_{U_t}) \rightarrow TK_{U_t}$: For all attributes x present in A_{U_t}, the algorithm calculates $s_x = e(g^a, K_{3,x})$. Then, the algorithm randomly selects $\mu \in \mathbb{Z}_p^*$ and computes transformation key TK_{U_t} as:

$$TK = \{\forall x \in A_{U_t} : I_x = H_1(s_x)\}$$

$$TK_{1,x,U_t} = (D_{1,x})^\mu, \quad TK_{2,x,U_t} = (D_{2,w})^\mu$$

$$TK_{U_t} = \{TK, \{TK_{1,x,U_t}, TK_{2,x,U_t}\}_{x \in \{1,2,...,l\}}\}$$

Once TK_{U_t} is generated, it is forwarded to CSC. Since the attributes are masked, I_x is used as an index for attribute x.

$PDecrypt_{out}$ $(PP, CT, TK_{U_t}) \rightarrow CT'$: Once the CSC receives the transformation key TK_{U_t} from user U_t, the verification of indices of attributes in attribute set with respect to access policy in the ciphertext is done. Upon successful verification, the

CSC partially decrypts the ciphertext. The algorithm calculates the value of P_x, for each row x of τ corresponding to attribute $\rho(x)$ as:

$$P_x = \frac{e(TK_{1,x,U_t}, C_{2,i})}{e(TK_{2,x,U_t}, C_{3,x})}$$

$$= \frac{e(g^{r_t} \cdot H(\rho(x))^{r_x}, g^{\lambda_x})^\mu}{e(g^{r_x}, H(\rho(x))^{\lambda_i + \omega x})}$$

$$= \frac{e(g,g)^{r_t \lambda_x \mu}}{e(g, H(\rho(x))^{r_x \omega_x \mu})}$$

Then, it finds constants $\{c_i \in \mathbb{Z}_p\}_{i \in I}$ in such a way that $\sum_{i \in I} c_x \tau_x = (1, 0, ..., 0)$, where τ_x is row x of τ and calculates the value of P as:

$$P = \prod_{x \in I} P_x^{c_x} = e(g,g)^{r_t s \mu}$$

The CSC calculates $K_S = e(g^a, SK_{CSC}) = e(g^a, H(ID_{CSC})^\gamma)$ and $C_1' = C_1/K_S = Me(g,g)^{\alpha s}$. Then CSC sends $CT' = (C_0', C_1', P)$ where $C_0' = C_0 = h^s$.

$Decrypt_{out}(PP, CT', SK_{U_t}) \to M$: In outsourced decryption, the users will get partially decrypted data from CSC which are easy to decrypt by lightweight devices having low computation power and limited resources. The algorithm retrieves the plaintext by calculating:

$$\frac{C_1'}{\left(\frac{e(C_0', K)}{(P)^{\frac{1}{\mu}}}\right)} = \frac{C_1'}{\left(\frac{e(h^s, g^{\frac{\alpha + r_t}{\beta}})}{(e(g,g)^{r_t s \mu})^{\frac{1}{\mu}}}\right)}$$

$$= \frac{C_1'}{\left(\frac{e(g^{\beta s}, g^{\frac{\alpha + r_t}{\beta}})}{(e(g,g)^{r_t s})}\right)}$$

$$= \frac{C_1'}{\left(\frac{e(g,g)^{(\alpha + r_t)s}}{(e(g,g)^{r_t s})}\right)}$$

$$= \frac{M \cdot e(g,g)^{s\alpha}}{e(g,g)^{s\alpha}} = M$$

This concludes the formal construction of our proposed CP-ABE cryptosystem.

4 Experimental Results and Performance Analysis

To test correctness and measure the efficiency, we have prototyped our proposed CP-ABE-based scheme using Charm framework [12]. Charm uses hybrid design which uses performance intensive mathematical operations to reduce code complexity and

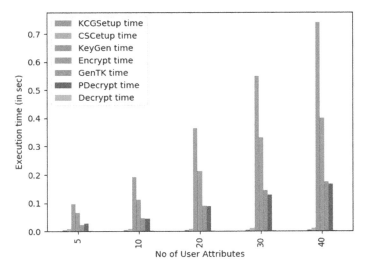

Fig. 4 Execution time versus number of user attributes

development time. For the purpose of implementation, we translate our proposed system to an asymmetric domain as Charm libraries use asymmetric groups. The experimentation is performed on a system with an Intel Core i3 CPU M 380 @ 2.53 GHz x 4 processor, 8 GB RAM, and Ubuntu 18:04:4 LTS operating system installed on it. Charm latest version 0.43 is used with Python version 3.4 as a dependency. Experimentation is performed by changing the number of user attributes and the number of access policy attributes. We observed the average execution time of seven different algorithms of our proposed cryptosystem: KCGSetup, CSCSetup, KeyGen, Encrypt, PDecrypt, GenTK, and Decrypt time.

Figure 4 depicts the average execution time of the aforementioned algorithms when changing the number of user attributes. The execution time of KGCSetup and CSCSetup algorithm is almost constant while the execution time of other algorithms increases linearly with addition of user attributes.

Figure 5 shows the variations of encrypt algorithm's execution time on changing the number of access policy attributes. The time taken to encrypt the data increases linearly with increasing the number of attributes in access policy.

The experimental results indicate that the execution time of our cryptosystem is within seconds for large number of user attributes and access policy attributes. Moreover, we also observe the performance of our proposed cryptosystem does not vary significantly with variation in data size. Therefore, our scheme is useful for secure storage and access of large volume of data for EHR applications.

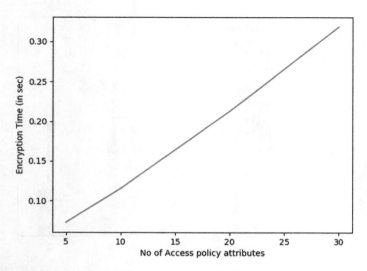

Fig. 5 Execution time of encryption versus number of access policy attributes

5 Conclusion

Our proposed CP-ABE cryptosystem obfuscates the access policy embedded in the ciphertext and supports outsourced decryption. Hence, our cryptosystem is apt for effective implementation of cloud-based EHR applications. We have used bilinear groups of prime-order and matrix-based LSS scheme that enhances computational efficiency of the cryptosystem. In addition, use of LSS scheme increases the expressiveness of our proposed CP-ABE cryptosystem. The outsourced decryption phase is characteristically built in the construction, that reduces the overhead at user end. The experimental results suggest that the proposed cryptosystem performs reasonably well, thereby may be employed as a substitute to standard CP-ABE schemes in cloud-based EHR systems to provide usable and secure medical services with reduced costs. Our cryptosystem do not support multiple domain authorities and traceability of malicious users which can be the desirable features of a EHR-based cryptosystem. Our future work is to include the aforementioned characteristics in the scheme to enhance its usability.

References

1. Beimel, A.: Secure schemes for secret sharing and key distribution, Ph.D. dissertation, Faculty Computer. Sci., Technion-Israel Inst. Technol., Haifa, Israel (1996)
2. Seo, J.H., Cheon, J.H.: Beyond the limitation of prime-order bilinear groups, and round optimal blind signatures. In: Theory of Cryptography Conference, Berlin, Springer, LNCS, vol. 7194, pp. 133–150 (2012)

3. Sahai, A., Waters, B.: Fuzzy identity-based encryption. In: Proceedings of the Advances in Cryptology-EUROCRYPT, vol. 3494, pp. 457–473, LNCS (2005)
4. Goyal, V., Pandey, O., Sahai, A., Waters, B.: Attibute-based encryption for fine-grained access control of encrypted data. In: Proceedings of the ACM Conference Computer and Communications Security (ACM CCS), pp. 89–98, Virginia, USA (2006)
5. Bethencourt, J., Sahai, A., Waters, B.: Ciphertext-policy attribute-based encryption. In: Proceedings of the IEEE Symposium Security and Privacy, Oakland, CA (2007)
6. Ibraimi, L., Asim, M., Petković, M.: Secure management of personal health records by applying attribute-based encryption. In: 6th International Workshop on Wearable, Micro and Nano Technologies for Personalized Health (2009)
7. Narayan, S., Gagn, M., Safavi-Naini, R.: Privacy preserving EHR system using attribute-based infrastructure. In: Proceedings of the ACM Workshop on Cloud Computing Security Workshop (2010)
8. Li, M., Yu, S., Zheng, Y., Ren, K., Lou, W.: Scalable and secure sharing of personal health records in cloud computing using attribute-based encryption. IEEE Trans. Parallel Distrib. Syst. (2013)
9. Qian, H., Li, J., Zhang, Y., Han, J.: Privacy-preserving personal health record using multi-authority attribute-based encryption with revocation. Int. J. Inf. Security **14**, 487–497 (2015)
10. Joshi, M., Joshi, K., Finin, T.: Attribute based encryption for secure access to cloud based EHR systems. In: IEEE 11th International Conference on Cloud Computing (CLOUD) (2018)
11. Sethi, K., Pradhan, A., Bera, P.: Attribute-based data security with obfuscated access policy for smart grid applications. In: International Conference on Communication Systems and NETworkS (COMSNETS) (2020)
12. Akinyele, J.A., Garman, C., Miers, I., Pagano, M.W., Rushanan, M., Green, M. and Rubin, A.D.: Charm: a framework for rapidly prototyping cryptosystems. J. Cryptogr. Eng. **3**, 111–128 (2013)

Viral Internet Challenges: A Study on the Motivations Behind Social Media User Participation

Naman Shroff, G. Shreyass, and Deepak Gupta

Abstract Of the many genres of content available, "viral Internet challenge" has become popular in the recent years and the existing research is centered on specific viral Internet challenges only, such as the ice bucket challenge. To address the gap in scientific evidence on the general motivation of the social media users to engage and co-create viral Internet challenges, our study was conducted in the context of popular social media apps in India. Based on a review of literature, a conceptual model was proposed and empirically tested. The influence of personal factors, outward motivations and inner motivations were also mapped in this study. Finally, we also investigated the influence of gender roles we play—specifically masculinity/femininity. It was observed that important drivers of the intent to participate in Internet viral challenges were the respondents' need to belong to a group, lack of immediacy of rewards and the ability to showcase personal strengths mattered. The association of influencers with the viral challenge significantly increased the likelihood of participation. The most interesting finding was the impact of gender identity in driving participation in viral Internet challenges. Those who identified higher on masculine traits were more likely to display an intent to participate and it was the converse for feminine traits. This effect was independent of the impact of gender. Our analysis suggests an interesting and hitherto unexplored dimension of viral Internet challenges—there is something in them that saliently appeals to our inherent masculinity. This result bears further investigation.

N. Shroff (✉) · G. Shreyass · D. Gupta
Amrita School of Business, Amrita Vishwa Vidyapeetham, Coimbatore, India
e-mail: namanshroff.93@gmail.com

G. Shreyass
e-mail: shrerenga95@gmail.com

D. Gupta
e-mail: dgshobs@gmail.com

© The Editor(s) (if applicable) and The Author(s), under exclusive license to Springer Nature Singapore Pte Ltd. 2021
T. Senjyu et al. (eds.), *Information and Communication Technology for Intelligent Systems*, Smart Innovation, Systems and Technologies 196, https://doi.org/10.1007/978-981-15-7062-9_30

1 Introduction

We have been exposed to some of the most popular trends, videos and memes to reach our devices, but one of the most intriguing elements is the phenomenon termed as "viral content." Before the Internet, such "virality" was common in the domains such as fashion, sports, film, music and more, building a sense of community [1]. With the advent of digital platforms, the media changed the way the message was passed on by the users [2].

By connecting the physical and digital worlds together, the appeal of these activities varies but the motivation was to understand the influencing factors across the timeline of viral Internet challenges. As digital communications have worked their way into society, so it too have this spontaneous type of viral trends [3]. From emails to short clips to videos, the viral Internet challenge is well a part of the Internet ecosystem.

Most importantly, these challenges perform the best by eliciting emotions and reactions from their participants [3]. The origins of viral Internet challenges are many: Influencers, social cause, brands, common place social media happenings, etc. And with the increasing screen time per user, the ability of viral content to reach more users is greatly magnified and utilized by the viral Internet challenge creators for commercial purposes and otherwise. As a result, they have become a powerful medium of communication for marketing entities to connect with segments of the society.

But the interesting thing is that while viral communication has been covered extensively with empirical models to aid research [4, 5], the question of why social media users interact and participate in viral Internet challenges is surprisingly under researched, with the few existing studies focusing on specific viral challenges rather than the phenomenon in general—the novel focus of this study. Based on an extensive literature review, a conceptual model was developed to map the likelihood of participation in viral Internet challenges in the near future, followed by a pan-India survey. The data was analyzed using ordered logistic regression. In the following sections, we present the model along with a discussion on the analysis of the survey results.

2 Literature Review

2.1 Community in Social Media

The exploration started with the effect of community in influencing participation in a viral Internet challenge. It was found that audience size influenced what participants were willing to share; a closer circle meant the sharing was influenced by its level of informativeness but diminished as the level of closeness got diluted [6]. Further exploring users' motivations to participate in viral communication on social media, existing research suggests that heavy users of social media are more likely to participate in a sharing event, influenced by social pressure [7]. Thereby, necessitating a study into the wants of social media users to integrate and differentiate [8].

2.2 Mirroring Attributes in Social Media

This also highlighted the role of self-esteem and self-image because when sending content to a very narrow audience, it was the usefulness of the content that mattered. Broadcasting, however, increased protective self-presentation, decreasing people's willingness to share content that made them look bad [6]. Moving from this degree of social influence to more outward social influences, we found that popular attention by a crowd of common users in the early stages results in large-scale coverage, while the participation of opinion leaders at the early stage does not necessarily lead to the same level of popularity or coverage [9].

2.3 Reading Signals in Social Sharing

But influencers are a big part of the viral Internet challenge trend in the Indian context which prompted us to weave in this variable. Research showed that virality or social transmission of content was about motivation to share according to the recipient and more about the transmitter's internal states [10], another study posited that a basic human motive to self-enhance leads consumers to generate positive word of mouth but transmit negative word of mouth regarding others' experiences [11]. This moved our focus onto the decision-making abilities which were split up between rational and hedonistic tendencies as outlined by the literature on the decision-making process. This is divided into two stages: whether content should be considered and when in agreement, whether they want to interact with it [12]. Figure 1 shows the conceptual model for the likelihood of participation in viral Internet challenges that emerged from this extensive review of literature.

3 Conceptual Model

See Fig 1.

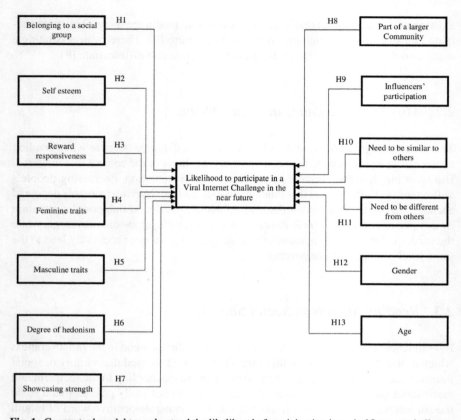

Fig. 1 Conceptual model to understand the likelihood of participation in a viral Internet challenge

4 Hypothesis

H1: People who feel the need to belong in a social group are more likely to take part in a viral Internet challenge in the near future than those people who do not feel the need to belong in a social group.

H2: People who have high self-esteem are more likely to take part in a viral Internet challenge in the near future than those people who have a low self-esteem.

H3: People who have higher responsiveness toward rewards are more likely to take part in a viral Internet challenge in the near future than those people who have a low responsiveness toward rewards.

H4: People who have feminine traits are less likely not to take part in a viral Internet challenge in the future than those people who don't have feminine traits.

H5: People who have masculine traits are more likely to take part in a viral Internet challenge in the future than those who don't have masculine traits.

H6: People who have a higher degree of hedonism are more likely to take part in a viral Internet challenge in the future than those who have a lower degree of hedonism.

H7: People who like to showcase their strength are more likely to take part in a viral Internet challenge in the near future than those who do not like to showcase their strength.

H8: People who feel they need to be part of a larger community are more likely to take part in a viral Internet challenge in the future than those who feel they need not be a part of a larger community.

H9: People who attach more importance to participation of an influencer are more likely to take part in a viral Internet challenge in the future than those who attach no importance to participation of an influencer.

H10: People who have a higher need to be similar to others are more likely to take part in a viral Internet challenge in the future than those who have a lower need to be similar to others.

H11: People who have a higher need to be different from others are more likely to take part in a viral Internet challenge in the future than those who have a lower need to be different from others.

H12: Gender has an impact on participation in a viral Internet challenge in the future.

5 Methodology

5.1 Data Description

The data ($n = 259$) was collected through the means of a pan-India cross-sectional sampling to reach the target segments; an initial quota sampling and subsequent snowball sampling were employed.

Age groups: Based on this measure, our sample size was divided into I (less than 18 years old), II (between 19 and 24 years old), III (between 25 and 30 years old)—these three segments making up out target group owing to their level of exposure to viral Internet challenges. The other age groups included IV (between 31 and 36 years old) and V (above 36 years old). The major emphasis was on the age groups II and III and minor emphasis was on the age groups I, IV and V. This was on the assumption that the emphasized age groups tend to engage with viral Internet challenges and therefore were the target segment.

Demographics: Based on this measure, our sample size and consequent responses were divided almost equally into male—53.7% (139) and female—46.3% (120). 125 (48.2%) of our responses were from major population centers of Bengaluru, Chennai, Delhi, Hyderabad, Ahmedabad, Pune, Mumbai and Kolkata.

Employment status: Based on this measure, our respondents were expected from and classified into students, employed full time (40 or more hours per week), self-employed and other employment options.

5.2 Questionnaire Development

An online survey was developed for the purpose of understanding the attitudes and behavior of Indians across four age groups toward viral Internet challenges. The first section was on the respondents using social media and their awareness of viral Internet challenges.

The next two sections were on peoples' reasons for their choice of engagement with viral Internet challenges including their thoughts and feelings on the action.

The following sections were about their self-description in adjectives, perceptions toward task difficulty, social belonging, the need for social integration/differentiation and affinity to the attributed of a viral Internet challenge to ascertain their individual selves. The succeeding section measured their tendency to make decisions through either emotional or rational means and closed with questions on age, gender, employment status and location.

6 Results and Discussion

By running ordered logistic regression in Stata, we were able to find out that belonging to a social group increased the odds on the likelihood of participation in a viral Internet challenge in the near future by 5%. This means that people who had a greater sense of belonging in the social group they were a part of, were significantly more open to taking part in a viral Internet challenge in the near future (Table 1).

It was observed that influencer participation increased the odds by 54.3% on the respondents' likelihood of participation. This shows that an external factor like an influencer doing a viral Internet challenge heavily influenced our respondents to participate in a viral Internet challenge. The need to showcase own strengths too significantly increased the odds of participation in the future—they rose by 34.7%.

It was also observed that reward responsiveness of the respondents decreased the likelihood of participation in a viral Internet challenge in the near future by 9.8%. This meant that respondents who were not driven by the need for immediate rewards were more likely to participate.

An interesting observation was the correlation of masculine and feminine traits to the likelihood of participation in a viral Internet challenge. These traits were measured through the Bem's sex role inventory scale (BSRI-12). Those exhibiting masculine traits were 4.6% more likely to take part in a viral Internet challenge in the near future per unit change in masculinity, and conversely the presence of feminine traits decreased the likelihood of participation in the future by 4.4%.

Table 1 Results of ordered logistic regression model

Variables	Odds ratio	z	P > [z]	95% conf. interval	
Belonging in a social group	1.05*	2.09	0.037	1.003	1.098
Self-esteem	1.082	0.79	0.428	0.891	1.313
Response to reward responsiveness	0.902**	−3.82	0.000	0.855	0.951
Being a part of larger community	1.038	0.41	0.682	0.869	1.239
Being a part of social cause	0.935	−0.74	0.460	0.783	1.117
Effect of influencers	1.543**	5.09	0.000	1.306	1.823
Showcasing own strength	1.347**	3.25	0.001	1.125	1.613
Feminine trait	0.956*	−2.35	0.019	0.921	0.993
Masculine trait	1.046*	2.34	0.019	1.007	1.085
Hedonic	1.016	1.12	0.263	0.988	1.046
Need to be similar to others	1.033	0.84	0.399	0.958	1.114
Need to be different from others	1.01	0.19	0.852	0.913	1.117
Age	1.249	1.13	0.258	0.850	1.837
Gender	0.778	−0.98	0.326	0.491	1.266

Note *If the P value ($P[Z]$) is less than 0.05 and greater than 0.01 ($0.01 < P[Z] < 0.05$)
**If the P value ($P[Z]$) is less than or equal to 0.01 ($P[Z] \leq 0.01$)

Regression showed that self-esteem, degree of hedonism, need to be similar to others and need to be different from others, gender and age showed no impact on likelihood of participation.

7 Limitations and Future Research

This study focused on understanding social media users' actions toward viral Internet challenges which were based on the motivations underpinning their choices. Further studies could focus on understanding attitudes toward and the drivers of the specific choices between viral Internet challenges and the role of brand marketing in the same context of spurring participant from the users. Our study only focused on sample derived from Indian population, a study with different nationalities could be done for a better global understanding of the research question.

8 Conclusion

8.1 Objectives and Summary

The aim of a viral Internet challenge is to engage social media users and allow them to participate so that the phenomenon gains coverage and momentum on the Internet space. In this research, we sought to understand the motivation behind the action and thoughts of social media users in an Indian context regarding their likelihood of participation in a viral Internet challenge.

8.2 Primary Findings

Our study showed that the effect of an influencer's participation in a viral Internet challenge served as a strong external positive impact on the likelihood of participation in a viral Internet challenge. The analysis also revealed that while their responsiveness toward immediate rewards translated to lower likelihood of intended participation, the respondents' need to showcase their strengths was a driving factor to do a viral Internet challenge. A strong sense of belonging to a social group significantly increased the intention to participate. The emergent picture then was of a social media audience who were strongly impacted by influencers in the societal context, traded immediate rewards for social currency and the chance to exhibit strengths. However, these factors could also be standalone yet significant aspects of a diverse set of audience.

8.3 Secondary Findings

Masculine traits significantly enhanced the willingness to participate in a viral Internet challenge and feminine traits in the respondents reduce the likelihood of participation, respectively. This result held independent of the use of gender as a control variable. This result suggests an interesting link between our gender role identities as masculine/feminine and the attraction of viral Internet challenges and merits further research.

Acknowledgements The authors would like to express their heartfelt gratitude to the faculty members and colleagues at Amrita School of Business for their motivation throughout the process of this business research. The authors would also like to thank all the respondents for their time and effort put forward to fill the survey.

References

1. Berger, J.: Word of mouth and interpersonal communication: a review and directions for future research. J. Consum. Psychol. **24**(4), 586–607 (2014)
2. Berger, J., Iyengar, R.: Communication channels and word of mouth: how the medium shapes the message. J. Consum. Res. **40**(3), 567–579 (2013)
3. Berger, J.: Arousal increases social transmission of information. Psychol. Sci. **22**(7), 891–893 (2011)
4. Reichstein, T., Brusch, I.: The decision-making process in viral marketing—A review and suggestions for further research. Psychol Mark. **36**, 1062–1081 (2019)
5. Camarero, C., San José, R.: Social and attitudinal determinants of viral marketing dynamics. Comput. Human Behav. **27**(6), 2292–2300 (2011)
6. Barasch, A., Berger, J.: Broadcasting and narrowcasting: How audience size affects what people share. J. Mark. Res. **51**(3), 286–299 (2014)
7. Tiago, M.T., Tiago, F., Cosme, C.: Exploring users' motivations to participate in viral communication on social media. J. Bus. Res. **101**, 574–582 (2019)
8. Baumeister, R.F., Leary, M.R.: The need to belong: Desire for interpersonal attachments as a fundamental human motivation. Psychol. Bull. **117**(3), 497–529 (1995)
9. Zhang, L., Zhao, J., Xu, K.: Who creates trends in online social media: the crowd or opinion leaders? J. Comput. Mediated Commun. **21**(1), 1–16 (2016)
10. Berger, J., Milkman, K.L.: What makes online content viral? J. Mark. Res. **49**(2), 192–205 (2012)
11. De Angelis, M., Bonezzi, A., Peluso, A.M., Rucker, D.D., Costabile, M.: On braggarts and gossips: a self enhancement account of word of mouth generation and transmission. J. Mark. Res. **49**(4), 551–563 (2012)
12. Guadagno, R.E., Rempala, D.M., Murphy, S., Okdie, B.M.: What makes a video go viral? An analysis of emotional contagion and Internet memes. Comput. Hum. Behav. **29**(6), 2312–2319 (2013)

Early Flood Monitoring System in Remote Areas

John Colaco and R. B. Lohani

Abstract The flood is one of the most dangerous natural disasters in the world. Due to this natural disaster, many humans as well as animals have lost their lives. This flood occurs in many ways such as tsunamis, extremely heavy rainfall, etc. In this paper, Internet of things-based wireless system for continuous alert and monitoring system is introduced. The rate of flow and high rise in water level detected using water flow sensors, pressure level sensors, and rain sensors. This system having sensors node placed at places such as rivers, dams, drainages, pipes. The data collected by sensors communicated over Wi-Fi to the control room where alarms get generated when received sensor values exceed beyond particular threshold and authorities need to take necessary preventive action and also an alert message sent over mobile device of the authorities.

1 Introduction

The most protruding of the world's monsoon systems are Indian monsoon system. This mainly affects Indian people and water bodies surrounding them. This monsoon comes from the northeastern parts of India during the coolest months and reverses its direction to come from the southwestern parts of India during the warmest months of the year. This brings the lot of rainfall during the month of June and July. Weather forecasts are moist, cloudy, and warm all over India most of the time. Rainfall ranges between 400 and 500 mm depending upon the states especially in Goa where rainfall lavished with on balance 2813 mm of rainfall per year, or 234.4 mm per month. The excessive rainfall gives rise in water level and so flood. The Table 1 depicts the

---

J. Colaco (✉) · R. B. Lohani
Department of Electronics & Telecommunication Engineering, Goa Engineering College, Ponda, Goa, India
e-mail: j_7685@yahoo.com

R. B. Lohani
e-mail: rblohani@gec.ac.in

Table 1 Water level status of Goa dams

Name of the dam	Max water level (m)	Threshold level (m)	Current status water level (m)	% of Dam filled
Selaulim	41.15	20.42	37.14	67
Anjunem	93.20	61.50	88.82	77
Chapoli	38.75	22.00	36.21	79
Amthane	50.25	29.00	48.05	73
Panchawadi	26.00	14.70	22.30	43

findings of the survey conducted by the researchers about status of dam's in Goa [1] as it is observed by the authors that the water level in the dams in Goa is giving serious threat of flood as the water level of the dams is exceeding threshold level (dead level), and the same is analyzed by plotting graphs as shown in Fig. 1. Floods arise in all depths, from inches to many feet. The velocity of flow of floodwater is sometimes extraordinary and deadly. Within no time, extremely excessive rain increases water level into a 30-foot-high that overpowers everything in its way or path. Flooding happens due to the overflowing of rivers, dams streams, lakes, or oceans. In remote areas of Goa, an alarming system is required to take preventive measures by the government authorities especially disaster management cell and be more ready to overcome its effects. Wireless sensor networks (WSN) have proved that its worth in monitoring environmental disasters and physical quantities. That is due to this technology that remarkable capability of quick capturing, processing, and transmission of life-threatening data in real time with high resolution. One of the most important applications of wireless sensor networks is environmental disasters. In this system, Raspberry Pi, rain sensors, water-level sensors used to make flood prediction and warn authorities, and also, pressure-level sensors perform liquid-level measurement by submerging sensor at a certain fixed depth under the water surface

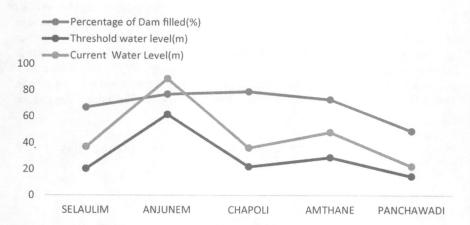

Fig. 1 Graphical analysis of water level status of dams in Goa

and sound quick warning in nearby villages to quickly transmit information o about possible floods using Internet of things (IOT). The water sensors are used for the measurement of water level at various locations. Also, rain sensors are used for the measurement of rainwater level in remote areas. These sensors generate their information over the IOT using Raspberry Pi. After detecting the conditions of flooding, the system makes predictions about the time required to flood in a particular area and warns the villages/remote areas that could be affected by it. The system also estimates the time required for flood to reach them and allocates a time to people for evacuation accordingly. In order to obtain the real-time water level of water bodies and also the flow rates, sensors are used. This data is transferred to the controller section over Wi-Fi. The Raspberry Pi3 updates the sensor data graph on a Web site which provides the weather data from the Internet. Also, an alert is provided if the values are not within the safe levels. The alert includes a buzzer sound and also an SMS to the concerned authority.

2 Related Work

In [2], authors alerting the authorities only after flood occurrence thereby monitoring the water level in sewage system in urban areas which causes severe damage as that happened in Chennai. Also, the rainfall outline of the area and the forecast for the area is considered. This information enables the authorities to be more prepared in the eventuality of a flood. In [3], artificial neural network algorithm such as radial bias function is used to predict flood, and final, the result is displayed on Android mobile application. In [4], the system helps in the transformation in management of city and improves life of citizens. This will also help in making the public awareness among the various people regarding the change of climate in the city. The monitoring system in real time established to check the level of water changes and warn the user through SMS or using GSM module. In [5], Eagle Layout Editor 5.4 software is used to design real-time flood monitoring system which only monitors changes in level of water. In [6], sensors have been used for measuring rate of flow of water, perception, and water level. The data unit processes through GPRS node. Here long-distance communication and monitoring, the various GDUs are a challenge. In [7], based on the integrated hydrological as well as meteorological data flood warning and monitoring are issued. In [8], the system is designed to measure only the speed of water and height of river. The proposed system is easily installed in a remote area having limited cellular data network infrastructure which makes system more reliable and accurate.

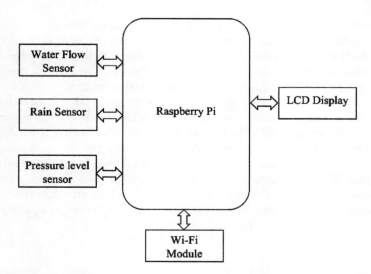

Fig. 2 Block diagram of data acquisition and transmission unit

3 Overview

This paper is divided into two sections: one section is data acquisition and transmission and other section is data receiving and monitoring. In the first section, data with respect to the amount of rainfall, height of water level, and rate of flow of water detected using respective sensors and the same transmitted through Wi-Fi to the control room of authority which is the receiving station.

In the second section, the detected data received through Wi-Fi to the said control room and alarm generated if the received values are exceeding the threshold level, and then the concerned authorities have to take urgent necessary action. The received sensed data displayed on the created Adafruit server, and using python script, the graph is plotted. Weather forecast data will be also displayed, and the information is displayed.

3.1 Data Acquisition and Transmission Unit

See Fig. 2.

3.2 Data Receiving and Monitoring Unit

See Fig. 3.

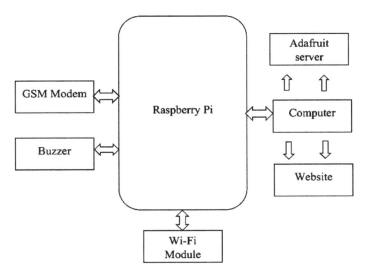

Fig. 3 Block diagram of data receiving and monitoring unit

3.3 Flowchart

See Fig. 4.

4 System Details

4.1 Water Flow Sensor

This water flow sensor is also known as area velocity flow meter which is used to measure water flow through open channels, full pipes(partial), and surcharged pipes which are ideal for the wastewater, stormwater, effluent, industrial wastewater, and irrigation water [9]. It has an ultrasonic sensor which is submerged to measure both water level and velocity of water flowing in the channel continuously. This sensor transmits a sound wave that reflects back from water surface and measures the time it takes for the echo to return and determine the distance to water [10]. The sensor has resistant to abrasion, corrosion, and fouling because materials exposed are plastic; hence, the sensor resists abrasion, corrosion, and fouling. It does not have moving parts and no ports, orifices, and electrodes. The flow meter configuration is done with the standardized submerged velocity-level sensor along non-contacting ultrasonic level sensor separately, for extremely aerated fluids or those with extreme concentration. In this model, sensor AVFM 6.1 area-velocity flow meter is used to determine both velocity and level and measures flow in a pipe or open channel.

Fig. 4 Flowchart

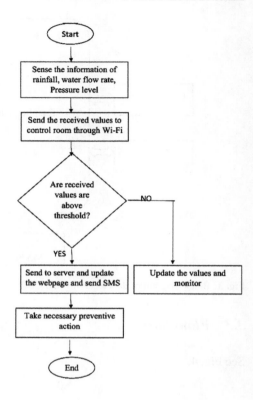

This configuration is very simple. First enter the diameter of pipe or dimensions of open channel, and then it will compute and display flow volume automatically. Inside the pipe/channel, the ultrasonic sensor is mounted and on the bottom of a channel having a mounting bracket of stainless steel and one screw into the bottom of the pipe or channel. The sensor is sealed with no openings storing time and date-stamped flow values at intervals between 10 s to 60 min. Reports of daily flow are created automatically where average, minimum, maximum and total flow rates are displayed. Log files and daily flow reports can transfer to flash drive such as USB by connecting to the USB output of loggers. The velocity-level sensor measures flow in partially or full and surcharged pipes up to 10 psi (pressure). A proper sensor selection of mounting location gives the best performance and maintenance-free operation. Using AVFM 6.1, one can measure forward flow velocity up to 6 m/s(20 ft/sec) as well as reverse flow up to 1.5 m/s (5 ft/sec) (Figs. 5 and 6).

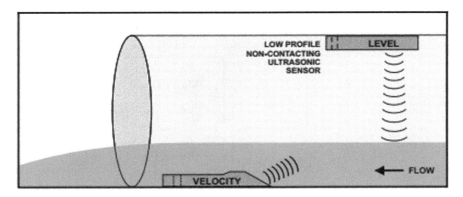

Fig. 5 General view of flow and velocity detection

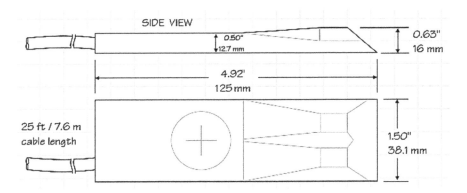

Fig. 6 Dimensional view of velocity sensor

4.2 Rain Sensor

Rain detection sensor is used in detecting rainwater. On the rain board what happens is that when raindrop falls on it, it could be used as a switch and also for measuring the intensity of rainfall. This raindrop sensor used has two modules: one is control module which relates the value of analog and converts it to a value of digital and other module has rain board that detects the amount of rainfall [11]. Further, these sensors are being used in the automobile sector to automatically control the windshield wipers, home automation systems also in the agriculture field. Interfacing the raindrop sensor with a microcontroller such as Raspberry Pi and Arduino is simple. This rain board module has been connected along with the control module of the raindrop sensor. When the surface of rain board module's gets exposed to rainwater, the surface of the rain board module will become wet, and it will offer less resistance to the supply voltage. Hence, the minimum voltage will appear at the non-inverting terminal of

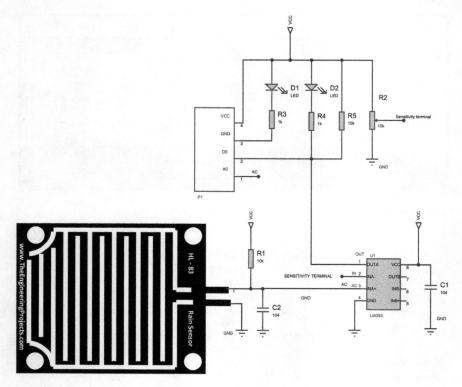

Fig. 7 Circuit diagram of a raindrop sensor module

operational amplifier. The comparison between both inverting and non-inverting terminal voltages is being done by comparator. Now if the condition is such that the output of the Op-Amp will be digital HIGH and digital LOW under any case (Fig. 7).

4.3 Pressure Level Sensor

The pressure sensor used is Seametrics PT2X Sensor which is a smart sensor. The sensor networks were with all Seametrics Smart Sensor family. It has data logger integrated and temperature as well as pressure sensor [12]. This sensor is used for monitoring water level in the tank, well, groundwater, sea tide as well as for slug testing and pumping. It has digital intelligent sensor which is based on a microprocessor intended to measure and record temperature, pressure and time and battery-operated circuitry of low power. Measurement of pressure is done with a very durable and stable piezo-electric media, isolated pressure element which has combination of 16-bit analog-to-digital converter and hence provides extremely stable pressure and accurate input into the microprocessor. It has industry standard digital RS485 interface device which records high of temperature/pressure level and time data, operating with low power. This pressure sensor built with stainless steel or titanium and fluoropolymer, thus providing high-accuracy readings in corrosive field conditions, the field using a portable computer and DataGrabber.

4.4 Raspberry Pi

Raspberry Pi 4 model has good features such as high-performance Quad core Cortex-A72 (ARM v8) with 64-bit System on chip (SoC) at 1.5 GHz 1 GB, Broadcom BCM2711, 2 GB or 4 GB LPDDR4-3200 SDRAM [13]. It enhances the speed of processor, memory, multimedia performance and connectivity compared to the previous-generation Raspberry Pi model while retaining similar power consumption, backwards compatibility. It has dual-display which supports at high resolutions up to 4 K, hardware video decoder up to 4 gb of RAM Gigabit Ethernet, USB 3.0, Bluetooth 5.0 dual-band 2.4/5.0 GHz wireless LAN. It has a good quality 2.5 A power supply and is used if downstream USB peripherals consume less than 500 mA overall.

4.5 Adafruit Server

Adafruit server is a cloud service. Uploading, monitoring and displaying the information are done through Internet and make system IoT enabled. It is primarily used for and retrieving and storing data. It has functions such as displaying data in real time, online, make systems Internet-connected: Adafruit.io can visualize and handle multiple feeds of data. Integrated dashboards on Adafruit IO allows to log, graph, chart, gauge, and display data and view dashboards from anywhere around the world. The feeds are the core of Adafruit server. They can handle the data you have uploaded and meta-data the sensors push to Adafruit IO [14].

5 Conclusion

The current system has ability to save as much people as possible in the remote areas by alerting the government authorities especially the disaster management cell about the critical situations of flood level in real time, and hence, the system can issue a warning by sounding an alarm and also by SMS in the events flood. However, the limitations can be water level errors because of floating objects flowing through the ultrasonic sensor. There could delay in taking preventive action from authorities due to the remote areas.

6 Future Scope

For future research, the system can be modified using various soft computing techniques such as fuzzy logic, artificial neural networks, artificial intelligence and analyze using MATLAB and mobile application can be also developed to continuously observe and monitor the data real time.

References

1. https://goawrd.gov.in/dam-levels
2. Devaraj Sheshu, E., Manjunath, N., Karthik, S., Akash, U.: implementation of flood warning system using IoT. In: 2nd International Conference on Green Computing and Internet of Things (ICGCIOT). IEEE, Bangalore, India (2018)
3. Ni, K., Ega, K., Muhammad, A.M., Casi, S: Floods prediction using radial basis function (RBF) based on Internet of Things (IoT). In: IEEE International Conference on Industry 4.0, Artificial Intelligence, and Communications Technology (IAICT). IEEE, Bali, Indonesia (2019)
4. Pradeep Kumar, N., Ravi Kumar, J.: Development of cloud based light intensity monitoring system using raspberry pi. In: International Conference on Industrial Instrumentation and Control (ICIC). IEEE, Pune, India (2015)
5. Baharum, M.S., Awang, R.A.,. Baba, N.H.: Flood monitoring system. In: International Conference on System Engineering and Technology (ICSET). IEEE, Shah Alam, Malaysia (2011)
6. Shen, S., Chen, P.: A real-time flood monitoring system based on GIS and hydrological model. In: 2nd Conference on Environmental Science and Information Application Technology. IEEE, Wuhan, China (2010)
7. Limlahapun, P., Fukui, H.: Flood monitoring and early warning system integrating object extraction tool on web-based. ICCAS-SICE, IEEE, Fukuoka, Japan (2009)
8. Nasution, T.H., Siagian, E.C., Tanjung, K.: Soeharwinto, Design of river height and speed monitoring system by using Arduino. 10th International Conference Numerical Analysis in Engineering, vol. 308. IOP, Banda Aceh, Indonesia (2018)
9. https://www.inmtn.com/water/water-flow/flow-sensors/
10. https://www.waterworld.com/technologies/flow-level-pressure-measurement/article/162 10884/ultrasonic-sensors-for-water-level-measurement
11. https://components101.com/sensors/rain-drop-sensor-module/
12. https://www.rshydro.co.uk/water-level-sensors/pressure-level-sensors/pressure-transducers/sdi-12-pressure-temperature-sensor/
13. https://www.raspberrypi.org/products/raspberry-pi-4-model-b/
14. https://iotdesignpro.com/projects/control-raspberry-pi-gpio-with-adafruit-io-to-trigger-an-led

A Survey on Securing Payload in MQTT and a Proposed Ultra-lightweight Cryptography

Edward Nwiah and Shri Kant

Abstract Internet of things foresees devices that are connected globally, accessing and sharing pertinent information to make life easier and better. This highly interconnected global network seeks to advance business productivity, government efficiency and agriculture growth. However, these fantastic opportunities also present a number of significant security challenges. Quite a number of researches have gone into the domain of Internet of things (IoT). MQTT which is the short form for Message Queuing Telemetry Transport is an application layer protocol that has been proposed to smoothen publish and subscribe procedures in exchanging data between client and the server. MQTT, however, was designed without proper security imbibition. In this paper, recent security mechanisms for enhancing the security of the MQTT protocol have been studied and analysed comparatively. Based on the challenges that surfaced from the survey, this research proposes an ultra-lightweight cryptography, Hummingbird-2 as a novel security solution. This is done to achieve bandwidth and memory efficiency, quality of service of data delivery.

1 Introduction

IoT is an emerging technology that extends the power of Internet to encompass computers, smartphones other devices, processes as well as the environments. Such connected devices help in collecting and sending information to and fro. IoT offers a viable means in simplifying forthcoming Internet concept, in a way domains which

---

E. Nwiah (✉)
M. Tech CSE—Networking & Cybersecurity Department of Computer Science, Sharda University, Knowledge Park III, Greater Noida, Uttar Pradesh 201310, India
e-mail: nwiahkikii@gmail.com

S. Kant
Department of Computer Science, Sharda University, Knowledge Park III, Greater Noida, Uttar Pradesh 201310, India
e-mail: shri.kant@sharda.ac.in

include smart cities, smart homes, public health, energy management, agriculture, smart transportation, smart grids, waste management, which embedded with sensors, actuators, electronics and network connectivity which enables such domains to communicate [1]. Though the connectivity is capable of transforming various sectors of life, there is still a challenge that potential hackers might cease this opportunity to exploit these technologies in order to cause harm with these critical infrastructure. [2].

Gartner mentioned in its report predicting massive sales of IoT devices by 2020 and indicating approximately twenty-one billion devices connecting to each other by some said date which is 2020. The rate at which systems, devices and services are capitalizing on the IoT environment is really creating great opportunities, and its relevance to the society is highly substantial. However, the security aspects are not marching parallel with the innovations it brings on board, thereby creating mayhem and economic risk [3]. Security in IoT devices and systems raises concern as developers put much emphasis on customer convenience, functionality, compatibility requirements, etc., rather than security. About 600% attacks were launched against IoT devices from 2016 to 2017 according to Symantec, and this indicates that threats and security issues are rising each day [4].

According to the Vice President of Gartner Research, Bob Gill, "By 2020, more than 25% of identified attacks in enterprises will involve the IoT, although the IoT will account for less than 10% of I.T. security budgets". IoT presents variety of challenges which includes security dangers to IoT devices, operating systems and platforms, and they are connected to. In this regard, competent technologies are needed to safeguard IoT devices and platforms from the hands of hackers tempering on salient data [5].

Billion devices exchange information through to the Internet and are therefore evident that these devices are prone to security attack. Protocols in IoT systems differ from traditional Internet protocols and as a result work comfortably with constrained devices. Some of these constraints include low power consumption, limited bandwidth, low computational capabilities and small memory size. A lot of protocols have been implemented and are already standardized in the Internet of things.

In view of this Message Queuing Telemetry Transport (MQTT) protocol came into the limelight. The main focus of MQTT was to cater for devices with low bandwidth, high latency or unreliable network. MQTT is entirely lightweight protocol since it offers a lean header structure and also simple to implement. This research therefore proposes Hammingbire-2 to provide extensive security to MQTT since it was not developed with security mechanisms.

1.1 Objectives

The following objectives have been enumerated by the researcher in this survey:

1. To analyse existing security techniques that have been implemented in securing MQTT protocol.

2. To develop an system to secure payload in an IoT environment using MQTT.

1.2 Problem Statement

IoT systems comprise of "things" such as temperature and humidity sensors, radio-frequency identification (RFID), motion sensors which are constraint devices with limited memory, low power consumption, limited computational capacity and limited bandwidth. There are various protocols proposed for IoT devices which MQTT is one of them. This protocol is devoid of strong security for transferring messages, the basic authentication mechanism data are transmitted in clear text which is not secure. This protocol depends on another protocol for its security which is Transport Layer Security. This TLS is a weightier and expensive with regard to its computation. Therefore, a lightweight encryption for payload is suggested as optimal option [6]. This paper proposes an ultra-lightweight cryptography to secure payload in MQTT in an IoT-based network.

1.3 MQTT Protocol

MQTT is a protocol created and works best for constrained devices due to its lightweight nature. It works on top of the transport layer protocol TCP/IP for sending messages. It was proposed to smoothen publish and subscribe procedures in exchanging data between client and server. It was invented by Arlen Nipper of Arcom (now Eurotech) in 1999 and Dr. Andy Stanford Clark of IBM [7]. IBM submitted this MQTT to OASIS specification body in 2013, and its standardization was successfully completed [8]. OASIS in 2014 approved version 3.1.1 of MQTT as a standard application layer protocol. [9]. MQTT possesses versatility features such as simplicity, openness, lightweight, low power usage and again designed to be easily implemented. These features render it effective in many circumstances, involving constrained environments which include machine-to-machine communication and IoT contexts where network bandwidth is highly rated (Fig. 1).

2 Review of Existing Techniques

A number of techniques have been proposed by many researchers in securing MQTT. This section presents an analysis of some techniques proposed by other researchers.

Bisne and Parmar [11] proposed a composite secure MQTT for Internet of things by means of dynamic S-box advanced encryption standard (AES) and attribute-based encryption (ABE). They presented a key result which offer confidentiality and access control of the exchange of information in an MQTT protocol. In their proposed

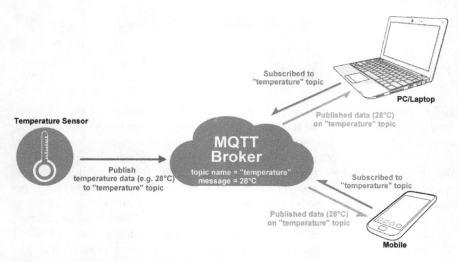

Fig. 1 Basic architecture of MQTT. Adapted from [10]

system, access is being provided by an external trusted authority which increases the overhead. Again this approach uses both private key and public key cryptographic solution and as a result doubles the decryption process by the subscriber which in turn increases the overhead.

Singh et al. [12]. In this paper, a technique called Key/Ciphertext-Policy Attribute-Based Encryption using lightweight elliptic-curve cryptography was used. CP/KP-ABE uses the bilinear pairing operations. The use of this scheme is computationally expensive. Again these operations are not appropriate for constrained devices.

Calabretta et al. [13] proposed a technique for MQTT using the AugPAKE. The proposed technique uses the AugPAKE algorithm for ensuring confidentiality. In this system, two tokens are used which provides authorization in accessing a topic and also authenticates how topics are being used.

Niruntasukrat et al. [14] proposed a mechanism for MQTT-based Internet of things known as the authorization mechanism. This ensures the authorization for single topic to be accessed. This security mechanism introduces a fourth element in the MQTT setup and that creates an overhead in the process of exchanged messages and the entire communication process.

Shin et al. [15] proposed a security framework for MQTT protocol. The proposed solution makes use of augmented password-based key exchange (AugPAKE) protocol. It is relatively useful because two separate session keys are used in the negotiation between the publisher and the broker, and the broker and the subscriber. This is done without the need for certificates either for validation or revocation. Their technique does not cater for the confidentiality of the data.

Matischek et al. [16] proposed a lightweight authentication mechanism for M2M communications in industrial IoT environment using Hash and XOR operations. The model is said to have low computational cost, communication and storage overhead, at the same accomplishing common authentication and above all agreement of session key but no mechanism for the data protection.

Almuhammadi et al. [17] in their proposed, common AES modes of operation are analysed. The comparison was done with regards to time taken for encryption and decryption and throughput with variable data packet sizes. In their analysis, they concluded that, for faster modes of operation ECB is considered compared to other modes of operation.

Thatmann et al. [18] the focus of their research was to use encryption for securing MQTT. Attribute-based encryption technique was used to publish/ subscribe message patterns in MQTT. Their security solution was used to compare different security mechanism previously studied. Their proposed solution is used to mitigate cluster communication security not end-to-end communications. Their work requires another entity called the group controller to communicate with MQTT network across HTTPS RESTful calls which is a tailback.

Peng et al. [19] they used identity-based cryptography (IBC) to develop a secured MQTT protocol. With this cryptography, a client, who accesses the public parameters of the system, also encrypts a message using the receiver's key which comprises name or email address. The decryption is derived from a central authority and sent to the receiver. The decryption is trusted because it also generates a secret for all users resulting several entities, IoT gateways and external administrators, acting as private key generators and administrating distinctive trust areas are introduced.

Another shortcoming of this technique is the introduction of IoT gateways and devices which handles two IBC-based private keys, and this in effect generates more overhead.

Upadhyay et al. [20] they proposed a solution that seeks to secure MQTT by the use of access-control lists (ACLs) included in the mosquito broker. With this technique, different usernames and passwords are needed for different data and which in turn causes more overhead.

Bhawiyuga et al. [21] in their security solution, to authorize access to a specific topic a token is required. To do this, another entity to carry out the access tokens is known as the authentication server. The introduction of the authentication server contributes to further overhead both in exchanging message and the management of the server.

Katsikeas et al. [22] in their research highlighted appropriate properties of MQTT as a lightweight protocol and its usage for industrial purposes. Different security techniques for MQTT were evaluated, noted among them was payload encryption with AES, payload encryption with AES-CBC, payload authenticated encryption with AES-OCB and link layer encryption with AES-CCM. It was concluded in their evaluation that all the stated techniques are good depending on the purpose.

3 Comparative Analysis of Literature Review

S. No.	Author	Year	Title	Tools/techniques	Contribution
1	L. Bisne and M. Parmar	2017	Composite secure MQTT for Internet of things using ABE and dynamic S-box AES	ABE and dynamic S-box AES	There are confidentiality and access control of data but the introduction of external trusted authority for access policy increases the overhead
2	Singh et al.	2015	Secure MQTT for Internet of things (IoT)	Key/Ciphertext-Policy Attribute-Based Encryption using lightweight Elliptic Curve Cryptography	CP/KP-ABE uses the bilinear pairing operations. The use of this scheme is computationally expensive These operations are not appropriate for constrained devices
3	Calabretta et al.	2018	A token-based protocol for securing MQTT communications	AugPAKE security protocol	Their technique ensures confidentiality. In this system, two tokens are used which provides authorization in accessing a topic and also authenticate how topics are being used

(continued)

(continued)

S. No.	Author	Year	Title	Tools/techniques	Contribution
4	Niruntasukrat et al.	2016	Authorization mechanism for MQTT-based Internet of things	OAuth framework	Their system ensures the authorization for single topic to be accessed. Again a fourth element is introduced in the MQTT setup and that create an overhead in the process of exchanged messages and the entire communication process
5	Shin et al.	2016	A security framework for MQTT	AugPAKE security protocol	Two separate session keys are used in the negotiation between the publisher and the broker, and the broker and the subscriber. Their technique does not cater for the confidentiality of the data
6	Matischek et al.	2017	A lightweight authentication mechanism for M2M communications in industrial IoT environment	Hash and XOR operations	Their proposal is a lightweight authentication solution which uses only hash and XOR operations

(continued)

(continued)

S. No.	Author	Year	Title	Tools/techniques	Contribution
7	Almuhammadi et al.	2017	A comparative Analysis of AES Common Modes of Operation		The comparison was done with regard to the time taken for encryption and decryption and throughput with variable data packet sizes and concluded that ECB mode is faster than other modes of operation
8	D. Thatmann et al.	2015	Applying Attribute-based Encryption on Publish Subscribe Messaging Patterns for the Internet of things	Attribute-based Encryption (ABE)	Technique used to mitigate cluster communication not end-to-end communications. Their work requires another entity called the "group controller" which communicate with the MQTT network through HTTPS RESTful calls which is a tailback
9	Peng et al.	2016	A secure publish/subscribe protocol for Internet of things using identity-based cryptography	Using identity-based cryptography	IoT gateways and external administrators are used, thereby increasing overhead

(continued)

(continued)

S. No.	Author	Year	Title	Tools/techniques	Contribution
10	Upadhyay et al.	2016	MQTT-based secured home automation system	Access Control Lists (ACLs)	Access Control Lists (ACLs) included in the Mosquito broker With this technique, different usernames and passwords are needed for different data and which in turn causes more overhead
11	Bhawiyuga et al.	2017	Architectural design of token-based authentication of MQTT protocol in constrained IoT device	Token-based solution	Different entity to carry out the access tokens known as the authentication server. The introduction of the authentication server contribute to further overhead
12	S. Katsikeas et al.	2017	Lightweight and secure Industrial IoT Communications via the MQ Telemetry Transport Protocol	Evaluation of different security options	Comparative analysis of different security mechanism to secure MQTT protocol

4 Research Questions

1. What are some of the challenges identified in the existing techniques?
2. Which cryptographic technique can be implemented to address the security challenges of payload being transmitted in MQTT?

5 Proposed Model

5.1 Logical Architecture

The logical architecture as shown in Fig. 2 depicts a conceptual design of the proposed model which identify the workflow between the components. The components of the architecture include publisher, key management system (KMS), broker and subscriber. This shows a typical MQTT architecture with its entities communicating through the broker. The role of the publisher is to publish a payload (temperature reading). The subscriber is the entity that is allowed to receive a payload on topics of interest. KMS is another entity which provides a management service by distributing the secret key to the clients (publisher and subscriber). The broker then serves as a communication medium between the publisher and subscriber. It receives payload from the publisher and transmits to the authenticated subscriber.

The security technique this research is proposing to address the challenges outlined is an ultra-lightweight encryption scheme known as Hummingbird-2. Hummingbird-2 was proposed as an improvement to Hummingbird-1 which combines stream cipher and block structure. It has 256-bit security and can recuperate the entire private key with at most 2^{64} offline computational effort under two related IVs. Nevertheless, Hummingbird-2 incorporates in its design some basic design of Hummingbird-1 [24]. This research proposes Hummingbird-2 because it offers privacy-preserving identification and mutual authentication protocol for constrained devices which includes industrial controllers, smart meters, RFID tags, wireless sensors, etc. Hummingbird-2 has very small hardware as well as software footmark and is convenient for providing security for constraint devices [25]. Hummingbird-2 has a 16-bit block size, 128- bit secret key and a 64-bit initialization vector.

Hummingbird 2 is used in this proposed model to ensure the security of payload in an MQTT-based IoT system. The focus of this proposed model is payload encryption

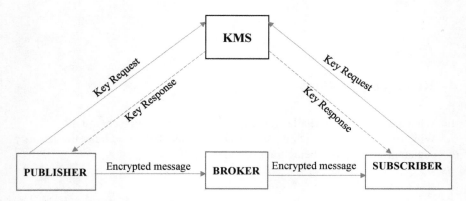

Fig. 2 Logical architecture of MQTT architecture

Fig. 3 Publish packet. Adapted from [23]

in MQTT. The payload is encrypted as part of the publish packets, which is transmitted to the broker. The Broker then forwards the encoded payload to the subscriber who has subscribed to the topic. Decryption occurs at the subscriber's end. All other data including client information and topic information are transmitted in a plaintext from the publisher to the subscriber as shown in Fig. 3.

5.2 Discussion, Analysis and Future Work

Hummingbird 2 is an ultra-lightweight encryption scheme used in this research. The model will be very difficult for hackers to manipulate the content since the payload is encrypted at the publishers end before transmitting to the broker for onward transmission the subscriber. Decryption is only done at the subscriber's destination. The system reads accurately and encrypted temperature readings from three different locations (IoT Lab, Networking Lab and My apartment) which show good progress of the model.

To ensure eavesdroppers are unable to follow the encryption pattern of plaintext by brute force attack, linear and differential cryptanalysis. The encryption is done in a way that same plaintext gives different ciphertext when encrypted multiple times.

The implementation of the model is at its final stage. We strongly hope, analysis after testing the model will yield great results as expected in the design.

6 Conclusion

In this research, a comprehensive comparative analysis was conducted on existing security techniques to secure MQTT protocol. The comparative analysis indicates most techniques that focus on authentication of users and few others on payload being transmitted in an end-to-end communication. Again most of the techniques introduce third parties which to some extent increase overhead depending on their operations.

The proposed security technique focuses on securing the payload being transferred from end-to-end in the MQTT-based system using Hummingbird-2 encryption algorithm.

Acknowledgements The work is possible with the help of many people. My sincere gratitude and appreciation to all those who contributed this project possible.

Firstly, I am extremely grateful to my research guide, Prof. Shri Kant Rastogi for his scholarly inputs and consistent encouragement. I could not have imagined having a better advisor and mentor for my project.

Secondly, a special thanks to my family especially Francis Nwiah, Dorothy Nyarko for their love, affection and financial support.

Finally, I give thanks to God for taking me through all the difficulties. I have experienced Your guidance day by day. I will keep on trusting You for my future. Thank you, Lord.

References

1. El-hajj, M., Chamoun, M., Fadlallah, A., Serhrouchni, A.: Analysis of authentication techniques in Internet of Things (IoT). In: Proceedings of the 2017 1st Cyber Security in Networking Conference (CSNet), Rio de Janeiro, Brazil, 18–20 October 2017, pp. 1–3
2. Commission on Enhancing National Cybersecurity. Report on Securing and Growing the Digital Economy, December 2016, 90 pp
3. Eddy, N.: Gartner: 21 Billion IoT devices to invade by 2020. Information week Nov. 2015
4. Internet Security Threat Report. Symantec Corporation, March 2018. https://www.symantec.com/security-center/threat-report
5. Gartner Insights on How to Lead in a Connected World. "Leading the IoT", https://www.gartner.com/imagesrv/books/IoT/IoTEbook_digital.pdf
6. Edielson et al. M2M Protocols for Constrained Environment in the Context of IoT: A Comparison of Approaches Dec. 2015
7. Rahul Gupta Banks. MQTT version 3.1.1. http://docs.oasis-open.org/mqtt/mqtt/v3.3.3/cos02/mqtt-v3.1.1-cos02.html
8. Oasis. [Online] Available: http://www.oasis-open.org
9. ISO/IEC20922, Information technology-Message Queuing Telemetry Transport(MQTT) v3.1.1, ISO, 2016. [Online]. Available: http://www.iso.org/iso/catalogue_detail.htm?csnumber=69466
10. ElectronicWings. [Online]. Available: http://www.electronicwings.com/nodemcu/nodemcu-mqtt-client-with-arduino-ide
11. Bisne, L., Parmar, M.: Composite secure MQTT for Internet of Things using ABE and dynamic S-box AES, 2017 Innovations in Power and Advanced Computing Technologies (i-PACT), pp. 1–5. Vellore (2017). https://doi.org/10.1109/ipact.2017.8245126

12. Singh, M., Rajan, M., Shivraj, V., Balamuralidhar, P.: Secure MQTT for internet of things (IoT). In: 2015 Fifth International Conference on Communication Systems and Network Technologies. pp. 746–751. IEEE (2015)
13. Calabretta, M., Pecori, R., Velti, L.: A token-based protocol for securing MQTT communications. In: 2018 26th International Conference on Software, Telecommunications and Computer Networks (SoftCOM). pp. 1–6. IEEE (2018)
14. Niruntasukrat, A., Issariyapat, C., Pongpaibool, P., Meesublak, K., Aiumsupucgul, P., Panya, A.: Authorization mechanism for MQTT-based Internet of Things. In: 2016 IEEE International Conference on Communications Workshops (ICC), pp. 290–295 Kuala Lumpur (2016). https://doi.org/10.1109/iccw.2016.7503802
15. Shin, S., Kobara, K., Chuang, C.C., Huang, W.: A security framework for MQTT. In: 2016 IEEE Conference on Communications and Network Security (CNS). pp. 432–436. IEEE (2016)
16. Matischek, R., Saghezchi, F., Rodriguez, J., Bicaku, A,. Maksuti, S., Tauber, M., Schmittner, C., Esfahani, A., Mantas, G., Bastos, J.: A Lightweight Authentication Mechanism for M2M Communications in Industrial IoT Environment. In: IEEE Internet of Things Journal (2017)
17. Almuhammadi, S., Al-Hejri, I.: A comparative Analysis of AES Common Modes of Operation. In: IEEE 30th Canadian Conference on Electrical and Computer Engineering (CCECE) (2017)
18. Thatmann, D., et al.: Applying attribute-based encryption on publish subscribe messaging patterns for the internet of things. In: IEEE International Conference on Data Science and Data Intensive Systems (2015)
19. Peng, W., Liu, S., Peng, K., Wang, J., Liang, J.: A secure publish/subscribe protocol for Internet of Things using identity-based cryptography. In: 2016 5th International Conference on Computer Science and Network Technology (ICCSNT), pp. 628–634. Changchun (2016). https://doi.org/10.1109/iccsnt.2016.8070234
20. Upadhyay, Y., Borole, A., Dileepan, D.: MQTT based secured home automation system. In: 2016 Symposium on Colossal Data Analysis and Networking (CDAN), pp. 1–4. Indore (2016). https://doi.org/10.1109/cdan.2016.7570945
21. Bhawiyuga, A., Data, D., Warda, A.: Architectural design of to- ken based authentication of MQTT protocol in constrained IoT device. In: 2017 11th International Conference on Telecommunication Systems Services and Applications (TSSA), pp. 1–4. Lombok (2017). https://doi.org/10.1109/tssa.2017.8272933
22. Katsikeas, S., et al.: Lightweight and secure Industrial IoT Communications via the MQ Telemetry Transport Protocol (2017)
23. HIVEMQ. [Online]. Available: http://www.hivemq.com/mqtt-essentials
24. Saarinen, M.J.O.: Cryptanalysis of hummingbird-1, fast software encryption, FSE'11. LNCS **6733**, 328–341 (2011)
25. Engels, D., Saarinen, M.J.O., Schweitzer, P., Smith, E.M.: The Hummingbird-2 lightweight authenticated encryption algorithm (2011)

Life Cycle Assessment and Management in Hospital Units Using Applicable and Robust Dual Group-Based Parameter Model

Vitaliy Sarancha, Leo Mirsic, Stjepan Oreskovic, Vadym Sulyma, Bojana Kranjcec, and Ksenija Vitale

Abstract Introduction: Complexity of hospital healthcare systems make it hard to manage using traditional statistical methods both in terms of describing their characteristics and in terms of the importance of qualitative parameters which have to be combined with quantitative ones. The aim of this work is to demonstrate robustness of dual group-based parameter model, following life cycle assessment and management principles (LCAM), with a goal to improve the efficiency of the Hospital Unit. Methods: This study was performed at the Department of Medical Biochemistry of the County Hospital. In the course of the study, we used elements of a LCAM principles combined with parametric analytics, structured interview with medical staff and direct content analysis (DCA). Study was conducted in two phases: data acquisition and analysis using dual group-based parameter model. During the first phase, the raw data were collected with total of 79 tests with adequate reagents data, test prices, number of tests delivered and duration of the sample processing, laboratory's staff working hours and wage data per period. In addition to the waste statistics, equipment, quality and energy expenses have been taken into account. Acquired data samples were pre-processed and adequately analysed. Results: We

---

This research did not receive any specific grant from funding agencies in the public, commercial or not-for-profit sectors.

V. Sarancha (✉)
School of Medicine, University of Zagreb, Zagreb, Croatia
e-mail: saranchavi@gmail.com

L. Mirsic
Algebra University, Zagreb, Croatia

S. Oreskovic · K. Vitale
School of Public Health, Zagreb, Croatia

V. Sulyma
Department of Traumatology, Ivano-Frankivsk National Medical University, Ivano-Frankivsk Oblast, Ukraine

B. Kranjcec
Laboratory of Medical Biochemistry of the Zabok General Hospital, Bračak, Croatia

© The Editor(s) (if applicable) and The Author(s), under exclusive license to Springer Nature Singapore Pte Ltd. 2021
T. Senjyu et al. (eds.), *Information and Communication Technology for Intelligent Systems*, Smart Innovation, Systems and Technologies 196,
https://doi.org/10.1007/978-981-15-7062-9_33

discovered the dual group-based parameter model elements consisted of quantitative "hard" and qualitative "soft" parameters. The "hard" parameters consist of: "tests", "reagents", "prices", "patients", "sample processing time", "waste", "equipment", "quality measurements" and "energy expenses" and "soft" parameters which include "financial conditions satisfaction", "working environment", "equipment functionality", "information flow", "communication" and "general organizational issues". Conclusion: Research proved applicability and robustness of dual group-based parameter model, following LCAM principles, with the goal to improve the efficiency of the Hospital Unit. Current methodology can be used to study a life cycle of a single process, chains of processes and can be successfully applied to assess the efficacy of Hospital Unit. As being the first study of LCAM, implementation in healthcare system in the Republic of Croatia as such can be used as a pilot project to support efforts in the well-recognized need to increase the efficiency of the healthcare units. Methodology can be extended as concept to any similar unit (e.g. laboratory, hospital, ambulance and operating theatre) or to healthcare system as a whole.

1 Introduction

Rapid industrial development and the extremely high level of the consumption of resources in a healthcare with the entailed pollution of air, water and soil have induced increasing interest in new optimization tools and methodologies. Despite the general problems in healthcare, common to every country in transition, due to its strategic location, mild climate, a plenitude of medical and rehabilitation facilities, reasonable prices for recreational activities and an increasing flow of tourists, Croatia has great potential with respect to development and implementation of such tools. Among them, we draw an attention to the life cycle assessment (LCA). Primarily, it was adopted for environmental studies with the purpose of resolving environment-related problems of corporations with a more holistic approach. LCA is a holistic view of interactions that covers a range of activities, from the extraction of raw materials from the Earth and the production and distribution of energy, through the use, and reuse and final disposal of a product. LCA is a relative tool intended to help decision-makers compare all major environmental impacts when choosing between alternative courses of action. The method was initially introduced in 1960s, developing until the late 90s, when it was implemented as a 14,000 series Standard of International Standards Organization [1]. A lifecycle approach takes into consideration the spectrum of resource flows and environmental interventions associated with a product or company from a supply, consumption chain perspective. In this context, we introduce additional key concept called life cycle management (LCM). LCM is a derivative and practical approach of LCA application to different sectors of economy. It emerged in a course of time and reflects the application of LCA to modern practice with an aim of managing the total life cycle of units, organizations, products and services towards more sustainable production and consumption. It is an integrated framework to address environmental, economic, technological and social aspects of products and services. LCM, like any other managerial pattern, is applied on a voluntary basis and can be adapted to the specific needs and characteristics of particular organization [2]. An illustrative example of the LCM application is the surgeon and nurse initiated Green Operating Room Committee. This is an internal medical staff initiative on the premises of one hospital in the USA. Routinely used consumables were replaced with recyclable and energy-efficient substitutes (single-use devices, reusable gel pads instead of disposable operating room foam pads.), resulting in the decreased amount of wastewater, in solid waste reduction, electricity, and great per year spending level reduction. As a result, the initiative provided significant opportunity to improve a healthcare unit's impact on the environment and generally save resources [3]. Following life cycle assessment and management (LCAM) foundations in a focus of this paper is the demonstration of applicability and robustness of novel dual group-based parameter model with the aim to assess and consequently improve the efficiency of the Hospital Unit. This paper is based on actual research which was performed at the Department of Medical Biochemistry (Hospital Unit), Laboratory of Medical Biochemistry of the General Hospital of Zabok (Croatia).

2 Methods

A general methodology was chosen a life cycle assessment (LCA) method and an approach that considers all aspects of resource assessment "from cradle to grave". According to that the data were acquired as follows: (i) documentation for all reagents and chemical substances needed for 79 test performance with the use of the official purchasing documentation such as invoices, order confirmations, warehouse receipts and product delivery statements, (ii) detailed calculation of the tests over the period of the study. It gives comprehensive information about the number of tests performed together with quantity of reagents used, (iii) step-by-step sample circulation process starting from the registration of a patient at the registration desk, over the whole processing cycle and until the final stage when the sample is validated. Duration of every stage was accurately calculated using a digital timer. We repeated the calculation of every process three times to get a time-based confirmation in order to calculate the average and decrease the level of any biases and incorrect results, (iv) estimation of the consumption of electricity, natural gas, water and fuel oil by the Unit, (v) quantity of hazardous waste and municipal waste per period of study, (vi) environmental parameters such as noise levels, temperature and humidity, (vii) data related to the staff of the laboratory. During the period of study, there were 25 employees at the laboratory. For the purposes of this study, International Standard Classification of Education (ISCED) was used [4], (viii) working hours of employees and gross salary expenses per period of study, (ix) anonymous questionnaire was used for data acquisition about level of satisfaction with working conditions: financial satisfaction, people/environment satisfaction (typical enquiry, responses ranged from 1 to 5). Every employee with an exclusion of those being on a sick or maternity leave was asked 2 questions: first was about satisfaction with financial conditions of his work and second-level of employee satisfaction with the environment conditions and colleague attitude. It was followed by additional qualitative data acquired with the help of structured interview in a form of "guided conversation". For this purpose, we chose three employees with different background and asked few open-ended questions in the same order: "Please reveal technical routine problems emerging on a daily basis"; "Please reveal organizational routine problems emerging on a daily basis"; "What would you recommend to improve the general functioning of the Laboratory?" The interviewing process was accurately recorded, and a precise transcript protocol was created. Afterwards, the codification of the interviews was done and assessment with conclusions was performed. For codification, we used the methodology of direct content analysis (DCA) [5]. For the purposes of this study, we missed out the metrics of environmental influences and paid attention primarily to revealing the key parameters that influence our system and can be potentially used for the optimization of the Unit. In the next phase, we performed computerized data mining to find main correlations among input data and to reveal the ways in which the identified parameters lead to optimization.

3 Results

After the acquired data samples were pre-processed and cleaned into data sets, we have revealed the following logical relations: (i) number of delivered tests combined with average price per test, (ii) seasonal characteristics through periods showing number of tests per period, (iii) correlation between test popularity and price, (iv) statistics of working staff hours per period, (v) correlation between staff wages and working hours, (vi) correlation between working hours versus working position, (vii) general waste data for total period of study, (viii) correlation of total tests and total staff working hours. Together with the assessment of acquired codes in result of the performance of DCA, we have found the key parameters that have an impact on the particular Unit. They are as follows: "tests", "reagents", "prices", "patients", "sample processing time", "waste", "equipment", "quality measurements" and "energy expenses". Current parameters have their number values and marked as "hard" (Fig. 1).

The second group of parameters is qualitative. These parameters crystallized from the responses obtained from: survey about environmental and financial conditions of staff, general satisfaction of staff and the structured interview. In our study, there were 22 employees interviewed. If we take 12 employees, which is a half of a total and constitutes the mathematical median value (read 50%), the survey about environmental and financial conditions and general satisfaction of the laboratory staff showed that 50% of employees rated their financial conditions as satisfactory (3 of possible 5) which mean that wages are relatively good for this kind of work/responsibility. The 8 and 9 employees that rated the situation 5 and 4 points, respectively, constituted 17 employees in total (i.e. 77%) who were more than satisfied with working environment (working conditions and relationships among the personnel). As a result,

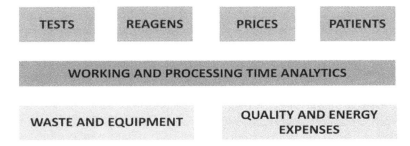

Fig. 1 Hard parameters. ©Vitaliy Sarancha, 2019

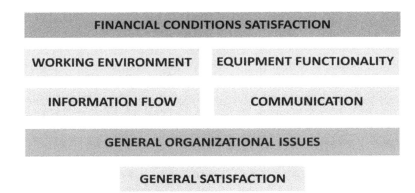

Fig. 2 Soft parameters. ©Vitaliy Sarancha, 2019

we defined parameters which are descriptive and we marked them as "soft". They are as follows: "financial conditions satisfaction", "working environment", "equipment functionality", "information flow", "communication", "general organizational issues" and "general satisfaction". As we mentioned above, these parameters are descriptive and in our opinion and are the key qualitative factors having obvious influence on the Unit functionality and efficiency (Fig. 2).

4 Discussion

Complexity of hospital healthcare systems makes them hard to manage using traditional statistical methods both in terms of describing their characteristics and in terms of the importance of qualitative parameters which have to be combined with quantitative ones. Research proved applicability and robustness of dual group-based parameter model, following LCAM principles, with goal to improve efficiency of the Hospital Unit. Dual group-based parameter model seems to be a convenient tool for analysing particular Unit through its quantitative and qualitative characteristics, influencing every step of the process or multiple processes. After proper identification and assessment of parameters, their influence can be modified, improved or decreased for the purposes of the optimization of the whole system. Optimization can be reached through the implementation of better managerial solutions according to LCAM principles some of them are improvement of ethical issues, introduction of more effective and less time-consuming internal procedures and manuals, working position-specific manuals and explanatory documents [6–8]. Analysis of the codes of DCA gives us the understanding of main problems occurring in the routine work of the employees of the laboratory, and this is the crucial point for making assumptions concerning the improvement and optimization of the work processes and relationships in general [5, 9]. With respect to state-owned institutions where the annual budgeting process is strictly predefined with limited resources, it is complicated to find an *ad hoc* solution concerning the purchasing of modern and updated computers and software but it is highly recommended to make all possible relocations, for example, to relocate PCs with high-speed processors and software to departments where their necessity is obvious and to decrease the level of purchasing of discounted and used appliances. Concerning the procedures, they should be implemented in such a way as to help to plan, organize, perform and control the processes but not in any way to create additional workload and bureaucracy, which can just complicate the processes and misuse the workforce. As a recommendation for a positive working environment, the engagement in the process of a professional time-planning trainer or coach can be considered, and this can create effective models of behaviour for work under pressure and in stressful conditions [1]. As it can be seen, one of the important issues is to pay attention to the best qualities of individuals, to get as many as possible from the existing human resources, to find triggers for staff motivation and the will to bring the best results, to find a proper model for different backgrounds and qualifications and to arouse the enthusiasm of the employees. The presented methodology can be used to study a life cycle of a single process, chains of processes and can be successfully applied to assess the efficacy of Hospital Unit. As being the first study of LCAM implementation in healthcare system in the Republic of Croatia, as such can be used as a pilot project to support efforts in the well-recognized need to increase the efficiency of the healthcare units. "Hard" parameters can be grouped, modelled, analysed using modern IT software [10–12]. It makes the whole analysis much more precise and quantitative with possibility to be extended as a to any similar unit (e.g. laboratory, hospital, ambulance and operating theatre) or to healthcare system as a whole. Similarities and groupings of "hard" parameters are used in design of standards. With relation to healthcare, standards are designed to encourage healthcare organizations to improve internal quality and performance, minimize the risks, including measures to protect and improve the safety of patients, to promote a culture of continual improvement, support efficient exchange of information and data protection while benefiting the environment and become close to the etalon. Depending on the scope of responsibilities and areas of activity, every organization is able to choose voluntarily among standards to implement. ISO alone created about

1200 health standards that are grouped in families. Some of them such as Environmental Management ISO 14000, Occupational Health and Safety OHSAS 18000, Guidance on social responsibility ISO 26000, Environmental management 14000 are featured as much applicable to public health and healthcare. A family contains a number of standards each focusing on different aspects of a corresponding topic. According to 2012 ISO Press release, the most commonly used standard is Quality Management Standard ISO 9001 (which belongs to the family ISO 9000—Quality management systems). Due to its generic basis, it is applicable to all types of organizations. It enables a company to develop a quality management system (QMS) which implies the introduction of quality planning, quality assurance, quality control and quality improvement [13, 14] and is a perfect tool for measurement and determination of the ultimate way of development of health services. In addition obtained results prove that for effective LCM management, the ratio between price/cost and volume of performed tests is crucial. In terms of financial efficiency, it would be preferable for the laboratory to perform more tests at higher prices. Finding such a "sweet spot" throughout the distribution of tests versus price can be used for marketing purposes to find a "niche" and to target that segment to create sales growth. Because the General Hospital of Zabok is a state-owned institution and patients are mainly the subject to the rules of the state insurance policy, it would be difficult to increase the volume of more expensive tests in which the profit margin is higher; but in privately owned laboratories, this recommendation could be effectively implemented, resulting in higher profit margin and financial sustainability. On the other hand, it would be difficult to manage the volume of frequently ordered tests in a state-owned healthcare system unit because it would result in additional demand for new facilities, technical equipment and staff. Seasonal activity can be also helpful for process optimization. The Unit may postpone non-urgent tests to the next period and increase the efficacy while prioritizing the volume of urgent tests. This would increase liquidity during the "non-active" months and decrease the pressure during "active" periods. This point of view is supported by several publications addressing similar issues to access to clinics and duration of sample processing with optimization using Lean philosophy and Six Sigma principles [8, 15]. It may be helpful also for personnel working time planning and distribution and in managing the life cycle in the waste disposal activity. We determined that the main components in waste disposal during the study period are infectious waste and mixed communal waste. In terms of optimization, this may be used for planning of processing and recycling. Also, figures prove that employees with upper educational levels such as the head engineer, the holder of a master's degree in medical biochemistry, head of the department with scientific degree and so on, taking into account the timeframe of 6 month period of the current study, have higher gross salaries than their colleagues with bachelor's or lower technical schools qualifications. Actually, this inverse correlation follows the general trend in healthcare in Croatia [16]. The situation is different at privately owned institutions in which the owners and CEOs are in positions of authority irrespective of their level of education. Anyway, understanding of staff overall wages and analysis of wage to working hours ratio can help in the managing of staff expertise and HR management. Furthermore, in our study we cannot see any correlation between the tests performed and the working hours of employees. This is due to the specific features of a state-owned unit. In this particular situation, the staff has scheduled working shifts and the number of tests conducted is not important. We suppose that situation is completely different in case of private ownership where in general performance and working hours are positively correlated. It is difficult to compare results with any others in Croatia since it was not possible to find any study particularly relating to this topic in any scientific publication. General expenses data for total period are taken into consideration just as a quantitative factor. In terms of optimization, they may be used for planning of expenses and effective budgeting. An important option is to consider general or partial substitution with alternative energy sources such as solar or geothermal. The team's knowledge and skills share empower the staff members to work together to make successful changes. This includes seeking information and effectively using that information to design, validate, and feedback process improvements, regularly assess progress and learn from the efforts and mistakes of others. In our opinion, much more attention should be paid also to the psycho-emotional state of the staff and general satisfaction. The results attained prove that the working environment and staff satisfaction in the unit contributes to the whole system's efficacy and aids in the creation of a healthy and comfortable atmosphere, which consequently gives new opportunities for development and process optimization. As it shown in our direct content analysis, the efficiency of fulfilling routine tasks depends on the mood and satisfaction and will of our interviewees. Of the 33 codes which were distinguished in course of DCA—21 (63.6%) are marked positive and 12 (36.4%) are marked as negative. Those marked as positive mainly address such factors as communication inside the system, working environment; but on the other hand, those marked as negative concern technical issues, out of date computers, delays in processing, insufficiency of professionals for particular laboratory tasks. Results concerning the general satisfaction of staff obtained during DCA and those from evaluation of structured interview are different. We may conclude that DCA is a more precise method of evaluation than structured interview and enables us to go much deeper into details while looking for the answers. Understanding of staff's overall workload pressure can help managing staff expertise and HR management of the unit. Despite global digitalization and the regular use of virtual reality, we would like to emphasize the importance of regular communication and interviewing of a staff. We would distinguish three main reasons for that. First of all, for the effective fulfilment of working tasks it is necessary to reveal the unproductive and ineffective models of behaviour and attitude which have a direct influence on relationships inside the system, on the patients and on the efficacy of the performed responsibilities at the end. The second component is the one-time feedback of what goes well and what should be corrected, improved or illuminated. The third and more important component, which we would like to underline, is the distribution of knowledge and experience that has a crucial impact on the system in a whole. All these topics were revealed in the course of our LCAM performed with respect to the Unit, which may lead to confirmation of the effectiveness of such holistic approaches as LCAM for this kind of unit assessment and scientific studies. The introduction of such approaches, leading staff members, from a fragmented perception to awareness of the system as a whole is in our opinion a necessity for the effective existence of modern systems and units. In our case, we also started from the assessment of one process but later transposed it to the whole unit. The same principle can be used for the application of the method and consequent optimization on the whole hospital or even healthcare system of, for example, Croatia. We understand that it is difficult to completely rely on LCAM based on one sample processing and took as a reference point the data based on a summary of tests during a one-month or even a six-month period in a laboratory. However, with the approach of complex data acquisition and system analysis, we elevated the LCAM to the level of a healthcare system Unit—the laboratory in a particular case. We admit that the LCAM of a multitude of processes comprises the cumulative LCAM of the healthcare system unit and consequently see feasible results. Summarizing all proposed holistic concepts and tools, we understand that from the start a holistic view will in some way frustrate many professionals because it requires much higher level of responsibility with relation to the routine tasks but according to a well-known saying "We can't solve problems by using the same kind of thinking that was used when we created them"; this means that without new initiatives and tools which can provide leverage for changes it is impossible to improve the systems and modernize the world as a big multilevel and multitasking system. Vital for the success of the application of holistic philosophy is the common understanding of its necessity and common wish, will, attempt and drive to eliminate waste and create new surroundings for sustainable and continual growth. In addition, all described innovative solutions would be the drivers for creating more advanced work frames and guidelines for the development of a normative—legal framework; guidelines for setting limits for certain hazardous emissions, effects and impacts on human health, working procedures and behavioural models.

References

1. Curran, M.A.: Life cycle assessment: principles and practice. Scientific Applications International Corporation (SAIC). Environmental Protection Agency, Washington, USA (2006). EPA/600/R-06/060
2. Wormer, B.A., Augenstein, V.A., Carpenter, C.L., et al.: The green operating room: simple changes to reduce cost and our carbon footprint. Am. Surg. 1(7), 666–671 (2013)
3. Kneifel, J.D., Greig, A.L., Lavappa, P.D., Polidoro, B.J.: Building for environmental and economic sustainability (BEES). Online 2.0 Technical Manual. National Institute of Standards and Technology. Technical Note 2032 (2018 December). https://doi.org/10.6028/nist.tn.2032
4. Parlament of Croatia. Zakon o zdravstvenoj zaštiti. Narodne novine. 2018 Nov 14; N100/18
5. Allen, B., Reser, D.: Content analysis in library and information science research. Lib. Inf. Sci. Res. 12(3), 251–260 (1990)
6. OHSAS 18001:2007. Occupational health and safety management systems. Requirements; 2007 July. Available at: http://www.producao.ufrgs.br/arquivos/disciplinas/103_ohsas_18001_2007_ing.pdf
7. Ramesh Aravind, T., Saravanan, N.: Guidelines to implementation of health and safety management systems through OHSAS 18001:2007. Int. J. Res. Develop. Technol. 2018 Apr;(4) (2018). ISSN(0):-2349-3585

8. Schweikhart, S.A., Dembe, A.E.: The applicability of lean and six sigma techniques to clinical and translational research. J. Invest. Med. **57**(7), 748–755 (2009)
9. Hsich, S.: Three approaches to qualitative content analysis. Qual. Health Res. **15**(9), 1277–1288 (2005). https://doi.org/10.1177/1049732305276687
10. Potter, J., Hepburn, A.: Qualitative interviews in psychology: problems and possibilities. Q. Res. Psychol. **2**(4), 281–307 (2005). https://doi.org/10.1191/1478088705qp045oa
11. Curran, M.A.: Life-cycle based government policies. Int. J. LCA **2**(1), 39–43 (1997)
12. Azapagic, A.: Life cycle assessment and its application to process selection, design and optimisation. Chem. Eng. J. **73**, 1–21 (1999)
13. Ministry of Health. Order On the introduction of standardization in health care. 1998 Jan 19-12/2 (original in Russian). Computer Technologies in Medicine 1998. Available at: http://www.ctmed.ru/DICOM_HL7/mz12_98.html
14. Boll, V.: The development of a uniform system of standardization in healthcare of Russia. (original in Russian). Russian Entrepreneur. (J.)**8**(80),148–152 (2008)
15. Murray, M., Bodenheimer, T., Rittenhouse, D., Grumbach, K.: Improving timely access to primary care: case studies of the advanced access model. JAMA **289**(8), 1042–1046 (2003)
16. Lazibat, T., Burcul, E., Bakovic, T.: The application of quality management system to Croatian Healthcare. Poslovna izvrsnost. Zagreb. 2007;2. UDC: 614(497.5): 658.562, JEL: I38

A Novel Approach to Stock Market Analysis and Prediction Using Data Analytics

Sangeeta Kumari, Kanchan Chaudhari, Rohan Deshmukh, Rutajagruti Naik, and Amar Deshmukh

Abstract Stock market is one of the most versatile sectors in the financial system. Historically, stock market movements have been highly unpredictable as there are many factors that affect it. Immense research about financial history, as well as the current trends, must be done before thinking of investing in a particular company. For the prospective buyer to get a better sense of future, predictive models of machine learning can be used to predict whether the stocks will go high or low and how much. While no one can say with certainty that a stock will go up in value, taking the time to evaluate the past few years of the company's growth can give some insight into the possibility. Employing traditional methods like fundamental and technical analysis may not ensure the reliability of the prediction. In this paper, we attempt to take financial historic data for the stock and try to get more detailed insights and derive parameters that we will use to predict the stock behavior.

1 Introduction

Financial time series change dynamically and selectively. Investors are looking at the stock market as an emerging sector in the world of trading. Many of them are currently involved in the stock market directly or indirectly, and this creates the need for investors to know about stock trends. Thus, nowadays, they are more curious to know insights into the stock price. Predicting the stock market is a challenging field in the finance domain and also in the engineering and mathematics field. Due to the rise in the financial market, stock market prediction algorithms came into consideration from the academic and business side. Finding the right time to enter/exit/hold for a particular stock is very difficult.

S. Kumari (✉) · K. Chaudhari · R. Deshmukh · R. Naik
Department of Computer Engineering, Vishwakarma Institute of Technology, Pune 411037, India
e-mail: ec.sangeeta@gmail.com

A. Deshmukh
Connecticus Technologies Private Limited, Pune, India

In past, stock market predictions were made only by financial experts; however, improvement in the fields like machine learning, deep learning, and data analytics, computer data experts has also shown an interest in predicting the stock market. Machine learning is popularly used in stock market prediction systems that help financial analysts in decision making. Researchers have used data from various stock markets, news to build and train machine learning models. Self-learning and the self-organizing ability of these algorithms can help predict the volatility for a particular stock [1]. Enhancement of the Internet has increased the speed of delivering news all over the world. Many websites like Bloomberg, Economic Times provide financial news to investors all over the world. Many investors have observed that news may have an impact on the price of the stock. Negative/positive news on a particular company may affect the share price. The study proposed by Li et al. proposes a multi-kernel learning technique that incorporates two information sources, i.e., news and stock price. [2]

It is observed that the stock market behavior can be predicted based on historical data, e.g., the stock price of Ashok Leyland always falls in May. Based on the inference of numerical analysis, traders/brokers can decide when to buy/exit/hold for a particular stock.

Analyzing the stock market trend will require large records from past years. The data that will be used in this paper to build the inference model will be the Maruti Suzuki India Limited of NIFTY 50 index National Stock Exchange from the financial year 2003--2020. The data consist of daily closing price of the stock. There are many methods to predict the price in the share market like technical analysis, fundamental analysis, time series analysis, statistical analysis, etc. But none of these prove to be consistent. The numerical analysis has given us many important financial statistical measures such as standard deviation, upswing, downswing, and probabilistic measures. As machine learning algorithms do not consider many financial statistical measures and rely only on data it may lead to poor predictions, we propose a novel approach to which uses numerical analysis to predict stock behavior. This will help the broker/trader to predict the market trend as well as give clear insights into the data and thus, helping them to grow their profits.

2 Literature Survey

Many data mining and artificial intelligence techniques are proposed to predict the price of stock. The most well-known publication in this area is by Venkata Sasank [3] the stages are all well mentioned right from the data collection, preprocessing till how to implement sentiment analysis. The method proposed by these authors helped to build the base of the project and complete the initial procedure. Algorithmic strategies even play an important role which was very well explained in Jadhav and Wakode [4].

Machine learning is shown as a prominent resource and also a powerful tool that we can use while predicting the stock price. Machine learning is very much data-centric and supervised learning approach is used in many of the prediction models [5]. Thus, understanding data is very important; data analysis is not an easy task.

Soni et al. have analyzed stock prices along with its some attributes and then compared multiple machine learning models results for predicted stock price [6]. This shows multiple derived parameters from the data can be used to predict the stock prices.

The researchers from many fields especially from the mathematics field are currently trying to find the insights into data; the most key focus area for the research is how we can process the data over so many years and how the external entities affect the data analysis. For example, Nicola has provided a rigorous discussion on dynamic change in stock price and how investors found it difficult to handle [7].

The volatility of the stock price is one of the most important aspects for prediction. Volatility indicates the risk associated with the stock that is the fluctuation of the stock stated by Kumar et al. [8].

Factors like opening price, closing price, high and low price for the stocks are shown major measures to define historical volatility of the stock stated by Singh and Thakral [9]. The approach suggested by Singh et al. shows that we can get insights into the stock market based on statistical analysis which can reflect the stocks behavior, its performance over the years and can develop an ideal portfolio to invest in stock with time-tested strategy [9]. The strategy suggested by the Xing et al. shows that there are some hidden relationships between the stock prices which we can identify using hidden Markov model [10]. This method is pretty accurate but for the shorter period prediction.

To deal with the changing nature of the stock price triggered by the multiple factors [11], nonlinear methods are considered to predict the stock price. And the frequent thought comes is to make use of nonlinear regression analysis [12].

As all these techniques are mostly dependent on data, we can transform the data into meaningful data which can give insights to the behavior of the stock.

The stock-related data and its prices are generally available to a large extent, so we can use this data for data analysis to make some profitable predictions.

3 Numerical Analysis

3.1 Closing Prices

A closing price is the final price for the day of a particular stock in a given trading day. It represents the latest stock price till the next trading day starts.

3.2 Percentage Change

It is a mathematical concept defining change in stock price over time. It is used to define price change of particular stock.

Percentage Change = ((New Price − Old Price)/Old Price) ∗ 100

We use it to compare the current price of stock with respect to previously dated stock.

3.3 Standard Deviation

Standard deviation in statistics is a dispersion of data concerning its mean and measured as the square root of variance. In finance, it is applied to a weekly/monthly/quarterly/half-yearly/yearly rate of return of an investment to highlight the volatility of that investment. The volatility of investment determines the risk associated with stock, i.e., high volatility of stock indicates greater fluctuation in the stock price positively or negatively. When the standard deviation is high, it implies that the stock price is more volatile. When the standard deviation is low, the stock price is less volatile that means less risk is associated with the stock.

3.4 Mean

Mean in statistics is the average or centric value of data. In finance, mean return, also known as expected return, defines how much we can gain from stock on weekly/monthly/quarterly/half-yearly/yearly basis.

$$\text{Formula}: \bar{x} = \left(\sum x_i\right)/n$$

where:

x_i is all values in dataset,

n is the number of items in the dataset (Fig. 1).

3.5 Upswing and Downswing

Upswing defines the maximum hike you can get after investment in a stock. Downswing defines the maximum decline after investment.

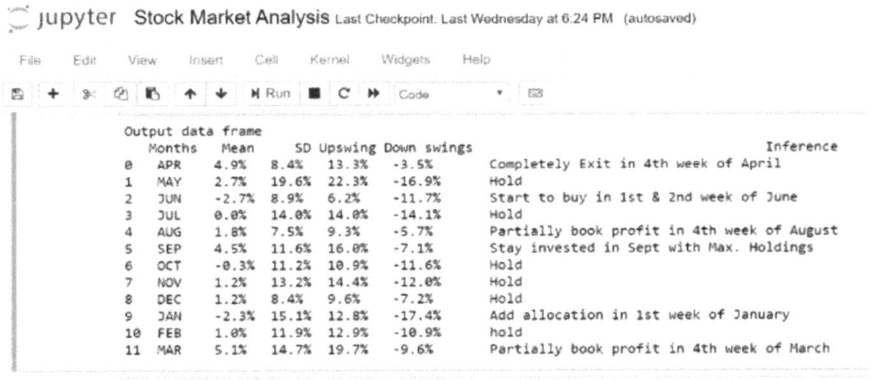


```
Output data frame
   Months   Mean     SD Upswing Down swings                                       Inference
0     APR   4.9%   8.4%   13.3%   -3.5%      Completely Exit in 4th week of April
1     MAY   2.7%  19.6%   22.3%  -16.9%      Hold
2     JUN  -2.7%   8.9%    6.2%  -11.7%      Start to buy in 1st & 2nd week of June
3     JUL   0.0%  14.0%   14.0%  -14.1%      Hold
4     AUG   1.8%   7.5%    9.3%   -5.7%      Partially book profit in 4th week of August
5     SEP   4.5%  11.6%   16.0%   -7.1%      Stay invested in Sept with Max. Holdings
6     OCT  -0.3%  11.2%   10.9%  -11.6%      Hold
7     NOV   1.2%  13.2%   14.4%  -12.0%      Hold
8     DEC   1.2%   8.4%    9.6%   -7.2%      Hold
9     JAN  -2.3%  15.1%   12.8%  -17.4%      Add allocation in 1st week of January
10    FEB   1.0%  11.9%   12.9%  -10.9%      hold
11    MAR   5.1%  14.7%   19.7%   -9.6%      Partially book profit in 4th week of March
```

Fig. 1 Screenshot of Jupyter notebook output for stock market analysis

Table 1 Stock returns in month of April

Year	03–04	04–05	05–06	07–08
APR mean (%)	2.7	1.1	6.5	22.6

Table 2 Mean and standard deviation of April (2003–08)

Mean	Standard deviation (%)
8.23	9.84

Table 3 Upswing and downswing of April (2003–08)

Upswing (%)	Downswing (%)
18.07	−1.61

$$\textbf{Upswing} = \textbf{Mean} + \textbf{Standard Deviation}$$
$$\textbf{Downswing} = \textbf{Mean} - \textbf{Standard Deviation}$$

By looking at only on the past performance (Returns) of the stock on monthly, weekly & quarterly, we calculated mean & standard deviation & on the basis of that calculated upswings & downswings (Tables 1, 2 and 3).

3.6 Probability Analysis

By looking at only the past performance (returns) of the stock on monthly, weekly & quarterly, we considered the probability of how many years of which months, which quarters, which month of which week it was positive or negative.

Table 4 Mean of stock returns in April (2003–08)

Year	03–04	04–05	05–06	07–08
APR mean (%)	2.7	1.1	6.	22.6

Table 5 Monthly probability analysis for the month of April

Year	03–04	04–05	05–06	07–08
APR analysis	Positive	Positive	Positive	Positive

Table 6 Upswing and downswing calculations in April month

Upswing (%)	Downswing (%)
18.07	**−1.61**

3.7 Risk to Reward

By inspecting the value of mean, if value is negative, then the respective week/month/quarter considered as unfavorable for investment. Also, if the value of mean is positive, then respective week/month/quarter is considered as favorable for investment.

3.8 Analysis

Simple probability analysis can be done. By looking at only the past performance (Returns) of the stock on monthly, weekly & quarterly, we considered the probability of how many years of which months, which quarters, which month of which week it was positive or negative (Tables 4, 5 and 6).

4 Performance Analysis

4.1 Yearly Performance

Yearly performance of stock can be calculated as given formula (Table 7)

Table 7 Yearly performance

Year	Date (dd/mm/yy)	Price (₹)	Date (dd/mm/yy)	Price (₹)	% Change (%)
2003–04	1/04/03	4.83	31/03/04	12.66	162.1

Table 8 Quarterly performance

Quarter	Date (dd/mm/yy)	Price (₹)	Date (dd/mm/yy)	Price (₹)	Change (%)
June	1/4/03	4.83	30/6/03	6.42	32.9
SEPT	1/7/03	6.05	30/9/03	9.47	45.7
DEC	1/10/03	9.96	31/12/03	14.71	47.8
MAR	2/1/04	14.7	31/3/04	12.66	−12.5

$$\text{Yearly Performance} = (x - y)/y$$
$$\text{Where } x = \text{Closing Rate of the Year}$$
$$Y = \text{Opening Rate of the Year}$$

4.2 Quarterly Performance

Quarterly performance of stock can be calculated as given formula (Table 8)

$$\text{Quarterly Performance} = (x - y)/y$$
$$x = \text{Closing Rate of the Quarter}$$
$$y = \text{Opening Rate of the Quarter}$$

4.3 Monthly Performance

Monthly performance of stock can be calculated as given formula (Table 9; Fig. 2)

$$\text{Monthly Performance} = (x - y)/y$$
$$x = \text{Closing Rate of the Month}$$
$$y = \text{Opening Rate of the Month}$$

Table 9 Monthly performance

Month	Date	Price (₹)	Month end date	Price (₹)	% Change (%)
APR	01/04/2003	4.83	30/04/2003	4.96	2.7
MAY	02/05/2003	4.96	30/05/2003	5.74	15.6

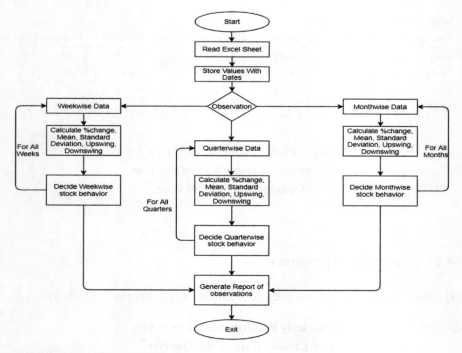

Fig. 2 Numerical analysis flowchart

4.4 Weekly Performance

Weekly performance of stock can be calculated as given formula below (Table 10):

$$\textbf{Weekly Performance} = (x - y)/y$$

$x =$ Closing Rate of the week
$y =$ Opening Rate of the week

Table 10 Weekly performance

Month	Week	Date (dd/mm/yy)	Price (₹)	Date (dd/mm/yy)	Price (₹)	% Change (%)
APR	1	1/04/03	4.83	4/04/03	5.09	5.3
APR	2	07/04/03	5.16	11/04/03	4.98	−3.3
APR	3	15/04/03	4.99	17/04/03	4.99	−0.1
APR	4	21/04/03	4.96	25/04/03	4.97	0.1

Algorithm for numerical analysis

This algorithm will find inferences for weekly, quarterly, monthly, and yearly basis for a particular stock.

Input: Price and date of stock

Output: Inference Model

1: priceanddate:= read(excel)
2: **for** (int i=1 ; i<length(excel); i++) **do**
3:       percentagechange:= function(newprice,oldprice)
4:       standarddeviation:= function(percentagechange)
5:       positiveinferece:= function(percentagechange)
6:       negativeinferece:= function(percentagechange)
7:       mean:= function(percentagechange)
8:       upswings:= standarddeviation + mean
9:       downswings:= standarddeviation − mean
end for
10: **if** standarddeviation(of next month/week) < 5% and upswing > 10% **then:**
11: Buy the Stock
12: **end if**
13: **if** standarddeviation(of next month/week) > 10% and downswing > 10% **then:**
14:      Completely exit from Stock
15: **end if**
16: **if** upswing:downswing difference is in small fraction(0.5%) **then:**
17:      Hold the stock
18: **end if**
19: **if** upswing>10% mean>5% **then:**
20:      Buy stock with max holdings (Add more allocation)
21: **end if**
22: **return**

5 Conclusion

This paper presents a novel idea using a numerical analysis technique based on the historical prices of stock to give buy and sell recommendations to traders/brokers. This technique can help a number of investors to extract insights of behavioral patterns for a particular stock. A lot of research has been carried out for stock prediction using machine learning techniques where the dataset is the key thing to achieve the correct inference. Also, the behavior of the model is dependent upon its structure and design. As most of the machine learning algorithms try to find some sort of patterns based on available data, analysis of stock prediction using statistics of data science is having an upper edge over machine learning and deep learning. The statistical factors of data science such as standard deviation, upswing, downswing, and probability analysis have been used in this paper to provide a precise insight into the stock market behavior even if the market is volatile in nature.

As a future scope, there is still scope for testing, improving, and optimizing the algorithm. Finally, considering vital factors such as GDP, Dollar Rate, news, and factors affecting stock can be used to improve accuracy.

References

1. Khan, W., Malik, U., Ghazanfar, M.A., et al:. Predicting stock market trends using machine learning algorithms via public sentiment and political situation analysis. Soft Comput (2019)
2. Li, X., Wang, C., Dong, J., Wang, F., Deng, X., Zhu, S. Improving stock market prediction by integrating both market news and stock Prices. In: Hameurlain, A., Liddle, S.W., Schewe, K.D., Zhou, X. (eds.) Database and Expert Systems Applications. DEXA 2011. Lecture Notes in Computer Science, vol. 6861. Springer, Berlin, Heidelberg (2011)
3. Pagolu, V.S., Reddy, K.N., Panda, G., Majhi, B.: Sentiment analysis of Twitter data for predicting stock market movements. In: 2016 International Conference on Signal Processing, Communication, Power and Embedded System (SCOPES), pp. 1345–1350. Paralakhemundi (2016)
4. Jadhav, R., Wakod, M.S.: Sentiment analysis of Twitter data for stock market prediction. Int. J. Advanced Res. Comput. Commun. Eng. Understand. Algorith
5. Pahwa, K., Agarwal, N.: Stock market analysis using supervised machine learning. In: 2019 International Conference on Machine Learning, Big Data, Cloud and Parallel Computing (COMITCon), pp. 197–200. Faridabad, India (2019)
6. Soni, D., Agarwal, S., Agarwel, T., Arora, P., Gupta, K.: Optimised prediction model for stock market trend analysis. In: 2018 Eleventh International Conference on Contemporary Computing (IC3), pp. 1–3. Noida (2018)
7. Darvas, N.: How I made $2,000,000 in the stock market. Publisher, Lyle smart (2001)
8. Hemanth Kumar, P., Patil, S.B.: Estimation & forecasting of volatility using ARIMA, ARFIMA and Neural Network based techniques. In: 2015 IEEE International Advance Computing Conference (IACC), pp. 992–997. Banglore (2015)
9. Singh, P., Thakral, A.: Stock market: Statistical analysis of its indexes and its constituents. In: 2017 International Conference On Smart Technologies For Smart Nation (SmartTechCon), pp. 962–966. Bangalore (2017)
10. Xing, T., Sun, Y., Wang., Q., Yu, G.: The analysis and prediction of stock price. In: 2013 IEEE International Conference on Granular Computing (GrC), pp. 368–373. Beijing (2013)
11. Huang, A.Y.: Asymmetric dynamics of stock price continuation. J. Bank. Fin. **36**(6), 1839–1855 (2012)
12. Li, Y., Cobourn, W.G.: Fuzzy system models combined with nonlinear regression for daily ground-level ozone predictions. Atmosp. Environ. **41**(16) 3502–3513 (2007)

Energy Trading Using Ethereum Blockchain

M. Mrunalini and D. Pavan Kumar

Abstract In the recent past, blockchain has gained more popularity and dragged the attention of both industry and academia. Blockchain is always misunderstood as Bitcoin by many people. Blockchain has the potential to implement many applications ranging from banking to digital Ids, e-commerce to immutable data backups. This paper implements energy transactions for smart grid using Ethereum blockchain. A smart contract is used to execute trading if necessary conditions are met. Blockchain is used to fuel the energy trading for peer-to-peer exchange of electricity. Blockchain enables the application more secure through its default feature immutability. Smart grid and its components are necessary to implement this trading platform. In this platform, people can buy/sell electricity whenever they require.

1 Introduction

Blockchain is one of the new technologies in today's era. Majority of the people have the misconception that blockchain is Bitcoin. Bitcoin is a cryptocurrency, and it is just an application on blockchain. Blockchain has the potential to implement various applications such as banking transactions, e-commerce transactions, cryptocurrencies, and digital Ids. Bitcoin is the first application on blockchain technology which was invented in 2008. An unknown author under the pseudonym Satoshi Nakamoto published a white paper "Bitcoin: A Peer-to-Peer Electronic Cash System." Nakamoto merged many cryptographic components in that paper to describe a peer-to-peer payment network known as the cryptocurrency [1].

Blockchain technology is used for trading between two untrusted parties. The main advantage of blockchain technology is high security and transaction immutability.

M. Mrunalini (✉) · D. Pavan Kumar
M S Ramaiah Institute of Technology, Bangalore, India
e-mail: mrunalini@msrit.edu

D. Pavan Kumar
e-mail: pavan_kumar_18@yahoo.com

© The Editor(s) (if applicable) and The Author(s), under exclusive license to Springer Nature Singapore Pte Ltd. 2021
T. Senjyu et al. (eds.), *Information and Communication Technology for Intelligent Systems*, Smart Innovation, Systems and Technologies 196,
https://doi.org/10.1007/978-981-15-7062-9_35

In blockchain, the data is stored in distributed ledger. In distributed ledger, there are multiple participating nodes. Each node contains a record of all the transaction data. The crucial component in the distributed ledger operation is to ensure that the entire network agrees collectively with the ledger's contents; this process is carried out as a consensus mechanism. The idea of a consensus system is to check the authenticity of information added to the ledger, i.e., the network is in consensus (agreement).

Different instances of the same data will be stored in a distributed database across several nodes where each network user node automatically updates itself. A blockchain is a collection of records stored in a block-like structure that continues to expand. The only procedure that can be carried out on such blocks is to add new data to the current block: it is impossible to modify or remove previously entered data on previous blocks, that is because the blocks can only be appended. In a distributed database, data can be easily manipulated. Typically, there are database administrators who are entitled to make changes to any part of the data or its structure where in the case of blockchain modifying/deleting data is almost impossible. Present transaction management is done only through third-party interfering and participating nodes cannot have data and data is not publicly accessible as well.

The blockchain is a relatively slow, very expensive database (in terms of computation) which provides excellent resistance to data hacking and manipulation. This is a method of write once read many (WORM). The blockchain is mainly used in the financial sector where the transaction includes only values and information but the actual asset (i.e., electrical commodities) when applying this in the energy sector (i.e., electricity) should be taken into consideration. However, blockchain platform provides following advantages:

- Multiple organizations which allow interoperability and ensure integrity of processes by standardizing data formats
- Reduces error, invalid transactions, and risk of fraud
- Low credit risk and capital exchange requirements.

2 Literature Survey

In [2], authors addressed several technical issues related to microgrids, such as the duplication of energy transactions in a microgrid causes a difference in power losses across all microgrid divisions. In [3] peer-to-peer energy market architecture, including payment, the authors have also shown how blockchain and smart contract can promote decentralized collaboration among non-trusting agents.

In [4] the author has suggested a competitive market between multiple peers for the energy scheduling process. In [5], by expanding the functionality of cryptocurrency exchanges to the Renewable Energy Industry, a project was introduced using blockchain-based software platform.

On the hybrid blockchain, consisting of private and public, the authors of [6] addressed. The authors suggest private blockchain verifies the transaction and serves the public to establish data integrity.

In [7], the author describes how to apply blockchain to decentralized energy service, which will offer the system more stability, versatility, and cost-effective operating process.

The research in [8] deals with the perspective of peer-to-peer energy sharing networks.

In [9] the authors discussed the application of blockchain technology on the microgrid with respect to the electricity market. They have also put forward a system for using blockchain for peer-to-peer electricity trading.

In [10] the authors discussed important problems such as congestion pricing and dispatch non-optimality.

In [11], the authors suggested a model where hybrid electric vehicles could be exchanged with electricity.

The authors addressed technical issues in implementing blockchain-based energy trading in [12]; issues relating to power loss are addressed in this study.

3 Proposed Model

3.1 Basic Definitions

Blockchain—The blockchain is an innovation which gives a stage to shared exchange that utilizes decentralized storage to store the transaction. In straightforward terms, it is a decentralized, distributed ledger where blocks are interconnected with hash esteems.

Smart grid—The smart grid is an electricity grid that consists of smart objects such as smart appliances, energy efficient resources, and renewable energy resources.

Smart contract— Smart contract is a piece of code with the terms of the deal between buyer and seller which executes automatically during the transaction.

3.2 The Architecture of the Proposed Model

In smart grids, it is crucial to have full control over electricity generation and distribution. Smart grids should have the abilities such as high security, hack proof, tamper-free data, and high reliability. From a structure viewpoint, a smart grid will probably consolidate new advances, for example, advanced metering, energy distribution, energy transaction, etc.

The domains within the smart grid: transmission, operations, service provider, generation, distribution, customer, and markets (removed here for brevity) that need to be considered for implementation.

A smart grid is a high-level grouping of appliances, networks, organizations, and individuals with common interests and engaging in projects of similar nature. To transmit, store, update, and process the necessary information within the smart grid, the various entities are needed. The entities in a particular domain also communicate with the other entities in other domains to allow smart grid functionality as shown in Fig. 1.

Advantages of implementing energy trading with blockchain are as follows:

1. Blockchains get more secure with more participating parties in the network
2. Blockchains improve trust between participants by having multiple points of verification (encrypting, proof of work, and hashing).
3. Blockchains create permanent records of data that cannot be edited or deleted.
4. Core objective in the system is designed to prevent double counting of assets, record ownership, and transfers.
5. Blockchains are transparent by design---where ownership or control of assets is public and transparent.

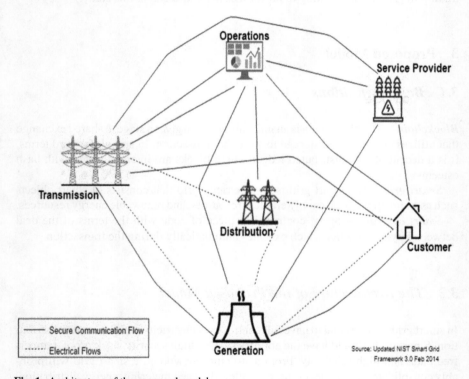

Fig. 1 Architecture of the proposed model

4 Implementation

The application is built using the Ethereum blockchain, and Remix IDE is used to write smart contracts using the Solidity programming language which is a contract-oriented programming language. TestRPC was used to simulate a blockchain network of connect nodes to execute smart contract in a virtual environment.

Ethereum is chosen due to the following advantages:

i. It supports cryptocurrency (ether)
ii. It supports smart contract
iii. It is a public blockchain.

It has few limitations too such as,

i. We cannot process much data as there is a limitation of the usage of parameters for functions in open source
ii. High computation require a high cost for processing data (ether, wei, gas).

This developed platform lets consumers buy and sell renewable electrical power directly from multiple entities which can be a domestic household or a commercial organization using a blockchain platform that includes trading energy like a commodity between users and as well as generating institute like government. In peer-to-peer trading manner, electricity transaction takes place directly from producer to consumer. Producer's source of generating electricity may be solar energy, wind energy, etc., which can be sold in the marketplace. By implementing this technology, the distribution of energy among multiple entities can be done efficiently with minimum loss of energy. The user transactions, buy/sell energy, are stored in the blockchain as distributed ledger.

5 Results and Discussion

Energy trading will be traded between producer and consumer. Energy generated from microgrids will be traded across consumer and producers of energy and primary power source from regular government electricity companies.

Following features have been implemented in the platform:

- Producers can sell their electricity in the platform which is generated by microgrids.
- The excess energy which is being generated is allowed to sell which is generated from multiple energy sources.
- Consumer can purchase electricity using the blockchain platform from the producer.
- The trading transaction is created as a ledger and stored on blockchain.

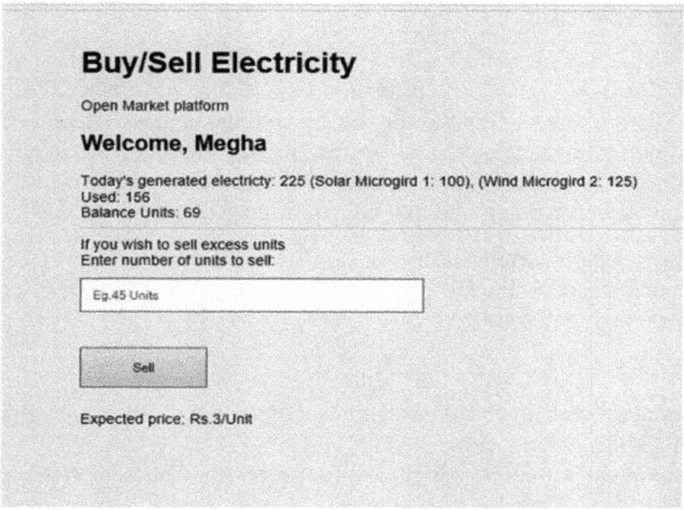

Fig. 2 Buy/sell interface

5.1 Screenshots

The screenshots of the application are presented below. Figure 2 shows the user interface for energy transaction.

The username will be displayed once he/she login then excess electricity which is generated by their microgrid will be shown and given the option to sell to open market. The transaction (buy/sell) is recorded in the blockchain after the authentication process.

In Fig. 3, list of consumers who is registered and able to consume electricity based on their need is shown.

Figure 4 shows list of producers. The list of producers is stored in the blockchain. The producers who have excess electricity generation are registered, and after authentication/verification, the producer will be registered in the blockchain.

Figure 5 shows people get to check a soft copy of their monthly used limit. The consumers who seek for electricity will make a request for the electricity. The producers who have excess power generation show their willingness to sell the electricity. After the mutual agreement between producer and consumer, the required electricity is supplied to the consumer. The consumer gets the electricity bill for the consumed electricity.

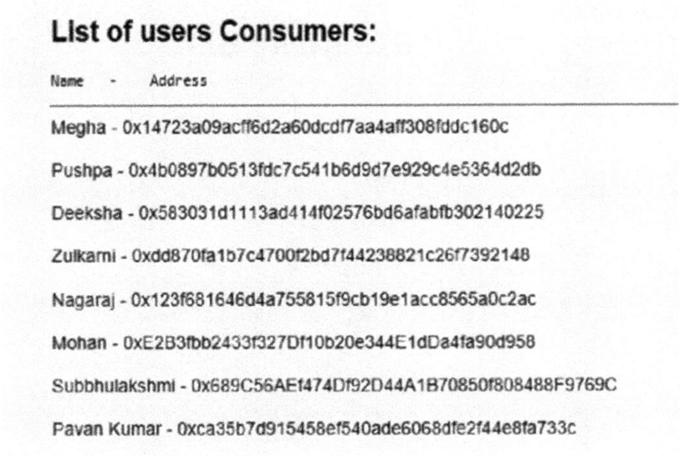

List of users Consumers:

Name - Address

Megha - 0x14723a09acff6d2a60dcdf7aa4aff308fddc160c

Pushpa - 0x4b0897b0513fdc7c541b6d9d7e929c4e5364d2db

Deeksha - 0x583031d1113ad414f02576bd6afabfb302140225

Zulkarni - 0xdd870fa1b7c4700f2bd7f44238821c26f7392148

Nagaraj - 0x123f681646d4a755815f9cb19e1acc8565a0c2ac

Mohan - 0xE2B3fbb2433f327Df10b20e344E1dDa4fa90d958

Subbhulakshmi - 0x689C56AEf474Df92D44A1B70850f808488F9769C

Pavan Kumar - 0xca35b7d915458ef540ade6068dfe2f44e8fa733c

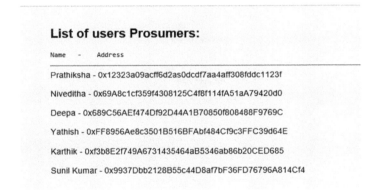

Fig. 3 List of consumers (from blockchain)

List of users Prosumers:

Name - Address

Prathiksha - 0x12323a09acff6d2as0dcdf7aa4aff308fddc1123f

Niveditha - 0x69A8c1cf359f4308125C4f8f114fA51aA79420d0

Deepa - 0x689C56AEf474Df92D44A1B70850f808488F9769C

Yathish - 0xFF8956Ae8c3501B516BFAbf484Cf9c3FFC39d64E

Karthik - 0xf3b8E2f749A6731435464aB5346ab86b20CED685

Sunil Kumar - 0x9937Dbb2128B55c44D8af7bF36FD76796A814Cf4

Fig. 4 List of producers (from blockchain)

6 Conclusion

Blockchain is going to revolutionize the business. In abroad, few countries have implemented smart grids on blockchain, where as in India it is yet to come. Simulation of Ethereum-based blockchain platform for peer-to-peer energy trading is discussed in this paper. The discussed model is an attempt to implement blockchain-based energy transaction. It has few limitations in implementation as it is built on free/basic version of Ethereum. Also, the implemented model is a basic prototype. Further, the model can be improved with authentication by adding verifiers to the system and integrating the prototype with hardware to control the physical energy transmission.

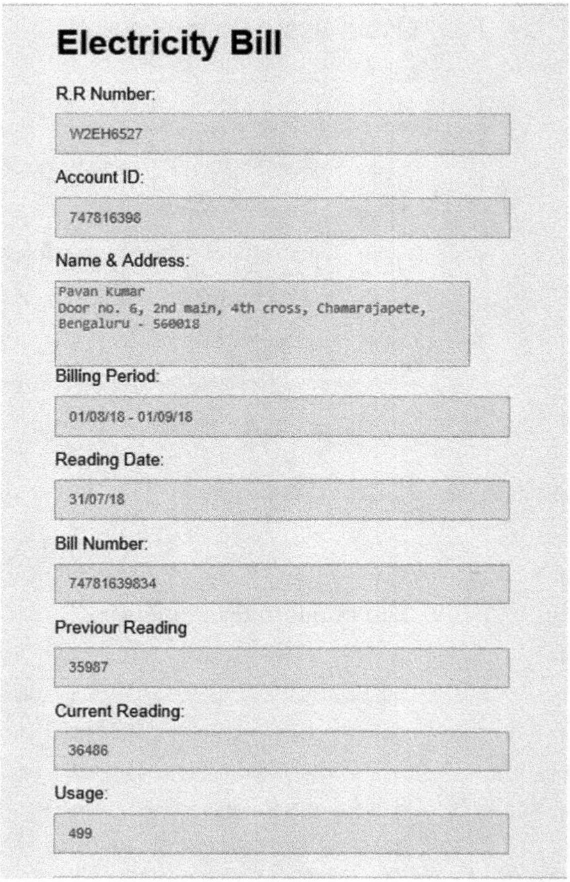

Electricity Bill

R.R Number:
W2EH6527

Account ID:
747816398

Name & Address:
Pavan Kumar
Door no. 6, 2nd main, 4th cross, Chamarajapete,
Bengaluru - 560018

Billing Period:
01/08/18 - 01/09/18

Reading Date:
31/07/18

Bill Number:
74781639834

Previour Reading
35987

Current Reading:
36486

Usage:
499

Fig. 5 Screenshot of electricity bill (from blockchain)

Acknowledgements The authors of this paper would like to thank "Department of Science and Technology, New Delhi" funding the project and thank the Principal and Management of MSRIT for their support and encouragement.

References

1. Nakamoto, S.: Bitcoin: A Peer-to-Peer Electronic Cash System. https://bitcoin.org/en/bitcoin-paper
2. Zizzo, G., et al.: A technical approach to P2P energy transactions in Microgrids. IEEE Trans. Indust. Inf. (99). 1–1, (2018, February)
3. Münsing, E., Mather, J., Moura, S.: Blockchains for decentralized optimization of energy resources in microgrid networks. In: IEEE Conference on Control Technology and Applications.: Energy, Controls, and Applications Lab, UC Berkeley, USA (2017)

4. Amanbek, Y., Tabarak, Y., Nunna, H. K., Doolla, S. Decentralized transactive energy management system for distribution systems with prosumer microgrids. In: 19th International Carpathian Control Conference (ICCC) (2018)
5. Mannaro, K., Pinna, A., Marchesi, M.: Crypto-trading: blockchain-oriented energy market. In: AEIT International Annual Conference (2017)
6. Wu, L., Meng, K., Xu, S., Li, S., Ding, M., Suo, Y.: Democratic centralism: a hybrid blockchain architecture and its applications in energy internet. In: IEEE International Conference on Energy Internet (ICEI) (2017)
7. Yang, T., Guo, Q.: Applying Blockchain technology to decentralized operation in future energy internet. In: IEEE Conference on Energy Internet and Energy System Integration (EI2) (2017)
8. Di Silvestre, M.L., Dusonchet, L.: Transparency in transactive energy at distribution level. AEIT International Annual Conference (2017)
9. Xue, L., Teng, Y.: Blockchain technology for electricity market in microgrid. In: 2nd International Conference on Power and Renewable Energy (ICPRE) (2017)
10. Sabounchi, M., Wei, J.: Towards resilient networked microgrids: Blockchain-enabled peerto-peer electricity trading mechanism. In: IEEE Conference on Energy Internet and Energy System Integration (EI2) (2017)
11. Kang, J., Yu, R., Huang, X.: Enabling Localized Peer-to-Peer Electricity Trading Among Plug-in Hybrid Electric Vehicles Using Consortium Blockchains. IEEE Trans. Ind. Inf. **13**(6) 3154–3164 (2017)
12. Di Silvestre, M.L., et al.: A technical approach to the energy blockchain in microgrids. IEEE Trans. Ind. Inf. **14**(11), 4792–4803 (2018)

Supply Chain Management for Selling Farm Produce Using Blockchain

Anita Chaudhari, Jateen Vedak, Raj Vartak, and Mayuresh Sonar

Abstract The blockchain technology holds an innovative and secured stand for a novel distributed and translucent business mechanism in businesses. The key goal of this research is to develop a framework for supplying farm produce in a system which monitors the profit of each node and maintains transparency using this technology. It helps the farmers to get the expected fair price of his product. If the price of the commodity of the particular product is increased beyond his/hers registered commission value, then fraud is detected easily in the system and the transaction will be terminated with an error message. The farmer can view the commission as well as the product price at each node (viz. packager, carrier, retailer) of supply chain. In addition to maintain transparency in the system, all these nodes can see the status of their products and transactions entered by every other node in the supply chain.

1 Introduction

Blockchain is a different result to the long-standing unruly of conviction between the users. It offers a manner to be called as trustless conviction. It permits us to conviction the crops of the scheme without believing anything within it. A blockchain practice uses Internet for functioning, on a point-to-point system of processors who maintains an undistinguishable replica of the ledger of communications, allowing

---

A. Chaudhari (✉) · J. Vedak · R. Vartak · M. Sonar
St. John College of Engineering and Management, Mumbai University, Palghar, India
e-mail: anitac@sjcem.edu.in

J. Vedak
e-mail: jateenvedak@gmail.com

R. Vartak
e-mail: rajvartak97@gmail.com

M. Sonar
e-mail: sonar.mayuresh260197@gmail.com

P2P authorize communications without a middleman though mechanism consent. Blockchain is nothing but a file—a collective and free ledger of communications that histories entirely communications since the origin chunk till today.

In the new beginning of technology, blockchain skill is a revolutionary idea in distributed information knowledge. Initially developed in place of chunk of Bitcoin's fundamental setup in 2008, its possible request spreads remote outside numerical coins and economic assets. The expertise is still in its opening stages and is yet to spread normal and initiative implementation. As the skill increased broader appreciation enlarged in recent centuries, there has been a burst of enhancements, original usage, and claims. The series of possible tenders of blockchain skill is limitless, since numeral coins to blockchain aided permissible agreements with the most proficient applications yet to be industrialized.

Around billions of harvests presence industrial average worldwide concluded compound source manacles that binge to complete chunks of the world. Supply chain information of how, when, and where these goods were created, manufactured, and used through their lifespan. Client, goods transportable complete an frequently huge system of venders, trailers, storing services. Dealers that contribute in creation, distribution, and sales, up till now in nearly each case these expeditions remains secreted to the end users. Supply chains are charming more intricate, more protracted, and universal. An occurrence on single sideways of the world can halt invention or transmission of a provision on the other sideways.

Supply chain prominence is a crucial corporate task, through maximum businesses taking tiny or not any evidence about their personal two- and three-tier suppliers. Entire supply chain clarity and perceptibility can benefit handling the stream of goods as of rare resources to business belongings, which allows novel types of analytics for actions, hazard, and sustainability.

2 Related Work

This section details the literature survey of different algorithms, technologies, and techniques used for farming using blockchain.

Food traceability structure is well-known, built on the blockchain skill. As proof of thought, they are initial with the basic method of food traceability structure using blockchain expertise. In further revisions, they would like to build up a whole food traceability scheme next, and they study the pragmatism and bound of the scheme in the impending [1]. Investigation provides a construction for a micro-economy platform for DIIs that founded on their early findings looks to be actual. One explanation they imagine would be to spread over cryptography to strongly sign messages, so their source is defensible to be genuine. Extra conceivable resolution is to use a scheme of observers that election scheduled the legality of succumbed information [2].

All-inclusive study on blockchain in cybersecurity examines the security matters we still face. In upcoming work, extra hard work will be finished to speaking the safety difficulties of blockchain and discovering better keys, including the ideals for blockchain security necessities and skill challenging, device for safeguarding the business stream, instructive safety actions in blockchain, etc. It examines the advantages that blockchain has got to cybersecurity and abridges current study and application of blockchain in cybersecurity linked areas [3]. Also discussed mitigation and compared different implementations to observe where the vulnerability is present and where it is not. Additionally, they discussed specific trades between public and private blockchains in terms of security. The results potentially indicate how to develop this fast-growing technology further and in a more secure way by expanding its disruptive nature to other application domains [4]. A completely distributed, blockchain-based traceability key for agri-food supply chain managing, gifted to unified assimilate IoT devices creating and overriding digital data with the chain. To efficiently evaluate Agri-BlockIoT, first, they defined a standard use-case in the given perpendicular domain, viz. from-farm-to-fork [5].

Theoretical model is proposed for supply chain using blockchain by referring different research articles [6]. To classify blockchain technology in food transparency and regulator. It helps to classify the present government of nutrition traceability, controller, skills and schemes that can be valued in deceitful blockchain keys [7]. To apply blockchain knowledge and schemes that can be practical in food traceability and regulator. Blockchain technology can profit the customers, manufacturers, and the management sections but also refining the proficiency of food supply chain's processing and flow. However, these technologies still stay in an idea, not pushing into practice. Based on the above motives, some proposals can be lent from the US applications and then propounded for building up the system in China [8].

The aim is to make a distributed record available for all investors in the supply chain. The plan determination makes an elementary outline for edifice an example or imitation by means of existing skills and procedures [9]. Grid upgrading has improved the distribution of clever Internet linked energy expertise that is possibly vulnerable to replicated threats [10, 11]. A context for blockchain built SCQI. Supply chain management is proposed that maintains all the records and information through developed software using blockchain [12]. Blockchain skill is likely to gaze in marketplace, this is actual inspirational for financial system to use this technology, in different domians to secure data. Author explained RTSM tactic to avoid risk of data manipulation [13]. The use of blockchain technology in the manufacturing of structures and components made of composites/prepregs has the potential to comply with any possible requirements involving tamper proof and provenance-tracking in highly regulated industries [14]. Modum.io is accessible, a start-up that practices IoT device leveraging blockchain expertise to affirm information immutability and community availability of temperature archives, while dropping operative charges in the pharmacological supply chain. The medicinal business has numerous compound and harsh conservational controller procedure (e.g., temperature and humidity) to promise excellence switch, and controlling obedience finished the carriage of medicinal crops [15]. Blockchain-based context for finding source of supply chain products. The proposed methodology gives unique contract solution which promises no member controls the blockchain [16]. To predicament the physical and cyber worlds, i.e., info warehoused trendy, the scheme can precisely replicate the position of the cargo [17]. Enlarged traceability, competence, and public fitness care validate more study and growth to engineering a combined blockchain- and IoT-centric initiative resolution [9].

System is developed to give security in food industry using food supply chain [18]. Emphasis scheduled the vehicle uniqueness and incorporation vehicle-to-vehicle, vehicle-to-infrastructure for independent driving and information conversation resolutions [19]. An organization should plan for appropriate research, using exposed topics and gaps in blockchain domain [20]. Proposed solution tracks goods and maintains all information [21]. Rice supply chain system using blockchain technology which will guarantee safety of rice during supply chain management processes and also obtainable a theoretical study on the blockchain technology and how many industries would get earned by applying this technology into their commercial practices [22, 24]. Crowd-validated, online transfer tracking framework that matches recent enterprise-based SC management keys [23].

Collects all farming related things information through IoT and proposed blockchain-based solution to secure agriculture field [25]. Blockchain expertise targets two main difficulties within the supply chain, precisely, data transparency and reserve allocation. Integration blockchain and supply chain knowledge will develop supply chains performance [26]. Created on the study of 181 supply chain consultants in India the projected prototypical was confirmed using mechanical equation exhibiting [27]. Nutrition supply chain traceability scheme for actual food sketching, shape a security controller scheme for food supply chain by mixing the aforementioned through overall supply chain hazard management procedures, and knowingly recover the routine of the food logistics firm [28]. Provides secure supply chain for agriculture food management. System provides tracking, sketching and information management of farming food [29].

3 Research Methodology

See Figs. 1 and 2.

Experimental Conditions

Users include the total number of clients registered on the Web page which may include either Farmer/Packager/Carrier/Retailer. Total roles include the total number of roles for a client; i.e., the client may have the role of a 1. Farmer or 2. Packager or 3. Carrier or 4. Retailer. Total Batches are the total number of chains added in the blockchain. A single chain includes a Farmer, a Packager, a Carrier and a Retailer (Fig. 3).

Every single batch/chain has a unique Product-id. Product name, the usernames of Farmer, Packager, Carrier and Retailer are also displayed. Also, the view button displays the details as well as transactions of the Farmer, Packager, Carrier and Retailer in proper format. It also shows the status of the product chain. The green

Supply Chain Management for Selling Farm Produce Using ... 371

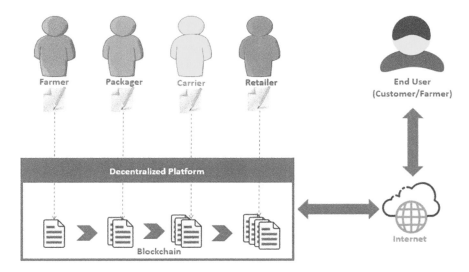

Fig. 1 Architecture diagram

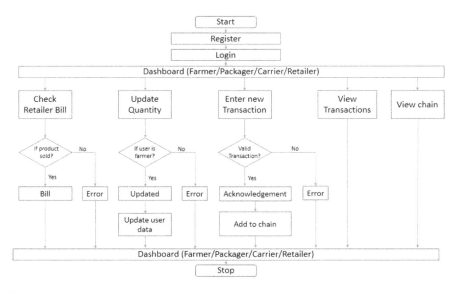

Fig. 2 Flow of proposed system

tick means the transaction is complete while red cross indicates the transaction is incomplete or not yet started. All Users Transactions includes all the transactions the sender hash, Sender username, Receiver username, the Product name with

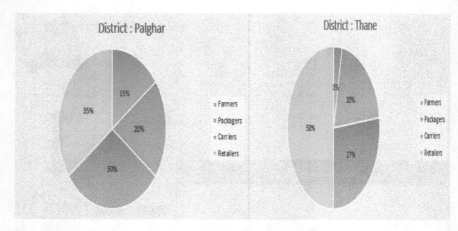

Fig. 3 Pie charts for total number of clients in different regions

id and quantity and at what price it was sold by the Farmer/Packed by the packager/transported by the Carrier/sold to the customer by the Retailer. The New Transaction helps the Farmer/Packager/Carrier/Retailer to create new Transactions based on the availability of product left with him/her (Fig. 4).

4 Conclusion

Using blockchain technology a transparent & tamper proof supply chain system for Farmer crop was made on a local platform. The results for the same are shown in results section. The system generates a bill at end which includes the commissioning prices & the total price at which the product was sold by the retailer in the market. This research will focus on secure transaction between famers and retailers. Framers should know how much amount they had sold farm produce and how much amounts customer paid for same thing while purchasing from retailer. Framers will get the actual value of farm produce. So farmers' conditions can be improved in India. In future we can trace the farm produce using IoT enable system. As Proposed research uses blockchain, so no one can modify transaction.

All Users Transactions:

New Transaction

Index	Sender Hash	Sender	Price	Receiver	Product Name	Quantity	Product id
1	32430d4434400105790a6a1ef011753791f3153a797bb7a6f6d2030099c8fe7c0	RV123	1000	JV123	Rice	50	FA12
2	7cb37c3195ba708b24bf1d5ff9cb8aa35de84b744a22d0e3e2d103a7f5730bde	JV123	1200	mayur1	Rice	5	FA12Tz3X
3	02a8d6e3d52b4c3d0df5d4f6c02de144deb06858d4f63469cd8710979a04fc9ed2	JV123	1200	mayur1	Rice	5	FA120gGO
4	3d14f65d93a5f3b91890e2788539b09cfe06c7db521e0e5ac72808f7002173318b	mayur1	1400	pp	Rice	5	FA12Tz3X
5	93d8fd06eadeb49b9779d9771d06deadf38a65808d05dd5ac5185a2a59fbcb545	mayur1	1400	pp	Rice	5	FA120gGO
6	2e179ff2c0e479f01f01a46dc8fe36c6b5ef889f6c6371b868f15d00407ceaae	pp	1400	enduser	Rice	5	FA12Tz3X
7	0863b06eca84498a8c9b6e7bc4c9b1015193adc71e1c3bb268debc5e4840e1f4	pp	1400	enduser	Rice	5	FA120gGO
8	f42039bedb4c8cf7b31f10236facf2a08e9a19196d0011386888d01bdc14eff2362	JV123	1400	mayur1	Rice	5	FA122ozJ
9	ae39052b859941b4b2cd3ce24ec0db0a856d0e16f879e98e2f4678d0e96c703df	JV123	1400	mayur1	Rice	5	FA123khN
10	d65d4060e173e58ac855cf7c86db52cb36a08ea6a8bbae00f15e774996c6af	JV123	1400	mayur1	Rice	5	FA1295JP

Fig. 4 List of transactions made

References

1. Lin, I.C., Shih, H., Liu, J. C., Jie, Y.X.: Food traceability system using blockchain (2017, October)
2. Kramer, J., van der Werf, J.M., Stokking, J., Ruiz, M.: A Blockchain-based micro economy platform for distributed infrastructure initiatives. In: 2018 IEEE International Conference on Software Architecture (ICSA), pp. 11–1109. Seattle, WA (2018)
3. Dai, F., Shi, Y., Meng, N., Wei, L., Ye, Z.: From Bitcoin to cybersecurity: A comparative study of blockchain application and security issues. 975–979 (2017). https://doi.org/10.1109/icsai.2017.8248427
4. Raimundas, M., Nicolas, M.: A Reference Model for Blockchain-Based Distributed Ledger Technology. Tartu 2017
5. Caro, M.P., Ali, M.S., Vecchio, M., Giaffreda, R.: Blockchain-based traceability in agri-food supply chain management: a practical implementation. In: 2018 IoT Vertical and Topical Summit on Agriculture—Tuscany (IOT Tuscany), Tuscany, pp. 1–4 (2018)
6. Francisco, K., Swanson, Rr: The supply chain has no clothes: technology adoption of blockchain for supply chain transparency. Logistics **2**, 2 (2018). https://doi.org/10.3390/logistics2010002
7. Martin & Servera, and Kairos Future. Blockchain use cases for food traceability and control.
8. Tse, D., Zhang, B., Yang, Y., Cheng, C., Mu, H.: Blockchain application in food supply information security. In: 2017 IEEE International Conference on Industrial Engineering and Engineering Management (IEEM), pp. 1357–1361. Singapore (2017)
9. Kim, M., Hilton, B., Burks, Z., Reyes, J.: Integrating blockchain, smart contract-tokens, and IoT to design a food traceability solution. In: 2018 IEEE 9th Annual Information Technology, Electronics and Mobile Communication Conference (IEMCON), pp. 335–340. Vancouver, BC (2018)
10. Nakasumi, M.: Information sharing for supply chain management based on block chain technology. In: 2017 IEEE 19th Conference on Business Informatics (CBI), pp. 140–149. Thessaloniki (2017)
11. Mylrea, M., Gourisetti, S.N.G.: Blockchain for supply chain cybersecurity, optimization and compliance. In: 2018 Resilience Week (RWS), pp. 70–76. Denver, CO (2018)
12. Chen, S., Shi, R., Ren, Z., Yan, Y., Shi, Y., Zhang, J.: A blockchain-based supply chain quality management framework. In: 2017 IEEE 14th International Conference on e-Business Engineering (ICEBE), pp. 172–176. Shanghai (2017)
13. Gao, B., Zhou, Q., Li, S., Liu, X.: A real time stare in market strategy for supply chain financing pledge risk management. In: 2018 IEEE International Conference on Industrial Engineering and Engineering Management (IEEM), pp. 1116–1119. Bangkok (2018)
14. Mondragon, A.E.C, Mondragon, C.E.C., Coronado, E.S.: Exploring the applicability of blockchain technology to enhance manufacturing supply chains in the composite materials industry. In: 2018 IEEE International Conference on Applied System Invention (ICASI), pp. 1300–1303. Chiba (2018)
15. Bocek, T., Rodrigues, B.B., Strasser, T., Stiller, B.: Blockchain everywhere-a use-case of blockchain in the pharma supply-chain. IFIP (2017)
16. Malik, S., Kanhere, S.S., Jurdak, R.: Productchain: scalable blockchain framework to support provenance in supply chains. In: 2018 IEEE 17th International Symposium on Network Computing and Applications (NCA), pp. 1–10. Cambridge, MA (2018)
17. Xu, L., Chen, L., Gao, Z., Chang, Y., Iakovou, E., Shi, W.: Binding the physical and cyber worlds: a blockchain approach for cargo supply chain security enhancement. In: 2018 IEEE International Symposium on Technologies for Homeland Security (HST), pp. 1–5. Woburn, MA (2018)
18. Feng T.: A supply chain traceability system for food safety based on HACCP, blockchain & Internet of things. In: 2017 International Conference on Service Systems and Service Management, pp. 1–6. Dalian (2017)
19. Perboli, G., Musso, S., Rosano, M.: Blockchain in logistics and supply chain: a lean approach for designing real-world use cases. IEEE Access **6**, 62018–62028 (2018)

20. Tribis Y, El Bouchti A, Bouayad H.: Supply chain management based on blockchain: a systematic mapping study. IWTSCE'18
21. Westerkamp M, Victor F, Küpper, A.: Blockchain-based supply chain traceability: token recipes model manufacturing processes. Axel (2018)
22. Vinod Kumar, M., Iyenger, N., Ch Sriman N.: A framework for blockchain technology in rice supply chain management plantation. 125–130 (2017). https://doi.org/10.14257/astl.2017.146.22
23. Wu, H., Li, Z., King, B., Miled, Z.B.,Wassick, J., Tazelaar, J.: A distributed ledger for supply chain physical distribution visibility (2017, November)
24. Vinod Kumar, M., Iyengar, N.Ch.S.N.: Blockchain: an emerging paradigm in rice supply chain management. IJERCSE (2018)
25. Lin, J., Shen, Z., Zhang, A., Chai, Y.: Blockchain and IoT based food traceability for smart agriculture. ICCSE'18 (2018, July)
26. Eljazzar, M.M., Amr, M.A., Kassem, S. S., Ezzat, M.: Merging Supply Chain and Blockchain Technologies
27. Sachin, K., Angappa, G., Himanshu. A. Understanding the Blockchain technology adoption in supply chains-Indian context. Int. J. Prod. Res (2018)
28. Alfred, T., Feng T.: An information system for food safety monitoring in supply chains based on ACCP. Blockchain and Internet of Things" Doctoral Dissertation
29. Feng, T.: An agri-food supply chain traceability system for China based on RFID & blockchain technology. In: 2016 13th International Conference on Service Systems and Service Management (ICSSSM), pp. 1–6. Kunming (2016)

Design and Construction of a Multipurpose Solar-Powered Water Purifier

Ayodeji Olalekan Salau , Dhananjay S. Deshpande ,
Bernard Akindade Adaramola, and Abdulkadir Habeebullah

Abstract Water is a vital resource in major aspects of human life especially for drinking, cooking, and for washing purposes. It is useful for energy generation, food production, and for cleaning. Lack of access to clean water not only leads to the spread of diseases, but robs people of the basic human necessities. This work presents the design and construction of a portable solar-powered ultraviolet (UV) water purification system. The water purifier system was designed and assembled to demonstrate the capabilities of solar power water treatment systems. The water purifier is designed to filter out dirt and kill bacterial contaminants restrained in the water. The purifier is powered by a solar panel (SP) of size 1 m × 1 m which collects energy from the sun to charge a 12 V battery. The energy obtained from the solar panel is used to power an ultraviolet test bulb. The results of the performance evaluation show that the pH content of the purified water drops from 7.16 to 7.10, while the conductivity and resistance show significant improvements. The developed system will help provide a clean supply of water to schools, homes, industries, offices, and hospitals located in rural areas, consequently reducing the rate of waterborne related diseases.

A. O. Salau (✉)
Department of Electrical/Electronics and Computer Engineering, College of Engineering, Afe Babalola University, Ado-Ekiti, Nigeria
e-mail: ayodejisalau98@gmail.com

D. S. Deshpande
Symbiosis Institute of Computer Studies and Research, Constituent of Symbiosis International (Deemed University), Pune, India
e-mail: dhananjay.deshpande@sicsr.ac.in

B. A. Adaramola · A. Habeebullah
Department of Mechatronics and Mechanical Engineering, College of Engineering, Afe Babalola University, Ado-Ekiti, Nigeria
e-mail: adaramolaba@abuad.edu.ng

A. Habeebullah
e-mail: habeebullah.abdulkadir@gmail.com

© The Editor(s) (if applicable) and The Author(s), under exclusive license to Springer Nature Singapore Pte Ltd. 2021
T. Senjyu et al. (eds.), *Information and Communication Technology for Intelligent Systems*, Smart Innovation, Systems and Technologies 196,
https://doi.org/10.1007/978-981-15-7062-9_37

1 Introduction

Water is one of the most important fundamental needs of human beings. Every person needs a minimum of 20–50 litres of clean water daily for drinking, cooking, and for cleaning purposes. Water pollution is a major problem nowadays; polluted water is very harmful to human existence. The World Health Organization (WHO) reports in [1] indicate that, for each year about two million people die as a result of preventable waterborne diseases, while one billion people lack access to clean drinking water. A staggering number of people have serious health challenges due to water-related diseases, mostly which are easily preventable. As stated by the United Nation (UN)—"universal access to clean water" is a basic human right; a necessary step towards rising living standards worldwide. Communities suffering from lack of portable drinking water are typically economically undeveloped [2].

In Nigeria, over 63 million people lack access to good drinking water [3]. However, most of those who lack access to portable drinking water have a source of water available nearby (stream, rivers, ponds, etc.). The major challenge with these sources is that the water is mostly contaminated and needs purification before use. This water is purified mostly by boiling and sieving; however, some micro-organisms such as "hyperthermophilic" can survive this extremely hot temperature. Furthermore, devices to purify water are very expensive for the common man to purchase and in fact, in most cases, the users do not have the technical know-how to operate these devices effectively. Therefore, this study presents the development of a simple purifier that combines a gravity fed filter system and a disinfecting UV-C light to deactivate harmful micro-organisms in water. For purification of water, UV water purification is one of the efficient ways to purify water. The UV rays kill deadly infectious agents in water and destroys infectious micro-organisms by attacking their genetic core (DNA) [4]. This method of disinfecting water using UV light is very simple, significantly effective, and environmentally friendly. UV systems can overcome 99.99% of deadly micro-organisms without adding any chemical agents or changing the taste of water or giving it odour. Examples are carbon block filters or reverse osmosis systems. The Electromagnetic spectrum range of UV light falls in between visible light and X-rays. Its frequency is about 8×10^{14} to 3×10^{16} hertz (Hz) or cycles per second, and its wavelengths is about 380 nm (1.5×10^{-5} inches) to about 10 nm (4×10^{-7} inches) [5]. A pictorial view of the Electromagnetic spectrum is shown in Fig. 1.

UV purification techniques have been used for many years in the health sector, cosmetic, beverage, and electronics industries, especially in Europe [6].

This study presents the development of a solar-powered UV water purifier for use in rural areas, refugee camps, and other areas where there is no access to clean water supply. A prototype was developed which could be improved and upgraded in the future for commercialization. The developmental stages are divided into three main parts, namely:

i. Design of the water purifier.
ii. Fabrication of the purifier using locally sourced materials.
iii. Testing and performance evaluation of the purifier.

Design and Construction of a Multipurpose Solar-Powered Water ... 379

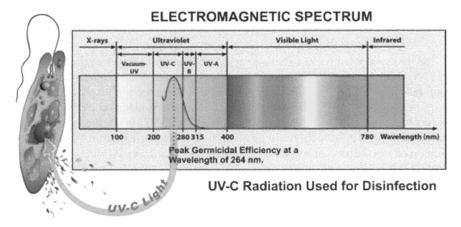

Fig. 1 Electromagnetic spectrum

The remaining sections of this paper are organized as follows: UV water treatment methods are presented in Sect. 2, while materials and methods are presented in Sect. 3, fabrication is discussed in Sect. 4, results in Sect. 5 and lastly, conclusion is presented in Sect. 6.

2 UV Water Treatment

UV water treatment process is a type of the purification unit setup which depends on numerous factors such as the source of the water and water quality. The purity of water may affect the transmitted UV light dosage. The quality of natural water is very important in selecting the type of water treatment device. Pre-filtration of water is highly recommended if the turbidity is 5 NTU or greater and/or the total suspended solids are greater than 10 ppm. Usually, it is recommended to install a 5–20 micron filter prior to a UV disinfection system [7, 8]. A typical UV bulb (UV-C screw type) is shown in Fig. 2.

Fig. 2 UV Germicidal bulb

There are three common types of Germicidal Lamps available, namely:

i. *Low Pressure Lamps (LPLs)*

The fluorescent lamp and LPLs are very similar having a wavelength of 253.7 nm (1182.5 THz). Mostly, the germicidal lamp looks similar to an ordinary fluorescent lamp but the tube of this lamp does not contains fluorescent phosphor. Instead of being made up of ordinary borosilicate glass, this tube is made up of fused quartz. These two modification combined together allows the 253.7 nm UV light produced by the mercury arc to pass out of the lamp unmodified. Whereas, in ordinary fluorescent lamps, it causes the phosphor to fluoresce, thereby producing visible light. Germicidal lamps still produce a little amount of visible light due to the other mercury radiation bands.

ii. *High Pressure Lamps (HPLs)*

HPLs are mostly similar to High-intensity discharge (HID) lamps. These types of lamps emit a broadband UV-C radiation instead of a single line. They are commonly used in industrial plants for water treatment because they produce very intense radiation but still produce very bright bluish white light.

iii. *Light Emitting Diode (LED)*

Recent advances in LED technology has led to the commercial availability of UV-C LED sources. The UV-C LEDs use semiconductor materials to produce light in a solid-state device.

3 Materials and Methods

Most times, rural areas that lack portable water also lacks electricity supply. Electricity supply is one of the important parts of the UV purifier. Designing a purifier to work with the grid is not feasible, thus the need to find a sustainable source of electricity. One of the most accessible sources of renewable energy is solar; therefore, we use solar power to generate the needed electricity. The purification process operates using the electricity generated from the sun. The material selection undertaken in any engineering project goes a long way in achieving its desired design goals. The design and construction of the system is achieved by using careful and scrupulous measures, working according to the design and its calculations in order to avoid errors, accidents, and system failure.

3.1 Components of the Purifier

This section discusses the components used in the development process.

i. *Cart*

This refers to a three-layered metallic shelf which holds all the components of the purifier (all the components of the purifier are attached to the cart). The cart is made from angle iron. The layers are covered with a 2 mm mild steel plate.

ii. *UV Chamber*

This is a cylindrical container which houses the UV Germicidal bulb. It receives filtered water from the filter and then exposes it to the UV bulb.

iii. *Tanks*

There are three tanks; the largest serves as the reservoir tank which holds the water coming directly from the source. It prevents dirt from getting into the system. The medium sized tank serves as the sedimentation tank which allows water already purified from the UV Chamber to settle. The small tank is the storage tank which holds water from the sedimentation tank; it has a tap where the purified water can be collected.

iv. *Filter*

The filter removes the major dirt from the water. The system uses a cartridge filter; the in and out section fits a 1-inch diameter pipe.

v. *Valves*

The purification system makes use of four ball valves which are used to control the flow of water in various stages.

vi. *Solar Power Source*

The UV system is powered by solar energy. In preference to other forms of energy, solar energy is readily available and eco-friendly. The solar power components used include: 12 V 50 W solar panel, 12 V 18 AH battery, 500 W inverter, and 10 A solar charge controller.

vii. *Plumbing*

This is the piping system that carries water to and from each component in the purifier. In the purification system, the entire piping system is made of plastic. In this work, a 1-inch plastic pipe was used throughout. Unions, back nuts, and a total of 4 ball valves were used. Seals and gums were also used in order to prevent leakages.

3.2 Design Calculations

This section presents the calculations used in the design process of the purifier.

i. *Ultraviolet Dosage*

The UV dosage gives an indication of the total amount of energy absorbed by micro-organisms present in water. The UV dosage (D) is given as:

$$D = I \times T \qquad (1)$$

where I is the effective intensity and T is the fluid retention.

ii. *Distance and amount of time needed to purify water*

UV bulbs usually emit light at a wavelength of 254 nm which cause DNA patterns in bacteria to be destroyed within a certain distance from the bulb. In [9], the equation for calculating this distance and the time needed to purify water is given as:

$$I(r) = \frac{PL}{2\pi r} e^{-aer} \qquad (2)$$

where:

$I(r)$ is the UV intensity at a distance r from the lamp (mW/cm^2), PL is the UV power emitted per unit arc length of the lamp (mW/cm), r is the Radial distance from the lamp (cm) and ae is the Base absorption coefficient of the water (1/cm) [10–12].

3.3 Solar Power Requirements

i. *Power Consumption*

The power consumption of UV Bulb is given as 15 Wh. If the bulbs rating is higher than this, the more energy it will consume. The UV bulb will need to run for an hour to purify the water contained in the UV purification chamber. Hence, it will require 15 Wh of energy to power a 15W UV bulb for 1 h.

ii. *Energy the Battery Stores*

Battery capacity is measured in Amp Hours (AH). This can be converted to Watt Hours (Wh) by multiplying the AH value by the battery voltage (12 V). The calculation is given as:

$$X \times Y = Z \qquad (3)$$

where:

X is the Battery size in AH
Y is the Battery voltage
Z is the Power available in Watt Hours.
Hence, a 18 AH battery is chosen. For the 12 V battery the Watt Hours value is

$$18 \times 12 = 216 \, \text{WH} \tag{4}$$

Considering the loss of energy through the DC/AC inverter (96% efficient), the battery can run a 15 W light bulb for 13.68 hrs:

$$\frac{216}{15} \times 0.96 = 13.68 \text{ hrs} \tag{5}$$

iii. *Solar panel power rating*

The power rating of the solar panels (SPs) is given in Watts and it indicates the maximum amount of power a SP can generate in bright sunlight. The amount of power it can supply to the battery is given by:

$$E = \alpha \times \beta \times \gamma \times \varepsilon \tag{6}$$

where:

α—solar panels rated power,
β—number of hours of direct sunlight,
γ—the percentage time estimate of direct sunlight received, and
ε—charge controllers efficiency.

Based on the selected components, the following estimations were used in the calculations: The SP would have 6 hrs of direct sunlight, and within this 6 hrs, clouds may obscure the sunlight. Therefore, it was assumed that 30% of the time, the panel will be covered by a shadow (70% in full sunshine). 85% efficiency is a typical value for a charge controller. Therefore, the energy generated is:

$$E = 50 \times 6 \times 0.70 \times 0.85$$
$$E = 178.5 \, \text{WH} \tag{7}$$

Hence, using a 50 W SP, it will generate 178.5 Wh of energy. This is almost enough to fully charge the selected battery.

3.4 Design Rendering

The purifier design was done using SOLIDWORKS 2015 Engineering Package. The 3-dimensionial design, assembly drawing, and 2-dimensional design concerning the purifier are shown in Fig. 3.

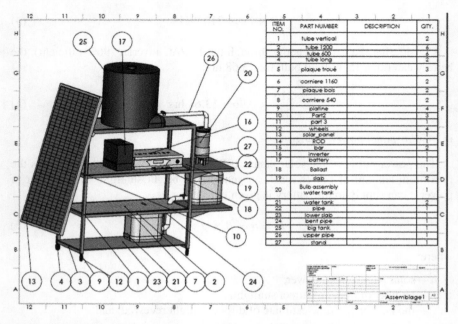

Fig. 3 Assembly drawing of the UV purification system

4 Fabrication

The fabrication of the different components of the water purifier are discussed in this section.

4.1 The Cart

The cart has a height of 6 ft (1.830 m) and a width of 3 ft (0.915 m). It was made using an angle iron and mild steel plate which was welded together to form the stand. The carts holds all the components of the purifier. The fabrication procedure used is very general and can be applied in the construction of carts of any size. The cart is then painted to prevent rusting and also to make the constructed purifier look neater.

Fig. 4 Tanks arranged on cart

4.2 Plumbing

The diameter of the pipes and valves used were 1-inch. This diameter was selected because the smaller the pipe, the higher the velocity and pressure. The fabrication process is described as follows:

Holes slightly larger than 1 inch were drilled into the tanks using a scriber. When the holes were made, the pipes were then fitted with back nuts and then fixed into the holes. The tanks were then arranged on the cart as shown in Fig. 4.

4.3 The UV Chamber

The UV chamber is where disinfection of micro-organisms takes place. It holds the UV bulb which emits UV light into water. Water comes into the chamber from the filter and subsequently leaves for the sedimentation tank. The fabrication process for the UV chamber is given as follows:

The UV Chamber is painted as shown in Fig. 5 in order to prevent the UV rays from causing danger due to the harmfulness of UV radiation. The UV bulb was then attached to the cover of the container as shown in Fig. 6.

Fig. 5 Spray painting of the UV chamber

Fig. 6 Installation of UV bulb

The UV Chamber is then mounted on the Cart and then connected to the solar source to supply power to the system.

5 Results

The water before and after purification was tested with a pH meter and resistivity cell to ascertain its level of purity. The pH meter was used to measure hydrogen-ion activity (acidity or alkalinity) of the water, while the Resistivity Cell was used to measure the conductivity and resistivity of the water. Table 1 shows the values obtained from the measurement, before and after the water purification process.

The values obtained which are presented in Table 1 indicate the level of purity. The pH value before purification was 7.16, while after purification it was 7.10. It can be deduced that water after purification has a pH value which is closer to 7 which is the pH value of pure water. In addition, the conductivity and resistance of the water after purification shows that there is an increase and a later decrease in values. This indicates that fewer number of impurities lead to a raise in conductivity and gives lower resistance.

The efficiency of the water purifier is deduced from the percentage of the water computed by the formula:

$$\% \text{ of purified water} = \frac{\text{volume collected}}{\text{volume added}} \times 100 \qquad (8)$$

Table 1 Level of purity of water sample before and after purification

Purification stage	pH value	Conductivity (μs/cm)	Resistivity (Ω)
Before purification	7.16	161	30,866
After purification	7.10	164	30,402

6 Conclusion

This study presented the design and construction of a solar-powered water purifier for the purpose of water purification. The development was achieved by the construction of an Ultraviolet-ray chamber which was used to disinfect filtered water by killing all the bacteria, viruses, and fungi within it. Unpurified water is made to enter a storage tank in which it is then transported into the cartridge filter to remove unwanted dirt and germs. After filtration, the water moves into the UV chamber where it is disinfected to remove micro-organisms and then passed to a Sedimentation tank, and then to the final storage tank. The results of this study show that the developed water purifier can be used for numerous purposes and guarantees provision of adequate clean drinking water without affecting the nutrients in the water.

References

1. The WHO and UNICEF Joint Monitoring Programme for Water Supply and Sanitation: Global water supply and sanitation assessment 2000 report. World Health Organization, Geneva (2000)
2. Nur, G.N., Sadat, M.A.: Design and construction of solar water purifier. In: International Conference on Mechanical, Industrial and Materials Engineering, RUET, Rajshahi, Bangladesh (2017)
3. Keleher, J.: Design of a solar powered water purification system utilizing biomimetic photo-catalytic nanocomposite materials (2016). [Online] Available at: https://cfpub.epa.gov/ncer_abstracts/index.cfm/fuseaction/display.abstractDetail/abstract/10517/report/0
4. Enzler, S.M.: History of water treatment (2018). [Online] Available at: http://www.lenntech.com/history-water-treatment.html
5. Oram, B.: Water Research (2011). [Online] Available at: https://www.water-research.net/index.php/drinking-water issues-corrosive-water-lead-copper-aluminum-zinc-and-more
6. Atlantic Ultraviolet Corperation (2010). [Online] Available at: https://ultraviolet.com/what-is-germicidal-ultraviolet/
7. Clarke, S.H.: Ultraviolet light disinfection in the use of individual water. US. Army Public Health Command, Virginia (2011)
8. Phalak, M., Kurkure, P., Bhangale, N., Deshmukh, V., Patil, M., Patil, M.H.: Solar powered reverse osmosis water purifier. Int. J. Res. Eng. Appl. Manag (IJREAM) **3**(1), 56–59 (2017)
9. Shaikh, S.K., Bhagwati, P.B., Mali, R.S., Gaikwad, S.C., Jadhav, P.B., Bongarde, S. J.: Mobile solar water purifier. Int. Res. J. Eng. Technol. (IRJET) **3** (2017)
10. Vivar, M., Skryabin, I., Everett, V., Blakers, A.: A concept for a hybrid solar water purification and photovoltaic system. Sol. Energy Mater. Sol. Cells **94**, 1772–1782 (2010). https://doi.org/10.1016/j.solmat.2010.05.045
11. Ismali, S.O., Ojolo, S.J., Orisaleye, J.I., Alogbo, A.O.: Design and development of dual solar water purifier. Int. J. Adv. Sci. Eng. Technol. Res. **2**(1), 8–17 (2013)
12. Oluwafemi, I., Laseinde, T., Salau, A.O.: Design and construction of a 0.5 kW solar tree for powering farm settlements. Int. J. Mech. Eng. Technol. **10**(6), 19–33 (2019)

Named Entity Recognition for Rental Documents Using NLP

Chinmay Patil, Sushant Patil, Komal Nimbalkar, Dhiraj Chavan, Sharmila Sengupta, and Devesh Rajadhyax

Abstract Information retrieval is the process of extracting a pertinent set of facts from a text or a document. The documents are of unstructured format, and thus, information retrieval techniques aim at organizing this data. Named Entity Recognition is one of the information retrieval techniques which classifies a particular word or a phrase in its appropriate class. NER can thus, also be used in extracting entities from legal documents, which would help in providing an effective way to represent these documents. This would reduce the task of a lawyer scrutinizing the document, multiple times, to look for the same set of information. NER systems can be developed with different approaches, one of which is utilizing an NLP library. However, these pretrained NLP libraries may or may not be suitable for a particular use case. Hence, in this paper, we depict an approach to analyze rental documents by custom training spaCy NLP library for tagging named entities such as a person, address, amount, date, etc. The system will provide an interface for the user to upload rent documents, and the result analysis will be stored for quick insights into the document.

C. Patil (✉) · S. Patil · K. Nimbalkar · D. Chavan · S. Sengupta
Vivekanand Education Society's Institute of Technology (V.E.S.I.T), Chembur, Mumbai, India
e-mail: 2016.chinmay.patil@ves.ac.in

S. Patil
e-mail: 2016.sushant.patil@ves.ac.in

K. Nimbalkar
e-mail: 2016.komal.nimbalkar@ves.ac.in

D. Chavan
e-mail: 2016.dhiraj.chavan@ves.ac.in

S. Sengupta
e-mail: sharmila.sengupta@ves.ac.in

D. Rajadhyax
Cere Labs Pvt. Ltd., Mumbai, India
e-mail: devesh.rajadhyax@cerelabs.com

© The Editor(s) (if applicable) and The Author(s), under exclusive license to Springer Nature Singapore Pte Ltd. 2021
T. Senjyu et al. (eds.), *Information and Communication Technology for Intelligent Systems*, Smart Innovation, Systems and Technologies 196,
https://doi.org/10.1007/978-981-15-7062-9_38

1 Introduction

Numerous legal documents are encountered by lawyers in daily business, which requires feasible solutions to be given to their clients or companies. It comprises vital information that can be used, such as deal value, multiple parties involved, disputed properties, loopholes, names, terms and conditions, etc. All of this needs to be stored, retrieved, and processed later to extract knowledge out of it for future investigation. However, the attributes mentioned above cannot be put into relational databases, which can be used for computational analysis. Moreover, the storage of legal documents requires the intervention of repeated manpower because of the natural humongous volume of the data. The data also comprises of rental documents in various formats long enough to be tedious to analyze manually.

The rental documents and contracts can be converted to a structured format with the help of recent advancements in Named Entity Recognition (NER). NER focuses on extracting entities from text, which is nothing but information retrieval. However, challenges are faced when working with rental documents with an Indian context to them. The text can often contain the first name of a person, which can also be the name of an apartment or a street, in an address. Mentioning of dates and their formats used also stands as a challenge. Further, for extracting named entities, there are numerous approaches present utilizing supervised models, semi-supervised models, unsupervised models, or deep learning approaches. Also, for NER, there are various tools available such as NLTK, OpenNLP, spaCy, StanfordNLP, from which we choose spaCy for a better accuracy of tagging entities. Thus, in this paper, we depict an approach to extract entities from rental documents using spaCy and integrating the entire system into a knowledge extraction engine—Cognie. The system will thus be useful for lawyers to analyze rental documents.

2 Related Work

Named Entity Recognition, being an important technique in information retrieval, many researchers have worked on it with varying purposes. Sun et al. [1] carried out a survey, based on various research papers from 1996 to 2017, on the approaches to NER and their target languages. They explored techniques such as Rule Based approach and different Machine Learning approaches, providing a brief overview of them. Chalkidis et al. [2] worked on describing several methods to extract elements from a contract. They explored written rules as well as linear classifiers; however, their work concluded with better results being obtained for a hybrid approach. Poostchi et al. [3] introduced a Named Entity Recognition Model for the Persian Language. Their system incorporated a number of steps. They developed two different corpora for entity extraction - an unsupervised corpus and another annotated corpus for supervised classification in the Persian Language. Partalidou et al. [4] developed a parts of speech (POS) tagger a named entity recognition model using spaCy for Greek language. Their POS tagger was a custom trained spaCy model, trained with 16 POS tags according to the guidelines of Universal Dependencies. The NER was trained on spaCy using Greek Wikipedia and Makedonia Corpus. Neumann et al. [5] developed a system for classifying entities in biomedical domain. They utilized the scispaCy library for this. Moreover, they also explored the dependency parser and POS tagger, which is trained simultaneously on the Genia Corpus. Their study also brought into light the performance of scispaCy when considering biomedical documents. Basaldella et al. [6] worked on extracting entities from a biomedical document by integrating a dictionary-based recognizer with a machine learning model. Yadav et al. [7] presented a survey on current developments in NER techniques which utilized Neural Networks. They studied different NER systems and categorized them, considering the way words are represented in a sentence. Corbett et al. [8] built three systems for recognizing named entities in the chemical domain. All the three systems were compared, the results were shown and the best one was chosen by them. Al Omran et al. [9] surveyed NLP libraries, which can be used for information retrieval from software documentation. Their research on Google's SyntaxNet, Stanford CoreNLP Suite, NLTK Python Library, and spaCy compared different aspects of how these performed on websites like Stack Overflow. They concluded that spaCy had the highest accuracy of tagging.

3 Proposed System

As shown in Fig. 1. Workflow for NER, the user will initially log on to the system and will provide the rent document as an input to the Cognie Platform. The document will then be passed on to the preprocessing module which will split the document into a list of sentences. These sentences are then given as an input to the Classifier and it produces a paragraph from the filtered sentences. From this, named entities are extracted which are then available to the user. The sentence classifier module is trained on a labeled set of sentences, while the entity extraction module is trained on an annotated dataset.

4 Methodology

The NER system was built in a number of stages. The dataset was prepared, which was passed to train the entity extractor and sentence classifier module, upon selecting the appropriate method and algorithm for each of them. Entity extractor module used spaCy while sentence classifier module was developed using a machine learning algorithm. The entire system is hosted on Cognie, which is developed by Cerelabs Pvt. Ltd. The user uploads documents to be analyzed on the system, and the extractor module extracts entities from the document. The output is available to the user in the form of visual representation.

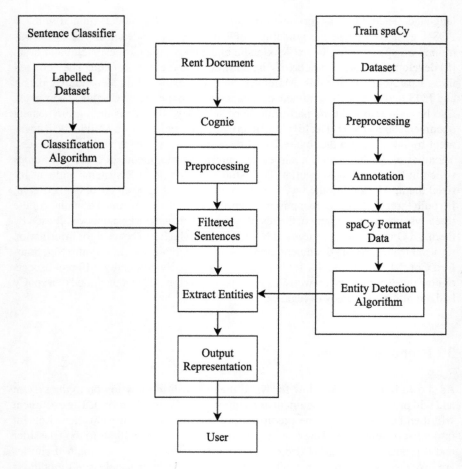

Fig. 1 Workflow for NER

4.1 Dataset Used

Dataset for training Sentence Classifier. A dataset for entity extraction of rental or legal documents with Indian context in it is not available to use directly. So, we handpicked news articles from various websites which had text in English language with Indian names, locations in India, and dates of various formats in them. Further, we inserted addresses of numerous shops, residential flats, and bungalows in the data. Similarly, other entities which are to be extracted from rent documents such as rent payment day, rent amount, etc. were inserted in the articles. Duplicates from the articles were removed and finally, it was ensured that there were an adequate number of entities of all the types. These articles are split into sentences and then labeled into two categories—the sentences which have entities in them and sentences which do not have entities in them.

Dataset for training spaCy. The entity extraction module uses spaCy, which is pretrained on OntoNotes 5 corpus. However, this pretrained model is not suitable for our system. Hence, spaCy is custom trained on our data. The news articles used above are utilized here too. Rather than splitting the articles into sentences, this data comprising of over 50,000 words was annotated for custom training using Dataturks. The annotation was then converted to a format required by spaCy for training its model. This was then passed along with appropriate parameters for training the model, which could extract the entities in the text passed to it.

4.2 The Cognie Platform

The "Learn and Recite" (L&R) system, Cognie, packs together Knowledge Discovery (KD), Knowledge Representation (KR) and Knowledge Query (KQ) modules. The Cognie KR system, is based on Description Logic (DL) presents a unique method of Type Representation. Both knowledge and programs are maintained as part of the KR module. The KD module is based on a declarative workflow system that relies on combinators and generics to implement process flows. Models such as ML and DL, are hosted inside the Cognie KD module. The KQ module is invoked by the Recite function and uses DL reasoning to search for knowledge.

4.3 Sentence Classifier

Various algorithms were considered for the classifier to suit our use case. We selected Bernoulli Naive Bayes algorithm as a classifier due to its higher recall than other classifiers. Recall is used as a key factor in choosing the classifier as the cost of false negative; i.e., a sentence containing entities is classified as a "No," is high. It results in a high recall low precision classifier module.

The document given by the user is at first fed as an input to the sentence classifier, which takes individual sentences from the rental document. These sentences are then classified into two classes by the trained model. When classifying, there is a slight bias toward classifying it as a "Yes" as a false positive sentence will have no entity extracted from it by the extraction module, but a false negative will result into an entity not being extracted, which is a problem for our case. Sentence Classifier module is used to increase the overall accuracy of the system as it avoids misclassification from spaCy by not passing the sentences which do not have entities in them.

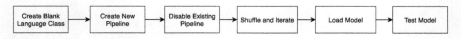

Fig. 2 spaCy pipeline

4.4 Extract Entities

spaCy, in its NER pipeline, incorporates Long Short Term Memory (LSTM) and Convolution Neural Networks (CNN). Further, spaCy allows us to add custom entities to a blank English language class and, in turn, train its deep learning model as shown in Fig. 2. It uses back propagation, to update its weights and at the same time, reduce error. For training the custom model, first a blank English language class. Then, we create a pipeline for the new class, disable the existing pipeline, and then feed the annotated data. The model can now be trained on multiple iterations. The resultant model is saved, which can be used by loading the custom model again. The sentences passed to the entity extraction module are searched for entities in them. spaCy marks appropriate text with appropriate labels and a precise boundary for that label. Various labels which were used for the system comprised of Names, Address, Payment Day, etc. All these entities are now ready to be represented to the user through Cognie.

4.5 Output Visualization

The program provided as input in Cognie is made of Sentence Extractor, Classifier and spaCy. Sentence Extractor receives (takes as input) rent document and produces as output the list of sentences in the document. Classifier receives the list of sentences and produces classified paragraph as in Figs. 3 and 4. This paragraph is passed on to spaCy for entity extraction and the final output is represented with different color highlightings for different entities as depicted in Fig. 5. This output is hosted on the server on a reserved port.

5 Conclusion and Future Scope

The developed custom NER model can accurately detect entities with their appropriate boundaries. Here, we have classified the rental document into tags such as Name, Address, Rent Amount, Payment Day for the rent, Date, Period of the agreement. Table 1 summarizes the results, where in the obtained overall precision is 92.80, recall is 89.58 and F1 Score is 91.16. The lawyer can thus have the document for a quick glance through Cognie Engine. Such analysis can also be carried out for various Terms and Conditions in a document to save the time of reading the document

Named Entity Recognition for Rental Documents Using NLP 395

Fig. 3 Extracted sentences

entirely to understand its intricate details. This can be done with help of a dependency parser which is capable of acquiring relationships in a set of text. Further, all of this can be integrated in Cognie's chatbot to whom a person can ask questions. The chatbot will reply to the question depending on the knowledge retrieved from the NER model and dependency parsers.

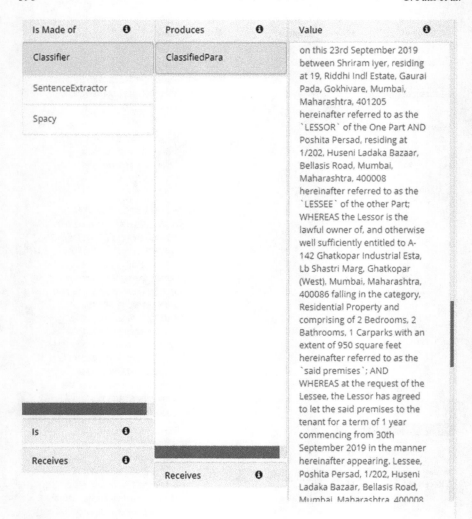

Fig. 4 Classified paragraph

Fig. 5 Extracted entities

Table 1 Results

Entity	Precision	Recall	F1 score
Address	89.18,918	82.5	85.71428
Payment day	90.90909	100.0	95.23809
Amount	95.45454	91.30434	93.33333
Name	95.83333	88.46153	90.47619
Date	90.47619	95.0	92.68292
Period	92.85714	100.0	96.29629
Day of week	100.0	83.33333	90.90909
Overall	92.80545	89.58333	91.16607

References

1. Sun, P., Yang, X., Zhao, X., Wang, Z.: An overview of named entity recognition. In: 2018 International Conference on Asian Language Processing (IALP), pp. 273–278. Bandung, Indonesia (2018)
2. Chalkidis, I., Michos, A.: Extracting contract elements. In: International Conference on Artificial Intelligence and Law (ICAIL). London, UK (2017, June 12–15)
3. Poostchi, H., Borzeshi, E.Z., Abdous, M., Piccardi, M.: PersoNER: Persian named-entity recognition. In: Proceedings of COLING 2016, the 26th International Conference on Computational Linguistics: Technical Papers, pp. 3381–3389
4. Partalidou, E., Spyromitros-Xioufis, E., Doropoulos, S., Vologiannidis, S., Diamantaras, K.I.: Design and implementation of an open source Greek POS tagger and entity recognizer using spaCy. In: 2019 IEEE/WIC/ACM International Conference on Web Intelligence (WI), pp. 337–341. Thessaloniki, Greece (2019)
5. Neumann, M., King, D., Beltagy, I., Ammar, W.: ScispaCy: fast and robust models for biomedical natural language processing. *arXiv preprint* arXiv:1902.07669 (2019)
6. Basaldella, M., Furrer, L., Tasso, C., Rinaldi, F.: Entity recognition in the biomedical domain using a hybrid approach. J. Biomed. Semant (2017)
7. Yadav, V., Bethard, S.: A survey on recent advances in named entity recognition from deep learning models. *arXiv preprint* arXiv:1910.11470 (2019)
8. Corbett, P., Boyle, J.: Chemlistem: chemical named entity recognition using recurrent neural networks. J. Cheminform
9. Al Omran, F.N.A., Treude, C.: Choosing an NLP library for analyzing software documentation: a systematic literature review and a series of experiments. In: 2017, 14th International Conference on Mining Software Repositories, p. 187–197 (2017)

Empirical Analysis of Various Seed Selection Methods

Kinjal Rabadiya and Ritesh Patel

Abstract Influence maximization is a core research issue in the field of viral marketing. Seed selection is one of the promising eras in influence maximization. It can be stated as to find a small set of initial influencers or early adaptors who can easily adapt the products as well as maximize the adoption through the diffusion model. There are two generic diffusion models, i.e., independent cascades and linear threshold models over the social network. In this paper, we have analyzed various algorithms of seed selection with two diffusion models. Based on the experiments, the best seed node-set can be formed by selecting nodes having maximum performance. All experiments show the performance of seed selection algorithms with four real-world collaboration networks datasets.

1 Introduction

1.1 Influence Maximization

Since the last decades, print media were the main source of advertisements that are less emotional than social networks. A social network is formed with trust, emotion, and satisfaction, etc. A social network can catch behavior, relationship, events, and interest of individuals as it is a significant part of human life. So this is a perfect combination that can be used to influence other people about product/social message awareness. Many companies use a social network as a platform for product promotions. The social network platform uses word-of-mouth effect [1, 31] to spread the advertisement messages with trust.

K. Rabadiya (✉) · R. Patel
U & P U Patel Department of Computer Engineering, CHARUSAT, Anand 388421, Gujarat, India
e-mail: kinjalrabadiya20@gmail.com

R. Patel
e-mail: riteshpatel.ce@charusat.ac.in

© The Editor(s) (if applicable) and The Author(s), under exclusive license to Springer Nature Singapore Pte Ltd. 2021
T. Senjyu et al. (eds.), *Information and Communication Technology for Intelligent Systems*, Smart Innovation, Systems and Technologies 196, https://doi.org/10.1007/978-981-15-7062-9_39

In the field of social network analysis, recently a major research works are based on influence maximization. Various seed selection algorithms give a different performance with all considered network datasets. All datasets have different properties so to study the performance and their properties we have conducted various experiments. Our research work focuses on the analysis of various seed selection methods that used to find highly connected with network elements and helps in influence maximization. This analysis helps researchers in finding more innovative and efficient seed selection algorithms. The social spread is maximum if there are strong pioneers to spread the message. So, the problem is to find initial seed users in a social network, which is known as seed selection. It is a recent interest of many companies and individuals who want to promote their products, service, ideas and opinions using social networks [26, 29].

In the network, there are mainly three types of nodes the same as people in society: people are unknown from some fact the same as the node is inactive, some people know something and nodes are aware, some people spread information in the network about something as active nodes do. Inactive nodes are nodes having no information about influence in the network, aware nodes have some of the information about influence but they are not enabled to spread information in the network and active nodes can spread information in the network. Over the time, some of the nodes in the network receive information from their neighbors and change their state from "Inactive" to "Aware." As time progresses, if the node gets more information from neighboring nodes then it changes its state from "Aware" to "Active." This process spreads influence in the network and inactive nodes become active. Every node gets a single chance to activate their neighbor nodes. For effective influence maximization, on the social network, there is a strong need for an efficient seed selection algorithm that can maximize the influence in the network [23].

We organized our paper as follows. In Sect. 2 we have gone through the related work on diffusion models and seed selection algorithms. Section 3 covers our experimental results and discussion and paper end with the conclusion in Sect. 4.

2 Related Work

Influence Maximization is the algorithmic problem was proposed by Domingos et al. [2] using the probabilistic method. The influence maximization problem is a discrete optimization problem that was formed by Kempe et al. [3].

The social network can be imagined as graph $G = (V, E)$ where peoples can be viewed as nodes (V) in the graph and their connection can be viewed as edges (E) in the graph [23]. Here, graph G mainly covers two details (1) information propagation model which decide the pattern of information diffusion in the network and (2) activation probability p or weight on edges. For a given a social network graph, a specific diffusion model and a small number k, influence maximization problem can be stated as under the diffusion model, the expected number of vertices get influenced

2.1 Linear Threshold Model

The linear threshold model works based on the threshold value of nodes [4, 5]. A larger threshold value represents that the node has less interest to change its state from inactive to active. In this model, the node changes its state from inactive to active only if the number of nodes that passes information to a given node crosses a threshold value, i.e., assume node v has threshold value 100. Node changes its influence count whenever an influenced neighboring node tries to influence the node v. When it reaches its threshold value 100 then node v will become an active node.

2.2 Independent Cascade Model

The Independent cascade model is based on the activation probability of other neighbor nodes and nodes have a single chance to get activated [4, 5]. Each edge has assigned probability p; it indicates that node u has a single chance to activate node v with probability value p. If edges have some weight assigned, then the weighted cascade model is taken into account.

2.3 Various Seed Selection Algorithms

As the seed selection algorithm is an NP-hard problem [31], there are multiple solutions available and they perform differently with all considered network datasets. In 2003, Kempe et al. proposed highest degree [3] algorithm for seed selection but sometimes it may not give optimal result because diffusion process only depends on out-degree of nodes. Highest degree algorithm also requires more computational power. In 2013, [6] Zhang et al. proposed a method that collects opinions from the network and spread opinion in the network. In 2015, Jayamangala et al. [4] proposed the work by considering various parameters of network and diffusion model. To spread maximum information the multiple-trial approach [7] was proposed by Lei et al. in 2015 for seed selection and information diffusion by considering Explore–Exploit strategies [7]. In 2015, Targeted Influence Maximization (TIM) was proposed by Pasumarthi et al. [8]. Social network consists of relationship information of users. By using this information in 2015 an expectation model was proposed by Lee et al. proposed [9]. In this model, it targets the users in the network based on a relationship between users. Similarly, opinion on anything can be counted as interaction with this product or message. So, in 2016, Galhotra et al. proposed holistic solution as an opinion-cum-interaction (OI) model [10] which was specifically designed for scalable and efficient influence maximization. Sometime event should be spread within specific time limit. The information should reach to maximum people within deadline. So, based on this requirement in 2016, Lamba et al. proposed a method to spread information in the network within a specific deadline [11].

In 2016, Cordasco et al. proposed a diffusion method in which each user's activity level [12] is considered. So, highly active users have high chances to get selected. Sometime, the specific message should reach a specific community so that the message can be useful for all people in the community. Based on this requirement, Lu et al. proposed a community based algorithm [13] in which synthetic and real networks were used. Each user in the network takes incentives for the promotion of a product or message [28, 32]. Company needs optimal budget allocation algorithm so maximum information spread can be achieved and budget should be as minimum as possible. Based on this theory, Kotnis et al. [14], Zhang et al. [15] and Lu et al. [16] proposed a seed selection method that can maximize spread in the network with optimal budget allocation.

There are various seed selection algorithm available which work on node properties like, random selection [17, 21], degree centrality [18, 21], Betweenness centrality [19, 24], Closeness centrality [20, 24], PageRank [17, 21] etc. They are as defined below:

(1) Random [17, 21]: From the network G, randomly seed set s is selected and then selected seed set is used for influence maximization.
(2) Degree Centrality [18, 21]: From the network G, the set of nodes gets selected having the highest number of connections as degree [30]. The node having more degree centrality influences more number of people. The node having more out-degree resistance to accept influence from the network [27].
(3) Betweenness Centrality [19, 24]: From network G, all nodes get selected which lies on the maximum shortest paths of the network.
(4) Closeness Centrality [20, 24]: From network G, all nodes get selected which have a small value for the network. The small value indicates that the nodes have a close connection with their neighbors.
(5) PageRank [17, 21]: It is the method of a link analysis algorithm in which the pages are ranked as per their importance in a Web graph.

3 Experiment Setup and Discussion

Here the experiments are implemented on a server with 2.50 GHz Dual-Core Intel® Core™ i5-3210 M and 4 GB memory. We have implemented four benchmark algorithms and compared them with each other for influence maximization. We have conducted our experiments on various benchmarks datasets as described below.

Empirical Analysis of Various Seed Selection Methods   403

3.1 Data Sets Description

We have performed extensive experiments on four real-world collaboration social networks.

- Zachary's Karate Club network [25]: This network was studied by Wayne W. It includes 34 nodes and 130 edges between nodes.
- Facebook Network [22]: Using the Facebook application, the survey was done to collect a friend list which is called circles.
- Collaboration networks [22]: AstroPh network, CondMat network are a collaboration network of Astro Physics, Condense Matter Physics, respectively. The data includes papers of the duration from January 1993 to April 2003.

3.2 Discussion

Figures 1, 2, 3 and 4 show the results of all the four simulated networks as well as real networks. Observations made from the experiments can be given as below:

For all network and seed selection methods, random seed selection does not perform well. As a random algorithm randomly select the nodes from the network as a seed and spread the information in the network but randomly selected nodes may not have a strong influence over the network. So, it degrades the performance of the diffusion process. As shown in Fig. 1 for the linear threshold model and

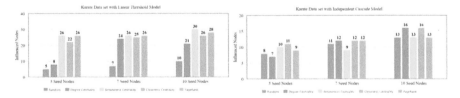

Fig. 1 Karate Dataset with 1 Random, 2 Degree Centrality, 3 Betweenness Centrality, 4 Closeness Centrality, 5 PageRank Seed selection using linear threshold model and independent cascade model

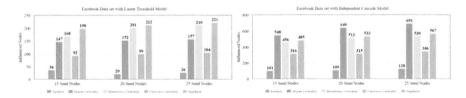

Fig. 2 Facebook Dataset with 1 Random, 2 Degree Centrality, 3 Betweenness Centrality, 4 Closeness Centrality, 5 PageRank. Seed selection using linear threshold model and independent cascade model

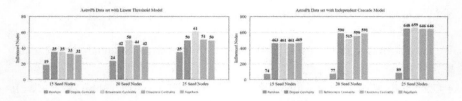

Fig. 3 AstroPh Dataset with 1 Random, 2 Degree Centrality, 3 Betweenness Centrality, 4 Closeness Centrality, 5 PageRank. Seed selection using linear threshold model and independent cascade model

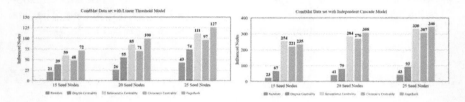

Fig. 4 CondMat Dataset with 1 Random, 2 Degree Centrality, 3 Betweenness Centrality, 4 Closeness Centrality, 5 PageRank. Seed selection using linear threshold model and independent cascade model

Independent cascade model, random seed selection doesn't generate good diffusion. So, for any type of network, it cannot be recommended for seed selection method. Degree centrality, betweenness centrality, and closeness centrality perform well on an average for all experiments.

As shown in Fig. 1, Karate dataset is a small network where betweenness centrality gives the best performance for linear threshold model whereas for independent cascade model degree centrality and closeness centrality gives the best performance. For the threshold model, based on performance the methods can be arranged in best to worst manner as (1) PageRank, (2) Closeness Centrality and then (3) Degree centrality. In the independent cascade model, remaining all three methods give an equal performance.

As shown in Fig. 2, for the Facebook dataset, the Independent cascade model performs better than Linear Threshold Model. Random seed selection performs worst among all seed selection methods for both diffusion models. For the Linear threshold model, PageRank gives the best performance and for Independent cascade mode, Degree centrality gives the best performance. From this, we can say that the Facebook dataset is dense and selected nodes have good contact to spread the information.

As shown in Fig. 3, for AstroPh dataset, the Independent cascade model performs better than Linear Threshold Model. Random seed selection performs worst among all seed selection methods for both diffusion models still perform better for Linear Threshold Model compared to Independent Cascade Model. For the Linear threshold model, Betweenness Centrality gives the best performance as the number of seed nodes increases. For the Independent cascade model, all seed selection methods give the best performance as they are very close to each other. From this, we can say that the Facebook dataset is dense and selected nodes have good contact to spread the information.

As shown in Fig. 4, for CondMat dataset, the Independent cascade model performs better than Linear Threshold Model. Random and degree centrality seed selection performs almost equal for both diffusion models. For the Linear threshold model, PageRank gives the best performance and for Independent cascade mode, Betweenness centrality, Closeness Centrality, and PageRank gives the equal but best performance. From this, we can say that the CondMat dataset is sparse and selected nodes do not have good contact to spread the information. So, from the above experiments common observations can be made as:

- When degree distribution follows poison degree distribution, i.e., most of the nodes have the same degree then different seed selection algorithms do not affect the performance.
- The nodes having a high degree of connectivity cover more number of nodes and put more impact on influence diffusion.
- The nodes having a high degree of connectivity take less time to cover more nodes compare to other methods.
- The set of nodes having high closeness centrality also takes less time to compare to other methods and gives better performance.

4 Conclusion

The social network generates vast data based on the various connections of people and their activities of the social network. Influence maximization can be used to promote the message and product using generated social network data. This paper provides a brief analysis of various seed selection methods in the field of influence maximization. This work can motivate the researchers for an initial study of seed selection problem and influence maximization and develops new matrices for seed selection methods. As the seed selection problem is an NP-hard problem so, there are high chances of the existence of a more scalable and efficient seed selection algorithm for influence maximization.

References

1. Eaton, D.J.: e-Word-of-Mouth Marketing, [Online]. Available: http://college.cengage.com/business/modules/eWOM_secure.pdf. Accessed 17 Jan 2020
2. Domingos, P., Richardson, M.: Mining the network value of customers. In: Proceedings of the seventh ACM SIGKDD international conference on Knowledge discovery and data mining, pp. 57–66. ACM (2001, August)
3. Kempe, D., Kleinberg, J., Tardos, É.: Maximizing the spread of influence through a social network. In: Proceedings of the ninth ACM SIGKDD international conference on Knowledge discovery and data mining, pp. 137–146. ACM (2003, August)

4. Jayamangala, H., Sheshasaayee, A.: A review on models and algorithms to achieve influence maximization in social networks. Indian J. Sci. Technol. **8**(29) (2015)
5. Chen, W., Lakshmanan, L.V., Castillo, C.: Information and influence propagation in social networks. Syn. Lect. Data Manage. **5**(4), 1–177 (2013)
6. Zhang, H., Dinh, T. N., Thai, M.T.: Maximizing the spread of positive influence in online social networks. In: Distributed Computing Systems (ICDCS), 2013 IEEE 33rd International Conference on, pp. 317–326. IEEE (2013, July)
7. Lei, S., Maniu, S., Mo, L., Cheng, R., Senellart, P.: Online influence maximization. In: Proceedings of the 21th ACM SIGKDD International Conference on Knowledge Discovery and Data Mining, pp. 645–654 (2015, August)
8. Pasumarthi, R., Narayanam, R., Ravindran, B.: Near optimal strategies for targeted marketing in social networks. In: Proceedings of the 2015 International Conference on Autonomous Agents and Multiagent Systems, pp. 1679–1680 (2015, May)
9. Lee, J.R., Chung, C.W.: A query approach for influence maximization on specific users in social networks. IEEE Trans. Knowl. Data Eng. **27**(2), 340–353 (2014)
10. Galhotra, S., Arora, A., Roy, S.: Holistic influence maximization: Combining scalability and efficiency with opinion-aware models. In: Proceedings of the 2016 International Conference on Management of Data, pp. 743–758 (2016, June)
11. Lamba, H., Pfeffer, J.: Maximizing the spread of positive influence by deadline. In: Proceedings of the 25th International Conference Companion on World Wide Web, pp. 67–68 (2016, April)
12. Cordasco, G., Gargano, L., Rescigno, A.A., Vaccaro, U.: Brief announcement: Active information spread in the networks. In: Proceedings of the 2016 ACM Symposium on Principles of Distributed Computing, pp. 435–437 (2016, July)
13. Lu, Z., Wen, Y., Zhang, W., Zheng, Q., Cao, G.: Towards information diffusion in mobile social networks. IEEE Trans. Mob. Comput. **15**(5), 1292–1304 (2015)
14. Kotnis, B., Sunny, A., Kuri, J.: Incentivized campaigning in social networks. IEEE/ACM Trans. Networking **25**(3), 1621–1634 (2017)
15. Zhang, B.L., Qian, Z.Z., Li, W.Z., Tang, B., Lu, S.L., Fu, X.: Budget allocation for maximizing viral advertising in social networks. J. Comput. Sci. Technol. **31**(4), 759–775 (2016)
16. Lu, Z., Zhou, H., Li, V.O., Long, Y.: Pricing game of celebrities in sponsored viral marketing in online social networks with a greedy advertising platform. In: 2016 IEEE International Conference on Communications (ICC), pp. 1–6. IEEE (2016, May)
17. Lu, W., Chen, W., Lakshmanan, L.V.: From competition to complementarity: comparative influence diffusion and maximization. arXiv preprint arXiv:1507.00317 (2015)
18. Freeman, L.C.: Centrality in social networks conceptual clarification. Soc. Netw. **1**(3), 215–239 (1979)
19. Beauchamp, M.A.: An improved index of centrality. Behav. Sci. **10**, 161–163 (1965)
20. Freeman, L.C.: A set of measures of centrality based on betweenness. Sociometry **40**, 35–41 (1977)
21. Brin, S.: Larry page. The anatomy of a large scale hypertextual web search engine. In: Proceedings of WWW7. Brisbane, Australia (1998)
22. Stanford University http://snap.stanford.edu/data/index.html#socnets. Last Accessed 21 Jan 2020
23. Rabadiya, K., Makwana, A., Jardosh, S., Changa, I.C.: Performance analysis and a survey on influence maximization. In: International conference on telecommunication, power analysis and computing techniques-2017. At Bharath University, Chennai (2017)
24. Bonacich, P.: Factoring and weighting approaches to status scores and clique identification. J. Mathe. Sociol. **2**(1), 113–120 (1972)
25. Network Repository. http://networkrepository.com/soc-karate.php. Last Accessed 21 Jan 2020
26. Saxena, B., Saxena, V.: Hurst exponent based approach for influence maximization in social networks. J. King Saud Univ.-Comput. Inf. Sci (2019)
27. Karampourniotis, P.D., Szymanski, B.K., Korniss, G.: Influence maximization for fixed heterogeneous thresholds. Sci. Rep. **9**(1), 1–12 (2019)

28. Han, S., Zhuang, F., He, Q., Shi, Z.: Balanced seed selection for budgeted influence maximization in social networks. In: Pacific-Asia Conference on Knowledge Discovery and Data Mining, pp. 65–77. Springer, Cham (2014, May)
29. Lee, C.W., Tang, Y.J., Kuo, J.J., Cheng, J.Y., Tsai, M.J.: The algorithm of seed selection for maximizing the behavioral intentions in mobile social networks. In: GLOBECOM 2017–2017 IEEE Global Communications Conference pp. 1–7. IEEE (2017, December)
30. Rui, X., Yang, X., Fan, J.,Wang, Z.: A neighbour scale fixed approach for influence maximization in social networks. Computing 1–23 (2020)
31. Chen, Y., Wang, W., Feng, J., Lu, Y., Gong, X.: Maximizing multiple influences and fair seed allocation on multilayer social networks. PLoS ONE **15**(3), e0229201 (2020)
32. Hong, W., Qian, C., Tang, K.: Efficient minimum cost seed selection with theoretical guarantees for competitive influence Maximization. IEEE Trans. Cybernet (2020)

A Comparative Study on Interactive Segmentation Algorithms for Segmentation of Animal Images

N. Manohar, S. Akshay, and N. Shobha Rani

Abstract Throughout this article, we are researching two distinct algorithms to separate animals from animal pictures. Since animals appear in a very complex background and often surrounded by greenery, segmenting an animal from its background is a very challenging task. Animal segmentation further helps the problem of animal identification and classification. Here the animals are segmented using graph cut and similarity region merging techniques. To determine the efficiency of our process, an analysis is carried out on our own dataset of 50 animal types, comprising 5000 images. The various performance measures such as Information Variation, Global Consistency Error, Probabilistic Rand Index, and Boundary Displacement Error are used for the purpose of assessment.

1 Introduction

Animal detection and classification has got plenty of applications for e.g. avoiding animal-vehicle collisions, avoiding animal intrusion into residential areas, monitoring animals in zoo, behavioral study of animals etc. Since detecting animals and classifying are done manually which consumes more time and a tedious task, automation is essential. As we know, generally animals appears in cluttered and greenery background, with different lighting condition and occlusion makes segmentation a challenging task.

N. Manohar (✉) · S. Akshay · N. Shobha Rani
Department of Computer Science, Amrita School of Arts and Sciences, Mysuru, Amrita Vishwa Vidyapeetham, Mysuru, India
e-mail: n_manohar@asas.mysore.amrita.edu

S. Akshay
e-mail: s_akshay@asas.mysore.amrita.edu

N. Shobha Rani
e-mail: n.shoba1985@gmail.com

© The Editor(s) (if applicable) and The Author(s), under exclusive license to Springer Nature Singapore Pte Ltd. 2021
T. Senjyu et al. (eds.), *Information and Communication Technology for Intelligent Systems*, Smart Innovation, Systems and Technologies 196, https://doi.org/10.1007/978-981-15-7062-9_40

In all computer vision systems, segmentation plays a significant function. In recent days incorporating interactive segmentation algorithms [1–8, 9, 10] are becoming more and more popular. Our goal is to present an interactive segmentation algorithm for segmenting animals. A consumer can provide segmentation feedback by showing few pixels (seeds) that are part of the target, and a few pixels that are part of the context. These seeds points will clearly indicate on what the user is interested in segmenting. The remaining part of the image will be segmented automatically satisfying the above input. This will give the user a desired segmentation results quickly and accurately.

We can see plenty of works carried on segmentation in the literature. Here we present some of the segmentation algorithm related to our work. Image-based segmentation utilizes Gaussian mixture model [11] in which only the image presence is used to add a pixel to a mark. In [12] Guru et al. proposed a threshold based segmentation for segmenting flowers. Mean shift based segmentation algorithms [13, 14], watershed algorithm [15] and super-pixel [16] based algorithm will divide the image into several regions. These algorithms are considered as low level segmentation algorithms, which can be used to perform subsequent high level operations. In [5, 17] Li et al. proposed a segmentation algorithm where they combined graph cut and watershed algorithm. Oliver et al. [18] Using an unattended approach utilizing contour information to predict the first number of clusters, not the final number of regions. The pixels are often divided into sections by a clustering algorithm. Finally, we have boundary regions pixels and middle regions pixels are another. Ning et al. [19], Ardovini [20], An integrated region-based segmentation algorithm using mean shift for initial segmentation was suggested. In [21] Matthias Zeppelzauer and [22] Mark Dunn Carries out color-based segmentation in wildlife videos utilizing mean-shift clustering.

From the literature it is evident that there are few attempts were made towards segmenting animals in wild life footages but most of them on very small dataset and with less complexity in the dataset. In this work, the efficiency of the algorithms are computed using Probabilistic Random Index, Variation of information, Global Consistency Error and Boundary Displacement Error.

2 Interactive Segmentation Algorithms

Within this research, we performed a comparative analysis of two separate digital segmentation algorithms, namely segmentation based on Graph Cut and Area Merging segmentation based on Maximum Similarity. In both the segmentation algorithm the consumer must initially feed some seed points showing area of interest, i.e. area of target and context. Remaining part of the segmentation will be carried automatically to get the final segmented result where the animal regions are merged into one region and other regions into non animal regions which can be easily discarded.

2.1 Graph Cut Based Segmentation

A graph-based adaptive segmentation is enabled here [23]. As it is an immersive segmentation, certain seed points are inserted by the user to show the entity and context which are known to be hard constraints. Such tough constraints can help to segment the animals automatically by determining a global equilibrium for all segmentation that satisfies the tough constraints. The cost function determines the boundary and area features of the soft constraints divisions [24]. The role of expenses is carried out as follows:

Let D be collection of arbitrary data components, N consists of all unordered neighboring pixel pairs, either 8 or 26 under the regular neighborhood framework. Let $M = (M_1, M_2, \ldots, M_p, \ldots, M_{|p|})$ be a binary vector, where M_p define assignments to pixel p in D. M_p can be either object or background. Calculates soft limitations which we add to the M limit and region properties: The meaning function $K(M)$:

$$K(M) = \lambda \cdot R(M) + B(M) \tag{1}$$

where

$$R(M) = \sum_{p \in P} R_p(M_p) \tag{2}$$

$$B(M) = B_{\{p,q\}} \cdot \delta(M_p, M_q) \tag{3}$$

and

$$\delta(M_p, M_q) = \begin{cases} 1 \text{ if } M_p \neq M_q \\ 0 \text{ otherwise} \end{cases}$$

The importance of the field characteristics of $R(M)$ against the boundary properties of $B(M)$ is seen in (1). The term $R(M)$ implies that penalties for assigning pixel p to object and background are levied by the entity, correspondingly Rp(object) and Rp(background). Rp can concentrate, for example, on how the strength of the pixel p fits into a defined strength model, say the histogram of the goal and the background. The word $B(M)$ indicates the restricting features of the segmentation M. $B\{p, q\}$ a fine for a discontinuity between pair p and q. $B\{p, q\}$ a fine. When $B\{p, q\}$ is high, and when $B\{p, q\}$ is close to null, p and q are identical. $B\{p, q\}$ costs are based on the gradient of local strength.

The goal of our program is to find the maximal global (1) of any section that meets the hard constraints of the customer.

The graph cut based segmentation has two stages: graph construction and graph cut. In the first step, we construct a graph of two terminals given a picture and in the second phase we determine the optimal global cut dividing the two terminals. The cut would provide two regions: target area and context area, from which the target of our concern, i.e. the subject, can be easily extracted from the picture supplied.

2.2 Graph Construction and Graph Cut

A graph is rendered in which the 'V' is $G = \{v, e\}$, and the nodes p(pixel) = D of the picture is 'v.' A source terminal (Source Src) and a source terminal (Sink Ter) are connected to two separate nodes. So $v = P$ is the edges of the margins: n-left (neighborhood link); t-left (terminal link), which contains two forms of indexed margins. In the graph we are creating,. pixel p consists of two connections, one connecting source terminal Src ($\{p, \text{Src}\}$) and a terminal Ter which is linked ($\{p, \text{Ter}\}$). In comparison, in N a relation binds a pair of pixels $\{p, q\}$. Accordingly,

$$e = N \bigcup_{p \in P} \{\{p, S\}, \{p, T\}\}$$

The weights of the edges are given by following table (Table 1):

$$\text{where } K = 1 + \max_{p \in P} \sum_{q:\{p,q\} \varepsilon N} B_{\{p,q\}}$$

The graph is now entirely designed. In the second stage of the algorithm, on graph G defined in [23] we find the minimum cost cut C. We select the minimum cost which is determined using Eq. (1) as the best cut of the segmentations. The following technique operates effectively after the addition of a new seed context points. Now, the point of concern can be conveniently removed from the context.

Table 1 Edge weights computation

Edge	Weight	For
$\{p, q\}$	$B_{\{p, q\}}$	$\{p, q\} \in N$
$\{p, \text{Src}\}$	$\lambda \cdot R_p$ (Background)	$p \in P, p \notin O \bigcup B$
	K	$p \in O$
	0	$p \in B$
$\{p, \text{Ter}\}$	$\lambda \cdot R_p$ (Object)	$p \in P, p \notin O \bigcup B$
	0	$p \in O$
	K	$p \in B$

2.3 Region Merging Based Segmentation

The dynamic fusion-based segmentation of the full similarity region is described in this portion. Initially the image is divided into regions during segmentation using rapid segmentation of shifts. Later, the segmentation would be performed automatically depending on the zone merging process [25], based on user feedback.

Region Merging. Initial segmentation is enough to reduce the amount of primitive components, not to enable the segmentation of each pixel. Here we use segmentation with Fast Shift [26]. Fast Shift segmentation has a greater benefit in maintaining the borders of artifacts, and is also computationally efficient.

After we have finished the initial segmentation we have many tiny areas. In order to conduct the process of merging the region, we must derive certain rules to describe the regions and also for the process of merging. In this work we have used color histogram as a feature descriptor to represent each region. The histogram of color is determined by RGB. Every region's histograms are measured in 16 rates by quantizing every color source. Thus, each area has $16 \times 16 \times 16 = 4096$ bins of feature space.

Any regions are listed as regions with evaluation and meaning. We have three regions after area marking, namely the selected portion of the object, the selected background zone, and the unselected sector. Now the main question is to combine the unmarked regions to either target area or context area using color histogram and a scale of similarity. In order to determine the similarities between the regions, we need to identify some similarities test such that the regions with identical character value can be combined. Here we have used Bhattacharya coefficient [26] to measure the similarity which is given by,

$$\rho(X, Y) = \sum_{k=1}^{4096} \text{Hist} X^k \cdot \text{Hist} Y^k \qquad (4)$$

In which HistX and HistY are the normalized histograms of area X and Y respectively.

Let's see Y as a neighboring area X, labelled with $SY = SiYi = 1, 2, 3,\ldots$, and set Y's neighboring regions. The correlation between Y and its neighboring regions is determined. If all calculated variations have an optimal similarity of X and Y, otherwise X and Y must be merged, i.e.

$$\text{Merge } X \text{ and } Y \text{ if } \rho(X, Y) = \max_{i=1,2,3,\ldots,y} \rho(Y, S_i^Y) \qquad (5)$$

To fully extract the object region, the merging process should be carried out iteratively using the aforementioned merging rule until all non-regions are properly labeled as either object or background. The entire cycle of merger takes place in two phases which are iterative processes which ends when no further merger is necessary.

The first stage is to combine all identified background regions with unidentified background regions adjacent to them. For each region $P \in M_{BKG}$ (where M_{BKG} denote marker background region), we form the adjacent regions $S_P = \{Q_i\} i =$

$1, 2, 3 \ldots r$. Now the similarity between P and each element Q_i in S_P, where $Q_i \notin M_{BKG}$ and $Q_i \in N$ (where N represent set of unmarked regions), is calculated i.e. $\rho(P, Q_i)$. P and Q are merged if they satisfy the merging rule (5), now we have new region P where $P = P \cup Q_i$. If they do not satisfy the above merging rule the region P and Q_i will not merge.

The merger method is an iterative cycle in which the designated history regions M_{BKG} and unknown area N are changed in increasing iteration. This cycle would end when no new combining regions are detected in any of the past regions.

The remaining unidentified regions are combined in the second stage of this process with the target area dependent on maximal similarity, as we did in the first stage with context regions. The resemblance between the unidentified N regions and the reference area M_{OBJ} is determined and integrated at this point if the unified law is complied with (5). This iterative process will stop when there is no new merging regions are available. Finally we extract only the region which contains animals.

3 Experimentation

3.1 Dataset

We also developed our own dataset to test the efficiency of the interactive segmentation algorithms. Our dataset is made up of 5000 photos of 50 separate classes of which 100 photos are taken per class. We have built our own dataset with multiple difficulties such as large intra and interclass differences, complicated context, differing posture, diverse lighting and different views. Figure 1 shows sample animals of different class. Figure 2 displays example pictures of animals with complex background.

Fig. 1 Sample images of animals from different class

Fig. 2 Sample images of animals with complex background and different illumination

3.2 Results

We conducted extensive experiments on our own dataset to check the efficacy of the segmentation algorithms. We have used the output of the proposed segmentation algorithm utilizing four separate parameters in this paper. The parameters used include: Probabilistic Rand Index (PRI) [27], Information Variance (VI) [27], Regional Consistencies Error (GCE) [28] and Boundary Displacement Error (BDE) [29]. In a picture the PRI values will be higher, while the values of VI, GCE and BDE should be smaller. The graph break based algorithm and zone combining dependent segmentation were evaluated on the basis of ground reality. The ground reality is generated by manually segmenting photographs of the species. Figure 3a–d shows the output of the two PRI, VI, GCE and BDE approaches respectively. From the

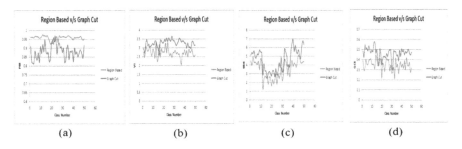

Fig. 3 Performance analysis of two interactive segmentation algorithms on our own animal image datasets in terms of **a** PRI, **b** VI, **c** BDE, **d** GCE

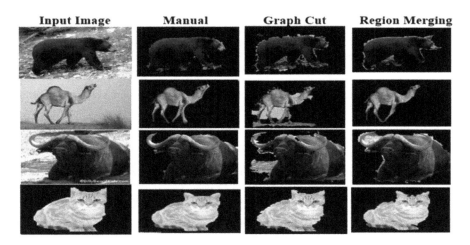

Fig. 4 Sample Segmented Images

graph we can easily note that the segmentation algorithm based on maximal similarities conducts the segmentation algorithm based on the graph split. Figure 4 displays sample pictures of segmented.

References

1. Kass, M., Witkin, A., Terzopoulos, D.: Snakes: Active contour models. Int. J. Comput. Vision **1**(4), 321–331 (1988)
2. Meyer, F., Beucher, S.: Morphological segmentation. J. Visual Commun. Image Represent. **1**(1), 21–46 (1990)
3. Felzenszwalb, P., Huttenlocher, D.: Efficient graph-based image segmentation. Int. J. Comput. Vision **59**(2), 167–181 (2004)
4. Yang, Q., Wang, C., Tang, X., Chen, M., Ye, Z.: Progressive cut: an image cutout algorithm that models user intentions. IEEE Multimedia **14**(3), 56–66 (2007)
5. Li, Y., Sun, J., Tang, C., Shum, H.: Lazy snapping. SIGGRAPH **23**, 303–308 (2004)
6. Levin, A., Rav-Acha, D., Lischinski, D.: Spectral matting. IEEE Trans. Pattern Anal. Mach. Intell. **30**(10), 1699–1712 (2008)
7. Carsten, R., Vladimir, K., Andrew, B.: Grabcut: interactive foreground extraction using iterated graph cuts. SIGGRAPH **23**, 309–314 (2004)
8. Blake, A., Rother, C., Brown, M., Perez, P., Torr, P.: Interactive image segmentation using an adaptive GMMRF model. In: Proceedings of the European Conference on Computer Vision, pp. 428–441. Springer, Heidelberg (2004)
9. Akshay, S., Apoorva, P.: Segmentation and classification of FMM compressed retinal images using watershed and canny segmentation and support vector machine. In: 2017 International Conference on Communication and Signal Processing (ICCSP), pp. 1035–1039. Chennai (2017)
10. Chaithanya, C.P., Manohar, N., Bazil Issac, A., Automatic text detection and classification in natural images, Int. J. Recent Technol. Eng (2019)
11. Das, M., Manmatha, R., Riseman, E.M.: Indexing flower patent images using domain knowledge. In: IEEE Intelligent Systems, vol. 14, pp. 24–33 (1999)
12. Guru, D.S., Sharath, Y.H., Manjunath, S.: Texture Features and KNN in Classification of Flower Images. IJCA, Special Issue on RTIPPR (1), 21–29 (2010)
13. Cheng, Y.: Mean shift, mode seeking, and clustering. IEEE Trans. Pattern Anal. Mach. Intell. **17**(8), 790–799 (1995)
14. Comaniciu, D., Meer, P.: Mean shift: a robust approach toward feature space analysis. IEEE Trans. Pattern Anal. Mach. Intell. **24**(5), 603–619 (2002)
15. Vincent, L., Soille, P.: Watersheds in digital spaces: an efficient algorithm based on immersion simulations. IEEE Trans. Pattern Anal. Mach. Intell. **13**(6), 583–598 (1991)
16. Ren, X., Malik, J.: Learning a classification model for segmentation. ICCV03, vol. 1, pp. 10–17. IEEE, Nice (2003)
17. Li, Y., Sun, J., Shum, H.: Video object cut and paste, SIGGRAPH **24**, 595–600 (2005)
18. Oliver, A,, Xavier, M,, Joan, B,P., Freixenet J.: Improving clustering algorithms for image segmentation using contour and region information. In: Proceeding of IEEE International Conference on Automation, pp. 387–397 (2006)
19. Ning, J., Zhang, L., Zhang, D., Wu, C.: Interactive image segmentation by maximal similarity based region merging. Pattern Recogn. **43**(2), 445–456 (2010)
20. Ardovini, A., Cinque, L., Sangineto, E.: Identifying elephant photos by multi-curve matching. J. Pattern Recogn. **41**(6), 1867–1877 (2008)
21. Matthias, Z.: Automated detection of elephants in wildlife video. EURSIP J. Image Video Process

22. Mark, D., John B., Neal, F.: Machine vision classification of animals. In: Mechatronics and Machine Vision 2003: Future Trends: Proceedings of the 10th Annual Conference on Mechatronics and Machine Vision in Practice
23. Boykov Y.Y., Jolly M.P.: Interactive graph cuts for optimal boundary and region segmentation of objects. In: Proceedings of International Conference on Computer Vision (ICCV-01), vol. 2. pp. 105–112. Vancouver, Canada (2001)
24. Greig, D., Porteous, B., Seheult, A.: Exact maximum a posteriori estimation for binary images. J. Royal Stat. Soc. Series B **51**(2):271–279 (1989)
25. Ning, J., Zhang, L., Zhang, D., Wu, C.: Interactive image segmentation by maximal similarity based region merging. Pattern Recog. **43**(2), 445–456(2010)
26. Kailath, T.: The divergence and Bhattacharyya distance measures in signal selection. IEEE Trans. Commun. Technol. **15**(1), 52–60 (1967)
27. Rubner, Y., Puzicha, J., Tomasi, C., Buhmann, J.: Empirical evaluation of disimilarity measures for color and texture. Compute Vision Image Understand. **84**, 25–43 (2001)
28. Rubner, Y., Tomasi, C., Guibas, L.: The earth mover's distance as a metric for image retrieval. Int. J. Comput. Vision **40**(2), 99–121 (2000)
29. Schmid, C.: Constructing models for content-based image retrieval, In: Proceedings of International Conference on Computer Vision and Pattern Recognition, vol. 2, pp. 39–45 (2001)
30. Issac, A.B., Manohar, N., Varsha Kumari Jain, M.: Segmentation for complex background images using deep learning techniques, Int. J. Recent Technol. Eng (2019)
31. Pantofaru C, Hebert, M.: Comparison of image segmentation algorithms, Tech. Rep. CMURI-TR-05-40, Carnegie Mellon University, vol. 14, pp. 383–39 (2005)

Implications of Quantum Superposition in Cryptography: A True Random Number Generation Algorithm

Dhananjay S. Deshpande, Aman Kumar Nirala, and Ayodeji Olalekan Salau

Abstract This paper addresses the problem of generation of true random numbers by using superposition of qubits in a quantum computer and its application in cryptography. We used QISKIT to create quantum circuits and executed those circuits on a publicly available IBM quantum computer to create an equal probability of getting all possible outcomes i.e. unbiased random output using a few limited random bits (qubits). Furthermore, we explain the physical and mathematical aspects of superposition and random number generation. The results of this study show that in theory, true random number generation is possible on quantum computers and has a lot of application in cryptography. In addition, we show how a truly random seed generated from one of the IBM quantum computers can be used to generate public and private encryption keys using RSA encryption.

1 Introduction

Quantum computing is a rapidly growing field that has the ability to transform the way computing is performed, existence of data, processing, and data security. Security based methods of encryption such as RSA are considered one of the strongest encryption methods. This method uses randomly generated values to generate encryption keys. Quantum computers which have the capability to generate true random

---

D. S. Deshpande · A. K. Nirala (✉)
Symbiosis Institute of Computer Studies and Research, Constituent of Symbiosis International (Deemed University), Pune, India
e-mail: amn1921010@sicsr.ac.in

D. S. Deshpande
e-mail: dhananjay.deshpande@sicsr.ac.in

A. O. Salau
Department of Electrical/Electronics and Computer Engineering, College of Engineering, Afe Babalola University, Ado-Ekiti, Nigeria
e-mail: ayodejisalau98@gmail.com

© The Editor(s) (if applicable) and The Author(s), under exclusive license to Springer Nature Singapore Pte Ltd. 2021
T. Senjyu et al. (eds.), *Information and Communication Technology for Intelligent Systems*, Smart Innovation, Systems and Technologies 196,
https://doi.org/10.1007/978-981-15-7062-9_41

numbers can be used for generating seeds that are truly random and unpredictable. However, quantum computers can also be used for quantum entangled encryption keys and decryption of encryption like RSA using superposition. This is because RSA encryption is based on the fact that finding the factors of a large composite number is difficult which is true for classical computers but not exactly for quantum computers [1]. This paper focuses only on random number (seed) generation for RSA encryption.

1.1 Background Information

The generation of random numbers, its algorithms, and randomness prediction is fast becoming an area of interest for hackers. Hackers always try to crack encryption algorithms by trying to predict the random number to neutralize the importance of the randomness of the number. But in reality, generating random numbers is a big issue for classical computers. The next hope for the generation of true random numbers is Quantum Computing [2]. In real sense, it is difficult to use a quantum computer for computation till date. However, IBM has provided a way for quantum computers to be used by research scholars. This study presents an empirical implementation of the experiment for generating true random number using an IBM Quantum Computer. Furthermore, the quantum computing is explained in terms of the physical and mathematical concepts required for the experiment and algorithm design. In addition, more details about the information of the mathematics and physics of superposition, and operations with quantum gates like Hadamard Gate and a bit of information about RSA encryption are provided in this paper.

1.2 Quantum Computing

A quantum computer is a device that makes direct use of the Quantum phenomenon like Superposition and Entanglement to perform computation [3]. A normal computer also known as the classical computer, stores information in bits (smallest memory unit of a classical computer) which has a value of either 0 or 1 at a particular time, whereas the quantum computer use quantum bits also termed as qubits to store the information. A qubit like a bit, stores the information as 0 and 1 but unlike a bit, it has the probability of having the value 0 or 1 at the same time i.e. Superposition.

1.3 Superposition

Superposition is a fundamental quantum property that states that a quantum system can be in multiple states at the same time and just like waves in classical physics, two or more quantum states can be superposed to give another valid quantum state [4]. Explained in a more abstract way, it can be defined as the ability of a quantum system to be in multiple states at the same time until it is measured [5]. This can be understood in a better way by understanding the Warner Heisenberg's Uncertainty Principle [6]. To obtain the property of a system, we need to be certain of its velocity and position at a particular time, but for a quantum particle, we can't be certain about its position and velocity at a set time i.e. if we are certain about the position, the velocity of the quantum particle is uncertain and visa-versa [7]. Thus, the particle is in the superposition of its velocity if we are certain about its position and visa-versa. So, in order to represent a superposition of a particular particle we use a probability density graph [8, 9].

To use any physical phenomenon as data bits in a computer, the physical phenomenon should have distinct binary outputs under several observations. In quantum mechanics, after the Stern-Gerlach experiment and discovery of spin of an atom (electron), this physical phenomenon was found in the deflection electrons in an inhomogeneous magnetic field due to the direction of the spin [10–12]. The spin of a particle is a vector whose magnitude is the same for the same particle but the direction varies every time. We can know the direction of the spin only at the time of observation and until then it is said to be in the state of superposition. The spin of an electron in the Stern-Gerlach kind of experiment has two Eigen states $|\uparrow>$ and $|\downarrow>$ i.e. up and down. Thus, we can use the direction of the spin of an electron as data bits or qubits when observed. The spin collapses to either of its two eigen states which can be mapped to be interpreted as $|0>$ or $|1>$. The $|0>$ and $|1>$ are two basis vectors $\begin{bmatrix} 1 \\ 0 \end{bmatrix}$ and $\begin{bmatrix} 0 \\ 1 \end{bmatrix}$ in the vector space respectively [13]. With the understanding of this simple concept of quantum memory in quantum superposition, we can formulate the superposition of both, the direction of the spin of the electron in Stern-Gerlach Experiment and the state of the qubit as:

$$|\Psi> = \alpha|\uparrow> + \beta|\downarrow> = \begin{pmatrix} \alpha \\ \beta \end{pmatrix} \qquad (1)$$

Equation of the superposition of an electron is given as:

$$|Q> = \alpha|0> + \beta|1> = \begin{pmatrix} \alpha \\ \beta \end{pmatrix} \qquad (2)$$

Equation (2) represents the superposition of the qubit [9].

where $|\alpha|^2$ and $|\beta|^2$ is the probability of the output being $|\uparrow>$ and $|\downarrow>$ respectively and $|0>$ and $|1>$ respectively. Therefore, we can also formulate the relationship between α and β as:

$$|\alpha|^2 + |\beta|^2 = 1 \tag{3}$$

where $P_\uparrow$ and $P_0 = |\alpha|^2$ and $P_\downarrow$ and $P_0 = |\beta|^2$ [9].

The state of the qubit and its superposition is not just defined using the spin of an electron but also by the (1) Photon polarization, (2) Impurity spins, (3) Trapped ions, and (4) Neutral atom.

1.4 Equal Probability and Randomness

To achieve this, we need to understand the Stern-Gerlach experiment a bit more in detail and how it is used to understand quantum superposition. In the Stern-Gerlach experiment, silver atoms travel through an inhomogeneous magnetic field and thus gets defected up or down due to their spin angular momentum. This represents two eigen states $|\uparrow>$, (electrons deflected up) and $|\downarrow>$, (electrons deflected down) of the spin in this experiment. This is shown in Fig. 1.

Figure 1 shows the Stern-Gerlach stepup of the experimental results. A similar setup can be seen in [12].

Now if we take the same bunch of atoms and the same setup rotated at 90°, we can see that the atoms get deflected towards left and right. But if the same bunch of atoms had spinned up and down in the first experiment, how then did they change their direction in the second experiment as left or right? This takes us toward the property of the quantum states which postulates that one quantum state can always be defined by another quantum state of the system and all quantum operations and transformations are reversible. Thus, we can certainly define the spin states $|\leftarrow>$ and $|\rightarrow>$ with respect to $|\uparrow>$ and $|\downarrow>$. Since in the first experiment, the atoms had

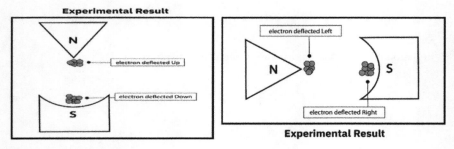

Fig. 1 Experimental results

the eigen states $|\uparrow\rangle$ and $|\downarrow\rangle$ the probability of it being $|\leftarrow\rangle$ and $|\rightarrow\rangle$ should be unbiased i.e. 50% [13]. The mathematical representation of this is represented below:

$$P_\rightarrow = \frac{1}{2} \text{ and } P_\leftarrow = \frac{1}{2} \qquad (4)$$

where Eq. (4) is the Probability of right and left spin.

$$|\uparrow\rangle = \alpha_1|\rightarrow\rangle + \beta_1|\leftarrow\rangle = \begin{pmatrix} \alpha_1 \\ \beta_1 \end{pmatrix} \qquad (5)$$

Equation (5) represents the superposition of the electron with the probability having a spin left or right when there is certainty of its spinning up. Equations (6), (7), (8), and (9) are the equations of the Superposition of the electron with the equal probability of having a spin left or right when there is certainty of its spin being up.

$$|\alpha_1|^2 = P_\rightarrow \qquad (6)$$

$$|\alpha_1| = \sqrt{\frac{1}{2}} \qquad (7)$$

Since $P_\rightarrow = P_\leftarrow$

$$|\alpha_1| = |\beta_1| \qquad (8)$$

$$|\beta_1| = \sqrt{\frac{1}{2}} \qquad (9)$$

It is observed from Eqs. (10) and (11) that the probability of the spin being left or right is half, i.e. unbiased.

$$|\uparrow\rangle = \sqrt{\frac{1}{2}}|\rightarrow\rangle + \sqrt{\frac{1}{2}}|\leftarrow\rangle = \begin{pmatrix} \sqrt{\frac{1}{2}} \\ \sqrt{\frac{1}{2}} \end{pmatrix} \qquad (10)$$

$$|\uparrow\rangle = \sqrt{\frac{1}{2}}|\rightarrow\rangle - \sqrt{\frac{1}{2}}|\leftarrow\rangle = \begin{pmatrix} \sqrt{\frac{1}{2}} \\ -\sqrt{\frac{1}{2}} \end{pmatrix} \qquad (11)$$

Thus, the classical output that we would get on collapsing the quantum superposition will be random. In order to perform the same operation on the quantum states to put them in a state of superposition with equal probability in quantum computers we use the Hadamard gate.

1.5 Hadamard Gate

Hadamard gate is a quantum gate that acts on a single qubit. It is represented as $[H]$ in the quantum circuit. The Hadamard gate is used to put a qubit into a superposition with equal probability of 0 or 1. The Hadamard gate maps the basic state of $|0>$ and $|1>$ as follows:

$$\boldsymbol{H}|0> = \frac{|0> + |1>}{\sqrt{2}} \tag{12}$$

Hadamard gate transforms $|0>$ to the superposition with equal probability [14]. Therefore

$$\boldsymbol{H}|1> = \frac{|0> - |1>}{\sqrt{2}} \tag{13}$$

Hadamard gate transforms $|1>$ to the superposition with equal probability [15].

The application of Hadamard gate on a qubit represents a rotation of Π about the axis $\frac{(\hat{x}+\hat{z})}{\sqrt{2}}$ which is eventually a combination of two rotations; Π about z-axis and $\Pi/2$ about y-axis in the Bloch Sphere model of the qubit.

The matrix form of the Hadamard gate and its effect on the state of a qubit gives us a much better understanding of the transformation. If a quantum gate acts on **n** number of qubits, the quantum gate has to be a $2^n \times 2^n$ unitary matrix. Thus, Hadamard gate is a 2×2 matrix. The derivation of the matrix form of Hadamard gate is given by Eqs. (14)–(19) [16].

$$\boldsymbol{H} = |0> \left[\frac{|0> + |1>}{\sqrt{2}}\right] + |1> \left[\frac{|0> - |1>}{\sqrt{2}}\right] \tag{14}$$

$$\boldsymbol{H} = \frac{1}{\sqrt{2}}[|0><0| + |0><1| + |1><0| - |1><1|] \tag{15}$$

$$\boldsymbol{H} = \frac{1}{\sqrt{2}}\left[\begin{bmatrix}1\\0\end{bmatrix}[10] + \begin{bmatrix}1\\0\end{bmatrix}[01] + \begin{bmatrix}0\\1\end{bmatrix}[10] - \begin{bmatrix}0\\1\end{bmatrix}[01]\right] \tag{16}$$

$$\boldsymbol{H} = \frac{1}{\sqrt{2}}\left[\begin{bmatrix}1 & 0\\0 & 0\end{bmatrix} + \begin{bmatrix}0 & 1\\0 & 0\end{bmatrix} + \begin{bmatrix}0 & 0\\1 & 0\end{bmatrix} - \begin{bmatrix}0 & 0\\0 & 1\end{bmatrix}\right] \tag{17}$$

Fig. 2 Quantum circuit

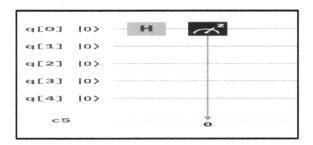

$$H = \frac{1}{\sqrt{2}}\left[\begin{bmatrix} 1 & 1 \\ 0 & 0 \end{bmatrix} + \begin{bmatrix} 0 & 0 \\ 1 & -1 \end{bmatrix}\right] \tag{18}$$

$$H = \frac{1}{\sqrt{2}}\begin{bmatrix} 1 & 1 \\ 1 & -1 \end{bmatrix} \tag{19}$$

Equation (19) is the matrix form of the Hadamard gate. In Fig. 2, we show the output of a quantum circuit with a Hadamard gate on a qubit. The sample is created on IBM Q Experience.

A quantum circuit with the Hadamard gate and measurement gate applied on the first qubit is shown in Fig. 2, while the output of the above circuit when executed on IBM-Q Experience is shown in Fig. 3. The Histogram, output shows a probability graph with $P_0 = 49.512\%$ and $P_0 = 50.488\%$. The absence of values on exactly 50% is due to the noise and de-coherence effect which is mentioned in the limitations section of this paper.

Fig. 3 Quantum circuit output (histogram)

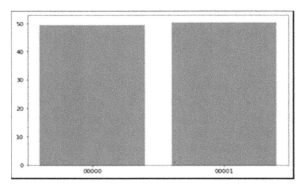

1.6 RSA Encryption

RSA (Rivest–Shamir–Adleman) is an algorithm used by modern computers to encrypt and decrypt messages. It is an asymmetric cryptographic algorithm. Asymmetric means that there are two different keys. This is also called public-key cryptography because one of the keys can be given to anyone, while the other key must be kept private. The RSA public-key cryptosystem was invented by Rivest et al. [8]. The RSA cryptosystem is based on the dramatic difference between the ease of finding large primes and the difficulty of factoring the product of two large prime numbers [17, 18]. In this section, we have focused on the generation of the key and actual encryption and decryption of data using quantum generated random number.

The RSA key generation algorithm is given as follows:

- Generate two large random primes, **p**, and **q**, approximately equal size such that their products $n = pq$ is of the required bit length, e.g. 1024 bits.
- Compute $n = pq$ and $\varphi = (p-1)(q-1)$.
- Choose an integer e, $1 < e < \varphi$, such that $\gcd(e, \varphi) = 1$.
- Compute the secret exponent d, $1 < d < \varphi$, such that $ed \equiv 1 \pmod{\varphi}$.
- The public key is (e, n) and the private key is (d, p, q). Keep all the values d, p, q and φ secret. We prefer sometimes to write the private key as (d, n) because you need the value of n when using d.

[*n* is known as *modulus*, *e* is known as *public exponent or encryption exponent or just the exponent*, *d* is known as *secret exponent or decryption exponent*]

True random numbers are used to generate the two large prime numbers p and q discussed in the first step of the algorithm. This makes the algorithm unpredictable.

2 Implementation of RSA Encryption with True Random Number in Python

This section presents the practical implementation of the steps of the algorithm presented earlier. We have used Qiskit and Python to make a working sample of an RSA encryption key generator. The IBM-Q Experience was used to run the quantum algorithm on the quantum computer "*ibmq*_burlington" that has 5 qubits to generate random seeds 0 and 31. These seeds were then used to generate random prime numbers between a given range. These prime numbers were then passed to the key generator, while the public and private keys were stored. The algorithm then takes the input message and then encrypts it using the private key and then decrypts it back to the original string.

Fig. 4 Visualized quantum circuit

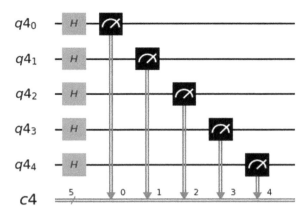

2.1 Algorithm

The section presents the designed algorithm used to generate the true random numbers and which was used for encryption of the string message. The steps of the algorithm are presented as follows:

- Import qiskit library and qiskit tools for visualization of circuit. Plot histograms for the probability density graphs for the output and job monitor for knowing the state of the job on the quantum computers on the cloud provided by the IBM-Q Experience.
- Import random, randint and seed python libraries for the generation of larger prime numbers based on the random seeds generated by the quantum circuit.
- Initialize your IBM-Q Experience account using the public key provided for your account to gain access to the quantum computer.
- Create an instance of 5 classical registers which will be used to read the values from the quantum registers and give classical outputs, i.e. 0 or 1.
- Create a quantum circuit using these registers.
- Apply Hadamard gates to all the quantum registers on the circuit to put them in a superposition with equal probability.
- Apply measurement gates to the quantum registers to measure the values of the qubits along the z-axis.

Visualize the circuit to be sure of connections and the gates. Figure 4 shows the visualized quantum circuit having 5 qubits with Hadamard gates and measurement gates.

- Initialize the IBMQ provider to get access to the quantum computers.
- Initialize the backend, i.e. the quantum computer provided by the IBMQ via IBMQ provider. Here we used "ibmq_burlington" which is a 5-qubit system.
- Define a function generate_random_seed () to generate and return random seed by executing the circuit on the quantum computer.

- Create a task object that would execute the circuit on the quantum computer.
- Call the job monitor to check the real-time status of your task on the quantum computer.
  - Create a result object to extract the results, i.e. the probability density from the task.
  - Create a variable *random_seed* to store the value with maximum probability from the density graph. This might not be technically correct because theoretically, all the possibilities should have equal probability but as we have mentioned earlier, due to noise and errors, the probabilities are not always equal.
  - Return *random seed*.
- Define a function 'prime (int)' to check and return True if a number is prime and False if it's not.
- Call the seed (int) function to set the seed for random number generator with the value returned by the generate_random_seed () function. This seed will be used to generate larger prime numbers.
- Initialize the variable p to store large prime number generated using the randint (min, max) function.
- Create a while loop with the condition of is prime (p) is False.
- If the condition is True, re-assign the value of p with a new random integer.
- Else, loop breaks; p is surely a prime number.
- Initialize the variable q to store large prime number generated using the randint (min, max) function.
- Create a while loop with the condition that prime (q) is False.
- If the condition is True, re-assign the value of q with a new random integer.
- Else, loop breaks; q is surely a prime number.
- Use the values of p and q to generate public and private keys using RSA-Encryption algorithm as seen in [19].
- Enter a string message to be encrypted.
- Use the private key to encrypt the string message.
- Store the encrypted string in the variable encrypted_message and print.
- Use the public key to decrypt the encrypted string stored in variable decrypted_message.
- Store the encrypted string in the variable decrypted_message and print. This will help check if the algorithm is encrypting and decrypting properly.

2.2 Output

In this section, the output of the designed algorithm is presented. A 5-qubit circuit was used for the random seed generation, i.e. between 0 and 31. The algorithm generated two random seeds, 21 and 5 which were then used to compute the prime numbers, i.e. $p = 53$ and $q = 79$. These values were then used to generate the private and

Implications of Quantum Superposition in Cryptography ... 429

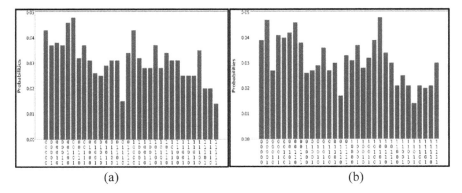

Fig. 5 Quantum circuit with random seed, **a** random seed = 5 and $q = 79$ and **b** random seed = 21 and $q = 53$

public keys. The public key generated is (3427, 4187) and the private key generated is (187, 4187). The private key was then used to encrypt a text "Aman Nirala". The encryption gave the output 79833614624022094339679438554633O746. We then decrypted the encrypted message with the public key and the output was the text "Aman Nirala". This shows that the RSA encryption algorithm encrypted the message using random numbers generated by IBM Quantum computer "*ibmq_burlington*" and also decrypted the encrypted message successfully. Figure 5 shows the Quantum Circuit with Random Seeds 21 and $q = 53$ and Random Seed = 5 and $q = 79$.

Thus the generated public and private keys using the value of p and q are:

public = 3427, 4187, private = 187, 4187, message = 'Aman Nirala'
encrypted_message = 79833614624022094339679438554633O746
decrypted_message = 'Aman Nirala'

This show that the text was encrypted and decrypted successfully using the private and public keys.

3 Limitations

This section presents a list of the limitations of the method and tools used in this study. Quantum computers being a very new technology has a number of limitations which affect the practical implementation of the proposed method of encryption. The algorithm presented in this paper is tuned and limited in many ways to fit the capabilities of the present quantum computers. Some of the limitations are:

1. **Limited and small number of qubits available**: Quantum computers provided by IBM-Q Experience have a maximum of 15 qubits (*"ibmq_16_melbourne"*) where others quantum computers have 5 qubits including *"ibmq_burlington"* which we used to execute this algorithm. This limited number of qubits and

connections between them limits the amount of data we can process at a particular time.
2. **Decoherence**: Quantum information in the qubits is very delicate. Any physical interference or noise destroys the quantum information of the qubit and collapses it to one of its classical forms. Quantum computers are thus constructed in such a way that interferences like electromagnetic radiations, heat, resistance, etc. are minimized. Even though they have many protective shields to prevent radiations, using superconductors like *Niobium* at 0.015 K to minimize resistance and heat, we can't shield out the noise completely and thus with time the qubits lose its quantum information; this is called *quantum decoherence* [20].
3. **Slow response from the IBM-Q Experience cloud server**: IBM-Q Experience is one of the most popular cloud platforms for quantum computing. There are only a limited number of quantum computers available and thus the computers remain busy most of the time and getting the response from the server with the result takes really long (especially the 15-qubit system *"ibmq_16_melbourne"*. So, we use of the 5-qubit system *"ibmq_berlington"*). This limited us to random seeds from 0 and 31.

4 Conclusion

In this study, we presented an algorithm for the generation of odd true random numbers and used them as seeds to generate prime numbers, which were later used for RSA encryption. The development of quantum computing and the core idea of superposition were discussed and Stern-Gerlach experiment was used to explain how the spin angular momentum of an electron in an inhomogeneous magnetic field gives us to eigen states which can be mapped as 0 or 1 in a quantum computer. A quantum state of the direction of the spin of an electron was put into a superposition having an equal probability of being either of its eigen states by rotating the Stern-Gerlach experiment setup by 90°. This lead to the discussion of the mathematics governing the operation and equations of superposition. We then introduced the *Hadamard Gate* and derived the matrix representation of the gate and how the Hadamard gate transforms $|0>$ and $|1>$ and puts them in a superposition with equal probability of being in either of its eigen states. Towards the end, an output of a quantum circuit executed on IBM-Q Experience was presented to show the effect of the application of Hadamard Gate on a qubit. Afterwards, the RSA encryption was introduced. The RSA algorithm decrypted the data successfully, showing the efficacy of the algorithm.

References

1. Colbeck, R.: Quantum and relativistic protocols for secure multi-party computation. PhD Thesis, University of Cambridge (2009)

2. Ma, X., Yuan, X., Cao, Z., Zhang, Z.: Quantum random number generation. *npj Quantum Information* (2) (2016). https://doi.org/10.1038/npjqi.2016.21
3. Jacak, J.E., Jacak, W.A., Jacak, L.: Quantum random number generators with entanglement for public randomness testing. *Scientific Reports* (10) (2020). https://doi.org/10.1038/s41598 019567062
4. Rouse, M.: What is superposition?—Definition from WhatIs.com (2017). [Online]. Available https://whatis.techtarget.com/definition/superposition
5. Bohr, N.: The quantum postulate and the recent development of atomic theory. *Nature* 580 (1928)
6. Heisenberg, W.: The physical content of quantum kinematics and mechanics. Quant. Theory Measur. 62–84 (1927)
7. Zettili, N.: Quantum Mechanics Concepts, and Applications. Wiley, New York (2009)
8. Michael, I.L.C., Nielsen, A.: Quantum Computation and Quantum Information, pp. 13–20. University Press, Cambridge, Cambridge (2010)
9. Ilan, Y.: Generating randomness: making the most out of disordering a false order into a real one. J. Transl. Med. 17 (2019). https://doi.org/10.1186/s1296701917982
10. Batelaan, J.J.S.: Stern-Gerlach effect for electron beams. Phys. Rev. Lett. **79**(23), 4517 (1997)
11. M. I. o. T.: Physics Department, The Stern-Gerlach Experiment. Massachusetts Institute of Technology, Massachusetts (2001)
12. Schmidt-Böcking, S.T.: The stern-gerlach experiment revisited. Eur. Phys. J. H **41**(45), 327–364 (2016)
13. Rieffel, P.W.: An introduction to quantum computing for non-physicists. ACM Comput. Surv. (CSUR) **32**(3), 300–335 (2000)
14. Shepherd, D.J.: On the role of Hadamard gates in quantum circuits. Quant. Inf. Process. **5**(3), 161–177 (2006)
15. Salemian, S.A.S.M.: Quantum Hadamard gate implementation using planar lightwave circuit and photonic crystal structures. Am. J. Appl. Sci. **5**(9), 1144–1148 (2008)
16. Poornima, A.U.: Matrix representation of quantum gates. Int. J. Comput. Appl. **159**(8), 2–5 (2017)
17. Rivest, R.L., Shamir, A., Adleman, L.: A method for obtaining digital signatures and public-key cryptosystems. Commun. ACM **21**(2), 120–126 (1978)
18. Preetha, N.M.: A study and performance analysis of RSA algorithm. Int. J. Comput. Sci. Mobile Comput. **2**(6), 126–139 (2013)
19. JonCooperWorks, A simple RSA implementation in Python, Github Gist (2013) [Online]. Available https://gist.github.com/JonCooperWorks/5314103
20. Coles, G.D.: Quantum algorithm implementations for beginners (2018 [Online]. Available *arXiv preprint* arXiv:1804.03719

Influences of Purchase Involvement on the Relationship Between Customer Satisfaction and Loyalty in Vietnam

Pham Van Tuan

Abstract This chapter has focused on influences of consumers' purchase involvement on the relationship between satisfaction and loyalty. The loyalty is based on three factors: commitment, trust, and word of mouth. A quantitative study has been conducted to determine the direction, magnitude, and effect of the relationship between satisfaction and loyalty; furthermore, to assess the influence of the purchase involvement, as a moderator variable, on the loyalty of Vietnamese consumers. The paper is inherited and developed from the author's doctoral thesis with an empirical research in Hanoi and Ho Chi Minh City in 500 samples. The paper aims at testing the proposed theoretical model and research hypotheses. The research shows that the satisfaction has a positive relationship with the loyalty' factors including commitment, trust, and word of mouth. The analysis of the impact of satisfaction on factors of loyalty says that the satisfaction makes the greatest impact on word of mouth (1.39) and the least on trust (1.043). At the same time, the paper also results that the purchase involvement increases the impact of satisfaction on loyalty. The impact is the most clearly demonstrated on commitment (0.498), while the satisfaction has less impact on trust (0.35) and word of mouth (0.01).

1 Introduction

The satisfaction has been considered as a central concept in marketing up to now. Increasing customer satisfaction is the strategic goal of many companies to gain a competitive advantage. Satisfaction is a multi-level concept and understood in various respects, including satisfaction with products themselves (or satisfaction with product features); with sales services (through interactions between salespeople and customers); or with after-sales services (through the service quality of the provider).

P. Van Tuan (✉)
Faculty of Marketing, National Economics University, Hanoi, Vietnam
e-mail: Phamvantuan@neu.edu.vn

Customer loyalty has always been extremely concerned by managers and marketers because of great benefits from such customers. The research by Hai [1] has shown that maintaining numerous loyal consumers over time will bring great benefits to businesses, specifically: (1) benefits from rates of high price; (2) ones from word-of-mouth communications and introduction to new customers; (3) ones from reduction of advertizing and marketing expenses; and (4) ones increase the quantity and frequency of use/purchase products over time. Inevitably, many theoretical and practical studies have been conducted over the past decades to find effective measures to build and maintain customer loyalty.

Many researches state that satisfaction results in loyalty, then loyalty results in repetitive purchase behavior [2, 3, 4, 5]. On the contrary, many demonstrate that satisfaction has a little relation with loyalty and repetitive purchase behavior [6].

The relationship between satisfaction and loyalty has not been clearly affirmed. The chapter also points out that the relationship between satisfaction and loyalty has been different in sectors and the strength of the relationship can be affected by many factors including commitment, belief, or engagement of consumers.

The relationship between satisfaction and loyalty is not simple and straightforward. Customers may be satisfied, but not loyal. What is the degree of customer loyalty to result in their repetitive purchase behavior? In fact, loyal customers are not all and even not all loyal customers carry out repetitive purchase behavior at businesses and any researchers and businesses would like to find answers for this issue. The main purpose of this study is to discuss and test the relationship between loyalty (through 3 manifestations of commitment, trust and word of mouth) and satisfaction. The other purpose is to discuss and test this relationship under the regulation of purchase involvement of Vietnamese customers.

2 Theoretical Framework

2.1 Relationship Between Loyalty and Satisfaction

This chapter is studied in three aspects of loyalty, including commitment, trust, and word of mouth, in the relation with satisfaction.

Oliver [6] has discovered that the relationship between loyalty and satisfaction has not been specifically defined in the corresponding theories. Kasper [7] has seen that many researches fail to find out differences between repetitive purchase and loyalty, and ones between true loyalty and false one in investigating the relationship with satisfaction. Recent researches have just focused on considering satisfaction as an independent variable and almost ignored the differences between types of satisfaction.

It appears two perspectives of evaluating the theory of relationship between satisfaction and loyalty. Firstly, satisfaction has been considered as a key factor of loyalty. Dixon et al. [3], Fornell [2] and Heitmann et al. [5] say that satisfaction has been strongly associated with loyalty to be positive and ready for introduction and word-of-mouth communication. Moreover, satisfaction affects consumers' future choices, which improve customer retention. Customers with satisfaction shall be loyal and intend to engage with organizations.

Meanwhile, others have argued that customer satisfaction has a highly positive influence on loyalty, but is not itself sufficient to create the loyalty [6]. Customers may be totally satisfactory, but also this satisfaction does not necessarily transform into loyalty. Satisfaction is one of the prerequisites to form the loyalty, and loyalty is influenced by several factors. Some researchers such as Reichheld et al. [8] and Suh and Yi [9] have displayed that even loyal and completely satisfied consumers are still attracted to situational factors such as competitors' coupons and discounts. Satisfaction, therefore, may not be the sole predictor of loyalty.

Rauyruen and Miller [10] proposed four important factors of loyalty, comprising service quality, commitment, trust, and satisfaction. According to Morgan and Hunt [11], the commitment requires consumers to strive to maintain relationships with retailers and suppliers. Trust can be determined by customers' belief in product features because consumers are provided with a sense of assurance of products. Consumers' trust in traders plays an important role in building the commitment. Suh and Yi [9] have indicated that the commitment is defined as an important intermediary factor in customers' decisions to purchase. Active word of mouth is a common approach to the loyalty when loyal customers become supporters of a service or product.

H1. Satisfaction and commitment have a linear relationship.
H2. Satisfaction and trust have a linear relationship

Hai et al. [1] state that satisfaction positively affects loyalty, willingly makes introduction and word-of-mouth communications, which means that satisfaction and word of mouth have a relationship. Therefore, the following research hypothesis is proposed.

H3. Satisfaction and word of mouth have a linear relationship.

2.2 Involvement

According to Tuan [12], the involvement is an individual difference variable that affects consumers' communication and decision-making behavior. This concept relates to other numerous marketing concepts, such as risk perception, information search, brand commitment, brand loyalty, brand similarity, leaders' opinion, brand change, advertising, and transmission of information.

The involvement is defined as the unobservable motivational states or interest in the product consumption. It is considered as a personal motivational state with an object or the influence of behavior to achieve related goals [12].

The environmental stimulations are situations of purchase involvement. For instance, sales promotions, such as discounts and coupons, are unexpected situations in consumer decisions. In consumer research, extrinsic stimulations are often used to assess levels of empirical participation. In contrast, personal intrinsic sources are relatively stable and long lasting. The ego involvement is often related to knowledge rooted in past experiences and stored in long-term memory.

Customers with a broad knowledge tend to make better purchasing decisions in the future, significantly expand efforts to find and process information, which increases their satisfaction [13]. Firstly, customers with ego involvement shall endeavor to avoid dissatisfaction after purchasing important products in life. Secondly, in high-risk situations, the high-involvement purchase takes customers much time and effort to make wise decisions after the careful consideration, investigation, and to bring about motivation of stronger satisfaction. Customers with high-involvement purchase tend to make better decisions and gain more satisfaction. The easy access to information provided by manufacturers can increase customer satisfaction. Customers with high level of involvement are specially promoted to experience the satisfaction with provided products.

The study by Suh and Yi [9] based on the approach and stability of theory, investigated the regulatory role of involvement in products in the relationship between customer satisfaction and loyalty.

Chen and Tsai [14] have also studied the impact of involvement, as a moderator variable, on the relationship between satisfaction and loyalty in China, a developing country with considerable similarities in consumer behavior and characteristics with Vietnam. The study has demonstrated that customers with high involvement tend to require more information than ones with low involvement do.

Manuel [15] has focused on purchase involvement (in case study) in moderating the relationship between satisfaction, trust, and commitment.

From overviews, the following hypotheses are proposed.

H4, H5, H6: The customers' purchase involvement has positive influence on the relationship between satisfaction and loyalty (commitment, trust, and word of mouth).

2.3 Research Model

From the above theoretical framework, three scales of loyalty, including commitment, trust, and word of mouth, are researched and tested to determine the relationship with satisfaction and repetitive purchase behavior of Vietnamese customers on ready-made shirts. The paper also tests the relationship between satisfaction and loyalty. Furthermore, this paper shall measure the relationships mentioned above by moderator variables of ego-involvement and purchase involvement to find out potential differences. The following research model has proposed scales to study relationships among satisfaction, loyalty and repetitive purchase behavior under the influences of the moderator variables of purchase involvement.

The involvement is defined as the unobservable motivational states or interest in the product consumption. It is considered as a personal motivational state with an object or the influence of behavior to achieve related goals [12].

Six research hypotheses have been divided into two hypothetical groups. In particular, the hypotheses from H1 to H3 are ones about the relationship between the factors without being influenced by moderator variables of involvement. The others from H4 to H6 are hypotheses related to purchase involvement.

3 Research Methodologies

The research has been conducted in two stages of preliminary surveys and official ones, in which both qualitative and quantitative methodologies are applied in preliminary surveys and the quantitative methodologies are applied in the official surveys. Specifically, in the qualitative preliminary survey, in-depth interviews are conducted with a focus group of experts to complete the draft scale 1 and create the draft scale 2. Then, quantitative preliminary survey was carried out with 200 samples and obtained 146 items ($N = 146$), in order to test the reliability of the scales by Cronbach's Alpha and EFA and to delete non-standard observed variables.

The completed questionnaire was used in the official quantitative survey in Hanoi and Ho Chi Minh City. Respondents are the entire customers of shirts under all various ages, residences, and occupation. The random sampling method shall be applied to collect data from two these cities. Samples are customers of ready-made shirts at the age of 18 or over. The interviews are carried out in districts with corresponding rates of residents. Specifically, research samples contain 58% of male customers and remaining 42% of female customers at five age groups (of 18–24, 25–34, 35–44, 44–54, and 55–64). The measurement model consists of 31 observed variables according to the principle of at least 5 elements per measurement variable [1]. Therefore, the initial number of calculated samples is $22 * 5 = 110$. However, in order to increase the reliability, the chapter plans to collect 1000 samples ($n = 1000$) and select 730 samples (by questionnaires). After checking and removing invalid respondents, 615 respondents are used for official analysis. The structural equation modeling (SEM) is used to test the scale model and theoretical model.

4 Research Results

4.1 Assessment of Preliminary Scale

SPSS 24.0 software presents the results of scale's reliability analysis (with Cronbach's Alpha) and the exploratory factor analysis (EFA) and suggests to remove observed variables to help the scales to accurately assess the concepts. According

Table 1 Reliability and average variance extracted of scales

No.	Scales	Number of observed variables	Reliability coefficient (Cronbach's Alpha)	Total average variance extracted (%)	Conclusion
1	Commitment (CK)	7	0.897	62.52	All scales' reliability coefficient are accepted
2	Trust (TT)	3	0.904	84	
3	Word of mouth (WOM)	4	0.852	69.37	
4	Satisfaction (STM)	5	0.805	56.42	

to the testing standards, Cronbach's Alpha coefficient must be ≥ 0.6 [1] and average variance extracted must be greater than 50%. The analysis results have been shown in the following Table 1.

4.2 Confirmatory Factor Analysis (CFA)

After the preliminary assessment of scales, AMOS 24.0 software is used to perform confirmatory factor analysis (CFA) of scales and test the suitability of the theoretical model and hypotheses. Testing standards include Chi-square with adjustable degree of freedom (CMIN/df); goodness-of-fit index (GFI); Tucker and Lewis index (TLI); comparative fit index (CFI); and root mean square error approximation (RMSEA). The model is considered to be suitable if Chi-square test with p-value ≥ 0.05. However, the disadvantage of the chi-square test is that it requires a sufficient sample size. The larger the sample size is, the larger the chi-square is, which reduces the suitability of the model. Therefore, the standards are based on both P-value and CMIN/df. In some practical studies, two occurrences of $\chi 2/df < 5$ (with sample size $N > 200$); or <3 (with sample size $N < 200$) indicate that the model is suitable [1]. The results of confirmatory factor analysis (CFA) of scales are presented in Table 2.

Table 2 Confirmatory factor analysis (CFA) of scales

No.	Items	Satisfaction	Loyalty
1	Chi-square/df	4.052	4.523
2	GFI	0.910	0.931
3	TLI	0.928	0.944
4	CFI	0.940	0.955
5	RMSEA	0.071	0.076

Table 2 shows that the entire GFI, TLI, CFI are greater than 0.9; chi-square/df < 5 and RMEA ≤ 0.08, which demonstrates that the model is suitable for the market data.

4.3 Testing the Theoretical Model

After assessing scales, the official theoretical model is tested.

The results of theoretical model test (in Fig. 1) are shown in the figure that chi-square/df = 4.460; GFI = 0.909; TLI = 0.926; CFI = 0,938; and root mean square error of approximation (RMSEA) = 0.075, which demonstrates that the model is suitable for the market data. Besides, the estimates indicate that all relationships are statistically significant ($p <1\%$), as specifically in Table 3.

The model and Table 3 show that of three factors of loyalty, the word of mouth has the most positive impact on satisfaction (with 0.638) and trust has the least impact (with 0.43).

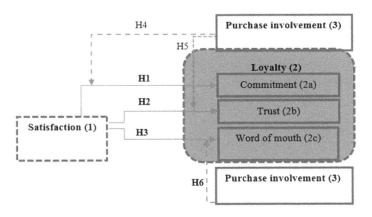

Fig. 1 Research model

Table 3 Testing relationships (standardized regression coefficients)

Relationship			Estimate	S.E.	C.R.	p	Label
Commitment	<–>	Satisfaction	0.514	0.052	9.903	***	
Trust	<–>	Satisfaction	0.434	0.049	8.854	***	
Word of mouth	<–>	Satisfaction	0.638	0.063	10.062	***	

In which, *estimate* average estimated value; *SE* standard error; *CR* critical value; *P* probability value; ***: $p < 0.001$

4.4 Testing the Hypotheses

The estimates in the Table 3 indicate that relationship weights are all positive and statistically significant ($p <= 0.05$), which demonstrates satisfaction with a brand of ready-made shirts has a positive relationship with commitment, trust and word of mouth (H1 = 0.514, H2 = 0.434, and H3 = 0.638). In general, hypotheses H1, H2, and H3 are accepted (Fig. 2).

To further clarify the relationship between satisfaction and loyalty to the product, hypotheses of H4, H5, are H6 are tested to determine the influence of the moderator variable of purchase involvement; Specifically,

H0: The moderator variable of purchase involvement has no positive influence on the impact of satisfaction on loyalty.

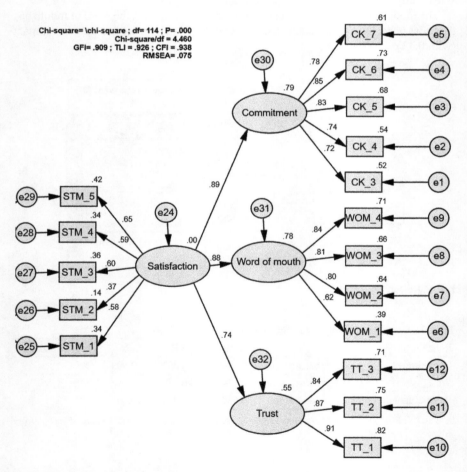

Fig. 2 Results of theoretical model test

H1: The moderator variable of purchase involvement has a positive influence on the impact of satisfaction on loyalty.

If the value of p-value <0.05, Ho is rejected and H1 is accepted, and the moderator variable has a positive influence on the impact of satisfaction on loyalty. If p-value <0.05, or vice versa, Ho is accepted and H1 is rejected.

The impact of satisfaction on commitment, trust, and word of mouth is under the influence by the moderator variable of purchase involvement.

Table 4 demonstrates that the influence of purchase involvement on the impact of satisfaction on factors of loyalty (commitment, trust, word of mouth) is statistically significant with $p < 0.05$. It proves that H1 is accepted and H0 is rejected. Thus, the hypotheses of H4, H5, and H6 are accepted, which means that the purchase involvement increases the impact of satisfaction on commitment (at most 0.498), trust (0.035), and word of mouth (0.001).

5 Discussions and Limitations of the Paper, and Recommendations for Further Researches

5.1 Discussions and Implications for Administrators

The findings of this chapter are completion of the theoretical model, tested hypotheses, corrected scales to create new scales for Vietnam's fashion, quantifying the intensity of the impact of moderator variable of purchase involvement on the relationship between satisfaction and loyalty. Noticeably, the word of mouth has been the most impacted (0.638) by satisfaction rather than trust and commitment. The affirmative finding in this paper is that satisfaction has the greatest influence on loyalty through word of mouth, which means that the more satisfied customers are, the more the customers will spread their satisfaction to customers and potential customers. This is the important focus paid high attention to by marketers for fashion products in general and fashion shirts in particular.

In general, the impact by satisfaction on loyalty is positive and proportional, but not all relationships are strong enough. This implies that the purchase behaviors in the fashion market are extremely complex, requiring proper marketing strategies and tactics. The paper provides references for administrators to know how to increase customer loyalty. Results of commitment, trust and word of mouth in this paper suggest administrators develop marketing plans and programs with the central focus on word of mouth for viral marketing to target customers. This is proper and appropriate in Vietnam market in which consumers have the "mob" mentality with a large proportion.

To further clarify the relationship between satisfaction and loyalty, the research results also show that the influence of the moderator variable on customers' purchase involvement is positive. The paper recommends that sale programs should be publicly announced and attached with "commitment" to attractive promotions to increase the

Table 4 Regression results for testing the hypotheses of H4, H5, and H6

Model	Commitment (CK)				Trust (TT)				Word of mouth (WOM)			
	Regression weight			p-value	Regression weight			p-value	Regression weight			p-value
	Unstandardized	Standard error	Standardized		Unstandardized	Standard error	Standardized		Unstandardized	Standard error	Standardized	
Constant	1.030	0.096		0.000	1.923	0.147		0.000	0.662	0.165		0.000
Satisfaction	0.273	0.029	0.303	0.000	0.459	0.031	0.510	0.000	0.671	0.032	0.660	0.000
STM*PUR	0.498	0.029	0.562	0.000	0.035	0.009	0.134	0.000	0.001	0.005	0.008	0.008
Multicollinearity VIF	1.764				1.078				1.037			
R^2	0.313				0.151				0.431			

Dependent variables: commitment, trust, and word of mouth (respectively)

loyalty (behavior loyalty), which significantly improves purchase behavior. This is considered as the false loyalty, which is explained that customers' purchase behavior only occurs at promotion programs. It is also quite true to the real situation in Vietnam when the income per capita is still modest.

5.2 Limitations and Directions for Further Researches

Firstly, the involvement is a very important and interesting subject in customer behavior research. This paper has just emphasized on purchase involvement, and not mentioned ego-involvement which aimed at further researches.

Secondly, this paper has focused only three factors of loyalty (including commitment, trust, and word of mouth) and purchase involvement to test relationships with satisfaction. The further researches may survey other factors of loyalty to more understand the researched relationships.

Last but not least, the word of mouth is referred to positive and advantageous communications and satisfaction hereof is general and overall satisfaction. Negative and disadvantageous word-of-mouth communications and partial satisfaction with each stage of sales processes has not been demonstrated in this paper and this is a direction for further research.

References

1. Hai, H.D.: Econometrics. Publisher of National Economics University (2019)
2. Fornell, C.: A national customer satisfaction barometer: the Swedish experience. J. Market. **56**(1), 6–21 (1992)
3. Dixon, J., Bridson, K., Evans, J., Morrison, M.: An alternative perspective on relationships, loyalty and future store choice. The International Review of Retail, Distribution and Consumer Research **15**(4), 351–374 (2005)
4. Hallowell, R.: The relationships of customer satisfaction, customer loyalty, and profitability: an empirical study. Int. J. Serv. Ind. Manag. **7**(4), 27–42 (1996)
5. Heitmann, M., Lehmann, D.R., Herrmann, A.: Choice goal attainment and decision and consumption satisfaction. J. Mark. Res. **44**(2), 234–250 (2007)
6. Oliver, R.I.: Whence consumer loyalty? J. Market. **63**, 33–44 (1999)
7. Bloemer, J.M.M., Kasper, H.D.P.: The complex relationship between consumer satisfaction and brand loyalty. J. Econ. Psychol. **16**, 311–329 (1995)
8. Frederick F. Reichheld and W. Earl Sasser, Jr.: Zero Defections: Quality Comes to Services. Harvard Business Review, 105–111 (October 1990)
9. Suh, J.C., Yi, Y.: When brand attitudes affect the customer satisfaction-loyalty relation: The moderating role of product involvement. J. Consumer Psychol. **16**(2), 145–155 (2006)
10. Rauyruen, P., Miller, K.E.: Relationship quality as a predictor of B2B customer loyalty. J. Bus. Res. **60**(1), 21–31 (2007)
11. Morgan, R.M., Hunt, S.D.: The commitment-trust theory of relationship marketing. Journal of Marketing **58**(3), 20–38 (1994)
12. Tuan, V.P.: Impact of satisfaction on loyalty and repetitive purchase behavior under the influence of fashion styles and involvement (2014)

13. Shaffer, R.R., Sherrell, D.L.: Consumer satisfaction with health-care services: The influence of involvement. Psychol. Market. **14**(3), 261–285 (1997)
14. Tsai, C.-F., Huang, H.T., Jaw, Y.L., Chen, W.K.: Why on-line customers remain with a particular e-retailer: an integrative model and empirical evidence. Psychol. Market. **23**(5), 447–464 (2008)
15. Sanchez-Franco, M.J.: The moderating effects of involvement on the relationships between satisfaction, trust and commitment in e-Banking. Facultad de Ciencias Economicas y Empresariales, Universidad de Sevilla, Avda. Ramon y Cajal, n1, 41018-Sevilla, Spain (2009)

A Study on Factors Influencing Consumer Intention to Use UPI-Based Payment Apps in Indian Perspective

Piyush Kumar Mallik and Deepak Gupta

Abstract Mobile payment is a system in which we do online transactions of money, using devices like smartphones or point of sale machine, etc. instead of cash. Mobile wallet or digital wallet is one of the popular mediums of mobile payment system. There is much research that focuses on mobile wallets and consumers' intention to use them. However, the mobile wallets are being increasingly replaced by UPI-based Payment Apps such as BHIM, PhonePe, and Google Pay, and little is known about the drivers for this adoption. Considering this gap, our study was conducted to understand the consumers' behavioural intention with respect to UPI-based Payment Apps. Based on the literature review a conceptual model was proposed using an extended UTAUT2 framework. This was tested using a pan-India survey with 224 valid responses. Ordered logistic regression models were run in Stata to understand the impact of factors like Performance expectancy, Effort Expectancy, Social influence, Facilitating conditions, Hedonic motivation, Price value, and User adoption on the intention to use UPI-based payment apps. Factors like Trust, Perceived security and Innovativeness were also included to get a better understanding of the consumer adoption perspective. It was found that Performance Expectancy, Price value, Trust and Perceived Security were significantly and positively related to behavioural intention. This is among the earliest studies in India to comprehensively model and empirically investigate factors driving the consumer adoption of UPI-based payment apps.

P. K. Mallik (✉) · D. Gupta
Amrita School of Business, Amrita Vishwa Vidyapeetham, Coimbatore, India
e-mail: piyushkumarmallik201@gmail.com

D. Gupta
e-mail: dgshobs@gmail.com

1 Introduction

Providing a digital solution for Monetary related transactions has always been a difficult job for financial institutions and service providers, and when it comes to offer these services to common people, the task is even more challenging. No wonder why even after almost one decade of the banking system's digitization in India, hard cash still dominates the market. It is not that effort has not been made—innovative technologies like ATM, Debit/Credit Card, Internet Banking, POS machine, Mobile wallet, etc. have been introduced to users but the situation could not improve because of lack of awareness, proper infrastructure and various reasons. However, post demonetization of Indian currencies in the year 2016, people started understanding the usefulness of these technologies and sudden increase in the use of these technologies was being witnessed by service providers.

The concept of mobile payment has been in the industry for almost more than two decades but it started gaining popularity only after the evolution of smartphones and availability of fast internet services at affordable prices. Many countries saw it as an opportunity to make their society and economy cashless.

In persuasion of cashless economy, the Government of India also took an initiative to launch a platform for digital or virtual cash transaction and came up with a Unified Payment Interface.

Unified Payment Interface or, UPI is a digital payment system launched by the Government of India in 2016. Through this payment system, users can connect more than one bank accounts and can operate them via one mobile app.

Apps like BHIM, Google Pay, PhonePe, Paytm, etc. which use UPI are providing services like, banking services, easy and secure fund transfer, merchant payment, Bill split and Bill payment, Peer to Peer collection request, Overdraft account and many other services at zero cost to its users. If we compare these services with services provided by mobile wallets and other medium of payment services like Debit/Credit cards, internet banking, etc. then UPI-based payment apps clearly outperforms and hence their increasing popularity. However, little is known as yet about the drivers of their adoption. This chapter seeks to fill that gap.

The flow of the rest of the chapter is as follows: In Sect. 2 we have explored the literature available on the digital payment system as well as M-payment system which helped us to build the background for our research; Sects. 3 and 4 depict the conceptual model and hypothesis, respectively on which analysis has performed. Section 5 talks about research methodology, questionnaire and method used for data collection. Section 6 discusses the results derived from data analysis whereas Sects. 7 and 8 focus on limitations, scope for future research and conclusion. Finally, acknowledgement and references have been given.

2 Literature Review

This study started with an intention to understand consumer perception about mobile wallets and whether they are familiar with them or not and if yes then what are the factors which motivate or prevent them to use it. According to literature, mobile payment is a system in which electronic devices like smartphones, point of sale machine (POS), etc. are used to perform commercial transactions [1]. Also, it is an evolved version of electronic payment system which helps users to do transaction of money, seamlessly and easily [2].

It was found that security and trust is one of the major concerns among them and it affects their behaviour towards M-payment systems [3]. Further, we tried to understand that how age affects users' intention to use, it was found that, youth are interested to adopt new technology but old people did not show much interest directly but they were influenced by their own social group for using new technology but in case of mobile payment system this relation could not be established for either youth or the older generation [4].

With reference to M-payment system as a means for performing daily transactions in different types of businesses by their owners, it was found that when it comes to merchant's intention to use mobile wallet technology, factors like compatibility and usefulness of service, customer value addition, influences their behavioural intention significantly [5]. In other words, for a merchant, usefulness of services means not only that it should be cost-effective but also it should be widely accepted by their customers, and if this payment system provides better convenience compare to other modes of payment then it will have a great influence on both customer as well as a merchant to use it.

With respect to individual or non-merchant users factors like innovativeness, stress to use (it measures the level of discomfort which an individual feel while using technology) and social influence were also explored and it was found that high stress negatively correlated with user satisfaction whereas social influence and recommendation are positively correlated with intention to use mobile wallet [6]. It was also found that early adopters give more importance to ease of use and think that they have enough understanding about M-payment system whereas late adopters are tech-savvy and use M-payment out of need [7].

We also tried to understand how user adoption and recommendation to use for digital wallets is being affected by different factors. Extant research reveals that perceived technology security, performance expectations and innovativeness have a positive impact on recommendation to use for digital wallets, suggesting that if the user feels that the payment system is secure and is meeting his expectation then he is more likely to recommend it to others [8]. Effect of personality traits and behavioural beliefs on adoption behaviour was also explored and it was found that both of them are very significant for digital wallet adoption and use but their impact on pre-adoption stage and post-adoption stage is different [9].

Further, we explored literature in order to understand post-adoption behaviour of mobile payment user and what motivates them to use it continuously, it was found that

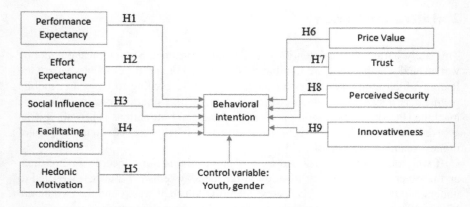

Fig. 1 Conceptual map to understand behavioural intention to use UPI-based payment apps

the quality of service and system is positively correlated with trust and satisfaction, respectively, which means better the quality of service and system greater will be trust and satisfaction in users for mobile payment system. Furthermore, quality of information and service affects the flow, which means if trust, flow and satisfaction are high then the user will keep using existing service [10].

Considering the above-mentioned literature we came to a conclusion that, though there are similarities between mobile wallets and UPI-based payment apps but the objective of these two-payment systems and the way they function is different and thus this research is focused on the user's intention to use UPI-based payment apps. Based on our literature review we propose an extended UTAUT2 model for conceptually modelling the factors that influence the adoption of UPI-based payment apps in India.

3 Conceptual Model

See Fig. 1.

4 Hypothesis

H1: Performance Expectancy is positively related to the intention to use UPI-based payment apps

H2: Effort Expectancy is positively related to the intention to use UPI-based payment apps

H3: Social Influence is positively related to the intention to use UPI-based payment apps

H4: Facilitating Conditions are positively related to the intention to use UPI-based payment apps
H5: Hedonic Motivation is positively related to the intention to use UPI-based payment apps
H6: Price Value is positively related to the intention to use UPI-based payment apps
H7: Trust is positively related to the intention to use UPI-based payment apps
H8: Perceived Security is positively related to the intention to use UPI-based payment apps
H9: Innovativeness is positively related to the intention to use UPI-based payment apps.

5 Methodology

- Data:

Total 224 responses were collected through an online survey from all over India. We divided age groups into eight categories as category 1–8, where first three categories (1–3) are our main target because they understand technology trends and can adopt it very quickly whereas for next three categories (4–6) we assumed that they may have some idea about UPI-based payment apps and may be willing to use it or might have already used it. The last two categories (7 and 8) were those above 45 (Table 1).

Demographics: Our sample size is divided between Male and Female as, Male—74.6% (187) and Female—25.4% (57). Most of the responses (more than 60%) were form Delhi, Mumbai, Kolkata, Chennai, Bengaluru, Hyderabad, Ahmedabad, Patna, Coimbatore and Kochi.

Employment Status: We have divided employment status into six categories. According to the survey response, 36.6% of people are employed, 54.5% are students, 6.3% are self-employed whereas rest of 2.6% are homemakers, retired or indulge in other activities.

Questionnaire: An online survey was conducted to understand consumers' intention to use for UPI-based Payment Apps.

Table 1 Different category of age groups

Category	Age
1	Under 18
2	18–25
3	26–30
4	31–35
5	36–40
6	41–45
7	46–50
8	Above 50

6 Results

By analyzing the data through ordinal logistic regression, it was found that Performance expectancy has a significant and positive impact on behavioural intention that is if people find the UPI-based Payment Apps useful in daily life than the odds of intention to use increase by 6.7% (Table 2).

It was also found that Price value has a significant and positive influence on the intention to use UPI-based Payment Apps—it increases the chance of adoption by 11.9%. It means if prices are becoming affordable then chances will be higher to use UPI.

It was observed that Perceived security has a direct influence on intention to use UPI-based Payment Apps; if UPI provides secure means of transaction then the odds of intention to use UPI-based payment apps increase by 9.9%.

It was also observed that Trust has a significant role in intention to use UPI-based Payment Apps, it means if people think that UPI-based apps are trustworthy then the odds will increase by 9.4% for intention to use UPI-based payment apps.

In case of the control variables, contrary to our expectations our results showed a negative relationship between Youth and the intention to use UPI-based payment apps.

It was also observed that Social influence doesn't play a significant role in intention to use UPI-based Payment Apps, similar to hedonic motivation, innovativeness and gender.

Table 2 Results of analysis based on ordinal logistic regression model

Variables	Odds ratio	Z	$P > [z]$	95% conf. interval	
Performance expectancy	1.067806**	2.37	0.018	1.011442	1.127312
Effort expectancy	1.021843	0.62	0.533	0.9547642	1.093634
Social influence	0.994385	−0.17	0.865	0.9317273	1.061256
Facilitating conditions	1.008656	0.29	0.775	0.9506769	1.070172
Hedonic motivation	1.042849	1.10	0.269	0.9680136	1.123469
Price value	1.11901***	2.86	0.004	1.036148	1.208498
Trust	1.094987***	3.19	0.001	1.035613	1.157764
Perceived security	1.09905***	2.84	0.004	1.029746	1.173019
Innovativeness	0.9919362	−0.24	0.808	0.9291906	1.058919
Youth	0.3835176**	−2.23	0.026	0.1648737	0.8921113
Gender	1.063931	0.21	0.832	0.5995572	1.887973

Note ***$P < 0.01$, **$P < 0.05$, *$P < 0.10$

7 Limitations and Future Research

This study focused on behavioural intention of people to use UPI-based Payment Apps in Indian context. Our analysis has shown that security has a positive correlation with the intention to use, which means if a UPI-based Payment Apps ensures full security during transaction or while making payments then it is highly likely that users will prefer to use that particular Apps compare to others. As we know with the increasing volume of UPI-based transactions, case of online fraud has also increased. Generally, these frauds happen in two forms, first as a theft of personal data like personal information, bank details, or data related to transaction history, etc. and in the second type of fraud, criminals use techniques like, Phishing, Malware, SIM Cloning, money Mule, etc. to steal money from a user's account. If we believe the data released by Reserve Bank of India in 2016–17 Rs. 42.3 crore and in 2017–18 Rs. 109.6 crore worth of cyber fraud happened, out of which one third was related to the online payment system.

Considering the above-mentioned facts, research can be done to explore the factors, which lead to these frauds and what should be done in order to protect the UPI-based Payment App users from these frauds. Our study is limited to the Indian context but the Indian government is planning to launch UPI abroad also so, further it can be explored the behavioural intention and motivational factors of people from other countries for using UPI.

8 Conclusion

The objective of our study was to understand what are the factors that affect behavioural intention of people to use UPI-based Payment Apps. We used an extended UTAUT2 model to understand how things like performance expectancy, trust, price value, security, etc. affect their decision to use UPI-based Payment Apps. Since UPI was introduced in India so the focus of our study was the Indian user only.

Our analysis showed that user performance expectancy and perceived security has a major impact on intention to use UPI-based Payment Apps. Price value also plays a significant role in behavioural intention, it means people are concern about the cost of using UPI. Finally, trust too has a positive impact on behavioural intention.

Acknowledgements The authors would like to express their gratitude to the institution for providing the opportunity to do this business research work, and faculty members for helping out throughout the journey. The authors also greatly appreciate the colleges and all the participants who gave their valuable time for the survey.

References

1. Au, Y.A., Kauffman, R.J.: The economics of mobile payments: understanding stakeholder issues for an emerging financial technology application. Electron. Commerce Res. Appl. **7**(2), 141–164 (2008)
2. Mallat, N.:. Exploring consumer adoption of mobile payments—a qualitative study. J. Strategic Inf. Syst. **16**(4), 413–432 (2007)
3. Shin, D.H.: Towards an understanding of the consumer acceptance of mobile wallet. Comput. Human Behav. **25**(6), 1343–1354 (2009)
4. Liébana-Cabanillas, F., Sánchez-Fernández, J., Muñoz-Leiva, F.: Antecedents of the adoption of the new mobile payment systems: the moderating effect of age. Comput. Human Behav. **35**, 464–478 (2014)
5. Singh, N., Sinha, N.: How perceived trust mediates merchant's intention to use a mobile wallet technology. J. Retail. Cons. Serv. **52**, 101894 (2020)
6. Singh, N., Sinha, N., Liébana-Cabanillas, F.J.: Determining factors in the adoption and recommendation of mobile wallet services in India: analysis of the effect of innovativeness, stress to use and social influence. Int. J. Inf. Manage. **50**, 191–205 (2020)
7. Kim, C., Mirusmonov, M., Lee, I.: An empirical examination of factors influencing the intention to use mobile payment. Comput. Human Behav. **26**(3), 310–322 (2010)
8. Oliveira, T., Thomas, M., Baptista, G., Campos, F.: Mobile payment: understanding the determinants of customer adoption and intention to recommend the technology. Comput. Human Behav. **61**, 404–414 (2016)
9. Yang, S., Lu, Y., Gupta, S., Cao, Y., Zhang, R.: Mobile payment services adoption across time: an empirical study of the effects of behavioral beliefs, social influences, and personal traits. Comput. Human Behav. **28**(1), 129–142 (2012)
10. Zhou, T.: An empirical examination of continuance intention of mobile payment services. Decis. Support Syst. 54(2), 1085–1091 (2013)

A Proposed SDN-Based Cloud Setup in the Virtualized Environment to Enhance Security

H. M. Anitha and P. Jayarekha

Abstract Virtualization provides the multitenancy feature in which many users reside in the same server. This is a benefit to share the resources among themselves. But it might cause security problem for each other. In order to monitor all the activities in the network, Software-Defined Network (SDN) is used. In this paper, a testbed involving the openstack is been explored and implementation details are provided. SDN has a controller to get the information about the entire network from the switches. A view on different controllers is seen in this work. Some of the existing testbeds are discussed. Case study with SDN is extensively studied to incorporate the security feature in network. Hence security is taken care of providing the integrity of the data stored in the physical machine and attacks from the neighbours is forbidden. Policies are maintained in the controller so that the network is safe with attackers. The SDN-based testbed setup presents detailed tools that are used and advantages.

1 Introduction

In the current scenario, Cloud Computing is an ongoing trend in the industries. It is used by different types of firms such as large and small. It offers services to users in any part of the world from anywhere. Services are offered at different levels. Cloud computing is defined as "A model for enabling ubiquitous, convenient, on-demand network access to a shared pool of configurable computing resources (e.g., networks, servers, storage, applications, and services) that can be rapidly provisioned and released with minimal management effort or service provider interaction" according to the NIST [1]. Cloud service providers offer three different kinds

---

H. M. Anitha (✉) · P. Jayarekha
B.M.S College of Engineering, Affiliated to VTU, Bengaluru, India
e-mail: anithahm.ise@bmsce.ac.in

P. Jayarekha
e-mail: jayarekha.ise@bmsce.ac.in

© The Editor(s) (if applicable) and The Author(s), under exclusive license to Springer Nature Singapore Pte Ltd. 2021
T. Senjyu et al. (eds.), *Information and Communication Technology for Intelligent Systems*, Smart Innovation, Systems and Technologies 196,
https://doi.org/10.1007/978-981-15-7062-9_45

of services Software as a Service (SaaS), Platform as a Service (PaaS) and Infrastructure as a Service (IaaS). Many users are moving towards cloud as they can access the services from anywhere and anytime. This flexibility has made many IT industries to stop backing up of data. Infrastructure as a service is the service which served to the users at the platform level and software level to deploy the applications. Any kind of facility required by the industries is leveraged by cloud providers which will reduce the investment cost. As per the users' payment and capabilities required, cloud providers offer the services. Virtualization has made a provision for the cloud providers to deploy the services to the users through the virtual machines. When the users request for the capability, cloud service provider leverages in the form of the virtual machines. Cloud service providers are equipped with high-end servers that are capable of granting resources to many users. Virtual machines are deployed to many users in the same physical machine. VMs communicate to each other using the network elements like switches and routers. This advantage is a security threat for users. There might be some non-trusted users in the network to gather or tamper sensitive information. In such scenarios to monitor such kind of problems in the network, Software-Defined Network technology can be applied.

The elastic and cost-effective property of the cloud has paved the way to software-defined network to manage the networks from centralized zone. SDN enables the central management of the switches and network devices. SDN controller [2] has an overlook of the entire network for the provisioning of resources to monitoring. This kind of monitoring and dynamically adjusting to the resource provisioning is difficult using traditional network. SDN controller is capable of monitoring the entire network and provides the guidelines in the network for further requests for resources. In this paper, SDN enabled cloud is proposed to monitor the security parameter in the network. Without the involvement of intrusion detection system information gathering can be done by switches of the network. Testbed set up is proposed to monitor the network using SDN. Openstack is an open-source cloud platform offering infrastructure as a service in the form of virtual machines based on the availability of the resources. Here SDN enabled cloud set up is discussed in order to address the security features. Section 2 describes the related work, Sect. 3 provides background, Sect. 4 describes some testbeds available, Sect. 5 discusses testbed set up for SDN-based cloud, Sect. 6 explains the case study of the SDN approach to enhance security and Sect. 7 concludes the chapter.

2 Related Work

SDN is a recent advancement in the IT industry wherein a wide variety of applications take advantage of it. The authors [3] have given the description of differences between the control plane and data plane. They have compared SDN with traditional networks with management involved requirements. Management functions can be implemented enterprises providing a wide range of services and control the datacentre traffic flows.

The authors [4] have given the correct picture of SDN-based network and explained how the network abstracts the underlying hardware. SDN adoption in the IT industry can benefit management, scalability and how dynamically they can adjust to variations in the network.

The authors [5] presented the sniffer reflector framework using SDN. They have discussed move target defense so that avoiding the attackers from collecting the information about the network. The framework contains a method to respond to attackers in a forged way and attackers do not know about these messages. Hence protecting the network from attackers.

One of the testbed based on SDN implementation [6] is incorporated in production environment. This testbed provides an opportunity for developers to exclude the outside traffic. The testbed is under the testing stage. The authors have presented different setups.

3 Background

Software-defined network is an emerging technology which can be incorporated with cloud to overcome the security challenges. Unlike traditional networks, the task of decision making and management of the network is done centrally. SDN controller has the entire view of the network.

SDN architecture [7] consists of three different layers as shown in Fig. 1 namely,

Application Layer: Many applications run in this layer which belongs to different users. Communication between the application layer and the control layer is carried out by northbound APIs.

Control Layer: SDN controller is the heart of the entire network which is located in the control layer. All the important decisions regarding the network are carried by the control layer. Decisions are based on the complete understanding of the network, unlike traditional network where independently decision is made. Since SDN is software-based, it is easy to manage resources and the traditional network is made up of hardware devices such as switches, routers and other physical devices and cumbersome to gather information. Controller can monitor the traffic flows in the network as it gets information from network devices using APIs. It's difficult in the case of traditional networks as every device information is not centralized. In order to communicate to application layer northbound APIs are used.

Data Plane Layer: In this layer, all networking components are residing in order to perform switching and packet forwarding in the network. Every operation is carried under the vigilance of the control layer. Communication between Data plane and control layer is carried out by the southbound APIs. Open flow is used for communication.

Open Flow: In order to communicate between data plane and controller, open flow is used. It is one of the first interface used directly between switches and

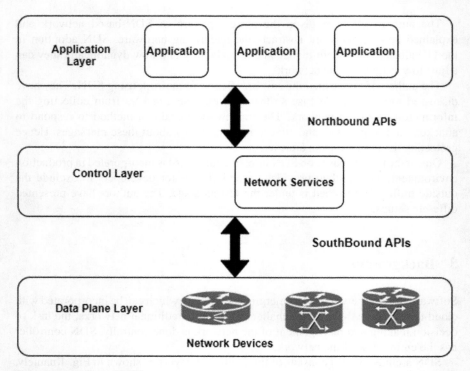

Fig. 1 Architecture of SDN

controllers so that required network alterations could be completed without any other elements' involvement [8]. Open flow is managed by the Open Network Foundation (ONF) which is open-source and used for SDN adoption among the users

Some of the open flow controllers [9] are described in Table 1.

4 OpenStack

Openstack provides the cloud computing platform [10] for the deployment of the private and public clouds. It was jointly started in the year 2010 by Rackspace and NASA. Later NASA contributed Nova. Nowadays most of the users from all the domains are using openstack to deploy the cloud. Openstack [11] provides infrastructure as a service by the use of well-defined APIs.

Openstack [11] offers the following services which are shown in Fig. 2.

Table 1 Description of openflow controllers

Controller	License	Description
Open daylight	Eclipse Public License (EPL)	It was released in the year 2014. It provides centralized monitoring of SDN network
OpenContrail	Apache 2	It is from Juniper networks. Virtual routers on top of hypervisor are required by the SDN controller. OpenContrail can be used as cloud computing platform. There are some of the important features of open contrail such as network virtualization, programmability, and big data
FloodLight	Apache	It is implemented in Java. It supports many varieties of networks. It is the most popular controller. It is used by applications such as open stack and floodlight virtual switch
Ryu	Apache	Ryu is an SDN framework offering the well-defined API software components. It allows creating applications required by the user
Flow visor	GNU general public license	Flow Visor provides network virtualization by creating many logical networks from a physical network. It only handles the switches and resources assigned

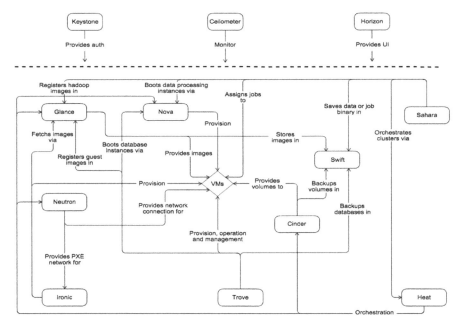

Fig. 2 Architecture of openstack [11]

a. Horizon: Horizon is dashboard. It is a user interface provided to interact with openstack services deployment of VM instances, allocating IP addresses and access control specification configuration.
b. Nova: It computes which is basically used to monitor the activities of VM such as launching, Migration, suspension, restart and shutdown of VM instances as per the requirement of the users.
c. Neutron: It is used for networking purposes. It offers Network connectivity as a service for some openstack compute. Offers as a set of well-defined APIs. It is pluggable so that support to many users can get benefit out of it.
d. Swift: It uses RESTful APIs to store the data. Its architecture is fault resistant provided with scalable environment.
e. Cinder: It gives persistent storage for running instances. It helps in managing many block devices.
f. Keystone: It offers key management for openstack services.
g. Glance: It provides the image management service. This service is used during deploying the VM instances.
h. Cellometer: It is metering service. Monitors the metering, pricing and scalability issues.
i. Heat: It provides orchestration services with multiple applications.

5 Testbeds

OpenSDNcore [12] is one of the testbed used by many researchers and vendors for developing their applications. It provides management and orchestration, more flexibility for network virtualization, etc. OpenSDNcore is an advanced implementation with openstack to manage the virtual networks. It uses openflow specifications to support telecom oriented features. Different kinds of applications can be developed in real-time as well as experimentation purposes. Mininet [13] is also a simulator provided for the users to implement their research problems. In a single system, Mininet can be tested for implementing the SDN applications. Real network scenarios can be implemented. Mininet has several advantages as it is simple to use, boots faster, and can scale well.

6 Proposed Test Bed for the SDN-Based Cloud

Laptop with 16 GB Ram and 1 TB HDD is considered for set up. In this setup, all in one is been implemented. Virtualbox [14] is used to deploy the setup with the SDN cloud. Vitualbox version 6.0 is used for setup.

Here in the proposed network [10] two nodes are considered namely the controller node and compute node shown in Fig. 3. One public network interacting with the internet. One provider network to manage the communication between VMs. The

A Proposed SDN-Based Cloud Setup in the Virtualized Environment ... 459

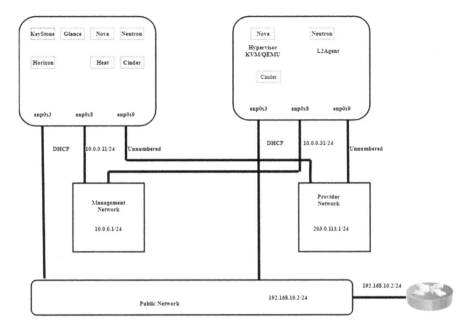

Fig. 3 SDN based cloud setup

controller node runs the services like keystone, glance, nova, neutron, and horizon. Compute node runs the hypervisor or virtual machine monitor to activities of VM instances. The hypervisor used KVM/QEMU. The memory provided for controller and compute node is 4 and 10 GB of minimum storage required. Opendaylight [15] is an open-source platform for SDN architecture. Opendaylight is used in the cloud environment to control and manage the networks. Openstack's most of the support is built into the opendaylight.

7 Case Study of SDN Setup Towards the Improvement of Security

In a university campus [16], many users bring devices such as mobile phones, tablets, laptops, and many more and start using for different applications. It's difficult for an administrator to monitor and leverage the secure service. What is the solution to this problem which is experienced not only on campus but also in corporates? The best solution is the introduction of SDN. The benefits of SDN in the campus are dynamically the services are offered, availability is enhanced, isolation of traffic and best usage of resources. With the network virtualization on campus, policies can be applied for the group or individual flows. There are different sets of people on the campus such as faculty, students, library management staff, book marts, doctors and non-teaching staff. Policies can be applied at the SDN controller for insiders as well as outsiders. Traffic bifurcation is achieved with respect to different sets of users in the campus. Controller with set of policies and rules are more advantageous in the campus network for management dynamically.SDN implementation in college campus gives several advantages with respect to security. The benefits are listed below:

a. Isolation of the traffic: Different logical networks will have different flows. Isolation is taken care and policies are set in the controller.
b. Security enhancement: With the switches following the rules framed by the controller security is enhanced. In traditional networks, normally the primitive methods such as firewalls are implemented.
c. Migration and support for devices: Many devices are used on the campus. With specific set rules and access policies, access to different networks is more easily compared to the traditional networks.
d. Multivendor interoperability: Different controllers, switches are used by vendors. In the case of switching to other networks, seamless interoperability is provided.

Hence, using the set up proposed enhances security in any network.

8 Conclusion

Many users use virtualization technology to access services from the cloud providers. SDN is the latest emerging technology in which users can benefit to monitor the cloud environment using a controller. Different types of controllers used commercially are explored. Well-known controllers such as Opendaylight, Opencontrail, Flowvisor, Ryu, and Floodlight are explored. Some of the available testbed setups like OpenS-DNCore and Mininet are provided here to analyze the cloud setup. The proposed work here uses Openstack as cloud computing framework that users could take advantage in creating the public or private cloud. In this paper, testbed set up using the openstack is proposed to get the benefits of SDN. Opendaylight is used with neutron in the openstack. Campus network case study is used to understand how SDN cloud can contribute towards security. In this case study, some of the features such as traffic isolation, support for devices and multivendor interoperability have contributed towards the enhancement of security in the cloud environment. There are some limitations of implementing SDN in the campus network though many benefits are highlighted. Staff are well versed with traditional approach and it is required to train them with SDN approach.

References

1. Mell, P., Grance, T.: The NIST definition of cloud computing recommendations of the National Institute of Standards and Technology. NIST Spec. Publ. **145**, 7 (2011)
2. Son, J., Dastjerdi, A.V., Calheiros, R.N., Ji, X., Yoon, Y., Buyya, R.: CloudSimSDN: modeling and simulation of software-defined cloud data centers. In: Proceedings of 2015 IEEE/ACM 15th International Symposium on Cluster Cloud, Grid Computing CCGrid (2015), pp. 475–484
3. Jammal, M., Singh, T., Shami, A., Li, Y.: Software-defined networking: state of the art and research challenges. Comput. Netw. **72**, 1–24 (2014)
4. Azodolmolky, S., Wieder, P., Yahyapour, R.: SDN-based cloud computing networking. In: International Conference on Transparent Optical Networks (2013), pp. 2–6
5. Wang, L., Wu, D.: Moving target defense against network reconnaissance with software defined networking. In: Lecture Notes on Computer Science (including Subseries Lecture Notes on Artificial Intelligence and Lecture Notes on Bioinformatics), vol. 9866, pp. 203–217. LNCS (2016)
6. Yang, C., Tsai, P., Huang, J.: Design and Development of a Large-scale Network Testbed on a Research and Education Network, pp. 85–90 (2016)
7. SDN Introduction. https://www.sdxcentral.com/networking/sdn/definitions/inside-sdn-architecture/. Accessed on 26 Jan 2020
8. What is Open Flow?. Available https://www.sdxcentral.com/sdn/definitions/what-is-openflow/. Accessed on 28 Jan 2020
9. McNickle, M.: Five-must-know-open-source-SDN-controllers. Available https://searchsdn.techtarget.com/news/2240225732/Five-must-know-open-source-SDN-controllers. Accessed on 26 Jan 2020
10. Briain, D.Ó.: Training Laboratory Guide, vol. 1 (2017)
11. Openstack. https://docs.openstack.org/install-guide/get-started-with-openstack.html. Accessed on 26 Jan 2020
12. SDNcore. Available https://www.opensdncore.org/. Accessed on 28 Jan 2020
13. Mininet. http://mininet.org/. Accessed on 28 Jan 2020
14. Virtualbox. Available https://www.virtualbox.org/. Accessed on 26 Jan 2020
15. Opendaylight. https://www.opendaylight.org/use-cases-and-users/by-function/cloud-and-nfv. Accessed 28 Jan 2020
16. O. N. F. Solution and B. September: SDN in the Campus Environment (2013)

A Light-Weight Cyber Security Implementation for Industrial SCADA Systems in the Industries 4.0

B. R. Yogeshwar, M. Sethumadhavan, Seshadhri Srinivasan, and P. P. Amritha

Abstract Industry 4.0 is expected to revolutionize the way industries are automated currently. Conventional Supervisory Control and Data Acquisition (SCADA) systems are connected to various remote terminal units that are geographically and functionally separated in an industry. This makes SCADA system vulnerable to attacks and the real-time timing requirements make it difficult to implement heavy security mechanisms and therefore, there is a need for light-weight security mechanisms exploiting the developments in the Industries 4.0. This paper presents a light-weight cyber security mechanism for Industrial SCADA systems using IoT hardware which acts as a firewall and protects the system from getting attacked by the adversary. Consequently, a light-weight cyber security measure is developed for the industrial SCADA system.

1 Introduction

The need for industrial automation is growing exponentially. With the emergence of the Industrial Internet of Things (IIoT), conventional automation system landscape is seeing a significant change. While the IIoT provides compact sensing, open networking, seamless connection of devices, and support of third-party applications,

---

B. R. Yogeshwar (✉) · M. Sethumadhavan · P. P. Amritha
TIFAC CORE in Cyber Security, Amrita School of Engineering, Amrita Vishwa Vidyapeetham, Coimbatore, India
e-mail: yogeshwar.br1@outlook.com

M. Sethumadhavan
e-mail: m_sethu@cb.amrita.edu

P. P. Amritha
e-mail: pp_amritha@cb.amrita.edu

S. Srinivasan
Berkeley Education Alliance for Research, University Town, Singapore
e-mail: seshucontrol@gmail.com

© The Editor(s) (if applicable) and The Author(s), under exclusive license to Springer Nature Singapore Pte Ltd. 2021
T. Senjyu et al. (eds.), *Information and Communication Technology for Intelligent Systems*, Smart Innovation, Systems and Technologies 196,
https://doi.org/10.1007/978-981-15-7062-9_46

it's also opening the conventional industrial automation architecture. It is considered as the major enabler for Industries 4.0 [1]. This means that a conventional industrial automation system which had a layered architecture based on ISA 95 with different layers being separated functionally into various organization regimes such as field-level control, supervisory control, real-time optimization, enterprise-level layers (see, [2] and references therein). This opening up makes it possible to perform plant-wide control and helps aggregating information across production lines, it also endangers the security of the system overall, as plant-wide information is available across different platforms, hardware, and services. Therefore, securing such systems becomes imperative. In this context, while IT security has been more related to the information integrity and attacks on it, the operation technology security has rather been in its infancy. Therefore, light-weight security mechanisms for OT security are required for industrial automation systems in the Industries 4.0. Our objective in this paper is to propose one such simple security mechanism for industrial automation systems and explain how the IIoT could be harnessed to provide the security mechanisms.

The ISA 95 architecture organizes the plant floor as field-level devices and controls. Typically, the programmable logic controller (PLC) is the workhorse for implementing field-level control actions. The SCADA stands for Supervisory Control and Data Acquisition. SCADA is a control system used in the industrial process and infrastructural process. In SCADA, the PLCs and RTUs are like microcomputers, where the communication happens using multiple objects like factory machines, devices, HMI and sensors. The SCADA software is the one which helps in sending the information from the objects to the computers and helps the employees to analyze the data for making the important decisions. It also helps the operators and displays the data. The SCADA systems are mainly proposed for industrial organizations. The public and private sector companies will always try to control, maintain, distribute the data and communicates with the issues of the system to make sure about the downtime as mitigation. The SCADA system is initiated in many different enterprises where the configurations from simple to larger complex installations can also be done.

2 Programmable Logic Controller (PLC)

The PLC is the one, which will get the input from the device, which is connected to it and the data gets processed. The output gets generated as per the instructions passed in the input parameters. The PLC will always monitor the input and output data and makes a record of the processing time and the data such as production capacity, temperature, starting, stopping process and alarms get triggered if any malfunction occurs. PLC is very adoptive and feasible for all the applications.

3 Distributed Control System (DCS)

A distributed control system [1] (DCS) is a control system, where it is fully computerized for processing of the plant with multiple loops for controlling it. It also distributes the controllers all over the system and the DCS does not have the central operating controller. This DCS will try to connect with the system, which has the controller, i.e. centralized. It can be located in a central control room or within the system. The DCS motive is mainly for reducing the cost of installation and helps in localizing the control functions of the plants by establishing the remote session to the plant. Not only this, but the DCS ideology also helped the industries to create large, huge systems with safety features. It brought the local system and the centralized system as an integration, due to this design integration risk was also reduced. Nowadays the working of the SCADA and DCS is the same, but the DCS is used only where the large processing plant with more security is required for the system and the remote access cannot be utilized.

4 Remote Terminal Unit (RTU)

A remote terminal unit (RTU) is a microprocessor that creates communication between human beings and the DCS or the SCADA. It helps the data to be transmitted to the machine or the system by using the messages from the main system to the connected devices. An RTU mainly monitors the sensor devices or the lower field devices and the parameters which are sent to the central monitoring system. It has a connection with the software, where the given input data and receives output data with a preferred communication protocol to know if any troubleshoot to be done.

5 Human–Machine Interference (HMI)

Human–Machine Interference is a device, which helps us in monitoring and displaying the status of the system. There are many factors that affect the machine operation such as temperature, pressure, flow of any liquid, etc. There is a control method (such as ON/OFF or PID) to maintain the temperature and see the desired temperature maintained for our process. HMI is used for controlling the machines for example switching on any motors or solenoid valves used in that machine. It is done through graphically made manual switches. Alarms trend can be monitored in the HMI and most of the time HMI interfaces with PLC and displays PLC data information and gives input to PLC.

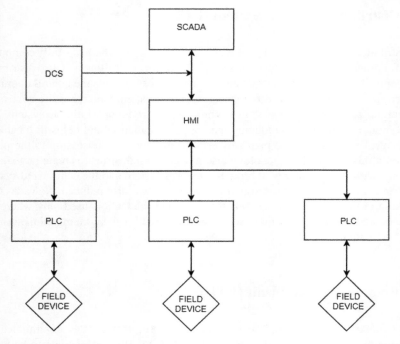

Fig. 1 Architecture of SCADA

6 Architecture of SCADA

This architecture is the general structure of the SCADA system, where SCADA will be at a higher level and the field devices are considered as the lower level. The DCS will help in the provisioning of the storage for the data generated. The PLC will act as a controller for the field devices as seen in Fig. 1 it follows the ISA 95 architecture [7].

7 Proposed System

7.1 SCADA System

The SCADA system [1] is the key operating mechanism for monitoring the operations. The system is completely automated, and the operations are handled online from different locations through the remote access local control module with standard protocols. This paper describes the implementation of security in the SCADA system. The Raspberry Pi is added to the system with all the security, where it will

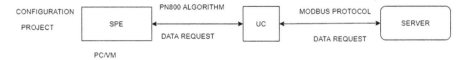

Fig. 2 System architecture

perform the operations of the firewall and protects the system from getting attacked by the advisory. This work can help in the development of a secure system.

7.2 Raspberry Pi

The Raspberry Pi is a series of small single-board computers developed in the United Kingdom by the Raspberry Pi Foundations to promote the teaching of basic computer science in schools and in developing countries. The Raspberry Pi is a configurable board with all the operations of a computer, which is smaller in size and available at a low cost. We will be installing the internal operating system for this Raspberry Pi and use the Ubuntu mate or Raspbian operating system to perform the required functions. The Raspberry Pi will have a few special features like in-built firewall in it, which will be activated for our security purpose. The firewall will be set up in such a way that it can monitor the logs, which is transmitted from and to the system.

The SPE will be system software functioning with the PN800 protocol and the UC will be the PLC of the system. The server will consist of the DCS simulator working with the Modbus protocol as mentioned in Fig. 2

7.3 SPE (S Plus Engineering)

The SPE is the software tool, where we are going to perform the engineering of the PLC, i.e. like how the process should function. It has a special environment where all the automation aspects can be performed on it. It provides a platform, where various data from multiple sources can be managed, if any changes are made at one point, and then it reflects the changes throughout the system. It also reduces the time for the architecture preparation and provides the functionality needed to engineer, configure, secure, communication, HMI systems to engineering platforms. We will be using this tool as mentioned in Fig. 2.

7.4 SPU (S Plus Operations)

The TWM is the simulator tool that works as a server. This tool will be helping us in retrieving the data at a faster rate. This tool also supports working with HMI, where multiple operations can be done. And it manages the information and alarms required for the system. This simulator will be working on the Modbus protocol.

7.5 PN800 Protocol

The PN800 protocol is also called as SAN (Single Attached Node) or DAN (Dual Attached Node). If two devices like PN800A and PN800B are present, then the protocol helps both the devices to communicate with the same IP address. The PN800 protocol uses the PRP (Parallel Redundancy Protocol) technology.

7.6 Modbus Protocol

The Modbus protocol [1] has the main functionality like Request and Response. It supports serial communication technology and provides high speed at longer distances and communicates with multiple devices with two-wire transmissions. The Modbus connected devices will not respond to the open calls of the request. The protocol has an error check header, where the device will check for the request from the master and the slave will send any exceptional error when the request from the master is wrong.

8 Working

We are having a working module with different things to ensure the secure SCADA system, i.e. like Firewall, Fail2ban, IPS [3], and TCPDUMP. This implementation helps to defend the attackers from the outside world.

8.1 Fire Wall

The implementation of the firewall [5] is always important for creating a secured system. Normally, the firewall is of different types like Packet Filtering, Application-Level Firewall, etc.

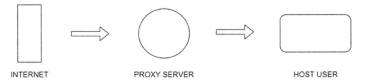

Fig. 3 Application level firewall

We implemented the application-level firewall as in Fig. 3 and this firewall will be connected to the internet using a proxy server as a mediator between the internet and the internal host machine. The functionality of the firewall is to restrict the access of the systems out of network range. We are going to use only particular ports and IP addresses to access the system or the plant from different locations. The firewall will help us to restrict others to access the system. We implemented IP tables with own set of rules restriction in the firewall and helps us to restrict the ports. In the same way, we restrict the IP address using the filters in the defender firewall.

8.2 Fail2ban

The fail2ban is also a kind of a firewall, where it helps in scanning the logs files which are coming to the system or the machine. It helps in the banning the IP address, where the malicious packets are trying to seek into the system even if too many attempts of the passwords or any exploits are detected [6]. The fail2ban will ban the IP address of the system with the help of the firewall, where the fail2ban will update the rules of the firewall. When we receive a packet if the packet is taking more time to be analyzed by the fail2ban, then we can suspect as a malicious packet and drop the packet, where this can be done by having a time out period in the fail2ban. This can help us in reducing malicious attacks. It also supports different services like apache, ssh, etc.

8.3 TCP DUMP

The TCP DUMP [2] helps us to set up the filters so that we can filter all the addresses need to access the system. We can also set the source and destination for the system and can use multiple filters at a time in the TCP DUMP with a specified port.

8.4 PLC Processing

The PLC used is SPC700, which generally needs a Master and a Slave. The master can be a computer as explained in Fig. 2. We have the flow of the circuits like FC222 to FC225 and using FC222 as INPUT, FC223 as OUTPUT, FC224 as DIGITAL INPUT, FC225 as DIGITAL OUTPUT. The DIGITAL INPUT/OUTPUT will be in the form of 0/1, which will help us to indicate the operation or functioning of the SPC700. Each FC circuit will consist of the IP ADDRESS, TYPE, and FC Code as seen in Fig. 4.

The PLC will be working on the PN800 protocol from the client to the PLC and the MODBUS Protocol from the PLC to the Server or the Field Device. The client is the one that is connected to the left side in Fig. 5. The gateway will be connected to multiple address with each individual address having their own inputs. There are multiple serial ports that will connect our client to one of the ports and multiple

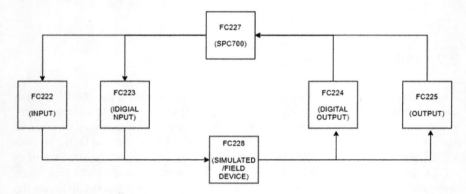

Fig. 4 Internal function of the PLC flow

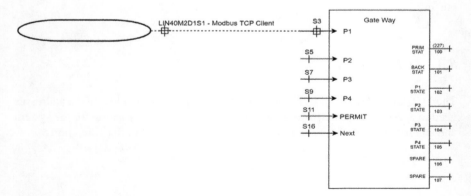

Fig. 5 Structure of the PLC gateway

A Light-Weight Cyber Security Implementation ... 471

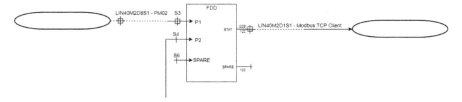

Fig. 6 The client and server connectivity flow

devices also can be connected to the PLC in different ports. A single DCS can be used for accessing the data of multiple devices with partitions in it.

In Fig. 6, we can find the server on the left side and the client on the right side. Here as a server, we are going to use a DCS simulator which will have default data in it. This will help to check the process and flow of the data accessing the system.

The client will have the IP address of the server system and the type of data can be specified or configured in the client phrase. The input will be given in the FCC222 as mentioned in Fig. 4. The input will be mentioned in the client-side and the control operations are passed to the PLC. The PLC will work on input and gets the address data to the client. Finally, the security [4] is implemented by assigning IP address and ports for all the devices like the client, PLC, and server. We have setup the fail2ban to analyses the logs of the packets that are communicated through the given network in the system. Securities aspects will be setup in between the PLC and the server. The fail2ban finds anything malicious, then it will make a rule in the filter list of the firewall and to make the firewall to block the IP address and hence not allowing the packet into the system. We can trigger alarm using this system if any malicious is detected in the network. Proxy server which is implemented in this system as mentioned in Fig. 3, will also have some set of rules. The IPS [3] is used to identify the purpose of the malicious packets and it is impacted. Using this data, we can also implement the AI (Artificial intelligence) into the SCADA system. We can bring a new propagation into the SCADA system and can be easily operated from any PLC with proper security, which can be future work.

9 Conclusion

The SCADA system is mainly used for the monitoring of the system's various processing of the infrastructure and process of the industries. We have implemented a flexible system for industrial usage with more security and the system can be accessible globally. This chapter gives more ideas on the security factors for the SCADA system with different aspects of the security components in a single working module. The Raspberry Pi is added to the system with all the security, where it will perform the operations of the firewall and protects the system from getting attacked by the advisory.

References

1. Muthukumar, N., Srinivasan, S., Ramkumar, K., Pal, D., Vain, J., Ramaswamy, S.: A model-based approach for design and verification of industrial internet of things. Future Gen. Comput. Syst. **95**, 354–363 (2019)
2. Muthukumar, N., Srinivasan, S., Ramkumar, K., Kavitha, P., Balas, V.E.: Supervisory GPC and evolutionary PI controller for web transport systems. Acta Polytechnica Hungarica **12**(5), 135–153 (2015)
3. Fillatre, M., Lionel, I.N., Willett, P.: Security of SCADA systems against cyber–physical attacks. IEEE Aerosp. Electron. Syst. Mag. **32**(5), 28–45 (2017)
4. Antonioli, D., Tippenhauer, N.O.: MiniCPS: a toolkit for security research on CPS networks. In: Proceedings of the First ACM Workshop on Cyber-Physical Systems-Security and/or Privacy (2015), pp. 91–100
5. Mo, Y., Chabukswar, R., Sinopoli, B.: Detecting integrity attacks on SCADA systems. IEEE Trans. Control Syst. Technol. **22**(4), 1396–1407 (2013)
6. Hu, Y., Li, H., Yang, H., Sun, Y., Sun, L., Wang, Z.: Detecting stealthy attacks against industrial control systems based on residual skewness analysis. EURASIP J. Wirel. Commun. Netw. **1**, 74 (2019)
7. Kalluri, R. Mahendra, L., Senthil Kumar, R.K., Ganga Prasad, G.L.: Simulation and impact analysis of denial-of-service attacks on power SCADA. In: 2016 National Power Systems Conference (NPSC) (2016), pp. 1–5

Conventional Biometrics and Hidden Biometric: A Comparative Study

Shaleen Bhatnagar and Nidhi Mishra

Abstract The biometric verification is necessary for the identification of human subjects for the purpose of security and authentication. Biometric data are very much unambiguous to each one, easily obtainable non-intrusively, no significant changes over a period of time and detectable by humans without much special training. Biometric traits provide a unique natural signature of an individual and it is accepted everywhere. Each biometric technique has its advantage and disadvantage. The methods based on conventional biometric traits such as the face, fingerprint, iris, signature, speech, etc., face problems of theft, imposter, and other security problems. Biometric template once theft cannot be used again. Hidden biometrics is a field where structural features of hidden body parts are used for biometric identification of human subjects. In this chapter, we will show a comparison of different hidden biometrics present and can be used for authentication.

1 Introduction

Prevalent security frameworks use secret codes, PIN, or tokens and these can be duplicated or lost without much effort. To build up a framework that guarantees more dependability, security, and foolproof identification it's mandatory to integrate biometric features in the framework. A biometric procedure of identification and conformation utilizes an individual's physical or behavioral traits to check their personality. Biometric features are particularly distinct to each human being, easily captured non-intrusively, and remain mostly unaltered for whole life expectancy and recognizable with no specific direction. It has a unique impression, which is all over acceptable [1].

S. Bhatnagar (✉) · N. Mishra
Presidency University, Bangalore 562106, India
e-mail: shaleenbhatnagar@gmail.com

N. Mishra
e-mail: nidhi.mishra@presidencyuniversity.in

© The Editor(s) (if applicable) and The Author(s), under exclusive license to Springer Nature Singapore Pte Ltd. 2021
T. Senjyu et al. (eds.), *Information and Communication Technology for Intelligent Systems*, Smart Innovation, Systems and Technologies 196,
https://doi.org/10.1007/978-981-15-7062-9_47

A new class of biometric came to presence barely any years back where hidden features of the human body like veins and DNA were utilized as the biometric characteristics for distinguishing between individuals [2]. Because of the huge variety in structure in body parts among humans and their stability under normal circumstances and these have become a usable source of recognition. As of late, a few researchers have claimed to utilize body parts like organs, i.e. brain, liver bones, etc. In this paper, we are giving a comparison between conventional and hidden biometric traits.

2 Advantages of Biometrics

Biometrics help us to replace "what you have" and "what you know" security proverb with the most important "who you are" by word, hence awarding one of the most substantial aids to the security world.

Enhanced security—Provide an advantageous and minimal effort extra level of security.

Diminish fraud by utilizing hard to forge mechanism. Get rid of issues that occurred due to lost IDs or forgotten pin by utilizing physiological traits.

Lessen password administration prices.

Lessen overhead of memorizing secret pin which can be shared or found.

Make it conceivable, naturally, to know WHO did WHAT, WHERE, and WHEN!

Allow noteworthy price savings or growing ROI in sectors such as Loss Avoidance or Time and Attendance. Unambiguously link a person to a transaction or event [3].

3 Challenges with Conventional Biometrics

Mainly there are two class of biometrics:

1. Physical biometrics
2. Behavioral biometrics

3.1 Physical Biometric

Physical biometrics works by utilizing the implicit physical properties of a person. It will in general be used for confirmation. Instances of physical biometrics include:

Facial Recognition—checking facial characteristics

Illumination. For example, a small change in lighting conditions has always been best known to cause a serious impact on its results. If the illumination tends to change, at that point; regardless of whether a similar individual gets caught with a similar sensor and with an almost identical facial expression and posture the outcomes that rise may show up very differently.

Background. It also plays a major role in identification. A facial recognition framework probably won't produce similar outcomes outdoors contrasted with what it creates indoors due to the area change. Additional factors like aging, back shadows on face, image resolution, and partial backlight can also influence the performance of the system.

Posture. Variation in posture creates a huge difference in Facial Recognition Systems. Point of view of the camera, movements of the head can constantly create variation in face appearance and create intra-class variations which make automatic face recognition a hard task.

Occlusion. Accessories (goggles, caps, veil, and so forth.), beard, mustache also interfere with the decision of a face recognition system. Their presence makes the subject different and it creates a problem with respect to accurate identification [4].

Expressions. Another critical factor that should be considered is different expressions of a similar person. This also makes accurate identification troublesome.

Fingerprint—analyzing fingertip patterns

Dry Prints. Because of low natural moistness in the skin, fingerprints can seem dry, damaged, or fragmented to electronic sensors. Due to this low-quality impression obtain during enrollment or irregular matching occur while identification. Because of environmental conditions, finger skin can be dry [5]. Dealing with certain materials or substances likewise absorbs the oils from the skin. For instance, substances like paper fabric, wood and chemicals like acetone thinners and cleaning agents all dry our skin (Fig. 1).

Wet Prints. Extra moistness can make our fingerprint impression seem blended together, this also results in low-quality impression during enrollment or irregular matching while identification. Extra moistness usually occurs due to sweating or by handling wet materials or substances like hand moisturizer and cosmetics, etc. (Fig. 2).

Scarred Prints. Scar prints have plastic-like nature. When it is dry, it doesn't picture appropriately, when it is wet, it would seem that a wreck around to the imaging framework [5] (Fig. 3).

Fig. 1 Dry print

Fig. 2 Wet print

Fig. 3 Scarred print

No Prints. Some professionals don't have an appropriate fingerprint impression to record like people who work as labor in mining or farming, and other professionals can cause extra fingerprint hardening, scarring, or wearing, rendering the impression unusable for fingerprint scheme (Fig. 4).

Fig. 4 Geometric shape of hand

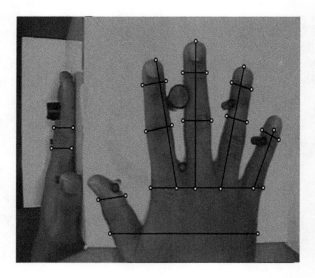

3.1.1 Hand Geometry—Analyzing the Shape of the Hand

Biggest advantage to use hand geometry traits is simple and easy to extract [6]. Ecological variables, for example, dry climate that causes the drying of the skin isn't an issue and normally thought to be less nosy than fingerprints, retinal, and so forth. But ornaments such as rings, limited dexterity (arthritis, etc.), etc. might cause a challenge in obtaining the hand geometry traits. For accurate identification, it's hard to use hand geometry as it is not very unique and not optimal for growing children. The data size of hand geometry biometrics is huge and is not optimal for practicing it in embedded systems.

3.1.2 Iris Scan—Measuring Attributes of Colored Ring of the Eye

Infants and few factors like long eyelashes, contact lenses, and even watery eyes will create trouble in reading from the scan. One more difficulty in implementing the Iris Scan Identification system, it's particularly difficult to take a reading of a handicapped person. The system can likely have 'failure to match' as number of attempts happen during the matching or verification period, despite the fact that the person is a similar individual as in the first time of enrollment. Head Tilt or Cyclotorsion [7], Template Aging in Iris identification are the big issues and in case if somebody has figured out how to get the photographs of your iris, it can be easily forged and you can't get another eye.

3.2 Behavioral Biometric

It generally works on the attributes which develop naturally with respect to time. It mainly helps in verification. Examples of behavioral biometrics are discussed below.

3.2.1 Speaker Recognition—Examine Vocal Behavior

One major problem in the speaker recognition method is irregularities in the various kinds of sound and their quality. One such issue, which has been the focal point of most research and publications in the field, is the issue of channel mismatch, in which the enrollment audio has been assembled utilizing one apparatus and the test audio has been delivered by an alternate channel.

3.2.2 Signature—Analyzing Signature Dynamics

The signature biometrics framework is broadly utilized in each confirmation area, however, it isn't achievable for users with highly inconsistent signatures. Furthermore, there frequently exist individuals whose signatures are exceptionally straightforward and can be copied easily. It degrades the performance of the biometric framework. Signature Recognition is prone to a lot more significant levels of error rate, especially when the behavioral characteristics of the signature are mutually inconsistent from each other.

3.2.3 Keystroke—Checking the Time Spacing of Typed Words

Keystroke dynamics checks behavior showed by an individual while typing on a keyboard [8]. User confirmation by way of keystroke dynamics is appealing for several reasons like: (i) it is not intrusive and (ii) it's comparatively cheap to implement since the main equipment required is the PC [9]. Other physiological biometrics such as fingerprints, retinas, and facial traits, all of which remain genuinely reliable over significant stretches of time, typing patterns can be rather erratic. Typing patterns also depends on the type of the keyboard being used, the keyboard layout, regardless of whether the person is sitting or standing, the individual's stance if sitting, and so on.

4 Hidden Biometrics

It can be viewed as a specific class of biometrics where the reason comprises of considering non-directly observable physical or behavioral traits of a person [10]. It utilizes information that is ordinarily utilized in the clinical field. Such techniques are powerful regarding spoofing. They need to be explored.

4.1 Vein

Vein Pattern Recognition is giving genuine rivalry to the most conventional biometrics. It is additionally viewed as what is known as an "Automated Physical Biometric."

4.1.1 Advantages of Using Vein Patterns

1. The structure of the vein patterns is remarkable among every person. Scientific studies have demonstrated that identical twins also have unique vein patterns.

With this degree of rich information, Vein Pattern Recognition has an extremely low degree of the False Rejection Rate.
2. The acceptance rate of Vein Pattern Recognition is regarded to be exceptionally high. A lot of this can be credited to the way that it is a non-contactless methodology, which requires very little human intervention. Subsequently, the issues of Civil Liberties and Privacy Rights infringement are for all intents and purposes non-existent.
3. In this verification, processing takes less than a second, and it is fastest than the other biometric systems. Results are affected by soil on the outside of the skin, cuts, wounds, scars, or even dampness or dryness on the fingertips or the palm.
4. It can also be utilized as a Multimodal framework just by itself. For instance, whenever required, the vein structure on both the fingertip and the palm can be caught and prepared by the same device.
5. The processing and storage necessities are low. Additionally, the mathematical algorithms required for unique feature extraction and template creation are considered to be very "light," when contrasted with those utilized in Iris Recognition and Facial Recognition.
6. It's hard to spoof a Vein Pattern Recognition device as continuous flow of blood is required in the veins for the raw images to be caught.

4.1.2 Disadvantages

7. Unlike the Iris or the Retina, there are chances that the vein pattern structure could change over the lifetime of a person, which creates the need for re-enrollment.
8. Vein Pattern Recognition can be significantly influenced by the negative effects of the ambient light from the external environment. As a result, this could have a negative impact on the quality of the raw images which are captured [11].

4.2 *DNA as Biometrics*

4.2.1 Advantages of Using DNA as Biometric

1. DNA is the main biometric that gives the chance of connecting family members to an unknown individual. Like fingerprints, DNA is one of only a handful, not many biometrics that can be "left behind", at a crime scene [12].
2. DNA testing is a generally experienced yet powerfully advancing technique that is getting broadly utilized and is familiar to the general population.
3. Ability to effortlessly store huge quantities of DNA brings the possibility of matches.
4. Advances in healthcare genomics offer guarantee that obscure forensic DNA samples can be described well enough to recognize the owners.

4.2.2 Disadvantage

It takes a lot of time in processing that's the main drawback to use DNA for verification. DNA is a profoundly useful information that can be utilized for negative methods whenever given the chance. In view of this, the nature of DNA and its utilization in the biometric condition can be undermining and an attack of the fourth amendment of the constitution—Guards against seizure of property explicitly in a non-criminal condition.

4.3 Brain Patterns as Biometric

Any piece of the skeleton can be considered and analyzed by utilizing brain signal and image processing algorithms. For instance, the skull, cervical vertebrae, humerus, ulna, radius can completely be modeled and important features can be extracted for using in hidden biometrics people vary from one another in thoughts, learning abilities and other cognitive ability learning builds up the mind and accordingly in long course of time large scale structure is framed having sulcus and gyrus structures on surface of the brain. Their numbers, size, and shapes are seen as one of a kind in a person [13]. The uniqueness of the human brain is one of the reasons for difference among the humans thus human brain structure can be used as a biometric authentication trait just like unique fingerprints.

The advantage is that neither the brain can be changed by the subject nor tends to be replicated by another person therefore this personality is constantly certifiable. Many feasible approaches are already available. Researchers can extend work in this direction to make a practical and more secure method for brain structure-based human identification.

The main drawback of this approach is its acceptability and its cost.

5 Conclusion

Biometrics appears to be secure superficially. After all, it always belongs to us. Biometrics is innately open. Consider it: our ears, eyes, and face are uncovered. We reveal our eyes whenever we look at things. We leave fingerprints wherever we go. Maybe someone is recording our voice when we are talking. Basically, there's simple access to all these traits. Our image is stored in more places than we realize. Not only does Facebook recognize our face, but every store we visit records and saves our image in its database to identify us and analyses our purchasing style. Truth be told, it's legitimate in 48 states to utilize software to identify anyone by using images taken without their assent for business-related work. Furthermore, law authorization organizations across the country can store our image without assent. The issue is identity management and security. If a hacker is breaching any of those databases, biometric ID can be forged easily. Biometrics gives another degree of security, but it's not foolproof thus it's necessary to switch to hidden biometric features for applications that are critical and require more security. In spite of the fact that right now obtaining the hidden biometric characteristics is still tedious and costly for users, we believe that for the high security required places where money is not the constraint hidden biometrics will be one of appropriate solutions in the future when the acquisition technique is developed.

References

1. Bhatnagar, S.: Cooperative biometric multimodal approach for identification. In: Satapathy, S., Joshi, A. (eds.), Information and communication technology for intelligent systems (ICTIS 2017), vol. 1 (2018); ICTIS: Smart Innovation, Systems and Technologies, vol. 83 (2017). Springer, Cham
2. Maheshwari, S., Choudhary, P.: Hidden biometric security implementation through human brain's artificial macro structure. Proc. Comput. Sci. **78**, 625–631 (2016)
3. Bhatnagar, S.: A reconsideration of schemes for fingerprint identification. Int. J. Adv. Res. Comput. Sci. **4**(3) (2013)
4. Li, S.Z.: Face recognition: technical challenges and research directions. In: Li, S.Z., Lai, J., Tan, T., Feng, G., Wang, Y. (eds.), Advances in Biometric Person Authentication. SINOBIOMETRICS 2004. Lecture Notes in Computer Science, vol 3338 (2004). Springer, Berlin, Heidelberg
5. Bhatnagar, S., Kumari, S.: Prevalent study of cooperative biometric identification system using multimodal biometrics. In: Fong, S., Akashe, S., Mahalle, P. (eds.), Information and Communication Technology for Competitive Strategies. Lecture Notes in Networks and Systems, vol. 40 (2019). Springer, Singapore
6. Amayeh, G., Bebis, G., Erol, A., Nicolescu, M.: Peg-free hand shape verification using high order Zernike moments. In: Proceedings of the IEEE Workshop on Biometrics at CVPR06, New York, USA (2006)
7. Rankin, D.M., Scotney, B.W., Morrow, P.J., Pierscionek, B.K.: Iris recognition—the need to recognise the iris as a dynamic biological system: response to Daugman and Downing. Pattern Recogn. **46**(2), 611–612 (2013). https://doi.org/10.1016/j.patcog.2012.08.008
8. Monrose, F., Rubin, A.: Authentication via keystroke dynamics. In: Proceedings of the 4th ACM Conference on Computer and Communications Security, pp. 48–56 (1997)
9. Gunetti, Picardi: Keystroke analysis of free text. ACM Trans. Inf. Syst. Security **8**, 312–347 (2005)
10. Nait-ali, A.: Hidden biometrics: towards using biosignals and biomedical images for security applications. In: 7th International Workshop on Systems, Signal Processing and their Applications, 2011
11. Aboalsamh, A.H.: Vein and fingerprint biometric authentication: future trends. Int. J. Comput. Commun **3**(4) (2009)
12. Maugards, D., Nait-ali, A.: Imaging for hidden biometrics. In: Hidden Biometrics, pp. 155–166. Springer, Singapore (2020)
13. Aloui, K., Nait-Ali, A., Naceur, M.S.: Using brain prints as new biometric feature for human recognition. Pattern Recogn. Lett. **113**, 38–45 (2018)

Performance Analysis of Various Trained CNN Models on Gujarati Script

Parantap Vakharwala, Riya Chhabda, Vaidehi Painter, Urvashi Pawar, and Sarosh Dastoor

Abstract In the age of digitization, it's worthy to preserve ancient scripts, researches, and documents. A digital format of the mentioned documents would help in preservation and replication and therefore, the chief goal of this research is to analyze the performance of various trained CNN (Convolutional Neural Network) models to recognize Gujarati scripts. As the ancient scripts are somewhat degraded in terms of their physical aspects and are sometimes unrecognizable, CNN plays a vital role in digitizing printed texts so that they could be edited in an electronic format, stored more compactly and used in various machine processes. The models used for experimentation includes LeNet (Learning CNN), DenseNet (Dense CNN), and VGGNet (Visual Geometric Group Network), which are used for Optical Character Recognition (OCR), and have obtained the results on our synthetically generated dataset. The dataset comprises more than 400 classes with an 80-20 division of train and test images respectively. The performance parameters include validation accuracy, validation loss and the number of epochs required to train the CNN model.

P. Vakharwala (✉) · R. Chhabda · V. Painter · U. Pawar · S. Dastoor
SCET, Surat, India
e-mail: pvakharwala32@gmail.com

R. Chhabda
e-mail: riyachhabda17@gmail.com

V. Painter
e-mail: vaidehi22698@gmail.com

U. Pawar
e-mail: urvashipawar511@gmail.com

S. Dastoor
e-mail: sarosh.dastoor@scet.ac.in

© The Editor(s) (if applicable) and The Author(s), under exclusive license to Springer Nature Singapore Pte Ltd. 2021
T. Senjyu et al. (eds.), *Information and Communication Technology for Intelligent Systems*, Smart Innovation, Systems and Technologies 196,
https://doi.org/10.1007/978-981-15-7062-9_48

1 Introduction

Character recognition is a continuously expanding field of computer vision. OCR has become an important and widely used technology for many practical applications like processing cheques in bank to avoid human involvement, digitizing paper documents, in retail and other industries. It is mostly used to classify optical patterns corresponding to alphanumeric or other characters. In our research, it is used to identify Gujarati scripts. Text recognition and conversation is a common application that is observed nowadays, but there hardly exists a solution for our native languages. One such language is Gujarati, on which much work of character recognition has not been accomplished with Machine Learning (ML). An application of text to speech on Gujarati language could be applied for the visually impaired members of our society. The dataset is synthetically generated for training of our model.

LeNet is a very basic network that could be used to classify small number of classes. DenseNet could be used on the pre-trained models of ImageNet. Therefore, it is expected that it will help in feature extraction, as well. All these networks vary in terms of number of parameters, error-rate and complexity, and hence a comparative analysis is important for its utility. A definitive network is in need because at the end, different models can be used for different applications.

The comparative analysis of various trained CNN models is done with the help of Tensorboard. All the networks are trained with a call-back to 'TensorBoard', 'Earlystopping' and 'ModelTest'. In our case, the model is trained on the dataset that was synthetically generated and the results obtained from each network are then measured with respect to their validation accuracies. The dataset consists of more than 400 classes. Hence, a model with capability to classify such high number of classes is quite crucial.

The remaining part of the paper has been organized as: Literature Survey is summarized in Sect. 2, Sect. 3 includes System Model; followed by comparative results in Sect. 4 and the summary of our work in Sect. 5.

2 Literature Survey

CNN is widely used in Deep Learning (DL) framework. But it is essential to figure out the type of network specific to a problem.

The basic architecture mainly used for character recognition is LeNet. A modified LeNet network is proposed consisting of two stages, one for recognizing the shape of the character, and other for dot recognition, with accuracy [1]. Further modification of LeNet architecture i.e. LeNet-5 is used for offline handwritten English character recognition. It possess special settings of the number of neurons in each layer and increasing the interconnectivity between the layers [2]. The same architecture trained with gradient based learning and backpropagation algorithm is implemented for classification of Malayalam character images with accuracy [3].

DurjoySenMaitra proposed that CNN could be used more efficiently as a feature extractor than a classifier [4]. Contradictory to that, handwritten character recognition for Arabic characters using CNN shows classification accuracy of 94.9% [5]. Embracing the qualities of CNN, this paper introduces DenseNet, which connects each layer to every other layer in a feed-forward fashion [6]. Also considering the size of the datasets, a modified ResNet-18 architecture is proposed for recognizing Bangla handwritten character recognition (HCR) and isolated Bangla HCR [7].

CNN has even provided an alternative for learning automatically the specific features with AlexNet as the base CNN model [8]. As CNN and ML is dependent on many variables, hence the best way to approach is to try and implement architecture which are best suited for our work. For the same reason, a performance analysis of CNN network models is shown for Synthetic Aperture Radar (SAR) image target recognition with accuracy more than 99% [9]. The similar performance analysis is shown in [10] to help the visually impaired people in which CNN with Computer Vision is combined to detect and recognize prohibition signs in real scenes with an accuracy of 95%.

CNN has also provided a new end-to-end approach to HCR with promising results. A streamlined version of GoogLeNet is shown for Chinese characters, which was 19 layers deep and had 7.26 million parameters [11]. One such challenging task of recognizing handwritten Devnagri characters is done with the help of AlexNet with around 16870 samples of 22 consonants with impressive results [12]. There are techniques which have been implemented for the recognition of Hindi characters using Deep CNN trained with backpropogation using thousands of hyper-parameters [13]. DenseNet has been used for the first time for text detection using FC-DenseNet (Fully Connected DenseNet) for semantic segmentation of the images [14]. The training of degraded patterns of image having characters are carried out by AlexNet for Kannada language, comprising of 497 classes and having accuracy of 91.3% for printed characters and 92% for handwritten texts [15].

Deep Learning (DL) has evolved in the field of mobile devices. CNN, being a powerful pattern recognizer, has been used for mobile computing with the use of image processing. The use of CNN has been validated by comparison with the existing models of OCR, GoogleNet being the best [16]. One application based on object detection algorithms like Haar Cascade and CNN, assisting a visually impaired individual to manage their activities independently is proposed in [17].

3 System Model

This section covers brief description about the various CNN models used for the comparative analysis. The various models used are LeNet, DenseNet and VGGNet. We have used a modified version of LeNet-5, a modified DenseNet-121 model as well as a modified VGG-19 architecture. The flow of our work is shown in Fig. 1.

For implementing architectures, we have used library TensorFlow and Keras. From these libraries, we have also called the callbacks for model performance

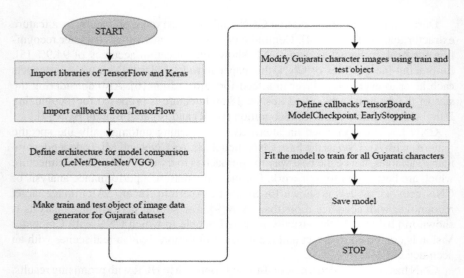

Fig. 1 Flowchart of implementation

comparison. The architectures of LeNet, DenseNet and VGGNet as defined using TensorFlow. For DenseNet and VGG-19, the models were imported from Keras applications. For importing the Gujarati dataset and handling class labels, Image Data Generator is used. The models were then fit using the generators obtained for all Gujarati classes and callbacks were associated. A callback is a set of functions to be applied at given stages of the training procedure. You can use callbacks to get a view on internal states and statistics of the model during training. The relevant methods of the callbacks will then be called at each stage of the training. The sample dataset is shown in Fig. 2.

Fig. 2 Sample generated dataset

Performance Analysis of Various Trained CNN Models ... 487

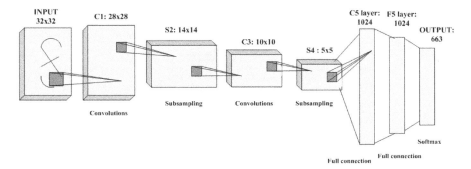

Fig. 3 Architecture of modified LeNet-5

Initially, the dataset was divided with 80-20 division. The dataset is synthetically generated using .TTF file (True Type Font) and tailor-made script. White Gaussian noise was added at random to increase the size of the dataset, and to provide robustness and versatility. The figure shows the images generated from Lohit, Padma, Aakar-medium and Rekha .TTF files. Various training models and their architectures are described in the following subsection:

A. **LeNet**

LeNet has a tendency to recognize one character at a time [1]. LeNet-5 is a general convolutional network for recognition of characters. Every unit in a particular layer of LeNet accepts input from a set of units in the small neighborhood in the previous layer [3]. The architecture of LeNet used for our Gujarati characters is shown in Fig. 3.

The images with image shape $32 \times 32 \times 3$ and 128 images per batch is fed into the first convolutional layer that has 32 layers, a kernel size of 3×3 and activation function ReLU (Rectified Linear Unit) which is followed by max pooling layer of 2×2. It is followed by another convolutional layer of 32 filters of size 3×3 and activation ReLU, followed by another max pooling layer of 2×2. The output is then flattened and given to two fully connected layers of 1024 units. The output is given to a softmax layer of 663 units. The model is trained for 50 epoch, but with an EarlyStopping callback with minimum delta of 0.0050 and patience 3. The history is locked with TensorBoard. The parameters related to the above discussion have been tabulated in Table 1.

B. **DenseNet**

To preserve the feed-forward nature of CNN, every layer receives additional inputs from all preceding layers and is passed on to all subsequent layers. This introduces $L * (L + 1)/2$ connections in every L-layer network. This approach is referred as Dense Convolutional Network (DenseNet) [6] as shown in Fig. 4.

The dataset was fed into the DenseNet architecture invoked from Keras with predefined weights of ImageNet. The top layer was excluded so that we could apply

Table 1 Parameters of modified LeNet-5 model

Layer		Feature map	Size	Kernel size	Activation
Input	Image	1	32 × 32	–	–
1	Convolution	32	32 × 32	3 × 3	ReLU
2	Max Pooling	32	30 × 30	2 × 2	ReLU
3	Convolution	32	15 × 15	3 × 3	ReLU
4	Max Pooling	32	13 × 13	2 × 2	ReLU
5	FC	–	1024	–	ReLU
6	FC	–	1024	–	ReLU
Output	FC	–	663	–	Softmax

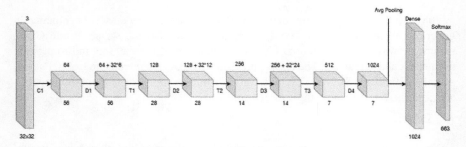

Fig. 4 Architecture of modified DenseNet-121

our own output layer. The input shape was 32 × 32 × 3. The DenseNet architecture is then given its final output layer with full connection dense layer with 1024 units followed by an output layer with activation Softmax. The EarlyStopping and training parameters were same as the previous model. The parameters used for this specific model are mentioned in Table 2.

C. **VGGNet**

VGGNet is a deep CNN developed at Visual Geometry Group of the University of Oxford. VGGNet enhances the robustness and generalization ability of the model. VGG-19 consists of 19 convolutional layers in sequence and has 5 configurations, A to E. We have used VGG-19 as increasing the network depth can effectively improve the performance of the model [9]. The modified VGGNet architecture used for our research in shown in Fig. 5.

The augmented dataset with an input image of 64 × 64 × 3, which was initially 32 × 32 × 3, was fed into VGG-19 which was invoked from Keras with predefined weights of ImageNet. The top layer was excluded, and hence three fully connected layers with units 512, 1024, 1024 respectively and activation ReLU followed by an output layer with activation Softmax were added. The EarlyStopping parameters

Table 2 Parameters of modified DenseNet-121 model [6]

Layer		Feature map	Size	Kernel size	Activation
1	Convolution	32	112 × 112	7 × 7 conv	ReLU
2	Pooling	32	56 × 56	3 × 3 max pool	ReLU
3	Dense block (1)	6	56 × 56	1 × 1 3 × 3	ReLU
4	Transition layer (1)	–	56 × 56	1 × 1 conv	ReLU
			28 × 28	2 × 2 avg pool	
5	Dense block (2)	12	28 × 28	1 × 1 3 × 3	ReLU
6	Transition layer (2)	–	28 × 28	1 × 1 conv	ReLU
			14 × 14	2 × 2 avg pool	
7	Dense block (3)	24	14 × 14	1 × 1 3 × 3	ReLU
8	Transition layer (3)	–	14 × 14	1 × 1 conv	ReLU
			7 × 7	2 × 2 avg pool	
9	Dense block (4)	16	7 × 7	1 × 1 3 × 3	ReLU
10	Classification layer	–	1 × 1	7 × 7 global average pool	Softmax
				1024 fully- connected	
				663 fully-connected	

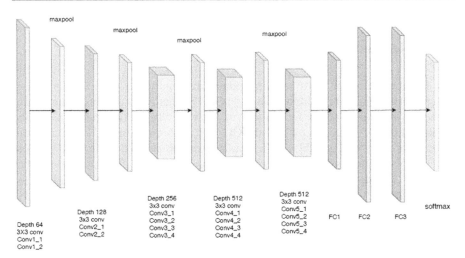

Fig. 5 Architecture of modified VGG-19

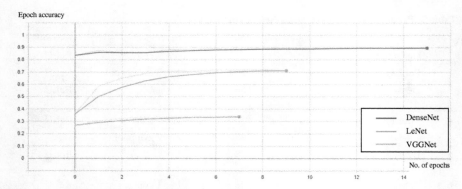

Fig. 6 Comparison graph of validation epoch accuracy

were also applied on this model. The flow of modified VGG-19 model is shown in Fig. 6.

Maxpooling: It is used to decrease the resolution and enhance the features. The largest value is the output. It is used against noise and distortion [17].

Softmax: This layer classifies and gives probabilities for each class label as shown in Fig. 5 [17].

ReLU: To increase the speed of training of the model, an activation function is used. ReLU is one such activation function which uses non-linear properties of the decision function and of the overall network [17].

4 Implementation Results

The architectures stated above were implemented using Python, Open CV for dataset generation and TensorFlow library based on GPU as a hardware platform. The dataset generated was used for the recognition of Gujarati characters. Architectures like modified LeNet-5, DenseNet-121 and VGG-19 have been compared using different parameters as shown in Table 3. The parameters compared for choosing the best suitable model for our purpose are validation accuracy, validation loss and number of epochs. The most prominent performance metric was proved to be validation loss, as once it starts to plateau, it starts to consider that the model is not working anymore, and hence the training was terminated after 3 epochs of change with a minimum delta

Table 3 Comparison of different CNN models

Parameters	LeNet-5	DenseNet-121	VGG-19
Validation accuracy	0.717	0.8978	0.3446
Validation loss	1.067	0.4373	2.17
No. of epochs	10	16	7

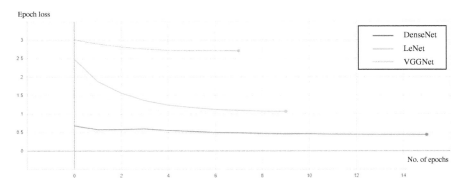

Fig. 7 Comparison graph of validation epoch loss

of 0.0050 in validation loss. Figures 1 and 2 shows the comparison graph of number of epochs versus validation accuracy and number of epochs versus validation loss respectively.

In LeNet, as we observe from Fig. 5, the validation loss starts to fall rapidly between first and fourth epoch and eventually plateaus after fifth epoch, which then settles on 1.067. Hence, the number of epochs it takes to reach validation accuracy of 0.717 is 10. In DenseNet, the initial validation loss does not vary much between first and fourth epoch, but as the accuracy increases, the training is not terminated and the final values arrived are 0.8978 validation accuracy with loss of 0.4373 after 16 epochs. In case of VGG-19, the initial epoch loss is very high, but it starts to decrease from first to fifth epoch and settles on 2.17. Without any further difference, the training is terminated at 7 epochs with validation accuracy of 0.3446 (Fig. 7).

5 Conclusion

In this paper, different architectures of Convolutional Neural Networks (CNN) are compared for the recognition of Gujarati characters. The synthetically generated dataset using .TTF font file has been used to train various CNN models. Various architectures used for the same include modified LeNet-5, DenseNet-121 and VGG-19. When processed on GPU, the callback Earlystopping used was configured to stop the training when there was a minimum delta of 0.0050, in validation accuracy over each epoch with patience of 3. Hence, all the models were terminated on their early stages as there was no significant change in validation losses and the models started to overfit. From the research, it can be concluded that due to extensiveness of Gujarati language, the shear amount of dataset and classes required are huge and hence, DenseNet turns out to be the better performer amongst the three. The validation loss was low from the very beginning and it achieved highest accuracy very rapidly but the reason it took most epochs is because the loss did not flatten out early on. Better results could be obtained by experimenting with various parameters such as learning rate, batch size, optimizer and activation function. We plan to fine tune these parameters in future work.

References

1. Al-Jawfi, R.: Handwriting Arabic character recognition LeNet using neural network. Int. Arab J. Inf. Technol. **6**, 304–309 (2009)
2. Yuan, A., Bai, G., Jiao, L., Liu, Y.: Offline handwritten English character recognition based on convolutional neural network. In: 2012 10th IAPR International Workshop on Document Analysis Systems, pp. 125–129. Gold Cost, QLD, (2012)
3. Kumar, K., Sachin, S., Anil, R.M., Soman, K.P.: Convolutional neural networks for the recognition of malayalam characters. In: FICTA 2014, pp. 493–500. Springer (2015)
4. Maitra, D.S., Bhattacharya, U., Parui, S.K.: CNN based common approach to handwritten character recognition of multiple scripts. In: IEEE 13th International Conference on Document Analysis and Recognition (ICDAR), pp. 1021–1025. Tunis (2015)
5. Elsawy, Ahmed, Loey, Mohamed, El-Bakry, Hazem: Arabic handwritten characters recognition using convolutional neural network. WSEAS Trans. Comput. Res. **5**, 11–19 (2017)
6. Huang, G., Liu, Z., Maaten, L.V.D., Weinberger, K.Q.: Densely connected convolutional networks. In: 2017 IEEE Conference on Computer Vision and Pattern Recognition (CVPR), pp. 2261–2269. Honolulu, HI (2017)
7. Albawi, S., Mohammed, T.A., Al-Zawi, S.: Understanding of a convolutional neural network. In: 2017 International Conference on Engineering and Technology (ICET), pp. 1–6. Antalya (2017)
8. Aloysius, N., Geetha, M.: A review on deep convolutional neural networks. In: 2017 International Conference on Communication and Signal Processing (ICCSP), pp. 0588–0592. Chennai (2017)
9. Shao, J., Qu, C., Li, J.: A performance analysis of convolutional neural network models in SAR target recognition. In: 2017 SAR in Big Data Era: Models, Methods and Applications (BIGSARDATA), pp. 1–6. Beijing (2017)
10. Motshoane, K., Tu, C., Owolawi, P.A.: Prohibition signage classification for the visually impaired using AlexNet transfer learning approach. In: 2018 International Conference on Intelligent and Innovative Computing Applications (ICONIC), pp. 1–5. Plaine Magnien (2018)
11. Zhong, Z., Jin, Z.X.L.: High performance offline handwritten chinese character recognition using GoogLeNet and directional feature maps. In: Proceedings of International Conference on Doctoral Analysis and Recognition (2015)
12. Sonawane, P.K., Shelke, S.: Handwritten Devanagari character classification using deep learning. In: 2018 International Conference on Information, Communication, Engineering and Technology (ICICET), pp. 1–4. Pune (2018)
13. El-Sawy, A., Benha, M.L.: Characters Recognition using Convolutional Neural Network (2017)
14. Behzadi, M., Safabakhsh, R.: Text detection in natural scenes using fully convolutional densenets. In: 2018 4th Iranian Conference on Signal Processing and Intelligent Systems (ICSPIS), pp. 11–14. Tehran, Iran (2018)
15. Rani, N.S., Chandan, N., Jain, S., Kiran, H.: Deformed character recognition using convolutional neural networks. Int. J. Eng. Technol. **7**, 1599. https://doi.org/10.14419/ijet.v7i3.14053
16. Weng, Y., Xia, C.: A new deep learning-based handwritten character recognition system on mobile computing devices. Mobile Netw. Appl. **7**, 1–10 (2019)
17. Shah, S., Bandariya, J., Jain, G., Ghevariya, M., Dastoor, S.: CNN based auto-assistance system as a boon for directing visually impaired person. In: 2019 3rd International Conference on Trends in Electronics and Informatics (ICOEI), pp. 235–240. Tirunelveli, India (2019)

Real-Time Human Intrusion Detection for Home Surveillance Based on IOT

Mohith Sai Subhash Gaddipati, S. Krishnaja, Akhila Gopan, Ashiema G. A. Thayyil, Amrutha S. Devan, and Aswathy Nair

Abstract Considering home security being more prominent nowadays, its proper surveillance and alerts at the right time warrant utmost importance. Our project focuses on an enhanced home security system that integrates the surveillance systems with powerful machine learning tools that guarantee a flawless responsible home safeguard system. To implement a real-time home security system for human intrusion detection, a remote monitoring video surveillance system based on Wi-Fi was developed. The video captured from the camera is further segmented and pre-processing techniques like Histogram of Oriented Gradients (HOG) and HAAR cascade algorithm are used for smartly eliminating false alarm due to animals. The former extract features from the input image whereas the latter is a cascade classifier that is used for identifying the objects in an image as propounded by Viola-Jones. The combination of these features is fed to SVM for multi-stage classification. The system then identifies the intruder automatically alerts the home resident and sends an alert message to the user's mobile using IoT. In this chapter, the proposed work is implemented using Arduino, PIR sensor, wi-fi camera and evaluated in Python and Matlab2019. The

---

M. S. S. Gaddipati (✉) · S. Krishnaja · A. Gopan · A. G. A. Thayyil · A. S. Devan · A. Nair
Department of Electronics and Communication Engineering, Amrita Vishwa Vidyapeetham, Amritapuri, Kollam, India
e-mail: mohithag13@gmail.com

S. Krishnaja
e-mail: krishnajas1502@gmail.com

A. Gopan
e-mail: akhilagopan890@gmail.com

A. G. A. Thayyil
e-mail: ashiemagathayyil@gmail.com

A. S. Devan
e-mail: amruthasdevan17@gmail.com

A. Nair
e-mail: aswathykn@am.amrita.edu

© The Editor(s) (if applicable) and The Author(s), under exclusive license to Springer Nature Singapore Pte Ltd. 2021
T. Senjyu et al. (eds.), *Information and Communication Technology for Intelligent Systems*, Smart Innovation, Systems and Technologies 196, https://doi.org/10.1007/978-981-15-7062-9_49

power of this video surveillance system is efficiently enhanced with the use of proximity sensors. Experiment results show that the integration of hardware with effective machine learning techniques can attain remote surveillance with high reliability and accuracy.

1 Introduction

One of the important problems that we face in our modern world is home security. The security of the situation plays a vital part in the present world. Numerous individuals use distinctive security systems to protect their assets from the entry of unauthorized persons. Security system helps people feel relatively secure while they are traveling or leaving their home for work. There are many cases reported all over the world due to the lack of home security. These security systems are generally constructed for the protection and safety of houses from theft, fire, managing electricity, odorless cooking gas, medical assistance, etc.

The security systems which are present now against robbery are so expensive because a fixed amount has to be paid to the administrator to store the recorded video but we won't get human movement often. The solution to this problem is an intelligent surveillance system, which can start recording video only after detecting a human presence. Many techniques have been adopted for the protection of the home from theft, such as video surveillance, cloud control, and other various kinds of applications. On the surface, these techniques might look fair, but this requires 24 × 7 power consumption and monitoring which is a tedious task. To overcome these limitations, we have made the camera triggered only when an intrusion is detected and for the rest of the time, the camera will be in the sleep state.

A PIR sensor is integrated into the system to trigger the system when motion is detected. Also, the incorporation of HOG features with SVM helps to avoid false alarm arising from animal movement. The system accuracy enhanced with the inclusion of HAAR cascade classifier and powerful Machine Learning (ML) technique, SVM. HAAR cascade classifier identifies the human facial features and this is being fed to SVM for multi-stage classification. At the first stage, SVM classifies human or non-human to ensure two-step reliability. If human presence is detected, then the machine learning tool evaluates the captured image with the trained home occupant's image to check whether the captured image is of an intruder or home occupant. If it is an intruder an alarm is raised, meanwhile an intrusion alert message is being sent to the concerned person's phone using Arduino and IoT. The proposed system is efficient in terms of less power consumption, simple architecture, user convenience, and prompt alert.

The chapter is organized as follows: Sect. 2 illustrates works related to the field, Sect. 3 elucidates system architecture, Sect. 4 describes the techniques adopted in this work, Sect. 5 explains the results and evaluations and finally Sect. 6 concludes the chapter.

2 Related Works

According to the studies conducted so far, there exist many systems that can detect human presence. There are home automation systems designed which are mainly based on sensors located at key points that work together with a main control or command panel installed in a convenient place somewhere near the home. Existing systems have some deficiencies such as high-cost, lack of intuitive User Interface (UI), lack of a good security system, and power inefficiency. Guo et al. explained a model that uses SVM with a binary tree recognition strategy. Two face databases are used to evaluate the multiclass face recognition using the proposed method. The binary tree extends naturally. The pairwise discrimination capability of the SVMs does the multi-class scenario [1].

Jin et al. explained about an intrusion detection system that uses WLID, a whole-home level intrusion detection system based on RSSI (Received Signal Strength Indicator) measurements of Wi-fi which can be an efficient way in detecting human indoor. The experimental results show that in a realistic home setting, WLID can achieve a consistent detection rate close to 100% [2].

Chowdhry et al. discuss the concept of a smart HAS with intrusion detection integrated to mitigate the harm done by burglary. Therefore, the current HAS incorporates a cloud server into the home appliances to view and monitor their status remotely. Intrusion detection uses the function descriptors Histogram of Oriented Gradients (HOG) and a Support Vector [3]. Xue et al. discuss UWB-based wireless sensor network to implement intrusion detection. After studying the characteristics of ultra-wideband (UWB) signals, convolution neural networks are used to automatically recognize the characteristics of UWB signals. The classifier Softmax or SVM is used to distinguish humans from animals [4]. Raghavachari et al. employed Frame Separation, Circular Hough Transform, and Oriented Gradient Histogram approaches are analyzed concerning various factors such as camera direction, illumination, occlusion, etc., for faster human counting. The success of such algorithms under various conditions shows the need for people counting algorithms more accurately and faster [5]. Jee et al. explained a process that uses color, binary data, and edge to detect pairs of eyes from input pictures, then extract faces from the identified pair of eyes. They used SVM to check both the candidate's face and eye pair [6].

Yousif et al. proposed a method for detecting humans from highly camera trapped images using background modeling and deep learning classification [7]. Xu et al. proposed a concept for the pedestrian identification algorithm for videos taken from fixed cameras on platforms. For each frame, the Harris operator is used to isolate corner points and the Normalized Cross-Correlation is used to align certain corner points with two neighboring frames. HOG functions are added to each picture and are used as an input to the SVM-HOG classifier [8]. Dewantara et al. discussed a model that identifies a human upper body utilizing a low-cost camera depth picture that is implemented on a PC [9]. Manohar et al. proposed an appropriate method for the identification and classification of animals focused on the texture attributes derived from the animals' local presence and texture. Animal classification is achieved by preparation and then two separate machine learning methods are evaluated, specifically k-Nearest Neighbors (k-NN) and Support Vector Machines (SVM) [10].

Ghatak et al. present the relative importance of IR and microwave sensor technologies and their conjunction with a wireless camera to establish a wall-mounted wireless intrusion detection device and describes the steps in which the intrusion information is gathered and sent to the central control station utilizing a wireless mesh network to interpret and process the data collected [11].

3 System Architecture

Figure 1 designed system is implemented by incorporating Arduino with PIR sensor and wi-fi video camera, that establishes a wireless connection network by transmitting captured videos through a wireless transmitter and alert through Blynk app to users mobile. The transmitted video is received at the wi-fi receiver which further processes the captured video (Fig. 1).

The object detection is done using a PIR sensor. It is an electronic sensor that measures radiating infrared (IR) light from the objects in its field of coverage [12]. It works based on IR rays which are emitted from the objects in its view. When it detects an object, the camera turns ON, which is an integral part of our work. The video is then transmitted for segmentation and pre-processing techniques such as HOG and HAAR classifiers, these are feature descriptors used in both computer vision and image processing. The object of interest is detected by segmenting the motion-captured from the video camera.

The features selected for classification purposes are mean and variance. These features are fed to the machine learning tool, SVM for the multi-stage classification. SVM is preferred over other ML techniques because of its optimal hyperplane separation, simple and faster computation that makes it a powerful ML tool for classification. The first stage mainly classifies humans from non-humans that avoid false

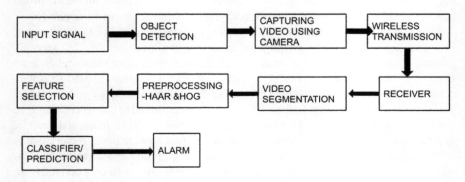

Fig. 1 Block diagram

alarm arising from animals. In addition to HOG classification, a double-checking is performed here, which increases the reliability and accuracy of the proposed system. This double-checking is performed using Machine learning tool which differentiates identified human with the family members. If the classifier detects it as a human, then a multistage classification is done using multi-stage SVM, which classifies intruders from family members or relatives. A data set of different images of family members was taken for training purposes and the same is fed to multistage SVM. When a human presence is detected, the classifier checks whether it belongs to a family member or an outsider. If it is an outsider or intruder, the system raises an alarm and warns the home occupants. Meanwhile using Blynk App, a cloud-based IoT platform via Arduino, an alert message is sent to the mobile of the concerned person indicating the presence of an unknown person.

4 Methodology

In this chapter, we have used two image processing techniques: HOG and HAAR for extracting features and machine learning algorithms-SVM for classification.

4.1 Histogram of Oriented Gradients

HOG is an image feature descriptor to describe the image based on the magnitudes and gradients directions. Using HOG we can extract histograms from all cells within the entire image patch, concatenating the histograms of all cells within the same block, normalizing the feature vector of each block, concatenating all feature vectors from all blocks. The fundamental steps involved in the implementation of HOG are as follows: [8].

4.1.1 Pre-processing and Calculating the Gradient Images

Pre-processing must be done for a patch of an image that has to be resized to 64 × 128 pixels (columns × rows).

The aspect ratio is fixed. Here the image is taken at a preferred size of 64 × 128 that will make the calculations easy.

Followed by this, calculate the vertical and horizontal gradients of an image by filtering it with the given Kernels (Fig. 2).

The following equation can be used to calculate gradient magnitude and direction:

$$\text{Magnitude}, g = gx_2 + gy_2 \qquad (1)$$

Fig. 2 The kernels

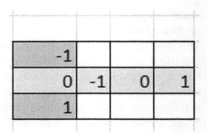

$$\text{Direction angle} = \arctan\left(\frac{gx}{gy}\right) \tag{2}$$

4.1.2 Calculating Histogram of Gradients (HoG) in 8 × 8 Cells

The histogram is a graph that shows the total number of pixels in an image at the different intensity in the image. It shows the frequency of each intensity value of an image. The input image is converted into gradients by dividing the whole image into 8 × 8 cells. There are 192-pixel values for an 8 × 8 cells. In total, this image contains a total of 128 numbers, i.e., 8 × 8 × 2 in which 2 indicates the gradient of patch (magnitude and direction) (Fig. 3).

Following this, we created a histogram of gradients on these 8 × 8 cells. The angles 20, 40, 60 till 160 corresponds to 9 bins of a histogram and calculates it for each cell. After that, gradient magnitude and gradient direction matrix of each cell were obtained. The histogram will be made using both for each cell in an image.

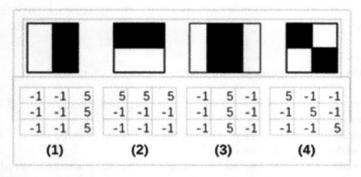

Fig. 3 The HAAR like features introduced in Viola Jones

4.1.3 Block Normalization

Block normalization is to be performed because some images will be in the light and some in dark. The gradient values will be different for sure in both the images. So to avoid this situation, we use the normalization of blocks.

4.1.4 Feature Vector

The final feature vector of the Histogram of Oriented Gradients will be calculated by the concatenation of feature vectors of all blocks in an image. So, in the end, we will receive a large vector of features [8].

4.2 HAAR Cascade Classifier

The HAAR Cascade classifier is based on the HAAR Wavelet technique for the computational classification of pixels in the picture into squares. It uses "integral image" to measure the observed properties [6]. The HAAR cascade classifier is based on the Viola-Jones detection algorithm, a method focused on machine learning where cascade function is learned from both positive and negative images and trained. HAAR Cascades uses an algorithm known as the Ada-boost learning which selects a few important features from a large set to give an efficient result for classifiers and later on uses cascading techniques to detect faces in the image.

Viola face detection algorithm-HAAR Features

HAAR Features are used to detect the features in the image. Here are some HAAR-Features. The first two (represent "edge features") are used to detect edges. The third is "line feature", while the fourth is "rectangle feature", mostly used to detect slanted lines.

The function results in a single value that is determined by subtracting the white rectangle pixel total from the black rectangle pixel amount. The above method uses 24×24 as the scale of the base window and scales the above attributes by 1 PX over the entire picture shift. If we find all potential parameters such as location and size, this will result in the calculation in a greater number of features. So we can avoid irrelevant features.

4.3 Support Vector Machines

We are using Support Vector Machines (SVM) as the main algorithm in the chapter. SVM is a binary classification method that finds optimum linear hyperplane based on the concept of minimal structural risk [5]. A weighted combination of training set components is represented by the decision surface. These training set components are called vectors of support, which define the boundaries between the group [13].

We have N examples

$$(x_1, x_2) \ldots (x_2, y_2) \ldots (x_N, y_N) \quad x_i \in R^N, y_i \in \{-1, 1\} \tag{3}$$

In the case of linearly separable data, the purpose of the maximum margin is to separate two classes with a hyperplane which maximizes support vector distance.

This hyperplane is called OSH (Optimal Hyperplane Separating) and is expressed in Eq. (4).

$$f(x) = \sum_{i=1}^{N} \alpha_i y_i \left(x_i^T x\right) + b \tag{4}$$

This solution is defined as a subset of support vectors that have a non-zero α_i.

5 Observations and Results

The entire project is integrated with the following hardware: PIR sensor, wifi camera, and wireless receiver, and Arduino. PIR sensors are placed at key points, near or in front of the door. PIR sensor and wifi camera are integrated into the Arduino board and when motion is detected by a PIR sensor, the camera gets triggered and starts capturing the video (Fig. 4).

The video is sent to a wireless receiver where the captured videos are further segmented. The images are segmented from the video at a sampling rate of 30 frames per second. As observed in Table 1, videos are captured at a fair speed and are sent for segmentation. Also, the time taken to send the alert message is 15–25 s. The segmented frames undergo the following steps.

5.1 Histogram of Oriented Gradient

The sampled images are fed to a HOG feature descriptor to extract the relevant features from the image. The aspect ratio is taken as 1:2 for preprocessing, hence the image size is taken as 64 × 128 for simple calculation. To make a real scenario, we have included background objects along with the person as shown in Fig. 6. Also, the image of a masked man, as in Fig. 6 was also taken into account, considering the fact that the intruder might have come with a mask. To both these images, HOG is applied and extracted the relevant features.

If we observe Figs. 5 and 6 it is understood that HOG applied to both the images give the same result, such that a person is identified with background noise. The

Fig. 4 Dataset used for training face of two human

Table 1 Observation table

Parameters	Results
Max detecting range of sensor	10–15 m
Response speed of PIR sensor	0.3–5 s
Segmented frames from video	30 frames per second
Motion detection speed	3.125–9 s for 100 images
Time required to send alert	15–25 s

second step in human identification is performed using SVM to authenticate the result obtained from HOG.

5.2 HAAR Cascade Classifier

Edge features and line features in HAAR are cascaded to identify the features of face since individually these features cannot accurately classify the pattern. Figure 7 shows that the segmented frames, when undergone HAAR cascade classifier, helps to extract face features such as eyes, nose, mouth, and face accurately.

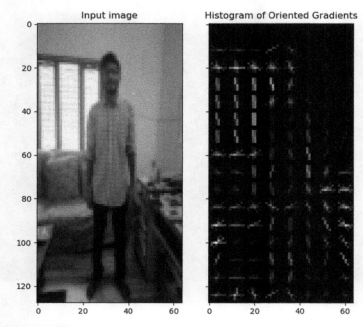

Fig. 5 HOG of a human

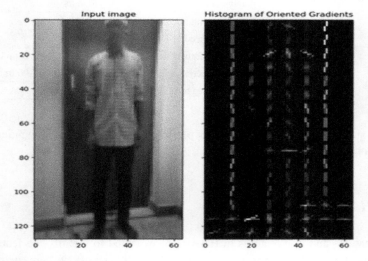

Fig. 6 HOG of a masked human

Fig. 7 HAAR classifier face detection

5.3 SVM

The first stage of SVM classifies the image between a human and a non-human. This stage of classification is considered as a double-precision checking in addition to HOG, which eliminates false alarms that arise from pets. 50 positive and 50 negative images of humans are taken and trained. We have created our own dataset from a wireless camera that consists of 40 images of each person as seen in Fig. 4.

Using the same camera, multiple images of each family member were gathered. To generate an SVM model, a complete set of features associated with each image, corresponding to each family member were trained. The second stage performed multiclass classification in which multiple persons are trained in the SVM model and labeled as class1, class2 and class3, and so on.

To train the SVM model 70% of data set is taken for training, 10 percentage for cross-validation, and 20% for testing. 40 images of each person in different positions and angles are trained using SVM and 20% of that are used as a testing dataset.

Here as a test sample, images of two persons were trained and accordingly confusion matrix was obtained as given in Fig. 8. The diagonal elements show truly predicted classes and off-diagonal elements give misclassified classes.

5.4 Accuracy

The accuracy of the system is calculated using the given Eq. (5)

$$\text{Accuracy} = \frac{(TP + TN)}{(TP + TN + FP + FN)} \quad (5)$$

Fig. 8 Confusion matrix of two humans

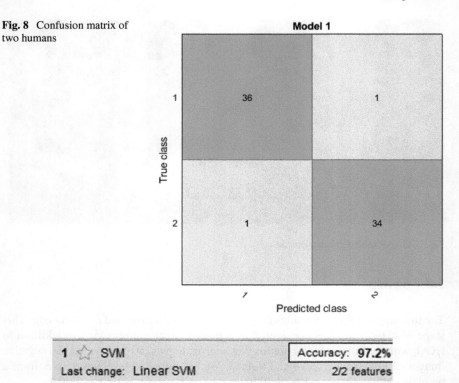

Fig. 9 Accuracy

where TP, TN, FP, and FN represent the number of true positives, true negatives, false positives, and false negatives, respectively. Here accuracy is obtained as 97.2% which is given in Fig. 9.

6 Conclusion

Real-time human detection system that uses the Histogram of Oriented Gradients, HAAR Classifier, and SVM is presented in this work. The system proposes a real-time security system for intrusion or unknown person identification. The integral part of the system, wifi video camera, captures the video only when PIR sensor detects motion, which efficiently reduces power consumption of the camera and other devices. The wifi video camera is in the sleep state unless PIR sensor detects motion and triggers the camera. Thereby, the system reduces the power consumption of the wifi video camera and other related hardware. This method upgrades the image segmentation than normal SVM based human intrusion systems, by the integration of HOG, HAAR, and SVM. The results obtained show that this method can be implemented in smart video-protection systems. As future work, the video is directly live streamed to authorize persons through suitable App when an unwelcome situation occurs.

References

1. Guo, G., Li, S.Z., Chan, K.L.: Support vector machines for face recognition. Image Vision Comput. **19**, 631–638 (2001)
2. Jin, Y., Tian, Z., Zhou, M., Li, Z., Zhang, Z.: A whole-home level intrusion detection system using WiFi-enabled IoT. In: 2018 14th International Wireless Communications & Mobile Computing Conference (IWCMC), pp. 494–499. Limassol (2018)
3. Chowdhry, D., Paranjape, R., Laforge, P.: Smart home automation system for intrusion detection. In: 2015 IEEE 14th Canadian Workshop on Information Theory (CWIT), pp. 75–78. St. John's, NL (2015)
4. Xue, W., Jiang, T., Shi, J.: Animal intrusion detection based on convolutional neural network. In: 2017 17th International Symposium on Communications and Information Technologies (ISCIT), pp. 1–5. Cairns, QLD (2017)
5. Raghavachari, C., Aparna, V., Chithira, S., Balasubramanian, V.: A comparative study of vision based human detection techniques in people counting applications. In: 2015 Elsevier Second International Symposium on Computer Vision and the Internet (VisionNet '15)
6. Jee, H., Lee, K., Pan, S.: Eye and face detection using SVM. In: 2004 IEEE Proceedings of the 2004 Intelligent Sensors, Sensor Networks and Information Processing Conference, Melbourne, Vic., Australia
7. Yousif, H., Yuan, J., Kays, R., He, Z.: Fast human-animal detection from highly cluttered camera-trap images using joint background modeling and deep learning classification. In: 2017 IEEE International Symposium on Circuits and Systems (ISCAS), pp. 1–4. Baltimore, MD, (2017)
8. Xu, F., Xu, F.: Pedestrian detection based on motion compensation and HOG/SVM classifier. In: 2013 IEEE 5th International Conference on Intelligent Human-Machine Systems and Cybernetics, Hangzhou
9. Dewantara, B.S.B., Ardilla, F., Thoriqy, A.A.: Implementation of depth-HOG based human upper body detection on a mini pc using a low cost stereo camera. In: 2019 International Conference of Artificial Intelligence and Information Technology (ICAIIT), pp. 458–463. Yogyakarta, Indonesia (2019)
10. Manohar, N., Subrahmanya, S., Bharathi, R.K., Sharath Kumar, Y.H., Hemantha Kumar, G.: Recognition and classification of animals based on texture features through parallel computing. In: 2016 Second International Conference on Cognitive Computing and Information Processing (CCIP), pp. 1–5. Mysore (2016)
11. Ghatak, S., Bose, S., Roy, S.: Intelligent wall mounted wireless fencing system using wireless sensor actuator network. In: 2014 International Conference on Computer Communication and Informatics, pp. 1–5. Coimbatore (2014)
12. Syazlina Mohd Soleh, S.S., Som, M.M., Abd Wahab, M.H., Mustapha, A., Othman, N.A., Saringat, M.Z.: Arduino-based wireless motion detecting system. In: 2018 IEEE Conference on Open Systems (ICOS), pp. 71–75. Langkawi Island, Malaysia (2018)
13. Ramanathan, R., Soman, K.P., Valliappan, N., Mathavan, S.P., Gayathri, M., Priya, R.: Generalised and channel independent SVM Based robust decoders for wireless applications. In: 2009 International Conference on Advances in Recent Technologies in Communication and Computing, pp. 756–760. Kottayam, Kerala (2009)

Token Money: A Study on Purchase and Spending Propensities in E-Commerce and Mobile Games

N. P. Sreekanth and Deepak Gupta

Abstract In businesses around the world, the Token Money system is getting adopted as an effective system in exchanging value between the company and customers. This system is most predominantly seen in mobile games, but is also now pervading the realm of e-commerce. For this research, the purchase and spending behaviors of Token Money in both mobile games and e-commerce are examined from the Token Money system's perspective. Primary research was conducted for 202 respondents across tier-1 and tier-2 cities in India. Results yield that Token Money system's characteristics such as the token's relevance, expiry and pricing play key roles in the purchase behaviors in both e-commerce and mobile games. Additionally, inclusion of social elements can have a significant impact on the same as well. This study contributes to both the theoretical and practical understanding of revenue generation in the above-mentioned areas.

1 Introduction

Mobile games, with their different models such as free, paid and freemium, are fast becoming the major contributor to revenue generation in the broader category of video games worldwide [1]. Much of this can be alluded to the success of freemium models where the game is provided for free but has in-app purchases that the player can purchase [2]. This model was arrived at only after a wide range of trial-and-error tests with various monetization techniques [3]. In the top-grossing mobile games, it is clear that a Token Money system is used for facilitating in-app purchases, rather than direct purchase. With the success of Token Money system, e-commerce is starting to adopt the same system as exemplified by Flipkart, the e-commerce giant in India, with

---

N. P. Sreekanth (✉) · D. Gupta
Amrita School of Business, Amrita Vishwa Vidyapeetham, Coimbatore, India
e-mail: sreekanthnp009@gmail.com

D. Gupta
e-mail: dgshobs@gmail.com

the introduction of what it calls SuperCoins with which one can avail several deals, enable premium features and even use to pay third-party vendors. Another major area where Token Money system can be seen and studied is the field of cryptocurrency [4]. For mobile games, prior research has examined how user's perceived utility of the Tokens is affected by variables such as sales, volume, frequency and the mobile market [5]. Research has also gone into how game elements affect grossing of the game [6].

There has also been research into identifying the moderating role of user's engagement level in the game on sampling effects, with a focus on the tactic of free item rewards to boost sales [7]. While the broader level of in-app purchases has been well researched, the deeper level, i.e., actual mechanism of Token Money, is an area left unexplored. Related to e-commerce, research has been done to understand the variables that affect transaction behaviors with virtual currency in Web2.0 environment [8]. Yet again, the impact of the Token Money system on the purchase behavior is left unexplored. This study aims to address this gap and understand how the Token Money system affects the purchase and spending propensities in both mobile games and e-commerce. The study also tries to identify if there is a correlation between purchase behaviors in mobile games and e-commerce due to the Token Money system.

2 Literature Review

In mobile games, monetization is done through one of three different models—free, paid or freemium. Our interest area lies in freemium models since they account for the highest grossing mobile games of all time in both Google Play Store and iOS App Store. In freemium model, the in-app purchases can occur via direct purchase or through the Token Money system. Prior research has been extensive in trying to understand the freemium model. For instance, it has been found out that for mobile social games, behavior Intention is positively impacted by perceived ease of use and perceived usefulness [9]. How social dynamics, in combination with user past performance, affects users' purchase propensity in freemium social games has also been explored [10]. The chapter consolidates several factors under the two main constructs: social dynamics and user past performance. The literature review on understanding the pricing of virtual goods and virtual currencies in free-to-play (F2P) mobile games found out that revenue generation is primarily dependent on game's design features and the game's challenge levels [11]. It has also been found out that there exists a relation between personality traits and mobile app purchasing tendencies [12]. Personality traits such as conscientiousness, emotional Instability, etc., are investigated to understand whether they affect in-app purchases. Moving on, the mobile apps as a revenue-generating business model, with a specific focus on the different app business, yielded how variables such as the App Store, Category, Usage and Consumer Attitude can affect the revenue of the different models [13].

To address the previously mentioned gap, we'll be looking at propensities in in-app purchases (for mobile games) and from website (for e-commerce) from the

Token system's perspective. Hence, the dependent variables will be set as purchase propensity and spending propensity so that both the purchase and spending of Token Money will be looked at separately, just as in the actual process of purchasing. The independent variables will include the Token system characteristics such as Token Attractiveness and Reward Attractiveness in addition to variables related to user's perception such as platform trust and perceived ease of use. To further strengthen the research, we'll also be looking at how individual characteristics and self-efficacy impact the above-mentioned propensities.

Research Question 1—How does the perception of Token Money system's characteristics affect the purchase and spending propensities in mobile games?

Research Question 2—How does the perception of Token Money system's characteristics affect the purchase and spending propensities in e-commerce?

3 Conceptual Model

See Fig. 1.

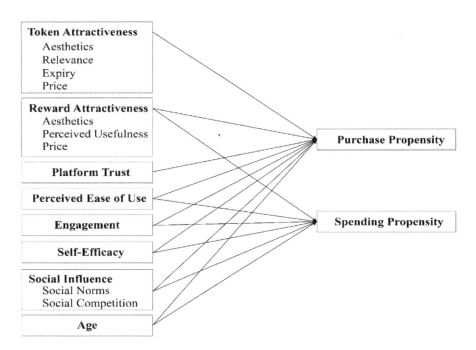

Fig. 1 Conceptual model

4 Hypotheses

H1 Higher the Token Attractiveness, higher the purchase propensity for Token Money.
H2 Higher the Reward Attractiveness, higher the purchase propensity for Token Money.
H3 Higher the Platform Trust, higher the purchase propensity for Token Money.
H4 Higher the perceived ease of use, higher the purchase propensity for Token Money.
H5 Higher the Engagement, higher the purchase propensity for Token Money.
H6 Higher the Self-Efficacy, lower the purchase propensity for Token Money.
H7 Higher the Social Influence, higher the purchase propensity for Token Money.
H8 Higher the Age, lower the purchase propensity for Token Money.
H9 Higher the Reward Attractiveness, higher the spending propensity for Token Money.
H10 Higher the perceived ease of use, higher the spending propensity for Token Money.
H11 Higher the Engagement, higher the spending propensity for Token Money.
H12 Higher the Self-Efficacy, lower the spending propensity for Token Money.
H13 Higher the Social Influence, higher the spending propensity for Token Money.
H14 Higher the Age, lower the spending propensity for Token Money.

5 Methodology

5.1 Data Description

A primary research was conducted with a structured online questionnaire and was distributed to people of 10+ years of age across tier 1 and tier 2 cities in India. Non-probabilistic sampling technique of purposive sampling was used. The total number of respondents was 212. Among the respondents, 176 were mobile game players and 182 were e-commerce users. The questionnaire initially focused on the respondents' familiarity with mobile games and asked to select the game they were most familiar with from a pre-populated list. Based on the game they selected, the respondents were shown relevant information on the game such as Tokens, Rewards, Pricing, etc. based on which they would answer the latter questions. Following that, the respondents would go through similar section but related to e-commerce. Table 1 shows the list of the most important constructs and scale items measured in this study.

With 176 overall usable responses, analysis was done for both mobile games and e-commerce. Odds ratio logistic regression tests were performed on the data to obtain the results.

Table 1 Constructs and sources of scales

Scale item	Adapted from
Token aesthetics, reward aesthetics	J. Blijlevens, Paul Hekkert et al. (2017)
Perceived usefulness	Fred D. Davis et al. (1989)
Platform trust	Schlosser, White and Lloyd (2006)
Perceived ease of use	Dabholkar (1994)
Engagement	Shiri D. Vievk, Sharon E. Beatty, Vivek Dalela and Robert M Morgan (2014)
Self-efficacy	Jan Smith, Benjamin Gardner, Susan Michie (2010)
Social influence	Stibe, A., and Cugelman, B. (2019)

6 Empirical Results and Discussion

Reported below are three separate odds ratio logistic regression models which together span the range of purchase and spending propensities in both mobile games (Table 2) and e-commerce (Table 3).

The results clearly show that for mobile games, Token's Relevance, Expiry and Pricing have significant impact on purchase propensity for Tokens. For both purchase and spending propensities, Reward's Aesthetics is positively significant. For e-commerce, Token's Relevance and Self-Efficacy have significant and positive

Table 2 Log odds ratio for mobile games

Independent variables	Purchase propensity	Spending propensity
Token aesthetics	1.059615	–
Token relevance	0.7150943**	–
Token expiry	0.7095511*	–
Token price	1.305111*	–
Reward aesthetics	1.160775***	1.086476**
Reward perceived usefulness	0.9719406	0.9641973
Reward price	1.046963	1.056962
Social norms	1.097016**	1.061248
Social competition	0.9881974	0.9792789
Self-efficacy	0.9992233	1.017087
Perceived ease of use (Token)	1.08203	1.002429
Perceived ease of use (Reward)	1.002102	0.9878315
Platform trust	0.9892534	3.326659
Youth (Age)	4.852812***	0.9220814***

Table 3 Log odds ratio for e-commerce

Independent variables	Purchase propensity	Spending propensity
Token aesthetics	0.740876	–
Token relevance	1.615832*	–
Token expiry	1.146533	–
Token price	0.7551428	–
Reward aesthetics	1.233875	1.067432
Reward perceived usefulness	1.597118	1.357131
Reward price	0.854866	0.9545191
Platform trust	1.764175	1.479295
Ease of use	0.8392007	0.654265
Variety of rewards	0.8503988	1.205951
Cash refund	0.6757828**	–
Self-efficacy	1.059027**	1.045113
Social norms	1.177381***	1.172509***
Social competition	0.9909505	0.9804468
Purchase propensity (mobile games)	1.422039	0.940414
Youth (Age)	2.282493**	2.633931***

Note $^*p < 0.1$; $^{**}p < 0.05$; $^{***}p < 0.01$ *for Tables 2 and 3*

impact on purchase propensity, whereas Cash Refund (ability to convert Token back to cash with a penalty) has a significant negative impact. Hence, it becomes clear that looking from Token system perspective can contribute to the current understanding of purchase behavior in mobile games and e-commerce. In both models, the effect of Social Norms is positive, signifying that users are often influenced by communities/fellow users into adopting the Token system. Moreover, across the board, Youth is positively significant meaning the younger the User, higher the chance they will adopt into the Token system.

An interesting result is that, breaking the "magic circle" of the game, thereby disrupting the immersive feeling experienced by the user by introducing reminders of real money seems to negatively impact purchase behavior. This is reflected by the negative significance of Token's Relevance on propensities in mobile games, whereas it has positive impact in e-commerce where an immersive feeling does not exist.

7 Implications for Theory and Practice

This study contributes to both the theory of understanding in-app purchases and the practice of increasing revenue generation via the Token Money system. As per the results, if freemium games are able to successfully implement and support communities, this shall greatly aid in the revenue generation using the Token Money system. For e-commerce, it is the relevant of Tokens as well as the social norms that mostly influence purchase of Tokens. Interestingly enough, the behavior with purchasing Token Money in mobile games has no impact on the purchase behavior of Token Money in e-commerce. With the increasing popularity of this system, combined with the underdeveloped body of law with respect to regulation of virtual currency, study of this system becomes imperative as the world moves forward into tokens and cryptocurrencies [14].

8 Limitations, Future Research and Conclusion

A limitation of this study is that the sample was limited to Indian users. Thus, future studies could extend the scope of this research geographically. An experimental study could be worth conducting to further explore the nuances of an effective Token Money system in other areas. This study, however, provides empirical proof that for both mobile games and e-commerce, it is worth adopting the perspective of Token Money system in order to better understand purchase behavior and revenue generation techniques.

References

1. Soh, J.O., Tan, B.C.: Mobile gaming. Commun. ACM **51**(3), 35–39 (2008)
2. Hanner, N., Zarnekow, R.: Purchasing behavior in free to play games: Concepts and empirical validation. In: 48th Hawaii International Conference on System Sciences, pp. 3326–3335. IEEE (2015)
3. Yamakami, T.: Revenue-generation pattern analysis of mobile social games in Japan. In: 14th International Conference on Advanced Communication Technology (ICACT), pp. 1232–1236. IEEE (2012)
4. Briere, M., Oosterlinck, K., Szafarz, A.: Virtual currency, tangible return: portfolio diversification with bitcoin. J. Asset Manage. **16**(6), 365–373 (2015)
5. Park, J.W., Yoo, C.S., Yang, S.B.: Analysis of users' utility on the virtual currency in mobile games: focusing on s mobile game. J. Inf. Syst. **27**(3), 141–160 (2018)
6. Alomari, K.M., Soomro, T.R., Shaalan, K.: Mobile gaming trends and revenue models. In: International Conference on Industrial, Engineering and Other Applications of Applied Intelligent Systems, pp. 671–683. Springer, Cham (2016)
7. Lee, J., Shin, D.H.: Positive side effects of in-app reward advertising: free items boost sales: a focus on sampling effects. J. Advert. Res. **57**(3), 272–282 (2017)
8. Shin, D.H.: Understanding purchasing behaviors in a virtual economy: consumer behavior involving virtual currency in Web 2.0 communities. Interact. Comput. **20**(4–5), 433–446 (2008)

9. Chen, H., Rong, W., Ma, X., Qu, Y., and Xiong, Z.: An extended technology acceptance model for mobile social gaming service popularity analysis. Mob. Inf. Syst. (2017)
10. Shi, S.W., Xia, M., Huang, Y.: From minnows to whales: an empirical study of purchase behavior in freemium social games. Int. J. Electron. Commerce **20**(2), 177–207 (2015)
11. Civelek, I., Liu, Y., Marston, S.R.: Design of free-to-play mobile games for the competitive marketplace. Int. J. Electron. Commerce **22**(2), 258–288 (2018)
12. Dinsmore, J.B., Swani, K., Dugan, R.G.: To "free" or not to "free": trait predictors of mobile app purchasing tendencies. Psychol. Market. **34**(2), 227–244 (2017)
13. Tang, A.K.: Mobile app monetization: app business models in the digital era. Int. J. Innov. Manage. Technol. **7**(5), 224 (2016)
14. Tu, K.V., Meredith, M.W.: Rethinking virtual currency regulation in the Bitcoin age. Wash. L. Rev. **90**, 271 (2015)

Data Augmentation for Handwritten Character Recognition of MODI Script Using Deep Learning Method

Solley Joseph and Jossy George

Abstract Deep learning-based methods such as convolutional neural networks are extensively used for various pattern recognition tasks. To successfully carry out these tasks, a large amount of training data is required. The scarcity of a large number of handwritten images is a major problem in handwritten character recognition; this problem can be tackled using data augmentation techniques. In this paper, we have proposed a convolutional neural network-based character recognition method for MODI script in which the data set is subjected to augmentation. The MODI script was an official script used to write Marathi, until 1950, the script is no more used as an official script. The preparation of a large number of handwritten characters is a tedious and time-consuming task. Data augmentation is very useful in such situations. Our study uses different types of augmentation techniques, such as on-the-fly (real-time) augmentation and off-line method (data set expansion method or traditional method). A performance comparison between these methods is also performed.

1 Introduction

Character recognition—a branch of pattern recognition, is one of the most active yet challenging fields. The handwritten character recognition process is more difficult compared to the recognition of printed ones, mainly due to variations in writing styles. The focus of our study is character recognition of MODI script. This script was used to write Marathi until 1950. It has been observed that very little work is done toward MODI script optical character recognition (OCR). Research and development are necessary to extract the information from MODI manuscripts, which are stored in various parts of the country and abroad. MODI script was developed in

---

S. Joseph (✉) · J. George
CHRIST (Deemed to be University), Bangalore, India
e-mail: solley.joseph@res.christuniversity.in

S. Joseph
Carmel College of Arts, Science and Commerce for Women, Goa, India

© The Editor(s) (if applicable) and The Author(s), under exclusive license to Springer Nature Singapore Pte Ltd. 2021
T. Senjyu et al. (eds.), *Information and Communication Technology for Intelligent Systems*, Smart Innovation, Systems and Technologies 196, https://doi.org/10.1007/978-981-15-7062-9_51

Devagiri in the twelfth century. It is observed that various libraries and temples in India still have a large collection of MODI script [1]. The focus of our study is the augmentation techniques for deep learning-based character recognition method. We have implemented the convolutional neural network (CNN) along with two different data augmentation techniques that classify and recognize the characters of ancient MODI script. The augmentation techniques used are on-the-fly (real-time) augmentation and off-line method (data set expansion method or traditional method). The performances of the two methods are then compared.

The basic character set of MODI script consists of 46 distinctive letters, of which 36 are consonants, and 10 are vowels. MODI was easy to write and was commonly used to write Marathi until 1950. The usage of the script was reduced when printing technology was introduced in India due to the difficulty in type-setting the script. Devanagari is now the official script used to write Marathi, MODI script is no longer widely used [2].

MODI character recognition is a very complex task due to variations in the writing style of individuals, shape similarity of characters and the absence of word stopping symbol in documents.

2 Literature Review

A review of the literature reveals that compared to the rest of the Indian languages, very little research work has been done toward MODI character recognition [1]. Sadanand et al. have implemented Zernike and Zernike's complex moments in combination with the Euclidian distance classifier for MODI script recognition and an accuracy of 94.78% was achieved [3]. Otsu's binarization method and Kohonen neural network classifier were used for MODI character recognition and an overall recognition rate of 72.6% was reported [4]. A two-layer feed-forward neural network and SVM were used for MODI character recognition and a 73.5% accuracy was reported [5].

It was observed that in recent years deep learning-based algorithms were effectively used in the character recognition of various scripts. Convolutional neural network (CNN) is one of the best performing deep learning models, which is implemented for character recognition by various researchers [6]. Arabic character recognition using CNN implemented by Najadat et al. reported an accuracy of 94.9% [7]. CNN architecture was experimented on Bangla character, by Keserwani et al. and an average accuracy of 99.76% was achieved [8]. Shobha et al. used CNN on deformed Kannada script and an accuracy of 92% was achieved [9]. Implementation of CNN for the recognition Hindi digit by Kumar et al. obtained an accuracy of 98.85% [10]. CNN-based handwritten digits recognition was performed by Hossain and Ali, and an accuracy of 99.15% was achieved [11]. CNN has also been used for Oriya numeral recognition and Telugu numeral recognition by Maitra et al., and the accuracy achieved was 97.2% and 96.55%, respectively [12]. An support vector machine (SVM)-based method is used in combination with restricted Boltzmann machine (RBM) for the recognition of handwritten multilingual numeral data set [13].

Deep learning-based algorithms are effectively used in various pattern recognition tasks, including character recognition. Convolutional neural network (CNN) is effectively implemented for character recognition and is one of the best performing deep learning models [6]. For Bangla character recognition, the method of feature fusion followed by spatial pyramid pooling has been used and had an accuracy of more than 99% [8]. CNN is implemented for deformed Kannada script recognition and achieved an accuracy of 92% [9]. For Hindi digits, the recognition accuracy of 98.85% has been obtained by using CNN with the help of the numerical database [10]. CNN-based method was implemented by Hossain and Ali for handwritten digit recognition using feature extraction and classification with an accuracy of 99.15% [11]. CNN has also been used for character recognition in Oriya numerical character recognition and Telugu numerical character recognition with an accuracy of 97.2% and 96.55%, respectively [12]. An SVM-based method is used in combination with restricted Boltzmann machine (RBM) for the recognition of handwritten multilingual numeral data set [13]. RBM is used for dimensionality reduction and was observed that complexity reduced using this method.

3 Methodology

The methodology proposed for the character recognition of MODI script involves the use of CNN. Data augmentation is performed on the data set before training the network. The step by step procedure of the methodology is shown in Table 1.

Table 1 Steps of the proposed method

Steps of the proposed method are as follows :
Step 1. Input the original image
Step 2. Apply Image Augmentation using :
(a)   on-the-fly augmentation (method1)
(b)   off-line method (method 2)
Step 3. Train the network using CNN
(a)   Apply a Convolutional filter on the input image with ReLU activation function.
(b)   Apply Max Pooling for down-sampling.
(c)   Repeat steps (a) and (b) with a varying number of filters counts and kernel sizes.
(d)   Apply a fully connected layer (two hidden layers) on the extracted features.
(e)   The output layer is constructed with 46 nodes to get the results for 46 characters.

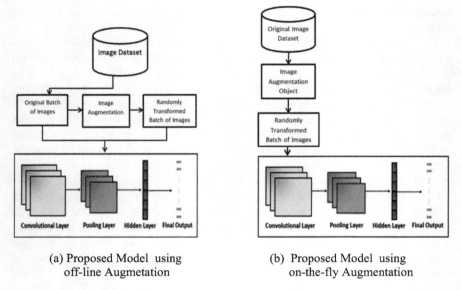

(a) Proposed Model using off-line Augmetation

(b) Proposed Model using on-the-fly Augmentation

Fig. 1 **a** Proposed model using off-line augmentation, **b** proposed model using on-the-fly augmentation

The original image (60 * 60 pixels) is given as input and the data augmentation method will be applied to it. In this experiment, we are using two different methods of data augmentations: off-line method as well as on-the-fly augmentation. Detail descriptions of both methods are given in the next subsection. After the data augmentation, the data is passed through the convolutional neural network for the recognition of characters, as shown in Fig. 1.

3.1 Data Augmentation

Data augmentation can be used in a situation where there is data scarcity. Artificially creating a new set of data from the existing data set is termed as data augmentation. Deep learning models generally require large amounts of data for accurate predictions and augmentation in useful for such cases. Several augmentation techniques include rotation, flipping, scaling, adding salt and pepper noise to the data set and many more. Data augmentation can be applied in different ways. There are two types of data augmentation methods that are commonly used while applying deep learning. In the first type, which is called the off-line method (also called as expansion method or traditional method), the data is augmented by transforming existing images into a new set of images by expanding an existing dataset. The problem with this type is that the model will not be able to generalize to the unseen data while encountering a small set of data for training. The second method is on-the-fly/real-time data augmentation methods. Instead of generating new data, the augmentation process will transform the existing data and the new set of data will be generated. Thus, the network sees new variations of the data at each and every epoch during the training of the data. In addition, this method also saves disk space. In this study, we are using both methods to compare the performance of the on-the-fly method for data augmentation.

3.2 Architectural Details of the Proposed Method

The architecture detail of the proposed CNN method is shown in Fig. 2. The classification model has ten convolutional layers, five pooling layers (2 * 2 sub-matrices) and three dense layers.

This architecture takes a greyscale image as input (60 * 60 pixel sizes). The input image is passed through the first convolutional layer (32 filters of window size 2 ×

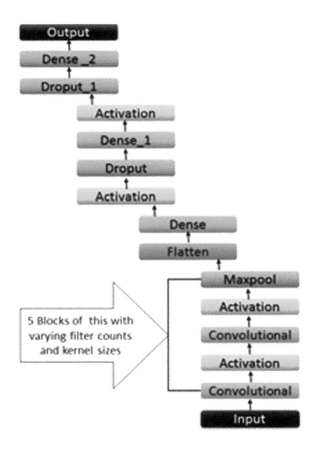

Fig. 2 Architecture of proposed CNN

2). The activation function used is ReLU. A pooling layer is subsequently used for down-sampling. The pooling layer has 2 * 2 sub-matrices (max pooling is used in this case). As a result, the image volume is reduced and filter samples are increased. The convolution and max pooling operations are repeated serially with varying numbers of filter counts. A flatten layer is added to convert the pooled feature map to a single column that is passed to the fully connected layers. The features are then passed through two dense layers of 256 nodes each. The ReLU activation function is used for the dense layers. Two dropout functions are added between the two hidden layers. The output layer is constructed with 46 neurons for the classification of 46 classes in the MODI character set. The sigmoid activation function is used in the output layer.

4 Experimental Study and Results

The experiment is conducted in the Python environment and the data set used is a basic MODI script character set. The MODI script has 46 characters: 36 are consonants and ten vowels. The initial data set used in this experiment is 4600 MODI characters. The augmentation is then performed on the initial data set. Off-line method of augmentation is used in the first experiment and the on-the-fly augmentation is used in the second experiment. Handwritten characters written by different persons were used for the study.

In the first experiment, the data augmentation method used was the off-line method. The final data set was created using rotation (20° and 40°, and also by adding Gaussian noise and salt and pepper noise). Thus, the data set was increased to 23,030 (the original set of data in addition to the data generated using five augmentation methods). The data set was divided 70:30 ratio (16,100 training samples and 6930 test samples).

In the second experiment, data augmentation is performed using ImageDataGenerator function (of Keras), which performs on-the-fly data augmentation. All the original images are just transformed at every epoch and then used for training; this way, the learned model tends to become more robust and accurate as it is trained on different variations of the same image.

The augmented data is then subjected to training using CNN. The proposed model has ten convolutional layers (as described in Sect. 3.2), so that more features can be extracted, which leads to better accuracy. The accuracy is predicted with the help of the recognition rate and error rate.

The recognition accuracy of both the experiments and the performance comparison is shown in Table 2. The first experiment (CNN with on-the-fly augmentation method) achieved an accuracy of 99.78%. The accuracy achieved in the second experiment (CNN with off-line augmentation technique) is 99.47%.

Thus, it is observed that the accuracy of the on-the-fly method is better. In addition to that the on-the-fly method also saves disk space as in this case, there is no need to store a very large amount of off-line augmented data. Thus, the overall performance of the on-the-fly method is better compared to the off-line method.

Table 2 Performance comparison of the two augmentation techniques

S. No.	Augmentation method	Classification method	Data set	Accuracy (%)
1	On-the-fly augmentation	CNN	MODI character set of 46 characters	99.78
2	Off-line augmentation	CNN	MODI character set of 46 characters	99.47

5 Conclusion

Handwritten character recognition for MODI script is a promising research area as very less work is done in this field. Deep learning methods are extensively used in character recognition as it gives better performance. In this study, we compared the performance of two augmentation techniques in CNN-based character recognition of MODI script. The CNN architecture for both methods is the same. The method with the on-the-fly data augmentation technique achieved better accuracy compared to the off-line method of augmentation. The on-the-fly data augmentation method equips the network to see a new set of data each time, and thus the efficiency of the system is increased.

Segmentation is one of the difficult tasks in MODI handwritten character recognition and in the future, we intend to work on the segmentation of handwritten MODI manuscripts.

References

1. Joseph, S., George, J.: Feature extraction and classification techniques of MODI script character recognition. Pertanika J. Sci. Technol. **27**(4), 1649–1669 (2019)
2. Joseph, S., George, J.P., Gaikwad, S.: Character recognition of MODI script using distance classifier algorithms. In: Fong, S., Dey, N., Joshi, A. (eds.) ICT Analysis Application Lecture Notes Networks Systems, vol. 93. Springer, Singapore (2020)
3. Sadanand, K., Borde, P.L., Ramesh, M., Pravin, Y.: Off-line MODI character recognition using complex moments. Procedia Comput. Sci. **58**, 516–523 (2015)
4. Anam, S.: An approach for recognizing Modi Lipi using Otsu's Binarization algorithm and kohonen neural network. Int. J. Comput. Appl. **111**(2), 28–34 (2015)
5. Besekar, D.N.: Special approach for recognition of handwritten MODI script' s vowels. Int. J. Comput. Appl. 48–52 (2012)
6. Cires, D.C., Meier, U., Gambardella, L.M.: Convolutional neural network committees for handwritten character classification. Int. Conf. Doc. Anal. Recognit. **10**, 1135–1139 (2011)
7. Najadat, H.M., Alshboul, A.A., Alabed, A.F.: Arabic handwritten characters recognition using convolutional neural network. In: 2019 10th International Conference on Information Communication Systems ICICS 2019, pp. 147–151, Jan 2019
8. Keserwani, P., Ali, T., Roy, P.P.: Handwritten Bangla character and numeral recognition using convolutional neural network for low-memory GPU. Int. J. Mach. Learn. Cybern. **10**(12), 3485–3497 (2019)
9. Shobha Rani, N., Chandan, N., Sajan Jain, A., Kiran, H.R.: Deformed character recognition using convolutional neural networks. Int. J. Eng. Technol. **7**(3), 1599–1604 (2018)

10. Kumar Reddy, R.V., Srinivasa Rao, B., Raju, K.P.: Handwritten Hindi digits recognition using convolutional neural network with RMSprop optimization. In: Proceedings of 2nd International Conference on Intelligent Computing Control Systems ICICCS, pp. 45–51 (2019)
11. Hossain, M.A., Ali, M.M.: Recognition of handwritten digit using convolutional neural network (CNN). Glob. J. Comput. Sci. Technol. **19**(2), 27–33 (2019)
12. Sen Maitra, D., Bhattacharya, U., Parui, S.K.: CNN based common approach to handwritten character recognition of multiple scripts. In: Proceedings of International Conference on Document Analysis Recognition, ICDAR, pp. 1021–1025, Nov 2015
13. Solley, T.: A study of representation learning for handwritten numeral recognition of multilingual data set. In: Lecture Notes Networks Systems, vol. 10, pp. 475–482. Springer (2018)

Improved Automatic Speaker Verification System Using Deep Learning

Saumya Borwankar, Shrey Bhatnagar, Yash Jha, Shraddha Pandey, and Khushi Jain

Abstract In this paper we have implemented and designed an accurate and robust automatic speaker verification (ASV) system. There is a constant need for improving the performance of the ASV because of its many advantages over other biometric means. In this work we have experimented with the VoxCeleb dataset. The VoxCeleb Dataset which has more than 1200 individuals and around 300–500 utterances each. We have proposed a new approach to authenticate users at a text independent level. Firstly the audio files are converted to spectrograms which are pre-processed and then are classified using Convolutional Neural Networks (CNN) and a model is created at a text independent level. The classification of the individual speakers are made on the bases of Spectrogram. Finally, the performance evaluation is done using the training and validation accuracy plots.

1 Introduction

Today there is a need for a more robust security system for the verification of an individual based on their bio-metric characteristics in places like airport, banks, financial transactions, etc.

S. Borwankar (✉) · S. Bhatnagar · Y. Jha · S. Pandey · K. Jain
Institute of Technology, Nirma University, Ahmedabad, Gujarat, India
e-mail: 17bec095@nirmauni.ac.in

S. Bhatnagar
e-mail: 17bec109@nirmauni.ac.in

Y. Jha
e-mail: 17bec128@nirmauni.ac.in

S. Pandey
e-mail: 17bec108@nirmauni.ac.in

K. Jain
e-mail: 17bec038@nirmauni.ac.in

© The Editor(s) (if applicable) and The Author(s), under exclusive license to Springer Nature Singapore Pte Ltd. 2021
T. Senjyu et al. (eds.), *Information and Communication Technology for Intelligent Systems*, Smart Innovation, Systems and Technologies 196,
https://doi.org/10.1007/978-981-15-7062-9_52

A biometric authentication system needs to compare a sample of the newly captured individual with the already existing individuals in its database. Biometric authentication can involve many biological features of an individual like their fingerprints, retina, palm print, etc. During the formation of the database samples are captured which are processed and stored for further comparison. This system can be used for either recognition (where the sample is scanned throughout the database) or verification (where the sample is used to authenticate the individual's identity) [1, 6]. Our approach works with the verification of the user.

Human speech has been considered as a 'classic' biometric, from more than four decades [16]. Speaker verification system extract speech features from the audio signal of an individual which represents their physical anatomy of the vocal cords, larynx etc * and their style of speaking [1].

Speaker verification deals with both speech recognition and speech identification. Speaker verification system decides if an individual is whom he/she claims to be. Speaker identification system decides if an individual is a specific individual or part of a group. In speaker verification, the individual firstly claims to an identity, if the system is a text dependent system then the individual is told to speak the known phrase while in text independent system the individual is free to speak any phrase and lastly the phrase spoken by the individual is used to authenticate his identity.

There are 2 types of speaker verification first being Text-Dependent ASV system and second Text-Independent ASV system.

1.1 Text-Dependent ASV System

Text Dependent ASV system works on authenticating a user based on a sample of the users voice on a known phrase. The ASV system validates the user based on the sample voice by comparing features of the sample with the previously built model of the users voice. In [12] the author has worked with Hidden Markov Model (HMM) to build a text-dependent ASV system. The author describes the system with 5 steps, first training world model on available known phrases of the dataset, secondly training client model with the client phrases from the dataset, then viterbi alignment is applied on each observation, then the normalization based on acoustic score is calculated and lastly the performance evaluation is done with the help of False Rejection Rate(FRR) and False Acceptance Rate (FAR).

1.2 Text-Independent ASV System

This paper deals with the designing of text-independent ASV system with the help of deep learning techniques. Text-independent ASV system is also a system used to validate the user, which works at a text independent level and there are no known phrases as in [12]. The user is free to say anything to the system and his voice will be validated with the help of a customized model. The designing of our model has 4 parts.

1. Data Pre-processing: The audio in the dataset needs to be processed before further computation can be carried out on the audio sample. So the audio is sliced in equal parts and is passed from a pre-emphasis filter. After which the audio can be segmented.
2. Dividing Data: The audio files are then split into different classes, after which training and testing folders are created based on split size.
3. Spectrogram: The audio files are then transformed into spectrogram to analyze the different frequencies present in the spectrogram.
4. Classifier: Then the spectrograms are passed to a classifier which classifies them and the output layers accepts or rejects the user based on the weights calculated.

1.3 Spectrogram

A Spectrogram is a representation of the range of frequencies present in a signal with respect to time, with this one can also see the energy variation with time. These graphs are 2D with the third dimension being the colour. Time runs from left to right, frequency increases in the vertical axis and the amplitude is represented in colour (dark blue colour being low amplitude and bright colours being high amplitudes). The 3D representation is often called a waterfall. Our proposed approach uses spectrogram as a feature to differentiate between speakers (Fig. 1).

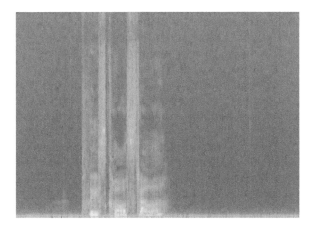

Fig. 1 Spectrogram

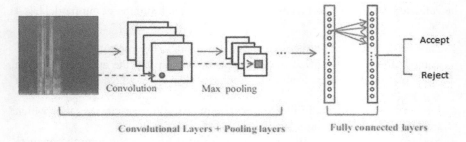

Fig. 2 CNN pipeline

1.4 Convolution Neural Network (CNN) in Speaker Verification

A convolutional neural network (CNN), is a deep learning algorithm that takes in an input image, calculates certain special features of the image and assigns importance to various features in the image so they can be differentiated from one another. Here in our case we have used CNN to classify among different individuals. The spectrogram are images and the classification of images can be done with the help of CNN. CNN has been successful in numerous studies [3, 4] to classify images (Fig. 2).

2 Relevant Work

Human voice can be used as an identity of a speaker. When the identity of a speaker is validated using his speech its called speaker recognition [1, 7].

The authors of [12] discuss a text-dependent Hidden Markov Model (HMM) based speech recognition. Their approach deals with the verification with the help of difference between log-likelihood which is derived using the viterbi algorithm. And the performance evaluation is done with the help of FAR and FRR. Which are standard performance analysis curves which help us judge the accuracy of our system.

In [10] the authors have worked with Support Vector Machines (SVM) to help build their speaker verification system, they have also discussed a hybrid model of SVM-GMM with the help of Kullback-Leibler kernel which makes decision based on the mean value of mixtures. They have also discussed about the effects of speech coding on speech recognition. The results were calculated in terms of Equal Error Rate (ERR) and they have discussed about how the ERR of 1.1% stays the same for SVM and GMM based techinques for the uncoded 16 Hz-sampled speech.

Recent development in the field of deep neural network (DNN) has been discussed in [19] wherein they battle environmental noise and background noise they have used Convolutional Recurrent Network (CRN) for speech separation which helps them with the speaker verification system so their system works better in a noisy environments. They have worked with the audio sample from 797 females and they have used the open sourced framework for Artificial intelligence (AI) called Pytorch [13]. Their proposed method gets an EER of 7.53 for an Signal to Noise Ratio (SNR) of −5 dB, 5.41 for 5 dB, 5.21 for 10 dB and an average EER of 6.04.

There are various research going on for development and increasing performance of Automatic speaker verification systems. ASV is an important biometric system with many advantages compared to other biometrics. Usually ASV are based on statistical models such as HMM, GMM, SVN and ANN.

This article showcases a system developed using Gaussian mixture model (GMM) [2, 16]. In [15] people have mainly focused on cepstral coefficients, Cheang 800 Yee and Abdul Manan Ahmad [18] proposed a study for ASV system in malay language using Artificial neural network (ANN) as classifier for MFCC, they have also used a sampled approach for ANN, their experiment shows highest accuracy using MLP, ANN showed better performance and used less data for training. Their ANN implementation gets an accuracy of 96% for a third run a3 layers, 97% on the third run with 9 layers and 98% with 3 runs with 12 layers.

Naoki Tadokoro, Tetsuo Kosaka, Masaharu Kato and Masaki Kohda [17] proposed an approach based on anchor model ASV with phonetic modeling. To improve performance phonetic modeling approach is preferred over GMM also to reduce speaker space dimension KL transform was proposed. Their paper also proposed the technique of anchor model speaker verification with modeling of phonetic features. They were able to get an EER of 2.50% with 762 speaker and 3048 utterances.

3 Dataset Details

The VoxCeleb dataset used in our experiments is a large scale public speaker database by the University of Oxford, UK[vox ki ref]. The whole dataset comprises of 2 parts first being VoxCeleb 1 and the other being VoxCeleb 2. In this paper we have experimented with the VoxCeleb 1 Dataset. It contains 1211 speakers with around 300 to 500 utterances for each speaker which comes to a total of 148,642 utterances in total. The dataset has utterances of minimum of 5 seconds and maximum upto 15 seconds. The total information about the VoxCeleb is given in Table 1. Next, we divide the dataset into training, testing and validation datasets, the training dataset consists of 70% of the total images, testing contains 20% of the total images and validation dataset contains 10% of the total images. Table 2 gives us a description about the training, testing and validation datasets.

Table 1 VoxCeleb dataset details

S. No.	Value
Number of speakers	1211
Number of utterances	148,642

Table 2 Split dataset details

S. No.	Value
Training set	118,913
Testing set	14,864
Validation set	14,865

4 Implementation Scenario

The implementation of the proposed algorithm start with converting of the training set, testing set and validation set samples to spectrograms with the help of Librosa library of python. The conversion of audio to spectrogram reveals more information about the audio signal. After the conversion the images are downscaled to 256×256 to reduce the computation and time complexity, next the rescaling of the images takes place so as to set the range of the pixel value between 0 and 1. This classification problem is a multi-class classification, the input being an RGB image and the output being acceptance or rejection. For the classification we have used ResNet architecture, the ResNet architecture [9] has been used after experimenting with various architectures, the problem of very deep neural networks has been solved with the introduction of a residual block (Fig. 3).

The vanishing gradient problem was solved by the ResNet which is caused when the backpropagation takes place and the multiplication of a number between 0 and 1 causes the initial layers to disappear. Identity shortcut connection skips the training of one or more layers and creates a residual block. The identity mapping that ResNet introduces is just present to add the output from the previous layer to the next layer. This helps the architecture train very deep networks which was not possible

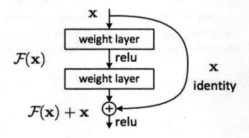

Fig. 3 Residual block [9]

before. The architecture outperforms ALexNet, VGG-16, Xception as the results with ResNet was consistent and most accurate with a specific database.

The architecture that has been used is a compact version of the ResNet. The first layer is a convolution layer with 64 filters followed by a batch normalization layer, then a stack of 3 residual blocks, with every layer in the residual block having 32, 32, and 128 filters respectively, basically every residual layer contains 1×1 convolutional layers followed by 3×3 convolutional layer with the third being another set of 1×1 convolutional layer also batch normalization and activation applied after every convolutional layer, with dimension reduction applied. After which 4 other sets of residual blocks, where the 3 convolution layers will have 64,64 and 256 filters respectively, with dimension reduction applied. Finally 6 sets of residual blocks where each convolution layer will have 128,128, and 512 filters respectively, with dimension reduction applied for the last time. Then an average pooling layer is applied with the last layer being a softmax classifier. The optimizer that is used is the stochastic gradient descent with a learning rate of 0.1 and momentum at 0.9. At the time of compilation of the model the loss function is set to categorical cross entropy as there are 1211 classes for classification, and the system parameters are updated after each epoch. The model has been trained for 35 epochs, with a batch size of 64.

5 Experiment Results

For evaluation of our proposed method we have calculated the accuracy and compared it with another algorithm based on ANN and applied it on the same dataset. The ANN architecture proposed in [18] has been proved to be effective, so the same algorithm has been used to calculate accuracy on the VoxCeleb dataset. Table 3 contains the results of the proposed method and the previous methods. The training and validation accuracy with training and validation loss has also be plotted in Figs. 4 and 5 respectively.

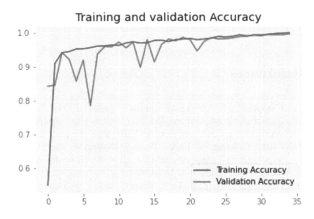

Fig. 4 Training and validation accuracy

Fig. 5 Training and validation loss

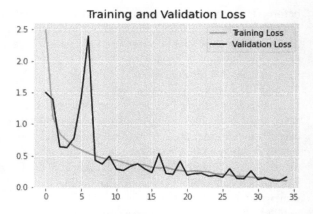

Table 3 Experimental results

Approach	Dataset	Accuracy (%)
Artificial Neural Network (ANN) [18]	VoxCeleb	95.0
Our approach	VoxCeleb	99.7

6 Conclusion

This paper has performed classification of audio using ResNet and has gained better performance than previous methods. The dataset of VoxCeleb has been collected and the evaluation carried out by our proposed method shows us that our proposed method can work with big datasets as well as provide robustness in our ASV system, moreover we can see that our propose method is much better and ANN used in the previous approach. We have developed an end-to-end ASV system that is capable of validating a user based on his voice and we can see that the accuracy of our system is about 99.7% proving to be better than the previous state of art methods like ANN which could only reach about 95%.

References

1. Campbell, J.P.: Speaker recognition: a tutorial. Proc. IEEE **85**(9), 1437–1462 (1997)
2. Campbell, W.M., Sturim, D.E., Reynolds, D.A.: Support vector machines using gmm supervectors for speaker verification. IEEE Sign. Process. Lett. **13**(5), 308–311 (2006)
3. Chen, Y., Jiang, H., Li, C., Jia, X., Ghamisi, P.: Deep feature extraction and classification of hyperspectral images based on convolutional neural networks. IEEE Trans. Geosci. Remote Sens. **54**(10), 6232–6251 (2016)
4. Chollet, F.: Xception: Deep learning with depthwise separable convolutions. In: Proceedings of the IEEE Conference on Computer Vision and Pattern Recognition, pp. 1251–1258 (2017)

5. Chung, J.S., Nagrani, A., Zisserman, A.: Voxceleb2: Deep speaker recognition. arXiv preprint arXiv:1806.05622 (2018)
6. Furui, S.: Recent advances in speaker recognition. Pattern Recogn. Lett. **18**(9), 859–872 (1997)
7. Furui, S.: Fifty years of progress in speech and speaker recognition. J. Acoust. Soc. Am. **116**(4), 2497–2498 (2004)
8. Gu, B., Guo, W.: Gaussian speaker embedding learning for text-independent speaker verification. arXiv preprint arXiv:2001.04585 (2020)
9. He, K., Zhang, X., Ren, S., Sun, J.: Deep residual learning for image recognition. In: Proceedings of the IEEE Conference on Computer Vision and Pattern Recognition, pp. 770–778 (2016)
10. Janicki, A.: SVM-based speaker verification for coded and uncoded speech. In: 2012 Proceedings of the 20th European Signal Processing Conference (EUSIPCO), pp. 26–30. IEEE, New York (2012)
11. Kataria, S., Nidadavolu, P.S., Villalba, J., Dehak, N.: Analysis of deep feature loss based enhancement for speaker verification. arXiv preprint arXiv:2002.00139 (2020)
12. Munteanu, D.P., Toma, S.A.: Automatic speaker verification experiments using HMM. In: 2010 8th International Conference on Communications, pp. 107–110. IEEE, New York (2010)
13. Paszke, A., Gross, S., Chintala, S., Chanan, G., Yang, E., DeVito, Z., Lin, Z., Desmaison, A., Antiga, L., Lerer, A.: Automatic differentiation in Pytorch (2017)
14. Ramoji, S., Krishnan, P., Ganapathy, S.: Nplda: A deep neural PLDA model for speaker verification. arXiv preprint arXiv:2002.03562 (2020)
15. Reynolds, D.A.: Experimental evaluation of features for robust speaker identification. IEEE Trans. Speech Audio Process. **2**(4), 639–643 (1994)
16. Reynolds, D.A., Quatieri, T.F., Dunn, R.B.: Speaker verification using adapted Gaussian mixture models. Digital Signal Process. **10**(1–3), 19–41 (2000)
17. Tadokoro, N., Kosaka, T., Kato, M., Kohda, M.: Improvement of speaker vector-based speaker verification. In: 2009 Fifth International Conference on Information Assurance and Security, vol. 1, pp. 721–724. IEEE, New York (2009)
18. Yee, C.S., Ahmad, A.M.: Malay language text-independent speaker verification using NN-MLP classifier with MFCC. In: 2008 International Conference on Electronic Design, pp. 1–5. IEEE, New York (2008)
19. Zhao, F., Li, H., Zhang, X.: A robust text-independent speaker verification method based on speech separation and deep speaker. In: ICASSP 2019-2019 IEEE International Conference on Acoustics, Speech and Signal Processing (ICASSP), pp. 6101–6105. IEEE, New York (2019)

Detection and Prevention of Attacks on Active Directory Using SIEM

S. Muthuraj, M. Sethumadhavan, P. P. Amritha, and R. Santhya

Abstract Active Directory is widely used in organizations to administer windows user accounts and related IT resources. It acts as centralized management to control windows based network. Attackers are focusing on compromising Active Directory Domain Services in order to take over the whole domain network. In this paper, we have studied about the detection of known attacks targeting on domain services from attacker end using SIEM and hence suggested prevention methods. SIEM's are widely used in many organizations by security analysts to monitor their network using event logs. The detection rules were designed and implemented in Splunk. The evaluations of rules and attacks are performed in a virtual environment. The proposed preventive measures will be able to resist against known attacks on active directory.

1 Introduction

Microsoft Active Directory (AD) is a common place for information about the objects that resides on a company network, such as users, groups, computers, printers, applications, and files. As a part of the windows server, Active Directory Domain Services (AD DS) enables admins to federate active directory user identities throughout windows based networks. In doing so, admins can leverage active directory as their core identity provider and connect users virtually to all of their Windows-based IT resources from one centralized location. Almost every organizations implement

---

S. Muthuraj (✉) · M. Sethumadhavan · P. P. Amritha · R. Santhya
TIFAC-CORE in Cyber Security,
Amrita School of Engineering, Amrita Vishwa Vidyapeetham, Amritapuri, Coimbatore, India
e-mail: muthu.6168@gmail.com

M. Sethumadhavan
e-mail: m_sethu@cb.amrita.edu

P. P. Amritha
e-mail: pp_amritha@cb.amrita.edu

R. Santhya
e-mail: r_santhya@cb.amrita.edu

network security protocols to defend against attacks by using firewalls or antivirus solutions fighting against malware. Still, some attacks are being developed to overcome those security mechanisms. Once the attacker is able to break-in the first line of security and gets their hands-on active directory then the consequences are fatal, which results in full domain compromise.

Active Directory is a place for the implementation of a directory service for Microsoft's Network Operating System (NOS). It is a term used to describe about the network environment in which different types of resources, such as user, group, organization and computer accounts are stored and managed centrally. This centralised management called Active Directory contains, network, application, or NOS information which are maintained by administrators and provided access to end-users. The directory services that provides access to this centralised management is called Active Directory Domain Services. Active Directory uses DNS domain names and NetBIOS domain names. The DNS domain name is also often referred to as a fully qualified domain name (FQDN). These two naming systems also apply to computer names and other objects in the AD. The basic physical component of AD is a domain controller (DC). The DC is a computer that runs a Windows Server OS and holds the AD DS role. The DC stores the ntds.dit database file, which is replicated with other DCs in the domain by using Directory Replication Service (DRS) remote protocol. The domain can have any number of DCs, but one DC can be authoritative for one domain only.

The Windows OS require every user to logon the computer using valid accounts. Such that they can access local and network resources. Authentication is a process of claiming one's identity by verification of rights to access resources. AD is the default management mode to store information and grant access to objects on domain-joined systems and therefore it is tied closely to authentication and authorization processes. Users are authenticated to windows-based computers through different types of logon process. Depending on how the logon process occurs, there are several scenarios defined:

- Interactive logon—A user logged on to this computer. (Local logon, Remote logon).
- Network logon—A user or computer logged on to this computer from the network.
- Smart card logon—A smart card followed by pin to get authorised.
- Biometric logon—Fingerprint.

In an interactive logon, a user typically enters their username and password in the entry dialog box. In case of network login functionality, windows systems include following authentication mechanisms:

- Kerberos version 5 protocol
- Public key certificates
- Secure Sockets Layer/Transport Layer Security (SSL/TLS)
- Digest
- NT LAN Manager (NTLM).

Kerberos allows users to access services on the network openly by simply requesting a service ticket to the Key Distribution Center (KDC). When clients request service tickets for a particular services from a DC, they use identifiers called Service Principal Names (SPNs). SPN is stored in AD in the ServicePrincipalName multi-valued attribute. SPN is constructed in the form of a service identifier, followed by the hostname, and optionally, a port number.

2 Literature Survey

In 2019, Kotlaba, Lukas mentioned about the detection of active directory attacks and how it can be implemented in an virtual environment in [1]. Since active directory is critical part in any organisation, the author have found out some of the attacks that can happen on active directory. We implemented the same environment and proposed preventives measures against active directory attacks.

In 2018, Wataru Matsuda, Mariko Fujimoto, and Takuho Mitsunaga [10] published about Detecting APT attacks against Active Directory. An advanced persistent threat (APT) is a type of persistent attack, where an attacker tries to exploit into an organization network and stay hidden until they accomplish their goals. When active directory is present in the network, then the attacker disguises themselves as a users of a legitimate domain administrator account which has the highest privilege in the AD. Windows system activities are collected using the windows event logs and used for analyzing attacks. If the attacker is using a legitimate account or built-in tool in windows then it is difficult to detect attacks. In this case, machine learning is used along with event logs to detect attacks which gives high precision rate even when the attacker uses legitimate account for attacks.

In 2017, M. Siva Niranjan Rajaa and A. R. Vasudevan discussed about Rule Generation for TCP SYN Flood attack in SIEM Environment [2]. Security Information and Event Management (SIEM) is a software solution that aggregates and analyzes activity from many different resources across your entire IT infrastructure. It helps in collecting event logs from various devices in the network, ordering them into a common event format and analyzing. Some attacks can be detected using a single event id or source. While some attacks require rules and other sets of algorithms for detection. The events collected are correlated and analyzed for changes in system behaviour. They have used RETE algorithm to formulate the rules which are stored in the database. An alert will be triggered when the attack matches the rules. This helps in detecting attacks from heterogeneous devices also.

In 2017, S. Sandeep Sekharan and Kamalanathan Kandasamy explained about profiling SIEM tools and correlation engines for security analytics [3]. In the case of large organizations, a huge amount of data will be generated. Organising and handling of those data are critical in IT. Therefore implementing centralized log management will improve the security and protection of data in an organization. Such IT sectors or any enterprises requires a high profiling tool to improve the level of security. SIEM tools are structured to handle such scenarios by various processes like collecting, analyzing, normalising and correlating the logs. Various SIEM tools and correlation engines are considered for a comparative study on those tools. There are open source SIEM tools available which are compared with popular tools and comparative study is made.

As it is said, in an advanced persistent threat (APT) attack, attackers intrude into an organization network and stay inside until they complete their goals by exploiting some sensitive information from the network [4]. If active directory is in place, then the attackers will try to get domain administrator account because it has the highest privileges to control domain users and files in the domain. The attacker uses several methods to obtain the account. CVE-2014-0317 is a vulnerability in active directory. There are other ways to steal credentials such as using tools like mimikatz to export the password dump and try offline password cracking. Once the attack is successful they create a backdoor to the server for future use. It is difficult to differentiate legitimate access and malicious access. There are several methods in place to detect such attacks but it is useful under specific conditions. In [4] they considered existing methods using a dataset and proposed a new detection algorithm with a precise detection rate.

Paper [5] discusses about protecting Security Account Manager (SAM). Securing information is paramount for the survival of any enterprise. The pillars of securing data are confidentiality, authentication, privacy, and integrity of the data. The most common practice of securing data is by setting up login credentials. There are different methods to ensure the security of login credentials still some attacks like brute-forcing and dictionary attacks lead to cracking the password. Here they enforced a security model to protect the SAM file used by the windows registry. Such that the system will be secured against such intruders.

3 Proposed System

In this paper, we have built and configured a virtual lab environment of a small domain to perform attack scenarios as in Fig. 1. The rules were created and tested on the data coming from this lab environment. Logs from all the monitored assets are sent to the physical machine where they are collected and indexed by a Splunk Enterprise instance. Splunk Enterprise instance is running on the host machine. On all monitored assets, there is Splunk Universal Forwarder deployed. The forwarder ingests logs on a remote system according to its settings and forwards them to the indexer. Since the designed rules rely on logs from Microsoft Windows, Active Directory and Sysmon, the following TAs are used:

- Splunk Add-on for Microsoft Windows
- Splunk Add-on for Microsoft Active Directory
- Add-on for Microsoft Sysmon.

The forwarder has all the permissions of this domain account and can collect all the data that the account has access to read. This setup is required for accessing

Detection and Prevention of Attacks on Active Directory Using SIEM 537

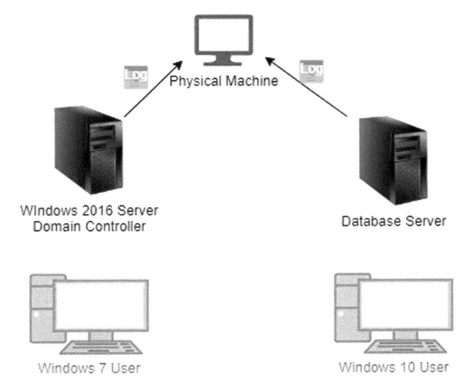

Fig. 1 Assets in virtual lab environment

the AD schema. Several designed detection searches require PowerShell logging. Module Logging and PowerShell Script Block Logging are enabled via Group Policy settings.We have suggested some prevention techniques to overcome these kinds of attacks. Figure 2 shows the proposed architecture. In this architecture we have implemented some attacks on virtual domain system using scripts and tools from linux system. These attacks are mainly focused on getting access to user account in the domain. And for detecting those attacks we have implemented SIEM in all the assets and forwarded logs to the physical machines. Based on the rules, Splunk will detect the attacks. If such attacks are possible then we have to take some preventive measures to overcome those attacks.

Every analytic story focuses on a certain attack scenario, and the following are the order of steps that would be potentially made by an attacker.

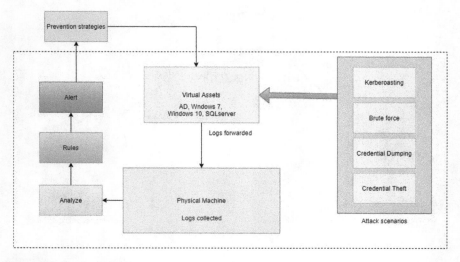

Fig. 2 Proposed architecture

3.1 Kerberoasting

The Kerberoasting technique is used to target Kerberos mechanism which is used to authenticate users who has access to protected network resources.

- **4769** kerberos service ticket was requested;
- **4770** kerberos service ticket was renewed;
- **4773** kerberos service ticket request failed.

Kerberoasting technique, involves the use of a valid domain user's authentication ticket (TGT) to request one or several service tickets using their SPNs [8]. Since the goal of an attacker is to crack the service ticket offline, tickets encrypted with weak cipher suites are vulnerable [9]. There are different ways to detect kerberoasting attack, because these types of attacks can be carried out by powershell, honeypot or suspicious service ticket request. Once attacker gets the hashed password, the adversary will perform offline password cracking to get service account password.

3.2 Brute Force Attacks

This attack is based on an elementary method of trial and error, its reliable detection is not so trivial. Brute force attacks involve attempts to log on by using explicit credentials, that is, a username and a password [7]. This implies that the source of information can be any logs auditing credential validation or account logon. There are certain Event id's to be considered for detecting brute force attacks 4624, 4625, 4648, 4740, 4768, 4771, 4776.

- **4625** logged on the computer where the logon attempt was made, with value of the field SubStatus = $0 \times C0000064$;
- **4678** recorded on a domain controller for Kerberos authentication ticket requests, with status = 0×6;
- **4771** recorded on a domain controller meaning that Kerberos pre-authentication failed, with Status = 0×18;
- **4776** auditing NTLM authentication attempts, with Status=0xC000006A and the generating host depending on the credentials used [1].

3.3 Credential Dumping

Credential dumping technique involves execution of tools which create processes, creating dump files in the file system, performing operations under special privileges, accessing particular network resources, or using specific applications [6]. The range of events that may contain valuable information is wide. The choice of monitored events depends significantly on the chosen detection method.

Detection of these tools can be based on monitoring process creation events. These events contain the name of the created process which can be checked to match names of known hacking tools.

3.4 Credential Theft

The techniques tricks windows authentication mechanisms to grant access to protected resources even for users who do not possess required privileges. However, the footprint of such activity looks like a legitimate authentication process. This implies that the spectrum of recorded events does not differ compared to the trail of standard authentication and resource access.

4 Prevention Strategies

1. Since kerberoasting attack is based on cracking passwords of service accounts it depends completely on strength of the password. The attack itself cannot be prevented, but selecting strong passwords can make it more difficult. Therefore service accounts should be treated with priveleged account and strong password should be preferred and password need to be changed at regular basis. The most effective method for preventing Kerberoasting is to use strong passwords that are at least 27 characters long with special characters. And to monitor the issuing of TGSs, the Audit Kerberos Service Ticket Operations setting must be enabled.

2. To prevent brute force attacks we can follow some basic methods like choosing strong username and password, disabling default administrator account and creating an administrator account with strong username makes it difficult for attackers to guess it, Remote access restriction to few users in network, account lockout policy helps in reducing number attempts can be made by attacker and Changing RDP port from default listening to some other port, by making changes in the registry.
3. For preventing credential dumping attacks never store your passwords in the system, check for reuse of users passwords against the breached passwords. Reviewing and auditing of use of NTLM should be done periodically. Also monitor for any suspicious process interacting with lsass .exe process.
4. To prevent credential theft use Multi Factor Authentication (MFA) for admin accounts, also create awareness among other users to set strong passwords and share knowledge about spear phishing and other scams. If any system is compromised isolate immediately and take precaution measures.

We have incorporated all these preventive measures in the system and we were able to resist attacks. These preventive measures are to reduce the chance of success rate of those attacks.

5 Conclusion

We have analyzed known attacks targeting Microsoft Active Directory and possibilities of their detection from windows security logs. The main task was to develop a set of detection rules, which would be able to detect the analyzed attacks by using windows security auditing. For many techniques, supplementing windows security events with PowerShell logs and Sysmon events, improved visibility and allowed building better detection rules. The future work can be on detecting attacks targeting multi-domain and multi-forest AD instances.

References

1. Kotlaba, L.: Detection of Active Directory attacks. Bachelor's thesis. Czech Technical University in Prague, Faculty of Information Technology (2019)
2. Siva Niranjan Raja, M., Vasudevan, A.R.: Rule generation for TCP SYN flood attack in SIEM environment. In: 2017 Procedia Computer Science, vol. 115, pp. 580–587, ISSN: 1877-0509
3. Sekharan, S.S., Kandasamy, K.: Profiling SIEM tools and correlation engines for security analytics. In: 2017 International Conference on Wireless Communications, Signal Processing and Networking (WiSPNET), Chennai, 2017, pp. 717–721
4. Fujimoto, M., Matsuda, W., Mitsunaga, T.: Detecting abuse of domain administrator privilege using windows event log. In: 2018 IEEE Conference on Application, Information and Network Security (AINS), Langkawi, Malaysia, 2018, pp. 15–20

5. Nair, H., Sridaran, R.: An innovative model (HS) to enhance the security in windows operating system—a case study. In: 2019 6th International Conference on Computing for Sustainable Global Development (INDIACom), New Delhi, India, 2019, pp. 1207–1211
6. Active Directory Security: How Attackers Dump Active Directory Database Credentials https://adsecurity.org/?p=2398
7. Techniques: Brute Force. https://attack.mitre.org/techniques/T1110/
8. Active Directory Security: Finding Passwords in SYSVOL & Exploiting Group Policy Preferences. https://adsecurity.org/?p=2288
9. Techniques: Kerberoasting. https://attack.mitre.org/techniques/T1208/
10. Matsuda, W., Fujimoto, M., Mitsunaga, T.: Detecting APT attacks against active directory using machine leaning. In: 2018 IEEE Conference on Application, Information and Network Security (AINS), Langkawi, Malaysia, 2018, pp. 60–65

Loki: A Lightweight LWE Method with Rogue Bits for Quantum Security in IoT Devices

Rahul Singh, Mohammed Mohsin Hussain, Milind Sahay, S. Indu, Ajay Kaushik, and Alok Kumar Singh

Abstract Internet of Things (IoT) is a platform that connects various devices through the Internet. Their abilities are limited in providing intelligent services. IoT devices are heterogeneous in nature and are able to connect sensors to simple devices that are constrained. Software and hardware attacks are a major concern in this domain. These attacks may lead to a breach of privacy, confidentiality, and malware attacks. To resolve this issue, a novel security algorithm is proposed in this paper. The proposed algorithm uses lattice-based cryptography method for encrypting gateways and cloud services and asymmetric key cryptography method for securing devices that protect the service gateways and low-energy IoT devices such as sensors. The proposed architecture uses asymmetric key cryptosystem that has nodes with a shared session key, and this session key is then further used to transfer messages. This protects the system from quantum attacks, eavesdropping, and public key attacks.

R. Singh (✉) · M. M. Hussain · M. Sahay · S. Indu · A. Kumar Singh
Department of Electronics and Communication Engineering,
Delhi Technological University, Delhi 110042, India
e-mail: singhrahuldps@gmail.com

M. M. Hussain
e-mail: mohdmohsin_bt2k16@dtu.ac.in

M. Sahay
e-mail: milind.sahay123@gmail.com

S. Indu
e-mail: s.indu@dce.ac.in

A. Kumar Singh
e-mail: aksingh@dce.ac.in

A. Kaushik
School of Computer Science and Engineering,
Galgotias University, Greater Noida, Uttar Pradesh, India
e-mail: ajay.kaushik@galgotiasunivesity.edu.in

© The Editor(s) (if applicable) and The Author(s), under exclusive license to Springer Nature Singapore Pte Ltd. 2021
T. Senjyu et al. (eds.), *Information and Communication Technology for Intelligent Systems*, Smart Innovation, Systems and Technologies 196, https://doi.org/10.1007/978-981-15-7062-9_54

1 Introduction

The use of IoT devices is growing with each passing day. In 2020, we have more than 20 billion IoT units connected, and this number is expected to grow by each passing day. IoT devices are low power devices and can transfer data over a network without requiring a human-to-human interaction. These interconnected devices leave the potential for hackers to attack them and use them to launch cyber attacks. Hence, the need for security standards in IoT devices is inevitable. Connecting more resource-constrained IoT devices to the network results in security threats [1]. IoT is shaping the future of the Internet, and while this happens, there is an urgent need to secure communication among IoT nodes against malware, viruses, and attacks. For a secure channel, each IoT node in the channel must implement specific cryptographic primitives such as key generation, public key encryption, public key decryption. It is important to note that most classical PKE systems rely on the high computation complexity they offer [2, 3]. Current cryptosystems, in addition to high computation complexity, rely on hard problems that are solved by quantum search algorithms leveraging polynomial time solution [4]. Thus, all of the classical cryptosystems will lose their security when quantum computing gains required computational power. Implementations of the RSA algorithm [5] and of ECC algorithms [6, 7] will be rendered unsuitable in the post-quantum era due to increased hardware complexity. Thus, there is an urgent requirement for cryptosystems that are quantum-resistant and rely on other hard problems. Some examples include shortest/closest vector problems in lattices [8], code-based cryptography, and learning with errors (LWE) [9].

2 Problem Formulation

Traditionally communication via cryptosystem such as RSA, Diffie–Hellman, Elliptic Curve Cryptography, etc., will be rendered obsolete shortly because the underlying hardness problem will be inadequate with the advent of quantum computers. Cryptosystems such as code-based cryptosystems, hash-based signatures, multivariate polynomial-based cryptosystems, and lattice-based cryptosystems are becoming widely accepted quantum-resistant PKE systems. However, the key sizes are plentiful, and thus, it cannot be used where computational resources are inadequate. There is a shift from traditional cryptography to lattice-based cryptography due to quantum resistance. Learning with errors (LWE) provides a quantum-secure algorithm for encryption/decryption and secure key exchange authentication for simple IoT devices. Thus, the LBC technique, which is lightweight and efficient, is more suitable for IoT applications. Therefore, a quantum resilient mechanism with modest key sizes and computational overhead is required.

3 Literature Review

Asymmetric key algorithms [10] have many benefits and are used for transmitting data almost in every communication network. They are based upon the computational complexity of some problems. NP-hard problems are chosen for achieving this purpose, such as integer factorization, discrete logarithm problems. Elliptical key cryptography [6, 7] is also popular because of the underlying hardness of elliptical curve problems. Most cryptographic problems are also based upon modular arithmetics, which are computationally expensive to decipher. However, with the advent of quantum computers, it is possible to crack these algorithms with the help of Shor's algorithm [11], and thus, newer cryptography algorithms were developed. Some of these new algorithms are:

3.1 Multivariate Cryptography

These are the cryptography algorithms based on multivariate polynomials over some finite field F. It is proven that solving of multivariate polynomials is considered to be NP-complete [12], and thus, they are preferred for post-quantum cryptography algorithms. They use very short signatures, which can serve the purpose of authentication in small devices such as IoT devices.

3.2 Lattice-Based Cryptography

These are the cryptography algorithms based on lattices during the construction or proving the hardness/security proofs of algorithms. A lattice is a set of all integer linear combinations of basis vectors b1,b2, …,bn Rn. SVP [8] is the most crucial lattice-based problem as to solve it, we have to approximate the minimum Euclidean length of a given nonzero lattice vector and cannot be solved efficiently even with a quantum computer. Some schemes have been under investigation for years such as the NTRU encrypt and no feasible attack have been found. Some other lattice-based algorithms have proof that their security reduces to a worst-case shortest vector length (SVP) problem. Some popular lattice-based cryptography algorithms are defined as follows:

1. **NTRU**: It works on the presumed difficulty of factoring polynomials in a polynomial ring into a quotient of two polynomials having tiny coefficients. Two algorithms are employed in it: NTRUEncrypt, used for encryption, and NTRUSign, used for digital signatures. Encryption and decryption, both require simple polynomial multiplications, and thus, the speeds are comparable to traditional asymmetric cryptography algorithms like RSA, etc. [13]. Studies have shown that

NTRU might have more secure properties than other lattice-based algorithms [14].

2. **Learning With Errors (LWE)**: LWE is a problem where a linear function is inferred over a finite ring from some given samples of the function. This scheme is much more optimized compared to other lattice-based cryptography algorithms with its public key size having a space complexity of $O(n^2)$. The ciphertext size also increases by a factor of $O(n)$, while previous algorithms had a complexity of $O(n^4)$ and $O(n^2)$, respectively. It is conjectured to be hard to solve as it reduces to the shortest vector problem [9].

3. **Lizard**: The variations of LWE had an issue that the parameter sizes were too large to be considered for use in practice. Thus, 'Learning With Rounding(LWR)' was introduced by the authors of [15], which is a derandomized variant of the LWE problem, which produces errors making use of a deterministic rounding mechanism in a smaller modulus without the summation of the additional errors [15]. If the number of samples is bounded, then its hardness is comparable to that of the LWE problems [15–17]. Lizard is a public key encryption scheme based on LWE and LWR, where the encryption process is carried out without Gaussian sampling by a procedure involving subset-sum followed by rounding operations [18]. It takes some advantage of sparse binary secrets. It provides better security than NTRU while achieving a comparable performance [18] (Figs. 1 and 2).

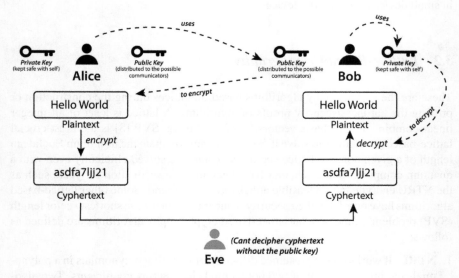

Fig. 1 Scheme showing how communications would take place

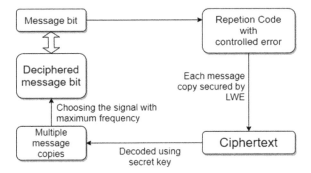

Fig. 2 Proposed work

4 Proposed Work

Asymmetric key cryptography [10] is a cryptosystem that uses private keys that are known only to users and pairs of keys—public keys which are disseminated widely. The generation of these keys depends on the underlying algorithm used in the proposed cryptosystems. Asymmetric key cryptography requires a private key to be confined to users, whereas public keys can be distributed widely without compromising security. Here, we propose an asymmetric key cryptosystem.

4.1 Standards and Notations

The following table defines the notations used throughout this text.

Notation	Description
Z_q	$Z \in [0, q)$, for a positive integer q
$x \leftarrow D$	x is sampled according to the distribution D, when D is a finite set, the sampling occurs with uniformity
G	Used to denote the Gaussian distribution set which generates a normal distribution from σ and μ
D^n	For an integer $n \geq 1$, it denotes a positive polynomial of dimension n
$bool(X)$	Function which converts its input into a $True$ if it is ≥ 0 and 0 otherwise
$RCS(x, i)$	It is rght circular shift function on x shifted by i numbers of size l

4.2 Proposed Algorithm

We propose a public key encryption system which uses three keys—two public keys to be distributed to clients and a secret key which is kept safe by the host in the network.

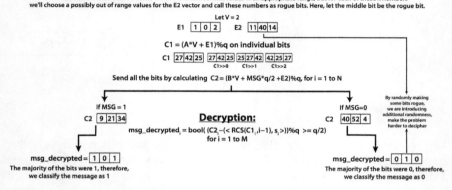

Fig. 3 Working of the proposed algorithm

We buildup on the LWE technique and combine it with our work of sending rogue bits to create a more robust LWE variant. Our work decreases the space complexity required for the public key and ciphertext in LWE from $O(n^2)$ to $O(n)$ by using a one-dimensional vector and then using the right circular shift arithmetic to calculate different public LWE search pairs using the same initial vector. This approach also helps us in reducing the computation time costs in processing of the ciphertext from the public keys. This is also illustrated in Fig. 3.

Setup: Choose positive numbers n, q, l, p such that $0 \leq n \leq q - 1$ where q is a prime integer having the range $[l/2, l]$ where $l = 2k$ and k is an integer denoting the number of bits required to store l and $p < n/2$. Choose a discrete Gaussian distribution G_σ having a parameter σ as the error distribution. Choose another positive number M, where $M < n$.

KeyGen: First, we generate A, a random public key polynomial $A \leftarrow R_q^n$. Generate a secret key S having dimension n, $S \leftarrow R_q^n$. Then, we generate an array of the same size but with small random error values, called the error term $E \leftarrow DG_\sigma^n$. Then, key $B \leftarrow R_q^n$ is calculated as

$$B_i = (< RCS(A, i-1), S > +E_i)\%q \in R_q \qquad (1)$$

where $i = 1$ to n. Output the public key as (A, B) and secret key as S. The secret key vector S remains constant for the system and must be kept safe and protected at all costs.

Encryption: Choose k indices in the encrypted message to be rogue bits; i.e., they signify the opposite bit and their places are chosen randomly for every message bit and let the set of these k indices be K and the set of remaining indices between 1 to n be I. Let the message bit be m and send the bit M times ($M/2 > k$). We generate an integer v and two other vectors $E1$ and $E2$ with random values in appropriate ranges $R1$, $R2$, respectively. For $i \in I$, $E1_i \in R1$ and $E2_i \in R2$. For $i \in K$, $E1_i, E2_i \in [0, l-1]$. Now, we generate two ciphertexts, $C1$ and $C2$,

$$C1_i = (A_i * v + E1_i)\%q \qquad (2)$$

$$C2_i = (B_i * v + m * q/2 + E2_i)\%q \qquad (3)$$

where $C1_i$, $C2_i$, $E1_i$ and $E2_i$ are i^{th} elements of the vectors $C1$, $C2$, $E1$ and $E2$ respectively.

Decryption: The ciphertexts $C1$ and $C2$ inherently represent the same message bit m along with the rogue bits at some random positions. Hence, when we solve the LWE problem for each element in the arrays, we can assign m to be the bit with the higher count. The decryption is done as,

$$m = bool(\sum((c2_i - <RCS(c1_i, i-1), s_i >)\%q \geq q/2) \geq M/2) \qquad (4)$$

Since the rogue bits will always be in the minority, thus, we use the majority principle to find the actual value of the message bit m.

5 Result

5.1 Hardness

Our proposed algorithm, Loki is an improvement of the LWE problem, which was proved to run in exponential time [9] with the ability to evade quantum attacks. It is also an extension of the LPN problem, which is widely believed to be hard.

Quantum attacks: Classical encryption schemes, for instance, elliptic curve cyptography based and symmetric cryptographic algorithms, cannot resist quantum computer attacks. The proposed scheme is based upon the worst-case hardness of SVP and SIS problems in lattices, and also, the security of the public key is impervious to quantum attacks.

Eavesdropping attacks: The security of the suggested algorithm is not only ensured by the used public-key encryption but by the usage of almost rogue message bits to increase the ciphertext randomness as well to prevent the unconditional attacks. Since the ciphertext is generated by processing the public keys with a randomly generated vector along with using specific parameters, it makes the message impossible to decode for the eavesdroppers.

Public key attacks: Our algorithm provides the public key security of learning With errors(LWE) as it is built on top of it. According to LWE, for $n \geq 1$ and $\epsilon \geq 0$, we define the 'learning from parity with error' problem as follows: find an unknown $s \in R_2^n$ when we are provided with an array of 'equations with errors.'

$$<s, a_1> \approx \epsilon b_1 (mod\, 2) \quad (5)$$

$$<s, a_2> \approx \epsilon b_2 (mod\, 2) \quad (6)$$

where $<s, a_i> = \sum_j s_j(a_i)_j$ is the dot product of s and a_i modulo 2 and a_i's are randomly chosen from a uniform distribution on R_2^n, and the probability of each equation being correct is $1 - \epsilon$. Here, the aim is to obtain s. This problem becomes much harder when we take any $\epsilon > 0$. An algorithm developed by [19] gave the first subexponential algorithm for the given problem. This approach is the best performing algorithm that operates with time complexity of $2^{O(n/\log n)}$. This high time complexity ensures safety from public key attacks.

5.2 Comparison with Other Schemes

The algorithm provided is compared with existing works such as NTRU, LWE, LIZARD [9, 13, 18] cryptography algorithms with respect to its performance in key generation, encryption, and decryption as seen in Figs. 4 and 5.

As per Fig. 4, the key generation time was the least for our proposed algorithm by a margin of 64.4%, 3359%, and 5693% compared to LWE, Lizard, and NTRU, respectively. The message encryption time was also the least for our work with a margin of 7596.6%, 1734.65% less than LWE, and Lizard, respectively. However, it was slower by a 1081% than NTRU in encryption time, yet it outweighs it when considering the total time taken, as shown in Fig. 5. Our algorithm has a higher ciphertext decryption time by a factor 96.5% when compared with LWE, Lizard algorithms, but our algorithm still outperformed them when the total time is under consideration. It performed comparably well with the total time taken less by 1170.5%, 587.12%, and 817.08% when compared to LWE, Lizard, and NTRU, respectively.

The key size required by different cryptography algorithms for providing the same levels of security is summarized by Fig. 6. Our proposed public key cryptography

Algorithm	Key Generation	Encryption	Decryption	Total Time
Loki(This Paper)	0.030459	0.033317	0.177953	0.241729
Lizard	1.043652	0.611252	0.006074	1.660978
LWE	0.500884	2.564362	0.006127	3.071373
NTRU	1.764524	0.00282	0.449521	2.216866

Fig. 4 Results of our algorithm compared to industry standards

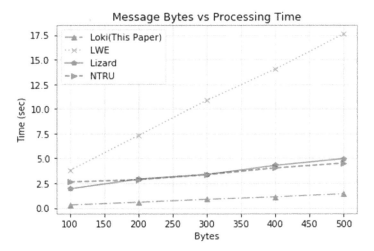

Fig. 5 Time comparison of our algorithm

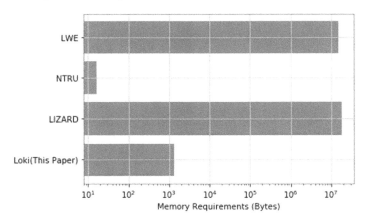

Fig. 6 Memory requirement comparison of our algorithm

system also requires smaller amounts of memory by almost $1.2 \times 10^6\%$ and $1.4 \times 10^6\%$ when compared to LWE and Lizard techniques, respectively. The least memory requirement is shown by the NTRU algorithm. However, this advantage is outweighed by the higher processing time that it takes when compared to the running times of our algorithm.

6 Conclusion

In this paper, we have proposed a new public key cryptography algorithm built on top of the LWE technique, which can be used in an IoT-based environment. It utilizes the majority principle over a repetitive transmission consisting of rogue error bits to encrypt the message bit. It was found to perform better than NTRU, LWE, and LIZARD in terms of computation efficiency. The underlying LWE problem ensures that our system is safe from public key attacks and eavesdropping. This system also provides increased security even in the Post-Quantum Era, with better performance and low bandwidth requirements.

References

1. Ling, Z., Luo, J., Yiling, X., Gao, C., Kui, W., Xinwen, F.: Security vulnerabilities of internet of things: a case study of the smart plug system. IEEE Int. Things J. **4**, 1899–1909 (2017)
2. Chaudhary, R., Aujla, G., Kumar, N., Zeadally, S.: Lattice based public key cryptosystem for internet of things environment: challenges and solutions, **10** (2018)
3. Patrick, C., Schaumont, P.: The role of energy in the lightweight cryptographic profile (2016)
4. Shor, P.W.: Algorithms for quantum computation: discrete logarithms and factoring. In: Proceedings 35th Annual Symposium on Foundations of Computer Science, pp. 124–134, Nov. 1994
5. Rivest, R.L., Shamir, A., Adleman, L.: A method for obtaining digital signatures and public-key cryptosystems. Commun. ACM **21**(2), 120–126 (1978)
6. Koblitz, N.: Elliptic curve cryptosystems. Math. Comput. **48**(177), 203–209 (1987)
7. Salarifard, R., Bayat-Sarmadi, S., Mosanaei-Boorani, H.: A low-latency and low-complexity point-multiplication in ECC. IEEE Trans. Circuits Syst. I Regul. Pap. **65**(9), 2869–2877 (2018)
8. Peikert, C.: Public-key cryptosystems from the worst-case shortest vector problem. In: Proceedings of the Forty-First Annual ACM Symposium on Theory of Computing, pp. 333–342 (2009)
9. Regev, O.: On lattices, learning with errors, random linear codes, and cryptography. J. ACM (JACM) **56**(6), 1–40 (2009)
10. Stallings, W.: Cryptography and Network Security: Principles and Practice. Prentice Hall, Upper Saddle River (1990)
11. Shor, P.W.: Polynomial-time algorithms for prime factorization and discrete logarithms on a quantum computer. SIAM J. Comput. **26**(5), 1484–1509 (1997)
12. Garey, M.R.: Computers and Intractability : A Guide to the Theory of NP-Completeness. W.H. Freeman, San Francisco (1979) (Johnson, David S., 1945)
13. Hoffstein, J., Pipher, J., Silverman, J.H.: NTRU: A ring-based public key cryptosystem. In: Buhler, J.P. (eds.) Algorithmic Number Theory, pp. 267–288. Springer, Berlin (1998)
14. Easttom, C.: An analysis of leading lattice-based asymmetric cryptographic primitives, pp. 0811–0818 (2019)
15. Banerjee, A., Peikert, C., Rosen, A.: Pseudorandom functions and lattices. In: Annual International Conference on the Theory and Applications of Cryptographic Techniques, pp. 719–737. Springer, Berlin (2012)
16. Alwen, J., Krenn, S., Pietrzak, K., Wichs, D.: Learning with rounding, revisited. In: Annual Cryptology Conference, pp. 57–74. Springer, Berlin (2013)
17. Bogdanov, A., Guo, S., Masny, D., Richelson, S., Rosen, A.: On the hardness of learning with rounding over small modulus. In: Theory of Cryptography Conference, pp. 209–224. Springer, Berlin (2016)

18. Cheon, J.H., Kim, D., Lee, J., Song, Y.: Lizard: cut off the tail! A practical post-quantum public-key encryption from LWE and LWR. In: Catalano, D., De Prisco, R. (eds.) Security and Cryptography for Networks, pp. 160–177. Springer International Publishing, Cham (2018)
19. Blum, A., Kalai, A., Wasserman, H.: Noise-tolerant learning, the parity problem, and the statistical query model. J. ACM (JACM) **50**(4), 506–519 (2003)

Implication of Web-Based Open-Source Application in Assessing Design Practice Reliability: In Case of ERA's Road Projects

Bilal Kedir Mohammed, Sudhir Kumar Mahapatra, and Avinash M. Potdar

Abstract Unreliable design practices on Ethiopian road construction projects have become a crucial matter concerning the late completion of road construction projects, high-cost overruns, and premature failures. The present research encompasses the revealed preference and stated preference data conducted through traditional printed questionnaires versus online Web-based cellular phone questionnaires for assessing design practice concerning design firms of road projects administered by Ethiopian Roads Authority. Online real-time application-based questionnaire has developed for the data collection and interpretation. Existing road design practices of the design firms regarding managing and evaluating information through planning, monitoring were assessed statistically and will helpful for those, involved in design management and the development of tools and practices to improve design reliability in engineering and construction industries.

1 Introduction

In the construction industry, particularly the road sector as often known have problems regarding to large slippage of calendar days of completion against the planned

---

B. K. Mohammed (✉)
M.Sc. in Road and Transport Engineering, College of Architecture and Civil Engineering, AASTU, Addis Ababa, Ethiopia
e-mail: Syambil0371@gmail.com

S. K. Mahapatra
Software Engineering Department, College of Electrical and Mechanical Engineering AASTU, Addis Ababa, Ethiopia
e-mail: sudhir.kumar@aastu.edu.et

A. M. Potdar
Road and Transportation, Department of Civil Engineering, College of Architecture and Civil Engineering AASTU, Addis Ababa, Ethiopia
e-mail: avinashmpotdar@gmail.com

duration, due to different inter-related cause and effect controversial project contract variables.

There are many researches have been attempted to directly assess the design associated issues and look into different features that lead to improving highway design outputs. Reviewing methods, procedures, and standards of previous experience were thoroughly examined to achieve the desired goal and design quality of road projects in this regard. Subsequently, this paper tries to investigate issues related to design reliability and frequent departure of the design elements that required to be updated for the highway projects.

Construction design is a sophisticated and important form of problem solving, where stakeholder's needs and requirements are conceptualized into a physical model of procedures, drawings, and technical specifications [1].

Reliability of road design has become a usual problem and continual concern that initiated researchers and stakeholders to study in Ethiopian construction industry. Unreliable design practice on road projects has become a critical issues which results in a considerable change to most projects in Ethiopia.

According to Wang and Wu [], definition of reliability-based design is a design trend considering the best design as one which performs as expected in the face of both expected and unexpected variations. Due to wide geographical settings with different soil and environmental characteristics, design information is very critical in road construction. The academic study conducted in Ethiopia on 24 projects by Wakjira [] shows 80% of the projects are experiencing the cost overrun ground investigation has been highlighted by based on an analysis of 24 large highway construction projects; they found that "the final cost was on average 35% greater than the tendered sum."

Design consultants are not performing their assignment in-house in doing the activities, i.e., many of their staffs are freelancer. These specialized professionals of the design consultant could have different backgrounds and working cultures, and they are usually located at different geographical locations. This is therefore, design stage failures are among the front lines for the construction failures and contractual implications caused by absence of integration among the design deliverables of independent works.

It is also important for the studies to be conducted on related to reliability of design practices that need to be taught at research institutions and universities to search for sound solutions to the existing/challenging situations for the upcoming projects in the country.

So that, this study will perform the design reliability assessment for the projects in the Ethiopian Roads Authority, in order to differentiate what are the root causes, to measure the road designs level of reliability, and identify areas where improvements are necessary.

2 Literature Review

2.1 General

There are many types of research have been attempted to directly assess the design associated issues and look into different features that lead to improving highway design outputs. Reviewing methods, procedures, and standards of previous experience to achieve the desired goal and design quality should be substantial. Subsequently, this paper tries to investigate issues related to design reliability and frequent departure of the design elements that required to be updated for the highway projects.

In the construction industry, 70% of the cost of the final product is spent on the design. It is a key activity where the customer's needs and requirements are conceptualized into a physical model of procedures, drawings and technical specifications []. Likewise, construction design is a sophisticated and important form of problem solving, where stakeholder's needs and requirements are conceptualized into a physical model of procedures, drawings, and technical specifications [].

Wang et al. [] definition of reliability-based design is a design trend considering the best design as one, which performs as expected in the face of both expected and unexpected variations, and it does so because the design is automatically not sensitive to changes in the design features and service environment. On the other hand, Halwatura et al. stated that "unreliable design is characterized by poor communication, lack of adequate documentation, deficient or missing input information, and poor information management, unbalanced resource allocation, lack of co-ordination between disciplines and erratic decision making []."

2.2 Design Practice and Contractual Complications

Poor co-ordination of design practice and information directly results in erroneous design causing complicated contractual issues. According to Lafford et al., design errors and variations are "inherent part of many construction projects and require deliberate effort to combat []." It is also possible to see the ranking of contractual problems faced due to design-related practices. Similarly, Koushki et al. mentioned that "the five most important factors agreed by the clients, consultants and contractors as causing project cost overruns, for instance, were schedule slippage, lack of project knowledge, underestimating and design errors followed in that order []."

Conversely, preceding studies were done that were published in the top-tier construction management journals failed to recognize design changes as a major cause of project cost overruns. Therefore, suggestions for future research are recognizing design changes as a major cause of delays and cost overruns in the construction project in Malaysia []. Fewer variation order involves if clients who spent more time and money on the design phase [].

In the Ethiopian construction industry particularly the road sector, the design-related problem is among those variables resulting in projects to be completed through un-conducive performances creating public grievances here and there in all around the country, variations, right of way claims, design problems (design risk) and scope change are identified as major factors leading to cost overrun among the eight potential factors which has been identified, from the desk study conducted by a senior researcher [].

3 Methodology

3.1 Overview of the Research Process

Owing to the desire of the researcher to undertake applied research, which aims at contributing knowledge toward solving a practical problem, the research started by investigating a practical problem. Experience, observations and results of previous studies from the ground for formulating the research problem statement and the research questions were measured.

The revealed preference (RP) and stated preference (SP) data conducted through traditional printed questionnaires versus online Web-based cellular phone questionnaires for assessing design practice concerning design and supervision firms of road projects administered by Ethiopian Roads Authority.

Based on the problem statement, first, an extensive review of the literature on the subject was undertaken. Next, an investigation on the existing project design management practice of consultants which have an agreement with Ethiopian Road's Authority for the design projects was carried out, with the view of discovering whether it matches what has been discussed in the literature or not. The results of the Web-based open-source application investigation gave a clear picture of the existing practice and assisted in identifying the major shortcomings and limitations of the design practice, which were used to propose the improvement interventions to achieve a reliable design of road projects.

3.2 The Study Approach and the Research Instrument

The research includes the development and implementation of online real-time application basis for data collection, assessment of existing practices of road design related with managing and evaluating information through planning, monitoring and controlling were performed in view of identifying shortcomings and limitations associated with each functions with the aim of improving and comprises contents of design practice and/or process.

The designed Web was interactive and user interface to make it easily accessible/convenient through a device in their hand. Besides, the developed questionnaire avoids reading, writing and reporting errors that the known paper-based questionnaire experienced.

The population of the study was limited to road design consultants which are working with Ethiopian Roads Authority, Design Management Directorate. Accordingly, the designed questionnaire with URL (eracivil.rf.gd) was forwarded/delivered to professionals of twenty-five (25) consultants, ten (10) counterparts of ERA and ten (10) other professionals of road design through their mobile phone and E-mail address; then, they can access through any browser like Opera, Chrome, Mozilla, Internet explorer network cages which can be accessed from any devices at hand. The participants have been well informed the objectives of the designed questionnaire and requested to filled properly and forward same in this respect.

3.3 Method of Analysis

The rating scale is one of the most common formats for questioning respondents on their views or opinions of an event or attribute. In this regard, respondents were requested to specify *an importance or level of influence factors (research variables) by rating them on a five-point scale, (0, 1, 2, 3 and 4)*. They are merely numerical labels indicating [0 = Not Important, 1 = Low Important, 2 = Medium Important, 3 = Highly Important, and 4 = Very high Important]. Accordingly, this statistical technique is intended to establish the rank of factors and same has been assigned.

$$\text{Importance index(II)} = (Wi * Fxi) * 100/(A \times n)$$

where

Wi weight given to ith response; $i = 0, 1, 2, 3, 4 \ldots$
Fxi Replies frequency
A highest weight value
N total number of responses (n responses).

The ranking format was used for analyzing question in which respondents were asked to place a set of attitudes in ranking order, indicating their importance priories or preferences.

3.4 Experiment and Results

3.4.1 Response Rates

According to the party, they represent

As mentioned in Sect. 3.5, the questionnaire form was distributed to 25 consultants, ten counterparts of ERA and ten other professionals of road design out of which 19 consultants, six counterparts of ERA and five other professionals of road design returned completed forms, representing a 76%, 60% and 50% response rate, respectively. The response rate regarding the total participants is 66.7%.

Based on position of respondent, they assigned

Among the participants/respondents, nine (9) or 36% are willing to disclose their professional assignment with their respective firm. In this regard, they have been assigned as Claim Expert, Counterpart Engineer, Project Manager, Highway Engineer, Resident Engineer, Quantity Surveyor, Material Engineer, Project Engineer and Office Engineer) and each of the assignment cover 11.11% per the response same.

3.4.2 Respondents Characteristics

From the responding 19 consultants, all of them participated in rendering Consultancy Service for Road Design Projects and Supervision of Works Contract. Besides, other respondents of ERA Counterpart Engineer (6) and other suggested professionals (5) are also well involved in design and supervision of road projects. An assessment was made to reveal the proportion of design projects among design and supervision service of road projects undertaken by the firms over the past five (5) years. In this respect, 90% of the respondent agreed that the minimum threshold portion of road design projects performed by the respective Firm was 50%.

3.4.3 The Research's Data Analysis

The start—up Meeting

A start-up meeting is important to identify key stages in the design process and introducing the designers and their personnel concerning objectives of the project. Further, start-up meetings would be held during investigation, feasibility design (surveying) and detail engineering design stages.

In this study, an assessment was made to investigate the design practice of consultants' decision for making meetings in the design process period. Accordingly, 27% of respondents practice a start-up meeting at project investigation stage. The 34% of respondents perform start-up meetings at feasibility design stage. The 39% of respondents perform start-up meetings at detailed engineering design stage. At the investigation stage, start-up meeting will generate discussion that ideas will germinate and that an understanding of the values, and capabilities of the client and the design team will be achieved. While start-up meetings at the feasibility design (Surveying) stage have a benefit to choose the range of specific studies necessary to improve.

Defining Tasks Practice

Detail engineering design is the immediate work detail after the feasibility study is completed. The detailed engineering designs are going to be regulated and permitted through an authorized office. An assessment has been made on the design consultants, 70.3% of respondents states that they follow each steps of the design process in order to meet the client need by defining the tasks. 21% of the respondents just focus on the feasibility (surveying) and detail engineering design stages while they receive a design project. 8.7% of respondents do the design process without defining the tasks; they just do the process as a whole. This brings the missing and orders less of tasks. For example, if they just start a design process without defining the statement of the need, finally they will miss the aim of design project.

Managing Information Production and Evaluation Practice

Information management (IM) is a system of controlling organizational qualities, responsibilities, confidentiality, co-ordination, and reliability of evidences. On the contrary, poor quality of information including; lack of consistency, duplication, and outdate information.

Information which is already produced by the contributors has been evaluated through design reviews in order to bring a completed production of design. In the managing and evaluation of information encompasses to be improved for the following activity;

(1) **Co-ordination of Information**

An assessment has been made in the co-ordination of information practice of consultants; from the respondents, 20% have good practice of co-ordination with regarded to information flow; this has been noticed by the period of interview and discussion with the professionals 72% of the respondents do not have consistent integrated co-ordination through each disciplines. This leads to an ambiguity between the works which are transferred through contributors.

The remaining 8% of respondents do works individually and they only submit works to one center while they finish. The next work is also distributed from the design head. This also plays a significant role for misunderstanding between the works of contributor. Many disciplines are engaged on the design process and they should have enough flow of information through them in order to meet the need of the client.

(2) **Design Reviews**

An assessment has been done that, from surveyed consultants, 15% of the respondents review the design at the schematic level only, 12% of respondents review design at the detail design stage, only 6% of respondents review the design at the final stage. The majority 35% of surveyed respondents review design at both stages, schematic and final stage of the design process. 6% of respondents practice the design review at the detailed and final design stages. The remaining 26% of respondents practice a design review effectively into three stages of designs, schematic, detail, and final design.

The survey shows that the design review practice of design consultants does not have a consistent or standard way in which stages must be done; this leads to happen un-clarity and also unnecessary items between information. It is suggested that most of the design reviews should be done at the early stages of implementation.

(3) *Thought of Maintenance*

The survey result shows that 60.5% of respondents follow the concepts of maintenance for the design projects in their firm and 35.5% respondents revealed that there is no concept of maintenance of design projects in their firm. The remaining 4% respondents totally ignore the concept of maintenance once they focused on the strength.

Causes for Design Changes

Control of change can be dealt with in three issues, by variations, by development of the design, and by request from the site for information. Variations are design revisions that occur when the scope of the work changes or when the client revises the requirement s for the road.

The survey result shows respondents give first rank for variations, with the important index of 86, give second rank for development of the design with important index of 78, and they give third rank for the request from site information as a factor for a change with the importance index of 77.

Factors to Produce Reliable Designs

The survey result displays, respondents give first rank for consistency, with the important index of 19.3, give second rank for esthetics with important index of 14.8, and they give third rank for engineering judgments as contributing factors for reliable design with the importance index of 12.5.

Factors for Unreliable Design Practice

This implies that the study on the reliability of design practice of road projects is timely concerns that require an improvement and/or minimize such occurrence of unreliability of designs in the future. The major contributing factor for unreliable design were grouped into employer related and design consultants related to make same easy for understanding and to point out which factors are more critical to employer and/or consultant.

(1) *Design Consultants Related*

Regarding consultants' performance with respect to reliable design practice, the survey finding revealed, as shown in figure below that lack of integration/co-ordination between designers of different disciplines has been ranked first with the important index of 20.5; low capacity/negligence of the design consultant has been followed as a second rank with the importance index of 18.5 while ineffective design review documents has got the third rank with the important index of 17.5 are among the top factors for unreliable design which are design consultant related. The consultants surveyed have problems of identifying the works and they have a trend of working in grouped manner; it has been assessed in literature review part that design review is a mandatory tool for insuring the quality design in the design process.

Lack of cost planning and control has ranked 4th with the importance index of 16.8. To this effect, there are financial deficiencies with the consultant leads to delays which finally bring unreliable design practice. Inefficient of definition of tasks has ranked 5th with an importance index of 16.0. In this regard, tasks should be identified and classified (by type, size and complexity) together with the schedule to produce complete information in the design process. Ineffective methods of programming (lack of proper project planning), inadequate evaluation of information, deviations in resource profile, inefficiency in co-ordination of information, and inadequate control of change have got ranks from 6th to 10th with the important index of 15.3, 14.5, 14.0, 13.5, and 12.3 respectively.

(2) **Employer Related**

Likewise, concerning employers' practice with respect to reliable design, the survey finding exposed that Duration of Design Projects has been ranked first with the important index of 20.5, the time interval to implement designed projects has been followed as a second rank with the importance index of 20 while both delays due to the reasons related with clients, and design submittals has been checked by Junior Engineers of the client/ERA have got the third rank with the important index of 14 are among the prominent factors for unreliable design which are related with employer.

Delays due to the reasons related with clients happen when clients do not provide enough information at the time of statement of the need/employer's requirement. In additions, there are gaps in client side technical staff requirement to review, check, and approve designs as design submittals has been checked by Junior Engineers.

Problems related with contract document, poor TOR development, and method of c administration have ranked from 5th to 7th with the important index of 13.5, 12.8, and 10.8, respectively. Besides, they also play a role for practicing of design management because it regulates the design process and helps as a bridge to meet the needs of client/employer's requirement.

4 Conclusion

The results revealed that factors contributing to the unreliable design practice are unfamiliarity with defining tasks, improper evaluation of information, inadequate planning/monitoring/controlling of design projects, insufficient time for project definition/design/documentation, lack of professionals' experience to check design submittals, design standards technical limitation to the designer's liberty, method of contract administration, poor TOR development, low capacity/negligence of the design consultant, and lack of integration/co-ordination between designers of different disciplines.

The construction design industry is characterized by the strong and multi-directional relationships and interactions that exist among the various stakeholders.

Hence, interventions or measures from the consultants' side only cannot result in improved outcomes, unless backed by pertinent or countermeasures from the other stakeholders.

5 Recommendation

Controlling professionals tangible productivity on a quantitative scale for intangible qualitative parameters involved in project reliability by developing Web-based interactive data repository to commensurate; communication among all on possible hierarchical position of design firm and client, computerized design and output standards library, and systems for controlling professionals' productivity and co-ordination of information through them.

References

1. Chileshe, N., Berko, P.D.: Causes of project cost overruns within the Ghanaian road construction sector. In: Proceedings 5th Built Environment Conference, p. 72. Durban, South Africa (2010). ISBN: 978-0-620-46703-2
2. Wang, W., Wu, Y.-T.J.: Southwest Research Institute, S. A. Deterministic Design, Reliability-Based Design and Robust Design. Southwest Research Institute, Sano Antonio (2016)
3. Wakjira, T.: Risk Factors Leading to Cost Overrun in Ethiopian Federal Road Construction Projects and its Consequences. Addis Ababa Institute of Technology, Addis Ababa (2011)
4. Hammond, J., Choo, H. J., Austin, S., Tommelein, I.D., Ballard, G.: Integrating design planning, scheduling, and control with Deplan. In Proceedings of the 8th International Group for Lean Construction Conference, Brighton, England (2000)
5. Halwatura, R.U., Ranasinghe, N.P.: Causes of variation orders in road construction projects in Sri Lanka. ISRN Construction engineering. Int. J. Constr. Eng. Management, pp. 1–7 (2013)
6. Kochan, A.: Boothroyd/Dewhirst—quantify your designs. Assembly Autom. **11**(3), 12–14 (1991)
7. Lafford et al: Civil engineering design and construction; a guide to integrating design into the construction process. Funders Report C534, Construction Industry Research and Information Association, London (2001)
8. Koushki, P.A., Al-Rashid, K., Kartam, N.: Delays and cost increases in the construction of private residential projects in Kuwait. In: Koushki, P. A.-R. (ed.) Construction Management and Economics, vol. 23, pp. 285–294. Kuwait: Kuwait (2005)
9. Horner, R.M.W., Zakieh R.: Improving construction productivity—a practical demonstration for a process based approach. Internal publication, Construction Management Research Unit, University of Dundee (1998)
10. Abdul-Rahman, H.: Impacts of design changes on construction project performance: insights from a literature review. In: Construction Management in Construction Research Association Conference and Annual General Meeting (p. 1). Kuala Lumpur, Malaysia: MiCRA (2015)

Influencer Versus Peer: The Effect of Product Involvement on Credibility of Endorsers

S. Rajaraman, Deepak Gupta, and Jeeva Bharati

Abstract The rise of internet and the penetration of social media has led to the increasing importance of online opinion leaders known as "Influencers" who influence purchase decisions. Another source of influence on purchase intentions is a peer, or a person a customer considers his/her equal such as a friend, a colleague, an acquaintance or a relative/family member. While there have been studies comparing the influence of celebrities and non-celebrities, as yet there is relatively little work on the comparative influence of online influencers and peers on purchase intention. The objective of this study was to identify which type of endorser people prefer, an online influencer or a peer, and determine who was more important when it comes to influencing the purchase decision of the customer. A conceptual model was developed from the literature to analyze the impact of factors such as expertise, trustworthiness, similarity and identification. The impact of the Big Five personality factors was also tested. The model was tested using ordered logistic multiple regression with a sample of 207 individuals living in Tier 1 and Tier 2 cities in India. The results show significant differences in the way influencer and peer credibility factors---trustworthiness and expertise---impact the purchase intention; and in the way, consumer product involvement moderates these influences.

1 Introduction

Behavior related to influencers or any kind of endorsers can be explained through the theoretical framework of social information processing theory [10]. Information about the behavior of others has an influence on current behavior, largely due to the need to emulate and due to peer pressure as well. One particular kind of social

---

S. Rajaraman (✉) · D. Gupta · J. Bharati
Amrita School of Business, Amrita Vishwa Vidyapeetham, Coimbatore, India
e-mail: koushik7496@gmail.com

D. Gupta
e-mail: dgshobs@gmail.com

information concerns the activity of a recommender or endorser who does the activity, who also persuades or invites another person to do it, on the grounds that "I do it, so can you." One such source of information is online influencers who have been found to influence purchase decisions. These influencers have a large online following on several online platforms and micro-blogging sites, such as YouTube, Facebook, Twitter, and Instagram where they share their content regarding products.

Studies have been made about the credibility of a peer as an endorser [9]. Comparative studies have been made between celebrities and influencers, and their respective weightages when it comes to influencing the purchase intention of the customers [8]. However, these studies have been done in one type of industry/product-FMCG—and they have dealt with the celebrity endorsement through one channel only, namely advertisements. The rise of social media has transformed the landscape of influencer [5]. Online influencers are not restricted by nationality and can reach even the millennial audience [6]. To the best of our knowledge, no study has been made by comparing an online influencer with a peer and more specifically, with the peer as an equal. This was the question we considered for the research study. In addition, there has been little research comparing the conditions under which influencers may play a more important role than peers or vice versa. In this study, peers are individuals who are considered as equals in the environment of the respondent and the influencer acts on any online platform.

From a review of the literature, it was found out that the factors which play the most vital role for online influencers when it comes to their weightage in purchase decisions factors were (a) expertise; (b) trustworthiness (which together contributed to the credibility); (c) similarity; and (d) identification. These factors were used to estimate the influence of both peers and online influencers. In addition, factors such as emotional connection, previous purchases made due to the word of mouth of the influencer/peer, level of satisfaction of the user and the degree of the product usage, were also considered as factors during the study. The Big Five personality factors (extraversion, agreeability, conscientiousness, emotional stability and openness to new experiences) were also measured. In the following sections, we present the model along with a discussion on the analysis of the survey results.

2 Literature Review

Endorsements and understanding how they work were the first components that needed to be understood. Therefore, the literature review initially started by asking the question---how do endorsements differ when they come from two different sources: celebrities and non-celebrities? It can be understood that people view credibility as a main component for buying a product [1]. Non-celebrities have a lesser impact, more specifically among households as well.

Among these, despite there being many qualities that the people look for in celebrities, expertise is the foremost one; rather, the customer's perception of a celebrity's perceived field of expertise. Apart from this, attractiveness was also considered to be another component which aided to the likeability of the source [2]. These factors not only affect the overall purchase intention, but when it comes to the customer, they also affect the decision-making process as well [3]; that was observed based on the time frame it took to buy the product. Brands seem to have recognized the impact celebrities seem to have on the customers. Therefore, a lot of popular brands have repositioned/changed themselves completely around the celebrity persona in order to raise brand awareness and to communicate effectively [4]. People look mainly for the way in which the celebrity "fits" with the product in question and literature suggests that this fitness has a significant influence on the celebrity influence [5].

One of the main reasons as to why these online influencers are being preferred compared to celebrities is their accessibility to the target audience as well as having a positive attitude toward social media, irrespective of many factors and thanks to their increasing usage [7]. There have been studies which have shown that by comparing them with the traditional celebrities, people's preference is toward influencers [8]. Both these types of endorsers have common factors through which their impact on brand credibility and further, purchase intention, can be measured. These similar variables can also be applied for peer endorsers as well [9]. This study also included product involvement as an influential variable. However, in this study, peers refer to a normal person, not necessarily an equal. Based on our literature review, a conceptual model was constructed using the demarcated variables. Product involvement was hypothesized to have a moderating influence on purchase intention.

3 Conceptual Model

See Fig. 1.

4 Illustrative Hypothesis

As seen in Fig. 1, there are 26 hypotheses for the study, 13 for influencers and 13 for peers, inclusive of moderation hypotheses for product involvement. The following are two illustrative hypotheses for influencers:

> H3: Trustworthiness of the influencer is positively related to the purchase intention of the customer for products recommended by the influencer.
> H11.a: The influence of trustworthiness on the purchase intention of the customer is <u>positively moderated</u> by the product involvement---the higher the product involvement, the greater the influence.

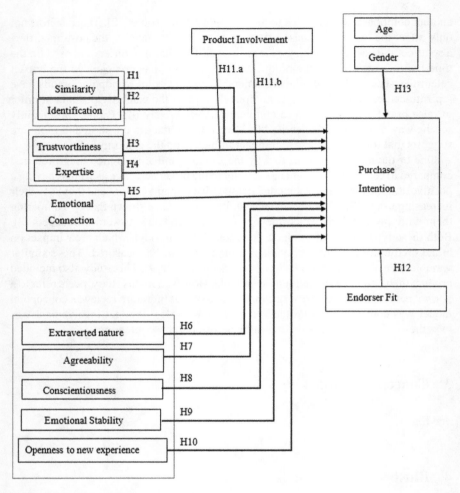

Fig. 1 Conceptual model

5 Methodology

5.1 Data Description

The sample size consists of 207 individuals, which was achieved through systematic sampling. There are specific target segments, and in order to reach them, quota-based sampling was done as well.

The sample groups were grouped into five categories based on age. The main age respondents are from the age group 19--24, at 56.1%, while the next group with a share of 29% belongs to the age range of 25--30.

Gender: Out of the respondents, 58.1% are male while the rest of 41.9% are female.

Location: The majority of the respondents were from the city of Chennai, at 24.1%, followed by Coimbatore (19.1%) and Bengaluru (18.5%). The rest are from cities such as Kochi, Trivandrum, Delhi, Mumbai, Hyderabad, Erode, etc.

Employment Status: Over 48% of the respondents are full time employees; 39.3% are students. The rest is a mixture of part-time employees and unemployed jobseekers.

5.2 Developing the Questionnaire

A questionnaire was developed for an online survey to understand the attitudes and behavior of the targeted age groups toward influencers and endorsers. The questionnaire was comprised of questions pertaining to the factors that are sought after in an endorser trustworthiness, expertise, identification and similarity.

Questions regarding product involvement, i.e., the level of importance the product (i.e., the one for which information is being sought after from the endorser) in question, were also included. Apart from that, the product's role in the purchase intention is also identified through direct questions, and the frequency of the product usage as well. The level of satisfaction the product has brought was also measured. With regards to the respondent, the "Big Five" personality test was conducted.

6 Results and Discussion

The results were derived from the variables using Ordered Logistic regression in Stata for 2 categories: one for influencers and one for peers. Table 1 shows results for influencer-based and peer-based endorsements and their impact on purchase decision.

For influencers, trustworthiness was positively and significantly related to the likelihood of purchase intention. The results for the moderation hypotheses suggested that the higher the involvement with the product, the greater the impact of expertise on purchase intention. However, the opposite impact can be observed for trustworthiness---the higher the product involvement, the lesser the influence of trustworthiness on purchase intention. Endorser fit mattered and there was a negative relationship with age as well as emotional stability.

For peers---unlike influencers---expertise had a significant and positive impact on likelihood of purchase intention. However, there was no significant moderating impact of product involvement seen for the influence of expertise on purchase intention. There was instead a significant interaction observed for trustworthiness and product involvement---the higher the product involvement, the greater the impact of trustworthiness on the purchase intention. In addition, emotional connect and endorser fit mattered.

Table 1 Effect of influencer and peer on purchase intention

Variable	Peer		Influencer	
	Odds ratio	$P > |z|$	Odds ratio	$P > |z|$
Expertise	**1.444****	0.014	0.821	0.115
Trustworthiness	0.877	0.354	**1.511****	0.006
Identification	**1.082***	0.075	0.949	0.276
Similarity	**0.916****	0.048	1.066	0.164
Product Involvement	0.788	0.395	1.030	0.928
Endorser expertise × product involvement	0.979	0.142	**1.032****	0.012
Endorser trustworthiness × product involvement	**1.026***	0.062	**0.972***	0.054
Emotional connect	**1.051***	0.053	1.037	0.239
Endorser fit	**2.262****	0.000	**1.297***	0.096
Gender	0.797	0.447	1.598	0.152
Age	0.821	0.494	**0.580***	0.094
Extraverted	0.937	0.301	1.036	0.618
Agreeability	0.927	0.297	1.074	0.375
Conesus	0.965	0.620	0.940	0.465
Emotional stability	1.103	0.165	**0.831****	0.020
Openness to new experiences	1.052	0.478	1.080	0.354

***$P < 0.01$, **$P < 0.05$, *$P < 0.10$

Thus, we see that for influencers, expertise matters more than trustworthiness when the product involvement is higher; however, when it comes to peers, the relationship between product involvement and trustworthiness reverses. These are interesting results and merit further study.

7 Limitations and Future Research

This study focused on the type of endorsers and the attitudes people tend to look toward in them. However, future studies can look more into the personalities of the people themselves to see how influencers are being sought. Similarly, the process in which information is being sought through these sources, especially during the information seeking phase of the consumer journey, has not been studied. That study in the future could also show how the various stages in the consumer journey could increase other factors such as product involvement and could affect other variables as well.

8 Conclusion

Information search has always been a main component of the customer journey. In this current era, since there is a vast flow of data flowing through multiple sources, brands need to ensure that product information comes from the right sources---sources that can lead to positive outcome. This is the main reason why there are different types of marketing even within influencers-endorsements, campaigns, etc.

People have always valued credibility of a source. Our study has shown that the variables affecting credibility-trustworthiness and expertise-influence purchase intention in different ways for influencers and peers. It has also been seen that if they have an emotional connect with the source, irrespective of its form, the stronger the purchase intention becomes. Our study has shown that this emotional connect is much more valuable and is much more present for a peer endorser. The results pertaining to the moderating influence of product involvement suggest important differences in the mechanism by which influencer and peer credibility is influencing product purchase.

References

1. Osei-Frimpong, K., Donkor, G., Owusu-Frimpong, N.: The impact of celebrity endorsement on consumer purchase intention: an emerging market perspective. J. Market. Theory Pract. **27**(1), 103–121 (2019)
2. Munasinghe, A.A.S.N., Bernard, D.T.K., Samarasinghe, H.M.U.S.R., Gamhewa, S.R.N., Sugathadasa, S., Muthukumara, T.C.: The impact of celebrity's field of expertise on consumer perception. Int. Rev. Manag. Market. **9**(2), 31 (2019)
3. Khamis, S., Ang, L., Welling, R.: Self-branding, 'micro-celebrity' and the rise of social media influencers. Celebr. Stud. **8**(2), 191–208 (2017)
4. Parmar, B.J., Patel, R.P.: A study on consumer perception for celebrity and non celebrity endorsement in television commercials for fast moving consumer goods. Global Bus. Econ. Res. J. **3**(2), 1–11 (2014)
5. Akar, E., Topçu, B.: An examination of the factors influencing consumers' attitudes toward social media marketing. J. Internet Comm. **10**(1), 35–67 (2011)
6. Malik, A., Sudhakar, B.D.: Brand positioning through celebrity endorsement-a review contribution to brand literature. Int. Rev. Manag. Market. **4**(4), 259–275 (2014)
7. Pradhan, D., Duraipandian, I., Sethi, D.: Celebrity endorsement: how celebrity–brand–user personality congruence affects brand attitude and purchase intention. J. Market. Commun. **22**(5), 456–473 (2016)
8. Schouten, A.P., Janssen, L., Verspaget, M.: Celebrity versus Influencer endorsements in advertising: the role of identification, credibility, and product-endorser fit. Int. J. Advert. **39**(2), 258–281 (2020)
9. Munnukka, J., Uusitalo, O., Toivonen, H.: Credibility of a peer endorser and advertising effectiveness. J. Consum. Market. **33**(3), 182–192 (2016)
10. Crick, N.R., Dodge, K.A.: Social information-processing mechanisms in reactive and proactive aggression. Child Dev. **67**(3), 993–1002 (1996)

The Influence of Fan Behavior on the Purchase Intention of Authentic Sports Team Merchandise

Anand Vardhan, N. Arjun, Shobhana Palat Madhavan, and Deepak Gupta

Abstract The ease of access to satellite TV and online content has led to an increase in popularity and growth of premier sports leagues and clubs across the globe. This has been accompanied by an increase in the sale of branded sports team merchandise which represents a major source of revenue for sports clubs. However, in the Indian context, little is known about what factors influence the purchase of authentic sports team merchandise by sports fans. The objective of this study is to understand how fan behavior influences the purchase of authentic or genuine sports team merchandise. Given the growing fans among Indians of international and national sports clubs, this study becomes very relevant. An online survey of 200 Indian sports fans living in Tier 1 and Tier 2 cities was conducted and the data analyzed using ordinal logistic regression. The study found that team identification had a positive influence on the purchase of authentic sports team merchandise as did the degree of fan involvement. This study contributes to the literature on sports marketing and offers practical implications for encouraging the purchase of authentic sports team merchandise.

1 Introduction

A number of sports fans of clubs such as Manchester United and Real Madrid are no longer restricted to Europe but have been increasing across the globe through

---

A. Vardhan (✉) · N. Arjun · S. P. Madhavan · D. Gupta
Amrita School of Business, Amrita Vishwa Vidyapeetham, Coimbatore, India
e-mail: anandvardhan96@gmail.com

N. Arjun
e-mail: arjun18.official@gmail.com

S. P. Madhavan
e-mail: m_shobhana@cb.amrita.edu

D. Gupta
e-mail: dgshobs@gmail.com

© The Editor(s) (if applicable) and The Author(s), under exclusive license to Springer Nature Singapore Pte Ltd. 2021
T. Senjyu et al. (eds.), *Information and Communication Technology for Intelligent Systems*, Smart Innovation, Systems and Technologies 196, https://doi.org/10.1007/978-981-15-7062-9_57

cyber-mediated relationships facilitated by the ease of access to the Internet and satellite TV [1]. This growth in popularity has contributed positively to the sports market that is expected to reach nearly $614 billion by 2022 [2]. While branded merchandise of sports clubs is an important source of revenue for sports clubs, the sale of counterfeits represents a major loss for these clubs and manufacturers [3]. The global market for counterfeit goods has grown exponentially over the last few decades to account for around 8 percent of world trade with annual sales estimated at US dollar 300 billion [3]. It is interesting to examine what factors would influence sports fans to purchase authentic or genuine sports team merchandise rather than purchasing the often lower-priced counterfeit sports team merchandise. To the best of our knowledge, the relationship between sports fan behavior and the purchase of authentic sports team merchandise has not yet been examined in the literature. The literature on counterfeits has mainly focused on luxury goods, music and software [4] with sporting merchandise receiving some attention [3]. This study seeks, therefore, to understand the factors influencing the purchase of authentic sports merchandise by sports fans.

2 Literature Review

Sports fans are defined as individuals who are interested in a sport and follow a particular sport, team or player [5]. Being a fan of a particular sports club is associated with emotional significance and has been found to contribute to both social identity and personal identity [5]. The behavior of sports fans has been widely studied. The success of a sports team, the quality of players, star players, influence of peers and friends, and the influence of family have been found to influence identification with a sports team [5]. Sports fans have been found to be emotionally engage with their team's successes and failures that in extreme cases can lead to aggression and violence [6]. The benefits of being a sports fan include social benefits such as a feeling of psychological well-being [7]. Purchasing and wearing authentic team merchandise of successful sports clubs can therefore contribute to self-esteem.

The theory of planned behavior [8] has been used to show how subjective norms, attitude and perceived behavioral control influence the purchase of counterfeit sports merchandise [3]. A study on Indian sports fans found that fan engagement can lead to a positive attitude toward the sponsor's brand [9]. However, in the Indian context, there has been no specific focus on the purchase behavior by sports fans of authentic sporting team merchandise.

3 Theoretical Framework and Conceptual Model

This study draws from social identity theory that suggests that individuals categorize themselves into groups in order to define their self-identity and make favorable comparisons within their social environment [10]. In the context of sports fans, sports fan identification has been found to be positively related to the prestige of the team and the length of association with the team [5]. Sports fandom, fan involvement, team identification, and psychological commitment are all expected to influence purchase intention [11–14]. Counterfeit purchase behavior has been found to be influenced by factors such attitude to counterfeits, materialism, and ethics; this has been termed as the counterfeit proneness [4]. Self-esteem and need to belong constructs have also been found to influence counterfeit purchase intention [4]. Gender and income were added as control variables. The conceptual model is shown in Fig. 1.

4 Conceptual Model

See Fig. 1.

5 Hypotheses

Based on existing literature and theory, we propose the following hypotheses:

H1: Higher the sports fandom, higher the purchase intention of authentic sports team merchandise

H2: Higher the team identification, higher the purchase intention of authentic sports team merchandise

H3: Higher the psychological commitment to a team, higher will be the purchase intention of authentic sports team merchandise.

H4: Higher the fan involvement, higher will be the purchase intention of authentic sports team merchandise.

H5: Higher the need to belong, higher the purchase intention of authentic sports team merchandise

H6: Higher the face consciousness, lower the purchase intention of authentic sports team merchandise

H7: Higher the counterfeit proneness, lower the purchase intention of authentic sports team merchandise

H8: Higher the self-esteem, higher the purchase intention of authentic sports team merchandise

6 Methodology

A quantitative approach using primary data was adopted. An online questionnaire was designed to collect primary data. The questionnaire comprised validated scales from the literature in addition to questions on intention to purchase counterfeit sporting

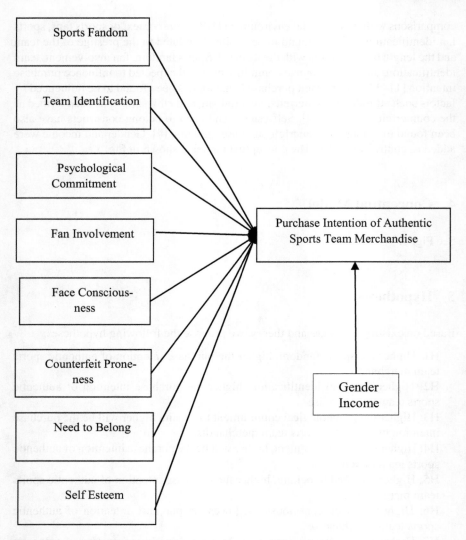

Fig. 1 Conceptual model

merchandise, types of sports clubs followed. Table 1 shows the sources of the scales used for measurement.

Non-probabilistic sampling technique of purposive sampling was used. Snowball sampling was also used. The total number of respondents were 223. Of these, only 200 were sports fans, 23 did not follow any sports. The data used for estimating the model was therefore from these 200 sports fans. Among these 200 fans, 144 fans followed foreign football clubs including Barcelona FC, Liverpool FC, Manchester United

Table 1 Constructs used in the model

S. No.	Constructs	Name of author--date--reference number
1	Sport fandom	Wann [12]
2	Team identification	Wann and Branscombe [11]
3	Psychological commitment to team	Mahony et al. [13]
4	Fan involvement	Capella [14]
5	Need to belong	Nichols and Webster [15]
6	Counterfeit proneness	Sharma and Chan [4]
7	Face consciousness	Sharma and Chan [4]
8	Self-esteem scale	Robins et al. [16]

FC, Real Madrid FC and Arsenal. Among the other respondents, 29 respondents mentioned being fans of Indian clubs IPL (cricket) and ISL (Indian Super League). Ordinal logistic regression was used to analyze the data using STATA software.

7 Analysis and Results

Table 2 shows the results of the analysis of the ordinal logistic regression.

As shown in Table 2, only three independent variables were found to have a significant influence on purchase intention. Hypothesis 1 and Hypothesis 4 were confirmed. In other words, team identification (H1) and fan involvement (H4) were found to have a positive influence on purchase of authentic merchandise ($p < 0.001$). The need to belong was found to have a positive influence on purchase of authentic merchandise at the $p < 0.1$ significance level.

8 Conclusions

The results of this study have important managerial implications especially given the phenomenon of growing fandom of Indians toward sport clubs and premier leagues. Both Indian and foreign sports teams could use the results of this study to increase team identification and fan involvement among Indian fans. Fan involvement has been found to depend on external displays of fandom [14]. Fan involvement can be therefore by increased among Indian fans by distributing vehicle stickers with the team's name, providing facilities at sports venues to paint faces, and encouraging fans to wear team colors. Fan engagement can also be increased by increased attendance at sports events [14]. Sports clubs can therefore offer discounts for season tickets

Table 2 Intention to purchase authentic sports merchandise

Variable	Odds ratio	Standard error	z	P > |z|	Confidence interval 95%	
Sports fandom	1.002967	0.01729	0.17	0.864	0.96965	1.0374
Team identification	*1.230633*	0.05510	4.63	*0.000***	1.1272	1.3435
Psychological commitment	1.015044	0.00982	1.54	0.123	0.99598	1.0345
Self-esteem	1.042955	0.12704	0.35	0.730	0.82145	1.3242
Counterfeit proneness	0.9963133	0.02427	−0.15	0.879	0.94986	1.0450
Face consciousness	1.048033	0.05416	0.91	0.364	0.94706	1.1598
Need to belong	*1.029986*	0.01727	1.76	*0.078**	0.99668	1.0644
Fan involvement	*1.487359*	0.14531	4.06	*0.000***	1.2283	1.8012
Female	1.286301	0.79606	0.41	0.684	0.38243	4.3264
Low income	1.281249	0.39726	0.80	0.424	0.69778	2.3526

***$p < 0.001$; **$p < 0.05$; *$p < 0.1$

to increase Indian fan involvement. Team identification can be increased through the help of influencers on social media. Using social media would also help those individuals and address their need to belong since social media platforms offer an opportunity to belong to a community. Sports teams could also facilitate the formation of social media fan clubs to address the need to belong and increase fan identification. Helping to build team identification and fan involvement would lead an increase in purchase of authentic sports team merchandise. This is the first study, to the best of our knowledge, that has identified these factors as being important to motivate the purchase of authentic sports team merchandise.

9 Limitations and Future Research

This study did not differentiate between fans of foreign clubs and Indian clubs. Future research could therefore analyze the difference between identification and involvement of fans and purchase intention with regard to foreign and Indian clubs among Indian sports fans. Future research could also test whether fan identification mediates the relationship between personality variables and purchase intention.

References

1. Katz, M., Baker, T.A., Du, H.: Team identity, supporter club identity, and fan relationships: a brand community network analysis of a soccer supporters club. J Sport Manag **34**(1), 9–21 (2020)
2. Business Research Company. https://www.thebusinessresearchcompany.com/report/sports-market. Last accessed 2020/3/15
3. Lang, B.: Using the theory of planned behaviour to explain the widespread consumption of counterfeit sports jerseys among American college students. J. Custom. Behav. **16**(4), 315–332 (2018)
4. Sharma, P., Chan, R.Y.K.: Counterfeit proneness: Conceptualisation and scale development. J. Market. Manag. **27**(5–6), 602–626 (2011)
5. Jacobson, B.: The social psychology of the creation of a sports fan identity: a theoretical review of the literature. Athletic Insight **5**(2), 1–14 (2003)
6. Toder-Alon, A., Icekson, T., Shuv-Ami, A.: Team identification and sports fandom as predictors of fan aggression: the moderating role of ageing. Sport Manag. Rev. **22**(2), 194–208 (2019)
7. Kim, J., James, J.D.: Sport and happiness: understanding the relations among sport consumption activities, long-and short-term subjective well-being, and psychological need fulfillment. J. Sport Manag. **33**, 119–132 (2019)
8. Ajzen, I.: The theory of planned behavior. Organ. Behav. Hum. Decis. Process. **50**, 179–211 (1991)
9. Pradhan, D., Malhotra, R., Moharana, T.R.: When fan engagement with sports club brands matters in sponsorship: influence of fan–brand personality congruence. J. Brand Manag. **27**(1), 77–92 (2020)
10. Tajfel, H., Turner, J.: The social identity theory of intergroup behavior. Psychol. Intergroup Relat. **5**, 7–24 (1986)
11. Wann, D.L., Branscombe, N.R.: Sports fans: Measuring degree of identification with their team. Int. J. Sport Psychol. (1993)
12. Wann, D.L.: Preliminary validation of a measure for assessing identification as a sport fan: the sport fandom questionnaire. Int. J. Sport Manag. **3**, 103–115 (2002)
13. Mahony, D.F., Madrigal, R., Howard, D.A.: Using the psychological commitment to team (PCT) scale to segment sport consumers based on loyalty. Sport Market. Q. **9**(1), 15–25 (2000)
14. Capella, M.E.: Measuring sports fans' involvement: the fan behavior questionnaire. Southern Bus. Review. **27**(2), 30–36 (2002)
15. Nichols, A.L., Webster, G.D.: The single-item need to belong scale. Person. Individ. Differ. **55**(2), 189–192 (2013)
16. Robins, R.W., Hendin, H.M., Trzesniewski, K.H.: Measuring global self-esteem: construct validation of a single-item measure and the Rosenberg self-esteem scale. Pers. Soc. Psychol. Bull. **27**(2), 151–161 (2001)

Purchase Decisions of Brand Followers on Instagram

R. Dhanush Shri Vardhan, Shobhana Palat Madhavan, and Deepak Gupta

Abstract Instagram is one of the fastest-growing social media platforms and businesses are increasingly using Instagram to promote their brands. However, in the Indian context, little is known on the role of brand promotion on Instagram on customer engagement. Even more importantly, it is not known whether brand strategies on Instagram lead to a positive purchase intention. This study analyzes the relationship between the customer experience of brand posts on Instagram and customer engagement. Furthermore, the study also examines the relationship between customer engagement and the purchase intention of Instagram users. The study investigates the influence of five categories of customer experience on three types of customer engagement. The five types of customer experience examined in this study are (i) entertainment; (ii) information; (iii) personal identity; (iv) utilitarian and (v) community. The three types of customer engagement are (i) affective; (ii) cognitive and (iii) behavioural. Primary data was collected through an online questionnaire from 259 respondents of which 150 responses were usable. The analysis was done using Partial Least Squares Structural Equation Modelling (PLS-SEM). Entertainment, information, personal identity, and utilitarian experiences were found to have a positive influence on affective engagement. Affective engagement in turn had a positive impact on purchase intention. Information, personal identity, and community were found to have a positive impact on behavioural engagement that in turn influenced purchase intention positively. Implications for marketers are discussed.

R. D. S. Vardhan (✉) · S. P. Madhavan · D. Gupta
Amrita School of Business, Amrita Vishwa Vidyapeetham, Coimbatore, India
e-mail: dhanushshrivardhan@gmail.com

S. P. Madhavan
e-mail: m_shobhana@cb.amrita.edu

D. Gupta
e-mail: dgshobs@gmail.com

1 Introduction

Instagram is one of the fastest-growing social media platforms with over 1 billion users per month and is increasingly used for brand promotion [1–3]. This is because around 200 million users visit at least one business profile daily and an estimated one-third of the most viewed stories are from businesses [1]. India is reported to have 69 million Instagram users [1]. The use of social media has been found to increase advertising effectiveness [3, 4]. Brands such as Nike, Calvin Klein, and Starbucks are leading brands on Instagram globally and reach millions of potential customers through this platform. The challenge for brands is to create Instagram posts that induce a positive customer experience that can help to convert viewers into customers. Customer experience is defined as customer beliefs or the impression a brand leaves upon the customer after a specific action, in this case, seeing a post [4]. Positive customer experience is expected to have a positive impact on customer engagement. Engagement refers to the degree and depth of customer interaction on a social media platform and denotes a psychologically-based willingness on the part of the customer to invest time and effort on the brand [4, 5]. Engagement with brand-related posts on social media is likely to lead to engagement with the brand [4]. Marketers must, therefore, make efforts to increase customer engagement on social medium platforms, with the ultimate goal being positive purchase intention.

While studies have examined the influence of customer experience on advertising effectiveness on social media [4], there has been no study on a brand's customer experience on customer engagement on Instagram in India. Customer engagement with particular brand posts on Instagram is expected to have a positive influence on the purchase intention of the brand. As yet, to the best of our knowledge, there is no study on the relationship between customer engagement with brands on Instagram and ensuing purchase intention. Purchase intention would be the main goal of most brands that use Instagram, so, finding which experiences induce which kind of engagement will push the brands towards a better decision on posts. Furthermore, there has been no study on how customers follow brands on Instagram, customer engagement, and purchase intention in India. This study fills this research gap by relating customer experiences to customer engagement and providing a clearer path towards inducing purchase intention.

2 Literature Review

Customer experience on social media sites such as Instagram tries to provide users with certain experiences or values [4]. Calder et al. classify these types of experiences into five types based on user gratification theory; the five types of customer experiences are [4]:

(i) Entertainment: viewing the site is regarded as a pleasure and as relaxing [4]
(ii) Information: viewing the site gives the viewer points for discussion [4]

(iii) Personal Identity: viewing the site contributes to the identity of the viewer [4]
(iv) Utilitarian: viewing the site helps improve purchase decisions [4]
(v) Community: viewing the site fosters a sense of connection with other viewers [4]

Customer experience, however, may not directly lead to purchase decision but is likely to be mediated by customer engagement [6]. Customer engagement has significant and growing importance for the management of brands [7].

Dessart classifies consumer engagement into three types: affective engagement, cognitive engagement, and behavioural engagement [5]. These are described as follow:

(i) Affective: Emotions experienced by a consumer with respect to his/her engagement focus [5]
(ii) Cognitive: Attention and mental absorption that a consumer experiences with respect to the focal object of his/her engagement [5]
(iii) Behavioural: The behavioural manifestations toward an engagement focus such as sharing and forwarding [5]

For instance, in the context of brands on Instagram, affective engagement [5] may be denoted by a consumer who gets happy when the brand replies to him or a consumer who enjoys interacting with the other fans in the comment section on Instagram. Cognitive engagement [5] may be denoted by a consumer who is so absorbed in the posts that she/he spends a lot of time browsing it or a consumer paying a lot of attention to the comments and replies of other consumers on the posts by the brand. Behavioural engagement [5] may be denoted by a consumer who shares his opinion about a product with the brand on Instagram or a consumer who asks advice or information from other members of the community about the brand.

Customer engagement has been found to have positive outcomes for marketers. Calder et al. found that personal and social–interactive engagement is positively associated with advertising effectiveness [4]. A study by Dessart found that high social media engagement increases brand trust, commitment, and loyalty [5]. The study also demonstrated the importance of individual traits in segmenting consumers for effective engagement [5]. For instance, highly informational and educational content is likely to suit highly involved consumers, whereas playful content would better suit customers who are not so involved [5]. It has been found that consumers who pay attention do not enjoy visiting a social media site might respond entirely differently to marketing efforts than low-attention who seek enjoyment from viewing the site [7].

3 Conceptual Model

The conceptual model was developed based on the constructs developed by Calder [4] and Dessart [5] and is shown in Fig. 1.

Independent variables

Dependent variables

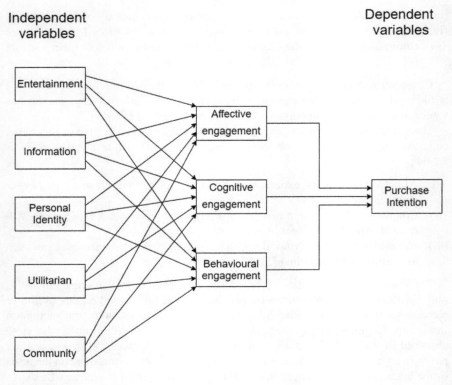

Fig. 1 Conceptual model

Calder developed five constructs of customer experience on social media namely, entertainment, information, personal identity, utilitarian, and community [4]. These are expected to result in three different types of consumer engagement, namely, affective, cognitive, and behavioural as developed by Dessart [5].

These three different types of engagement are expected to have a positive influence on purchase intention of the Brand posted on Instagram.

4 Hypotheses

Based on Fig. 1 we propose several hypotheses. Some illustrative hypotheses are listed below:

H1: Instagram posts on a particular brand that provide an entertaining experience to the viewers have a positive impact on affective engagement.
H4: Instagram posts on a particular brand that provide an informative experience to the viewers have a positive impact on cognitive engagement.

H7: Instagram posts on a particular brand that provide a sense of personal identity to the viewers have a positive impact on behavioural engagement.

H10: Instagram posts on a particular brand that provides utilitarian benefits to the viewers have a positive impact on affective engagement.

H15: Instagram posts on a particular brand that provides a sense of belonging in a community to the viewers have a positive impact on behavioural engagement.

H16: Affective engagement positively affects the purchase decision of the viewer.

H17: Cognitive engagement positively affects the purchase decision of the viewer.

H18: Behavioural engagement positively affects the purchase decision of the viewer.

5 Methodology

A quantitative approach using primary date was adopted. PLS-SEM was used to analyze the data using STATA software.

First, a structured online questionnaire was created to collect the data from the sample. The questionnaire was focused on the experiences related to brand posts on Instagram followed by questions on respondent engagement and purchase decision. Section 1 of the questionnaire concentrated on the experiences of the posts that particular brands provide. The types of experiences were Entertainment, Information, Personal Identity, Utilitarian and Community. Section 2 of the questionnaire focused on the engagement dimension: Affective Engagement, Cognitive Engagement, and Behavioural Engagement. Section 3 concentrated on Purchase Decisions and Sect. 4 was about demographics.

A pilot test was conducted to test the reliability and validity of the questionnaire. A non-probabilistic purposive sampling technique and snowball sampling was used in the survey. Before the questions on the subject, an initial question is asked to find if the participants were actually following any brands on Instagram. If they answer positively to that, they are asked to provide a few brand names they follow, to ensure if they have understood the context. Only after a positive response to these two questions, they are allowed to answer the rest of the questions. This purposive sampling was followed by snowball sampling. Respondents were asked to collect responses from other people who they knew followed brands on Instagram. The final data was collected from a sample size of 259 respondents who were demographically segmented based on their age, gender and income. However, only 150 responses were considered valid and were used in the analysis.

6 Analysis and Results

Table 1 shows the results of the analysis.

Table 1 Results of PLS-SEM

Variable	Affective engagement	Cognitive engagement	Behavioural engagement	Purchase intention
Entertainment	0.288*** (0.001)	0.348*** (0.003)	0.063 (0.516)	
Information	0.193* (0.064)	0.002 (0.991)	0.422*** (0.001)	
Personal identity	0.213** (0.028)	−0.027 (0.828)	−0.182* (0.093)	
Utilitarian	0.198** (0.029)	−0.029 (0.813)	−0.019 (0.860)	
Community	0.029 (0.779)	0.406*** (0.001)	0.413*** (0.001)	
Affective engagement				0.419*** (0.000)
Cognitive engagement				0.027 (0.809)
Behavioural engagement				0.234** (0.033)
r2_a	0.568	0.354	0.420	0.328

P values in parenthesis. ***P < 0.01, **P < 0.05, *P < 0.10

As shown in Table 1, our analysis shows that Instagram posts that provide an entertaining experience to the viewers engage the viewers through affective engagement and cognitive engagement. Instagram posts that provide an informative experience to the viewers, engage the viewers through affective engagement and behavioural engagement. Instagram posts that provide a sense of personal identity to the viewers, engage the viewers through affective engagement and behavioural engagement. Instagram posts that provide utilitarian benefits to the viewers, engage the viewers through affective engagement. Instagram posts that provide a sense of belonging in a community to the viewers, engage the viewers through cognitive engagement and behavioural engagement.

Only affective engagement and behavioural engagement positively impacts the purchase decision of the viewer. Cognitive engagement was not found to have a significant impact on purchase intention.

7 Conclusions

Our findings suggest that affective engagement of viewers is important for purchase intention. Affective engagement includes feeling enthusiastic about the brand and enjoying interacting with the brand. This is in line with prior studies that have shown the importance of affective engagement [8]. Marketers should, therefore, make efforts to increase affective engagement. Behavioural engagement that includes sharing ideas and content was also found to have a positive impact on purchase intention. Marketers can thus target those viewers who share ideas and content since these viewers are likely to be positively inclined to purchase the brand.

Our study shows that cognitive engagement does not have a positive influence on purchase intention. Cognitive engagement includes spending a lot of time thinking about the brand to the extent of forgetting everything else [5], This implies that even if viewers spend time on viewing posts this does not necessarily result in a positive purchase intention.

8 Limitations and Future Research

A limitation of our study is that this study did not differentiate between different product categories. Viewers may have different responses regarding purchase decisions to the posts that involve high involvement compared to posts that involve low involvement products. A good hypothesis to consider would be that positive purchase intentions related to brand posts may be higher for products that are low involvement. This could be an area for future research.

Another limitation is that we did not examine the difference between responses to product posts on Instagram versus posts on services. In the Indian context, brands such as Swiggy and Zomato that offer services are popular on Instagram. Future research could examine the differences between consumer responses to brand posts on products and services on Instagram.

References

1. https://blog.hootsuite.com/instagram-statistics, last accessed on 2020/3/17
2. Rietveld, R., van Dolen, W., Mazloom, M., Worring, M.: What you feel, is what you like influence of message appeals on customer engagement on instagram. J. Inter. Market. **49**, 20–53 (2020)
3. Anagnostopoulos, C., Parganas, P., Chadwick, S., Fenton, A.: Branding in pictures: using Instagram as a brand management tool in professional team sport organisations. Eur. Sport Manag. Q. **18**(4), 413–438 (2018)
4. Calder, B.J., Malthouse, E.C., Schaedel, U.: An experimental study of the relationship between online engagement and advertising effectiveness. J. Inter. Market. **23**(4), 321–331 (2009)
5. Dessart, L.: Social media engagement: a model of antecedents and relational outcomes. J. Market. Manag. **33**(5–6), 375–399 (2017)
6. Dessart, L., Veloutsou, C., Morgan-Thomas, A.: Consumer engagement in online brand communities: a social media perspective. J. Prod. Brand Manag. **24**(1), 28–42 (2015)
7. Dessart, L., Veloutsou, C., Morgan-Thomas, A.: Capturing consumer engagement: duality, dimensionality and measurement. J. Market. Manag. **32**(5–6), 399–426 (2016)
8. Casaló, L.V., Flavián C., Ibáñez-Sánchez, S.: Be creative, my friend! Engaging users on Instagram by promoting positive emotions. J. Bus. Res. (February 2020)

Deep Learning Based Parking Prediction Using LSTM Approach

Aniket Mishra and Sachin Deshpande

Abstract In daily life, parking vehicle is the most common issue arising due to the increase in a number of vehicles every year. Managing a parking system is quite a tough but important factor in minimizing the flow of traffic in cities. Research shows almost 30% of traffic is caused due to drivers watching for parking places. Many systems are installed to manage parking prediction but this system has less accuracy and cost of computation is high. The main intent of this proposed paper is to address these two major requirements and parking availability will be predicted during different time-periods like Weekday, Weekend, and Cultural events with peak hour attribute into consideration. In this paper, the model is designed such that the LSTM algorithm will be used to predict the parking space because it performs efficiently in case of time series prediction. The overall model works efficiently using LSTM for the project because the data-set contains sequential data of parking space information available for every single hour. Along with the prediction system, we are also developing a parking slot booking system, where the user of the app can book a parking slot well in advance. Also along with the prediction model, our system can give user live status of the parking slots available at a particular parking lot. The accuracy of parking prediction will be compared against the parking accuracy obtained using LSTM model.

1 Introduction

In cities, drivers have to stroll for the parking at the end of their destination by the search for a parking space, worsening the condition of traffic and exhausting there as well as others time and fuel. Parking with new Technologies refers to Intelligence and

---

A. Mishra (✉) · S. Deshpande
Department of Computer Engineering, Vidyalankar Institute of Technology, Mumbai, India
e-mail: amaniketmishra2@gmail.com

S. Deshpande
e-mail: sachin.deshpande@vit.edu.in

Communication of automation which is an explanation meant to improve parking search and its availability by providing information about parking locations and available slots and their authenticated or predicted availability. It is fairly difficult to know the parking availability and it becomes tricky for on-street parking. There is a need to nominate a machine learning based prediction of parking availability in distinct terms system architecture. The ecosystem is such that the components of our system will be linked in a specified way with ever other components for better performance by including the collected data and then data processing fundamental and application component.

The later phase of data processing, Long Short-Term Memory (LSTM) algorithm with its complete network is much needed, and is better to anticipate the parking availability. In case to boost prediction certainty, we consider lot more factors for developing countries which take into more attributes in an algorithm such as what time of day, climate, which day of the week, holiday, season and cultural events. The complete phase of processing of data is passed through the training phase of the design on the server to meet the requirements of high achievement in terms of computing. As the system designed provides an indicator for parking slots and service for many regions, huge data processing may cause the system cost to be rarely accepted. In our system design, we use archival data to train perception model for prediction. It uses the last added data and cleans and forward it to next phase of the model used on basis of newly entered data which is collected, rather than using the old data set with historical data again and again. There is a need of making the minimal count of computation cost and maintenance. Therefore, the system will update and store new data the results every specific hour so that the recent availability is known timely. When the updating process is accomplished, the guess results of the next specific minute will be stored in the database. The workflow would be cost-effective for lessening the computation cost, for the processes of refining data training and updating do not need to be in functioning state all the time. The aim is to make a proper flow of the proposed system so that the proposed system can conduct comprehensive experiments on the dataset gathered.

2 Literature Review

Harshitha Bura, Nathan Lin, Naveen Kumar, Sangram Malekar, Sushma Nagaraj proposed a technique using license plate detection and vehicle tracking. A customize network model is created for predicting the vacant slots in parking slots. On basis of data that will be stored will results in an increase in speed and accuracy of the algorithm using object detection model and linking connection between the cameras used on the ground and other cameras to achieve object tracking [1].

Johan Hakansson, Vijay Paidi, Hasan Fleyeh and Roger G. Nyberg Proposed the combination of different type of devices such as parking sensors, different technologies and applications using machine learning, convolutional neural network(CNN) or many other systems suitable for parking lots due to less cost and more resistance

Deep Learning Based Parking Prediction Using LSTM Approach   591

for different environmental conditions. In this Machine, vision is another technology that works by visual camera which acquires real-time parking slots information on open parking lots due to its minimal cost and maintenance [2].

Badii, C., Nesi, P., and Paoli uses four different machine learning algorithms to predict the parking slots of parking places and gates. The algorithms used are Artificial Neural Network with Support Vector Regression(SVM), Bayesian Regularization, Recurrent Neural Network(RNN), ARIMA Model(Auto-Regressive Integrated Moving Average). The system provides information about vacant parking slots into the drivers in advance. Drivers can decide which parking slots to be selected according to availability.

Julien Nyambal has presented a paper on an approach based on a realtime parking space classification based on Convolutional Neural Networks (CNN) using a framework known as Caffe and Nvidia DiGITS. The inputs and testing have been done using DiGITS and the output comes in a form of Caffe model which is used for predicting and checking occupied and vacant parking space. The system checks only for a specific area whether a parking space is containing a vehicle or not the paper uses CNN which provides a better success rate than other machine learning algorithms [3].

Dr. Sean McGrath, Dr. Eoin O'Connell has used Image Processing and Artificial Intelligence in this paper which is also used to make overall design of smart parking. Many cameras and ultrasonic sensors were used and linked with the different locations to recognize the plate numbers and ensure without ticket parking. Big data analysis and neural networks are used to integrate multiple technologies to provide the long-term and efficient solution and increasing the potential of space allocating the correct information of parking spaces to the users [4].

3 Proposed System

The main aim of the design is to make factual authentic predictions for the parking management system. The structural design of the system will compile the live data through Parking space report and user using Application based on Mobile and administration having access to the website of parking lot and use these data to anticipate the technology in public parking in a certain region. The process includes a huge amount of unsorted data to be processed then it is processed for data collection, data processing, Historical data and an API for users. The huge raw volume of data and computing attempt are a big threat, so system is linked with the server, which can contribute to high working of computation at a diminished cost and minimal maintenance (Fig. 1).

In the proposed system, we are developing Car parking locator system and Car parking availability prediction system. In this, admin have to manage application by web portal using web services over the internet. In this Android Application, all users have to log in, check parking lot, see parking availability prediction, book a parking slot, pay using wallet. To briefly describe, users have to Login Application by

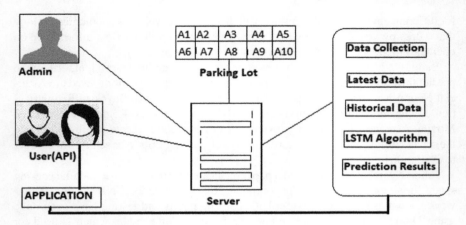

Fig. 1 System architecture

providing information such as Car number and person name. User can also check the predication of availability of parking based on Day, Time of Day, Holidays, Season, Climate, etc. User can also pre-book a parking slot at a particular time and the system assigns a parking slot where the user can directly go and park. The amount will be deducted from wallet according to Depending on their arrival to the parking slot and Out-Time for this Web Portal Application for Admin Login is made so that Admin and user can login by Identification document and password. Fill up Information like—Admin enters Car number, Person name, Parking Slot number.

Location—Find location By Entering Car number showing the location of Car. From this system, Car parking can be predicted and thus saves time from finding parking of vehicle.

4 Block Diagram

The following Block Diagram gives more understanding of the topic. In public parking there may be many parking levels, the user selects parking lot using a mobile application and then the system assigns a parking slot to the user according to availability of the parking space.

The user has a option of entering a "In time" and "Out time" to the parking spot and if any user goes without pre-booking of parking then also system will be updated with the remaining slots available.

Block diagram gives an overview of the proposed system (Fig. 2).

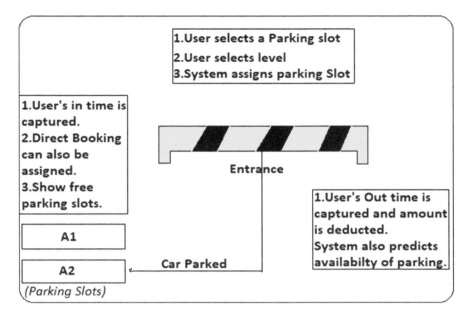

Fig. 2 Block diagram

5 Methodology

LSTM stands for Long Short Term Memory is an exceptional kind of RNN. LSTM is skillful in researching new concepts in the longer term. The main disadvantage of RNN is gradient vanishing and exploding problems which is addressed by LSTM through a concept called cell state which is the internal state of LSTM. Gates in the LSTM algorithm is able to count or eliminate the information stored in the cell state. These gates decide whether to retain or release information to the subsequent state. These gates are made of sigmoid layer outputs. These sigmoid layer generates two outputs 0 and 1. When 0 gets generated it represents no information gets passed through the gate and value is 1 it exhibits that all information will be passed through the cell state. The next phase is decider and finalize for the data to be selected and which of the information needs to be stored in the cell state.

This methodology executes into two levels. First step is called a sigmoid layer which concludes what values need to be amended. The second step is called a tan layer which creates a vector of new value that will be combined to the state. The final phase will compile both the two steps to create and manage and updated value. Vanishing and Exploding gradient, the major problems of RNN network is been address through LSTM network. LSTM network performs exceptionally in everlasting prediction and responds well to critical cases (Fig. 3).

$$*Cell : st\, c = bt\, f\, st - 1c + btl\, g(Ii = 1wicxti) \qquad (1)$$

Fig. 3 Flow diagram

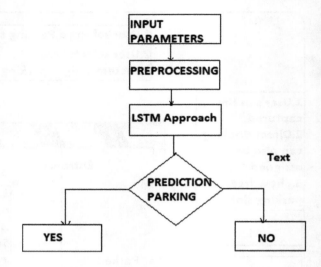

$$*Cell\,Output : btc = bt\,wh(st\,c) \qquad (2)$$

st c   (output)Cell at time t
bt f   (output)forget gate at time t,
st − 1 (output)Cell at time $t-1$,
Bt l   (output)Input Gate at time t, and
bt w (output)Output Gate at time t.

LSTM algorithm could be a network which is can be used for predicting the factual number of vacant parking slots and occupied in developing countries. To get the higher prediction report, we consider different possibilities which are a time of the day, Holiday, Climate, Season, day of the week then deployed the training process on the database to fulfill the need of the upper achievement rate. The following examined system provides a prediction of parking slots and this service is for several areas, therefore, a huge amount of information for processing and high computing.

In our structural work-flow, we use the old recorded archival data to an experienced model that is used for prediction. Then we use the new data to amend the design based on the experienced model on a regular basis, rather than accommodating the model with old archival data again. The model keeps the data and update of task which is executed every fixed hour. When the data of an archival is in a process and finishes its execution, then the prediction results are generated and is saved to the database of the next fixed hours. Due to which the structural workflow is prudent for the execution and also for saving computation and maintenance cost (Fig. 4).

In this section, we broadly discuss the principle of the LSTM algorithm and emphasize why there is a need for LSTM for the availability of parking prediction. Recurrent Neural Network (RNN) algorithm executes well in dealing with subsequent or time differing complications. However, it has a symbolic shortcoming,

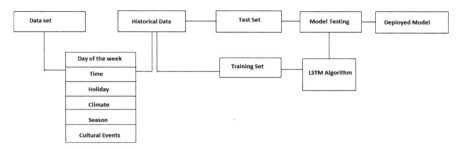

Fig. 4 Workflow

having the vanishing and exploding inclination problem by learning tasks involving long term inter-dependency.

The process starts after the cleaning of raw data from missing and null values. This process will be done in the Data Preparation phase and the cleaned data will be the input to the modelling phase. Once the data is transformed into the desired form. Feature selection process will be performed to extract the desired attributes. The attributes through which parking prediction will be done are extracted through this process. Attributes like Peak hour, Time based variables, Number of parking spaces, Date and time, etc. are some of the fields selected through this process. The dataset with selected attributes is then divided into two parts. Some portion of data will be provided to the training set and the rest of them will be sent to the Test set. Using the Trained set LSTM model will be built and will be trained for prediction. Then using the Test set of data the trained model will be tested. The tested model will be evaluated for accuracy with the current benchmark value.

6 Conclusion

The parking availability prediction is done on basis of LSTM network. In case to lift prediction rates, we consider lot more factors for developing countries which take into more attributes in an algorithm such as what time of day, climate, which day of the week, holiday, season, and cultural events. A lot of research has been done on the past using different attributes like Day, Time of day, and Holidays to evaluate the accuracy of parking prediction. The records of data that are acquired are processed from different sources in this research are to be checked and analyzed with the outcome of the sources which was the concatenated data of many of the trailing entered data's findings from different research to check for any improvement. Based on studies on LSTM, our prediction model will provide an economical and efficient flow. The Algorithm used will work according to the provided dataset and is complete is arranged and stored on the server with a dataset (Fig. 5).

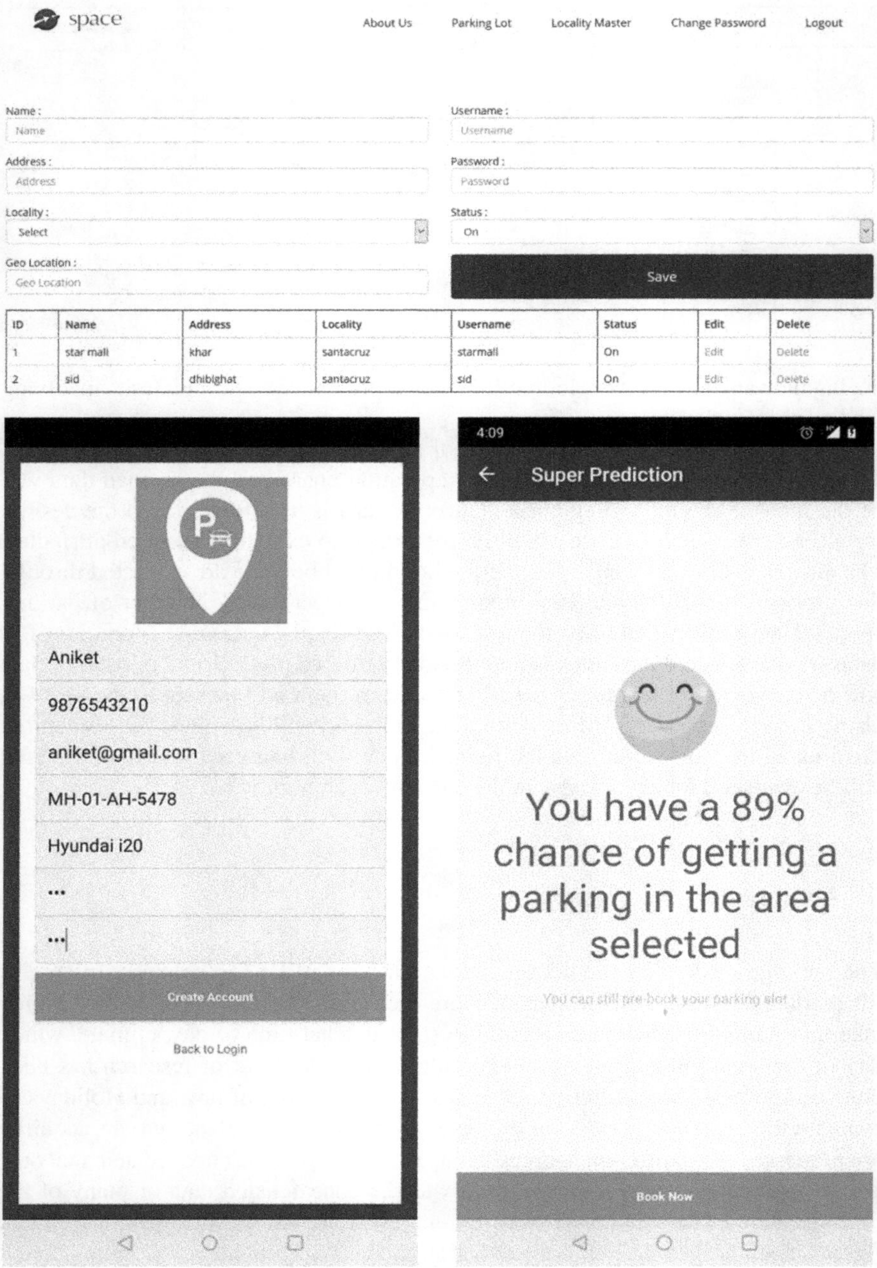

Fig. 5 Desired output

Acknowledgements I want to stretch out my genuine gratitude to all who helped me for the undertaking work. I want to earnestly express gratitude toward Prof. Sachin Deshpande for their direction and steady supervision for giving crucial data with respect to the effort likewise, for their help in completing this task work. I want to offer my thanks towards my folks and individuals from Vidyalankar Institute of Technology for their thoughtful co-activity and support.

References

1. Bura, H., Lin, N., Kumar, N., Malekar, S., Nagaraj, S., Liu, K.: Computer Engineering Department, San José, CA, USA An Edge Based Smart Parking Solution Using Camera Networks and Deep Learning
2. Paidi, V., Fleyeh, H., Håkansson, J., Nyberg, R.G.: School of Technology and Business Studies, Dalarna University, Borlange, Sweden. Smart parking sensors, technologies and applications for open parking lots
3. Nyambal, J., Klein, R.: School of Computer Science and Applied Mathematics University of the Witwatersrand Johannesburg, South Africa. Automated Parking Space Detection Using Convolutional Neural Networks
4. McGrath, S., O'sConnell, E.: Electronic & Computer Engineering Department, University of Limerick. Smart Parking System Using Image Processing and Artificial Intelligence
5. AlMaruf, M.A., Ahmed, S., Ahmed, M.T., Roy, A., Nitu, Z.F.: A proposed model of integrated smart parking solution for a city'. In *2019 International Conference on Robotics, Electrical and Signal Processing Techniques (ICREST)* (2019)
6. Alsafery, W., Alturki, B., Reiff-Marganiec, S., Jambi, K.: Smart car parking system solution for the internet of things in smart. In *2018 1st International Conference on Computer Applications Information Security (ICCAIS)* (2018)
7. Badii, C., Nesi, P., Paoli, I.: Predicting available parking slots on critical and regular services by exploiting a range of open data. IEEE Access **6**, 44059–44071 (2018)
8. Barata, E., Cruz, L., Ferreira, J.-P.: Parking at the UC campus: problems and solutions. Cities **28**(5), 406–413 (2011)
9. BLOG, C.: 'Understanding LSTM networks—colah's blog' (2019). http://colah.github.io/posts/2015-08-Understanding-LSTMs
10. Camero, A., Toutouh, J., Stolfi, D.H., Alba, E.: Evolutionary deep learning for car park occupancy prediction in smart cities'. Lecture Notes Comput. Sci. 386–401 (2018)
11. Chou, S.-Y., Arifin, Z., Dewabharata, A.: Optimal decision of reservation policy for private parking sharing system. In *2018 International Conference on Computer, Control, Informatics and its Applications (IC3INA)* (2018)
12. Das, S.: A novel parking management system, for smart cities, to save fuel, time, and money. In *2019 IEEE 9th Annual Computing and Communication Workshop and Conference (CCWC)* (2019)
13. Fan, J., Hu, Q., Tang, Z.: Predicting vacant parking space availability: an SVR method with fruit fly optimisation. IET Intel. Transport Syst. **12**(10), 1414–1420 (2018)

Beyond Kirana Stores: A Study on Consumer Purchase Intention for Buying Grocery Online

R. Sowmyanarayanan, Gowtam Krishnaa, and Deepak Gupta

Abstract In India, the use of online grocery shopping platforms—such as Big Basket, Amazon Pantry, Flipkart Supermarket, etc.—has been rising exponentially. The online grocery market value has been increasing steeply in recent years and is going to grow steadily, as the digital natives—the Millennial and the generation after them—grow older and start having families of their own. Given the importance of this growing market, there is surprisingly little research on the factors that are driving the adoption of online grocery in India today. Based on an extensive literature survey we first propose a conceptual model on the factors driving the online and offline grocery shopping behaviour of Indian consumers. We test the model using an online survey and a pan-India sample of 262 consumers. Our analysis suggests that the intention to purchase grocery online is significantly influenced by variety seeking, trust in seller, ease of use of online payments, convenience, purchase decision involvement, and preference for global products.

1 Introduction

Growing e-commerce business boosted by rising demand, an increasingly young earning population, and a generation that is willing to adapt faster has enabled the e-commerce industry to evolve itself to include even the daily used staple products such as grocery. With the increasingly time-crunched lifestyle of millennials and the growth of enhanced logistics and advanced warehousing techniques, both the demand and the supply for online on-time grocery has been rising exponentially.

R. Sowmyanarayanan (✉) · G. Krishnaa · D. Gupta
Amrita School of Business, Amrita Vishwa Vidyapeetham, Coimbatore, India
e-mail: sowmyanarayananr@gmail.com

G. Krishnaa
e-mail: gowtamkrishnaa@gmail.com

D. Gupta
e-mail: dgshobs@gmail.com

According to a recent research report,[1] the Indian online grocery market could grow with a compound annual growth rate of 55% in the period of 2016–21.

If the recent national lockdown on account of Covid-19 is any evidence that consumers are more inclined to use a single portal store for all their requirements with reliable delivery services. The market for online grocery is poised for sustainable and explosive growth.

Hence, it's more important than ever to understand the forces that are underlying the demand growth for the online grocery industry. This could help us identify the demand drivers and expectations of its consumer population. The analysis of these data could enable us to make the necessary tweaks that could refine and streamline its growth.

According to a RedSeer[2] report, the share of online grocery in the total retail market in India will grow to 7% by 2023. The most part of the current growth and market share of online grocery is currently driven only by the major metropolitan and Tier-I cities of India. In spite of rising growth rate and a number of earning digital natives, the number of adopters in Tier-II and Tier-III cities being very low is also a question that needs to be answered. However, there has been surprisingly little academic research on consumer demand drivers for online grocery in India.

This study aims to address this important gap in our knowledge by proposing and empirically testing a comprehensive conceptual model for factors that influence the purchase intention for online grocery products in India.

2 Literature Review

In grocery shopping, the consumers, in general, tend to purchase their groceries from any one of these three places or from a combination of these places—namely local stores, supermarkets or online stores. In order to understand the factors behind these purchase decisions an extensive literature review was conducted.

Trust in a seller is a pre-requisite for the repurchase intention of a consumer. In general, consumers with less or no online shopping experience perceive a notion of risk with e-commerce. Hence, it is important to study the influence of trust placed by consumers on online grocery sellers [1]. Furthermore, consumers who purchase special occasions or periods induced by a unique cause tend to stop using the online platforms once the occasion or period ends [2]. As the world progresses towards the cashless economy, the use of online payment has been drastically increasing as well. Many people find it easier and hassle-free to do transactions in online payments than through cash. Hence, it is crucial to understand its influence on purchase intentions in online grocery shopping.

---

[1]*"Online Grocery Market in India by Product Type, Consumer Behaviour, Competition Forecast and Opportunities, 2011–2021"*.

[2]*"India's online grocery retail market to touch $10.5 billion by 2023: Redseer"*.

Research suggests that online shopping is unique and different from the conventional retail environment, which is typically ranked for its accessibility by its consumers, unlike online retail which is mostly chosen for its convenience, product range or variety, or a time crunch at consumer's end. Therefore, it is essential to study the influence of convenience and variety on consumer's online grocery purchase intentions [3]. Lack of tangibility in online shopping raises the question of quality assurance and risk association by the hesitant non-users. Thus, the influence of need for touch on online grocery shopping intentions must be assessed [4].

Most grocery products are edibles and are associated with the health of the individual the purchase decision involvement is usually high. It is also vital to understand the relation between the level of commitment, one holds with their neighborhood or local grocery stores and his intention to purchase online. Further, to understand the product expectations of the consumers, we can analyze the product preference of consumers to purchase local or global versions of products with their online grocery purchase intentions. Thus, the influence of factors such as purchase decision involvement, commitment to local store, and product preference could also be studied to better understand the consumer's purchase intentions for online grocery. In this research, we propose and test a comprehensive conceptual model on factors that influence the purchase intention of consumers in an online grocery platform. In addition, we also look at the influence of relative preference for the neighborhood stores and local/global versions of products. The dependent variable is purchase intention for online grocery shopping and the independent variables are variety seeking, trust in online sellers, ease of use of online payments, convenience, purchase decision involvement and preference on global or local versions of products, among others.

3 Conceptual Model

See Fig. 1.

4 Hypotheses

H1-Greater the need for variety in shopping, higher the online grocery purchase intentions

H2-Greater the perception that the online payments are easy to use, higher the online grocery purchase intentions

H3-Greater the trust in the online seller, higher the online grocery purchase intentions

H4-Greater the purchase decision involvement in shopping, lower the online grocery purchase intentions

H5-Greater the need for touch in shopping, lower the online grocery purchase intentions

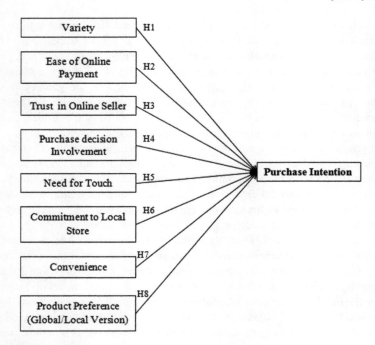

Fig. 1 Conceptual model

H6-Greater the commitment to local store, lower the online grocery purchase intentions

H7-Greater the need for convenience in shopping, higher the online grocery purchase intentions

H8-Greater the preference for global version of products, higher the online grocery purchase intentions

5 Methodology

5.1 Questionnaire Development

A dedicated primary research study was conducted with a structured online questionnaire to collect the data required for the research. The flow of the questionnaire started by understanding their familiarity in online grocery shopping and its frequency, the average cart value purchased, the preferred medium of purchase and questions analyzing the impact of the independent variables on their purchase intention of online groceries and their demographic information was collected towards the end of the questionnaire.

Table 1 Constructs and sources of scales

Constructs	References
Variety	Baumgartner and Steenkamp [5]
Online payment (Ease of use)	Venkatesh et al. [6]
Purchase intention	Swilley [7]
Trust in online seller	Thomson [8]
Purchase decision involvement	Mittal [9]
Need for touch	Peck and Childers [10]
Commitment to store	Cho [11]
Convenience	Noble [12]
Product preference (global/local version)	Zhang and Khare [13]

Table 1 shows the list of the most important constructs and scale items measured in this study.

5.2 Data Description

The non-probabilistic sampling technique of quota sampling was used in selecting the respondents and the questionnaire was distributed to people above 18 years of age across tier 1 and tier 2 cities in India. The total number of respondents who answered the questionnaire were 262. Among the respondents, 161 were male and 101 were female. About 236 respondents who participated in the survey had an undergraduate degree or higher

Figure 2 shows the percentage of female and male respondents of the survey. Figure 3 shows the educational qualification of the respondents.

Fig. 2 Male and female respondents

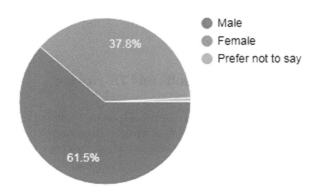

Fig. 3 Educational qualification of respondents

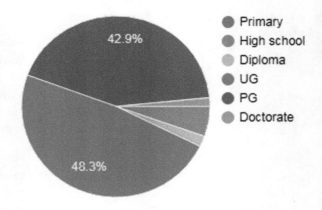

Table 2 Results of ordinal logistic regression model

Variables	Odds ratio	z	P > z	[95% conf. interval]	
Online payment (ease of use)	**1.125*****	3.1	0.002	1.043912	1.21085
Variety	**1.045****	2.07	0.039	1.00224	1.089077
Purchase decision involvement	1.014	0.46	0.648	0.955856	1.075209
Need for touch	1.002	0.14	0.889	0.970771	1.034791
Trust in online seller	**1.208*****	6.22	0	1.138068	1.28188
Commitment to local store	0.979	−0.77	0.441	0.926034	1.034026
Convenience	**1.114****	2.16	0.03	1.01029	1.22927
Product preference (global or local version)	**1.069*****	3.13	0.002	1.025303	1.115058
Gender	1.144	0.58	0.559	0.72866	1.79573
Age	1.123	1.19	0.234	0.927872	1.358852

***$P < 0.01$, **$P < 0.05$, *$P < 0.10$

Odds Ratio Logistic Regression tests were performed on the data to obtain the results.

Table 2 shows the results of the Ordinal Logistic regression model

6 Empirical Results and Discussion

From the results, it is evident that trust in the online seller, the availability of product variety, convenience, ease of use of online payment, and global product preference have a significant and positive impact on the consumer's purchase intention of online grocery.

That is, having trust in an online seller increases the likelihood of consumer's purchase intention of online grocery by 20.8%, and the perception of online payment as easy to use increases the likelihood of consumer's purchase intention of online grocery by 12.5%.

Similarly, seeking variety in purchase increases the likelihood of consumer's purchase intention of online grocery by 4.5%, and seeking convenience increases the likelihood of consumer's purchase intention of online grocery by 11.4%.

7 Conclusion

This study is one of the earliest in India to propose and test a comprehensive model which identifies the influence of factors such as variety, trust in the seller, ease of use of online payments, convenience, purchase decision involvement, and preference on the global or local version of products on the purchase intention of consumers in online grocery stores.

From the study, it is evident that factors such as trust in online sellers, ease of use of the online payment, product preference of global version over local version, variety and convenience have a significant and positive impact on the consumer's purchase intention of online grocery.

Hence, to promote online grocery the sellers must equally try to build trust as much as they are inclined to promotions. Also, providing online payment options in addition to payment by cash can significantly impact the purchase intentions of consumers. Further, the online platforms can ensure that the global version of the product is also displayed when a consumer searches for the local version of the same product.

Similarly, consumer's purchase intention is also significantly and positively impacted by the availability of variety and convenience provided by the online grocery platforms. Therefore, the firms can ensure that they never trade-off these constructs as consumer's purchase decision is also influenced by these factors.

8 Limitations and Future Research

This research attempts to contribute to the understanding of factors influencing the purchase intentions of consumers in buying groceries online. The study was only conducted among the Indian population, if a similar study is conducted with respondents world-wide, we can do cross-country comparisons. Also, comparative studies among the independent variables can be conducted to understand the influence and moderation they have on each other, and focus studies can be used to better understand the impact of situational parameters which plays a vital role in the purchase decisions of consumers.

Acknowledgments The authors are grateful for the support and inspiration provided by the peers and faculty members of Amrita School of Business which aided in the completion of this research study. We also like to thank our respondents for their time and efforts.

References

1. Mortimer, G., Fazal e Hasan, S., Andrews, L., Martin, J.: Online grocery shopping: the impact of shopping frequency on perceived risk. Int. Rev. Retail, Distrib Consum. Res. **26**(2), 202–223 (2016). https://doi.org/10.1080/09593969.2015.1130737
2. Hand, C., Dall'Olmo Riley, F., Harris, P., Singh, J., Rettie, R.: Online grocery shopping: the influence of situational factors. Eur. J. Mark. **43**(9/10), 1205–1219 (2009). https://doi.org/10.1108/03090560910976447
3. Morganosky, M., Cude, B.: Consumer response to online grocery shopping. Int. J. Retail Distrib. Manag **28**(1), 17–26 (2000). https://doi.org/10.1108/09590550010306737
4. Ramus, K., Asger Nielsen, N.: Online grocery retailing: what do consumers think? Internet Res. **15**(3), 335–352 (2005). https://doi.org/10.1108/10662240510602727
5. Baumgartner, H., Steenkamp, J.B.E.: Exploratory consumer buying behavior: conceptualization and measurement. Int. J. Res. Market. **13**(2), 121–137 (1996). https://doi.org/10.1016/0167-8116(95)00037-2
6. Venkatesh, V., Thong, J., Xu, X.: Consumer acceptance and use of information technology: extending the unified theory of acceptance and use of technology. MIS Q. **36**(1): 157–178 (2012). Retrieved April 12, 2020, from www.jstor.org/stable/41410412
7. Swilley, E.: Technology rejection: the case of the wallet phone. J. Consum. Market. **27**(4), 304–312 (2010). https://doi.org/10.1108/07363761011052341
8. Thomson, M.: Human brands: investigating antecedents to consumers' strong attachments to celebrities. J. Market. **70**(3), 104–119 (2006). https://doi.org/10.1509/jmkg.70.3.104
9. Mittal, B., Lee, M.S.: A causal model of consumer involvement. J. Econ. Psychol. **10**(3), 363–389 (1989). https://doi.org/10.1016/0167-4870(89)90030-5
10. Peck, J., Childers, T.L.: To have and to hold: the influence of haptic information on product judgments. J. Market. **67**(2), 35–48 (2003). https://doi.org/10.1509/jmkg.67.2.35.18612
11. Cho, J.: The mechanism of trust and distrust formation and their relational outcomes. J. Retail. **82**(1), 25–35 (2006). https://doi.org/10.1016/j.jretai.2005.11.002
12. Noble, S.M., Griffith, D.A., Adjei, M.T.: Drivers of local merchant loyalty: understanding the influence of gender and shopping motives. J. Retail. **82**(3), 177–188 (2006). https://doi.org/10.1016/j.jretai.2006.05.002
13. Zhang, Y., Khare, A.: The impact of accessible identities on the evaluation of global versus local products. J. Consum. Res. **36**(3), 524–537 (2009). https://doi.org/10.1086/598794

Car Damage Recognition Using the Expectation Maximization Algorithm and Mask R-CNN

Aseem Patil

Abstract In the course of possessing a car, even though you are very careful, it is almost inescapable to take a few scratches. Unkempt grasslands, traffic-scattered gravel, and impatient passersby in small car parks are all very difficult to keep clean paint. Sometimes we do not even know that our car has been scratched till the last moment. In this paper, I shall propose a method in identifying the scratches on a car by using a live stream from the Basler's camera using the expectation--maximization algorithm and the Mask R-CNN. Also, large datasets and domain features are used to best fit the data in order to perform this function. I have also introduced, trained, and tested several classification models for domain general databases, to categorize different types of car scratches from different angles and environments (i.e., either in the monsoon season, spring season, or the summer season), adding a limited time and data. After the model is trained, a validation accuracy of 96.65% is achieved with the test images.

1 Introduction

The catalyst behind computer vision applications, convolutional neural network (CNN), is rapidly developing with advanced and innovative architectures to solve virtually every perceptual system related issue of the world. Convolutional neural network (CNN) is one of the common techniques of artificial intelligence and machine learning that enables pattern recognition and image classification. They were designed to classify a large number of categories, using filtered packages with large databases, which include hundreds of visualizations for each category, such as AlexNet, to train its parameters. Nevertheless, CNN may have trouble classifying the images based on the categories to be educated and their identification, as in the case of medical images, as it is sometimes impractical to have collections of a significant

---

A. Patil (✉)
Department of Electronics Engineering, Vishwakarma Institute of Technology, Pune, India
e-mail: patilaseem98@gmail.com

© The Editor(s) (if applicable) and The Author(s), under exclusive license to Springer Nature Singapore Pte Ltd. 2021
T. Senjyu et al. (eds.), *Information and Communication Technology for Intelligent Systems*, Smart Innovation, Systems and Technologies 196,
https://doi.org/10.1007/978-981-15-7062-9_61

number of such pictures. We consider common types of damage, such as bumper denture, door denture, door glass break, broken head light, broken tail lamp, scratch and smash. There is no available dataset on the internet for the assessment of car damage recognition. Therefore, by collecting images from random websites on the internet and clicking photos of damaged vehicles in and around the city, I created my own dataset. Due to factors such as high inter-class similarity and little visible damage, the classification task is challenging. Segmentation algorithms inevitably make common errors, which cause any method using segmentation results to degrade its performance [8]. As a result, image systems designs have typically preferred to use global image properties that do not rely on exact segmentation. Segmentation of an image, however, enables reference to the image at the object level. As well as I think that this skill is important in order to retrieve images and progressing in objects. I have developed an image segmentation algorithm which provides segmentation which is reasonable and can provide better query performance than systems with global values although not always being optimal. To accurately segment each frame, we model the tone, structure, and position as a collective distribution of characteristics in a Gaussian mixture. Here, I shall use the EM algorithm to approximate this model's parameters, which gives a segmentation of the picture with the corresponding pixel-cluster representation [7]. If the image is separated into regions, the tone and structure features of each region will be identified. Throughout the query function, the user has direct access to the regions to see the complexity of the database image and to identify the key elements of the image that are relevant for the making of the database. Classical object recognition techniques generally rely on the clean segmentation or the design for fixed geometric artifacts, such as computer pieces from the rest of the scene. In natural images, there are no restrictions either: in objects like animals and automobiles, their shape, scale, and color are very subjective and their segmentation is imperfect. It is clear that identification of classical entities does not relate. More recent techniques can classify individual objects taken from a finite range (about 1000) but at general image processing no current strategy is successful, which involves both image optimization and image identification. The Gaussian mixtures are evaluated by an EM algorithm. The finite mix is a versatile and powerful probabilistic method for modeling. It can be used to provide a pattern recognition model-based clustering. Object recognition usually detects objects by making a bound box or laying out the object as a mask and identifying the object by training the model using the dataset in hand. However, this paper discusses the parameters of mask recognition of scratches of cars using the exact same method of identifying objects, but in our case we shall identify the scratches on a car by a database which is not fully complete. There are some pieces of the dataset that are missing and this is where the expectation--maximization algorithm comes into play.

2 Mask R-CNN and Its Concept

Mask R-CNN is a deep neural network designed to address machine learning or computer vision instance segmentation problems. In other words, in a picture or video, it can distinguish various objects. You give it a picture, it gives you boxes, classes, and masks on object bindings. The Mask R-CNN consists of two stages. First, it generates proposals for the areas where an object can be based on the image input. Second, the class is expected, the bounding box is refined, and a object pixel-level mask is created based on the proposal for the first point. The backbone structure is related to both levels [9]. The R-CNN mask is primarily a faster R-CNN extension. Faster R-CNN is commonly used for object detection purposes. The class mark and bounding box coordinates of each image object are returned for a given image. The faster R-CNN works are based on the following four steps,

1. First, faster R-CNN uses a ConvNet to retrieve image maps.
2. Such maps are then distributed through a regional proposal network (RPN) that returns the bounding boxes for candidates.
3. On these candidates' bounding boxes, we apply a RoI pooling layer so that all candidates can be of equal size.
4. Finally, the propositions are moved to an optimized layer to identify the border boxes for objects and output them.

As well as,

5. The RoI align network creates more than just one bounding box and integrates the bounding boxes into a fixed dimension.
6. Warped features then are applied to completely connected layers in order to render classifications by using the ConvNet predictor, and boundary box predictor is further refined by using the warped regression model.
7. Mask classifier enables the network to build masks without competition between the classes for any class

The R-CNN mask is different from normal object detection models like fast R-CNN in which it can also color pixels in the bounding box corresponding to that class, in addition to recognition of the class and its bounding box position. The working of the Mask R-CNN has been shown in Fig. 1.

A. *Backbone model used in the system*

Like the ConvNet we use in faster R-CNN to extract image charts, we use ResNet 101 for extracting characteristics from Mask R-CNN images. Therefore, the first step is to use the ResNet architecture to take the images and to extract characteristics. This is an input for the next step.

B. *Region proposal network*

From the previous step, we are now making the feature maps obtained and using the region proposal network or (RPM). It predicts whether or not an entity (or does not)

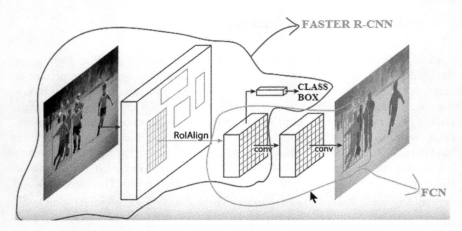

Fig. 1 Mask R-CNN architecture with a sample image

exists in that area. In this step, we obtain those regions or character maps that contain an entity in the model.

C. *Region of interest (ROI)*

There may be different types of the areas obtained from RPN. Therefore, we use a pooling layer and transform all regions into the same shape. These regions will then pass a completely connected network, which will predict the class label and bounding boxes. The steps are essentially identical to the faster R-CNN method until that point. Now the two systems have a gap [5]. However, R-CNN mask also generates the segmentation mask. We measure the area of interest first to reduce the calculation time. The intersection over union (IoU) is based on the ground truth boxes for all of the forecast areas. We can evaluate the IoU based on the following formula,

$$\text{IoU} = \frac{\text{Area of Intersection}}{\text{Area of the Union}} \quad (1)$$

Now, we find this is an area of interest only when the IoU is greater or equal to 0.5. Or else this particular area will be ignored. For every area, we only pick a set of regions where the IoU is higher than 0.5.

To summarize, R-CNN is basically the following three steps:

1. Build a set of bounding box proposals.
2. Run images in border boxes using a previously trained ConvNet and then an SVM to see the image object in a box.
3. Run the box using a linear regression model to provide tighter coordinates for the box after classifying the object (Fig. 2).

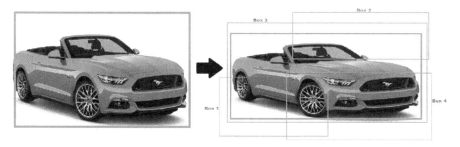

Fig. 2 Example considering the four regions of the RPN

From Fig. 1, Box 1 and Box 2 may not have the IoU of 0.5, while Box 3 and Box 4 may have a IoU of about 0.5. So, for this particular image, we may assume that Box 3 and Box 4 are of interest, while Box 1 and Box 2 will be ignored.

D. *Applying the segmentation mask*

After you have RoI based on the IoU values, the existing architecture can be supplemented by a mask branch. It returns for each area containing an item a segmentation mask, as we can see in Fig. 2. It returns a 28 × 28 mask for each area that is then scaled up. In Mask R-CNN, this is the last step to predict masks for all the image objects [2] (Fig. 3).

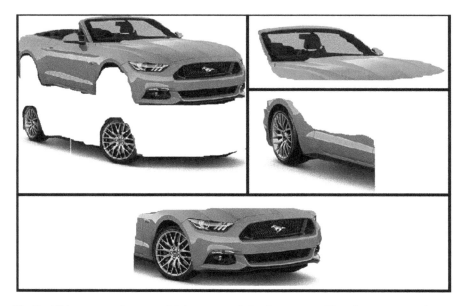

Fig. 3 All image parts in our model that were coded to be segmented have been segmented

3 The Expectation--Maximization Algorithm

An expectation--maximization (EM) algorithm is an iterative method for finding maximum likelihood in statistical models, where the models depend on unobserved latent variables. EM iteration alternates between performing an expectation (E) step, which creates a function for the expectation of the log-likelihood evaluated using the current estimate for the parameters, and a maximization (M) step, which computes parameters maximizing the expected log-likelihood found on the E step [3]. Then these parameter-estimates are used to determine the distribution of the latent variables in the next E step. For determining optimum likelihood estimates of parameters in the probabilistic models, the expectation--maximization (EM) algorithm is used in statistics, where the prediction is based on unknown latent variables. The concept behind the EM algorithm is intuitive and natural and hence extends to a variety of problems. But where there are several local maxima, the EM algorithm does not guarantee convergence to the global limit.

The expectation--maximization algorithm is used for the following two major reasons:

1. It can be used to fill the missing data in a sample.
2. It is always guaranteed that likelihood will increase with each iteration.

Here, we shall use the EM algorithm to help in the classification of data points even when the data sample is not fully enough for testing [4]. The block diagram of the following EM algorithm for the system can be described below:

The above diagram (Fig. 4) shows the working of the system using the expectation--maximization algorithm. During video stream conversion to frames, each frame passes through an iteration that is called the framewise constant iteration. Iterations will continue to take place till the very last frame is preprocessed and then the system shall exit the EM stage. In the stage above, the first frame is passed for preprocessing which includes either including a Gaussian mixture or adding a convolutional layer

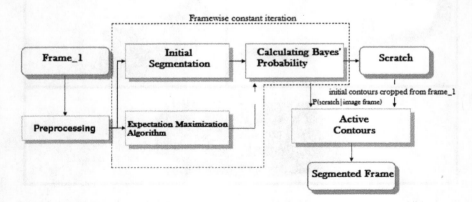

Fig. 4 Block diagram of the proposed system using EM algorithm

so as to make it easier for segmentation [6]. After the frame has been preprocessed, it sends the cache file of the image for initial segmentation of the image. Segmentation of the image is usually the first step used in computer vision applications. It includes clustering, edge detection, thresholding, and watershed techniques. In this system, we use clustering and edge detection as it will provide us a schema of the area we want to identify in the image. After the image has been segmented, the likelihood is calculated. Compared to supervised or unsupervised learning heuristics, the Naïve Bayes algorithm can deliver very precise classification results and likelihood with minimal training time. After the images' probability has been calculated, active contours are used to analyze the image and also for shape analysis. Active contours play a major role as they only identify and locate the homogenous coordinates of the specified location of the object present in that image, in this case a scratch or a dent on the cars' surface [1]. Before the image is sent to the R-CNN network, it is feature extracted using the following three steps:

1. Select an appropriate scale for each pixel and extract color, texture, and position features for that pixel at the selected scale.
2. Group pixels into regions by modeling the distribution of pixel features with a mixture of Gaussians using expectation--maximization.
3. Describe the color distribution and texture of each region for use in a query.

The segmented frame is finally sent to the R-CNN network for masking. Since the dataset is not fully complete, the system gets trained every time it makes a mistake or gives an error [4]. For example, the system may give you a hearty accuracy of 92.5% as the detection output after the first training cycle, however, after the training of the first cycle is complete, the system shall record a higher accuracy after the second training cycle is complete. Also, it shall increase every time you run the model. This helps in achieving a higher accuracy and precision percentage for the model.

$$P(c|x) = \frac{P(x|c) \times P(c)}{P(x)} \quad (2)$$

The above equation is known as Bayes' theorem. It is used to calculate posterior probability of the system, and in this case P (scratch|image frame), it is usually used when the dataset is really bulky. In Eq. (2), $P(c|x)$ is the posterior probability, $P(c)$ is the prior probability, $P(x|c)$ is the likelihood, and $P(x)$ is the prior probability.

In Fig. 5, the image is trained and tested using the EM algorithm. In the first image, the accuracy achieved is 92.51% after the first training cycle was complete. In the second image, the accuracy achieved is 93.11% after the first training cycle was complete. In the third image, the accuracy achieved is 93.66% after the first training cycle was complete. In the fourth image, the accuracy achieved is 93.83% after the first training cycle was complete, and in the final image, the accuracy jumps to 94.21%. As we can see, the accuracy increases after every images' iteration which proves the expectation--maximization algorithms' involvement, as can be inferred from Table 1.

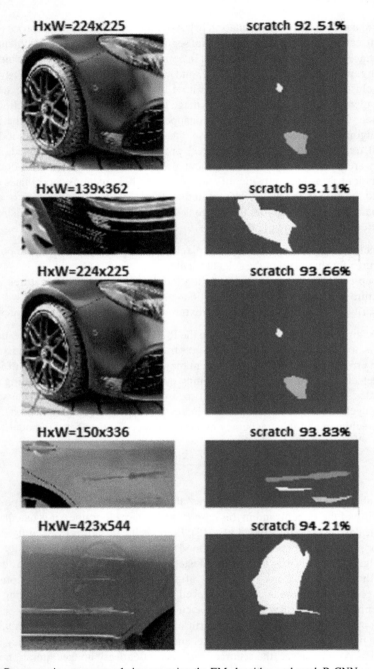

Fig. 5 Preprocessing some sample images using the EM algorithm and mask R-CNN

4 Results

Table 1 shows the accuracy achieved after the system had preprocessed the images using the EM algorithm. As we can see, the accuracy increases as we go down the chart.

Table 1 The following table shows the accuracy achieved after the system was preprocessed using the images by the EM algorithm. As we can see, the accuracy increases as we go down the chart

Frames	Average accuracy of ten frames (in %)
1–10	92.51
11–20	93.14
21–30	93.67
31–40	93.83
41–50	94.21
51–60	95.34
61–70	96.65

```
image ID: scratch.image52.jpeg (1) C:/Users/AseemP/Mask_RCNN/custom/val\image52.jpeg
Processing 1 images
image              shape: (1024, 1024, 3)    min:    0.00000  max:  255.00000  uint8
molded_images      shape: (1, 1024, 1024, 3) min: -123.70000  max:  141.10000  float64
image_metas        shape: (1, 14)            min:    0.00000  max: 1024.00000  int32
anchors            shape: (1, 261888, 4)     min:   -0.35390  max:    1.29134  float32
gt_class_id        shape: (1,)               min:    1.00000  max:    1.00000  int32
gt_bbox            shape: (1, 4)             min:  272.00000  max:  930.00000  int32
gt_mask            shape: (1024, 1024, 1)    min:    0.00000  max:    1.00000  bool
The car has:1 damages
```

Fig. 6 Scratch has been recognized successfully with an outstanding accuracy of 96.65%

5 Conclusions

A car scratch recognition model using incomplete image data as a training set has been used by the expectation--maximization algorithm and Naïve Bayes algorithm, which has been proposed in this paper. Because no unnecessary training method in the most artificial neural network architectures is typically needed for the Naive Bayes classification, the resulting classification system makes acceptable classification decisions, with a very limited computational effort. The system has used a minimal amount of data for training and has still given us a hearty accuracy of 96.65%. In the future, aim to automatically identify scratch categories based on regional characteristics. I shall also try to use the EM algorithm as a backup plan after the training is finished on the original and complete dataset so as to receive more accurate readings.

References

1. Dempster, A., Laird, N., Rubin, D.: Maximum probability from Incomplete knowledge via the EM algorithmic program. J. Royal Appl. Mathem. Soc. Ser. B **39**(1), 1–38 (1977)
2. Enser, P.: Query analysis during a visual info retrieval context. J. Document Text Manag. **1**(1), 25–52 (1993)
3. Flickner, M., Sawhney, H., Niblack, W., Ashley, J., Huang, Q., Dom, B., Gorkani, M., Hafner, J., Lee, D., Petkovic, D., Steele, D., Yanker, P.: Query by image and video content: the QBIC system. IEEE **28**(9), 23–32 (1995)
4. Belongie, S., Carson, C., Greenspan, H., Malik, J.: Color-and texture-based image segmentation using EM and its application to content-based image retrieval. In: Proceedings of International Conference Computer Vision, pp. 675–682 (1998)
5. BiguÈn, J., Granlund, G., Wiklund, J.: Multidimensional orientation estimation with applications to texture analysis and optical flow. IEEE Trans. Patt. Anal. Mach. Intell. **13**(8), 775–790 (1991)
6. Domingos, L.P.: Naive Bayes models for probability estimation. In proceedings of the 22th International Conference on Machine Learning, pp. 529–536 (2005)
7. Lewis: Naive Bayes at forty: The independence assumption in information retrieval. Lecture Notes in Computer Science, vol. 1398, pp. 4–15 (June 2005)
8. Mclachlan, G., Krishnan, T.: The EM algorithm and extensions. Wiley, Hoboken (1996)
9. Rane M., Patil A., Barse B.: Real object detection using tensorflow. In: Kumar, A., Mozar, S. (eds) ICCCE 2019. Lecture Notes in Electrical Engineering, vol. 570. Springer, Singapore (2020)

Deep Learning Methods for Animal Recognition and Tracking to Detect Intrusions

Ashwini V. Sayagavi, T. S. B. Sudarshan, and Prashanth C. Ravoor

Abstract Over the last few years, there has been a steady rise in number of reported human–animal conflicts. While there are several reasons for increase in such conflicts, foremost among them is the reduction in forest cover. Animals stray close to human settlements in search of food, and often end up raiding crops or preying on cattle. There are at times human causalities as well. Proficient, reliable, and autonomous monitoring of human settlements bordering forest areas can help reduce such animal–human conflicts. A broad range of techniques in computer vision and deep-learning has shown enormous potential to solve such problems. In this paper, a novel, efficient, and reliable system is presented which automatically detects wild-animals using computer vision. The proposed method uses the YOLO object detection model to ascertain presence of wild animals in images. The model is fine-tuned for identifying six different entities—humans, and five different types of animals (elephant, zebra, giraffe, lion, and cheetah). Once detected, the animal is tracked using CSRT to determine its intentions, and based on the perceived information, notifications are sent to alert the concerned authorities. The design of a prototype for the proposed solution is also described, which uses Raspberry Pi devices equipped with cameras. The proposed method achieves an accuracy of 98.8% and 99.8% to detect animals and humans, respectively.

A. V. Sayagavi (✉) · T. S. B. Sudarshan · P. C. Ravoor
Department of Computer Science & Engineering, PES University, Bengaluru, India
e-mail: ashwinisayagavi.work@gmail.com

T. S. B. Sudarshan
e-mail: sudarshan.tsb@gmail.com

P. C. Ravoor
e-mail: prash.ravoor125@gmail.com

© The Editor(s) (if applicable) and The Author(s), under exclusive license to Springer Nature Singapore Pte Ltd. 2021
T. Senjyu et al. (eds.), *Information and Communication Technology for Intelligent Systems*, Smart Innovation, Systems and Technologies 196,
https://doi.org/10.1007/978-981-15-7062-9_62

1 Introduction

There have been increasing reports of wild animals entering villages or towns, especially in settlements surrounding forest areas, endangering human lives. Intrusions by animals cause huge losses, be it in terms of crop loss or cattle being attacked. Increasing human population leading to decreasing forest cover is one of the leading causes for rise in human animal conflicts. Current methods to reduce such conflicts include installation of electric fences or have sentries watch for animals through the night. Electric fences cause severe injury to animals. Moreover, they require enormous initial investment and additionally have high maintenance costs. Recent developments in the field of of computer science enables use of technology to create low-cost solutions to such problems. Computer vision is one such technology which could potentially solve most of the associated problems.

Use of deep learning methods to classify images that contain entities of interest is gaining popularity. Deep convolutional neural networks (DCNN) are known to be accurate, and outperform all other existing methods in the task of image classification. Krizhevsky et al. [1], who submitted the winning entry for the ImageNet classification challenge, introduced a deep neural network-based solution for image classification. It is now considered a landmark achievement in computer vision, and has contributed to increased research in the field.

The main intent of this paper is to describe the design for a computer vision system, capable of detecting wild animals, and tracking their movement. DCNNs could be leveraged to detect the presence of animals in the captured images. In addition to detecting the presence of an animal, in order to effectively track them, and monitor their actions, it is also necessary to localize the animals within the image. This is the task of *object detection*. Object detection systems predict regions of interest within images, and in addition, classify entities within these regions. Thus object detection is the ideal choice for the system proposed in this paper.

This paper introduces a novel method of reducing human–animal conflicts through constant and automatic monitoring of vulnerable areas using a system of cameras. The proposed solution is accurate and cost effective and to an extent can be customized specifically for a particular region. Section 2 presents notable existing research work in the area. Section 3 describes the design of the proposed system and highlights the role of various components. Section 4 summarizes the important results of the study, followed by a brief discussion and scope for future enhancements in Sect. 5.

2 Literature Survey

The section is organized into two categories—some systems for animal detection have been reviewed, followed by methods of animal intrusion detection and prevention.

2.1 Animal Detection Using Computer Vision

Zhang et al. [2] describe a system for segmentation of animals from images captured through camera traps. The procedure employed uses a multi-level iterative graph cut to generate object region proposals and accurately recognize regions of interest. This is especially useful when the animal blends together with the background and is difficult to identify. These proposals segmented into background and foreground in the second stage. Feature vectors are extracted from each image using AlexNet [1] architecture, and combined together with the histogram of oriented gradients (HOG) to generate Fisher vectors. The system obtained an accuracy of 82.1% for animal and species detection.

Yousif et al. [3] combine deep learning classification with dynamic background modelling to evolve a swift and precise method for human and animal detection from highly cluttered camera trap pictures. Background modelling helps generate region proposals for foreground objects, which are then classified using the DCNN, resulting in improved efficiency and increased accuracy. The proposed system achieves 82% accuracy in segmenting images into human, animal, and background patches.

Kellenberger et al. in [4] use unmanned aerial vehicles (UAV) to monitor animals and prevent poaching. A two-branch custom CNN is built using AlexNet [1] as the backbone. The authors report a 60% accuracy over the dataset gathered from UAV images Kuzikus Wildlife Reserve park, Namibia.

Norouzzadeh et al. in [5] use the Snapshot Serengeti dataset [6] and apply deep-neural networks to detect and identify animals in camera trap images. The system consists of multiple parts (a) detection stage (whether there is an animal in the image), (b) species identification stage, (c) information stage, where the network reports additional data such as the count and attributes of the animals (standing, resting, etc.). An ensemble of nine models is used, and obtains a top-1 accuracy of 99.4% for the species identification task, and the overall pipeline accuracy was around 93.8%.

Parham et al. [7] propose a multi-stage pipeline for animal detection and recognition. The fundamental steps include animal classification, animal localization, and predicting animal characteristics, such as orientation. Animal localization is based on the YOLO [8] object detection model. The proposed system achieves an overall detection accuracy of 76.58% over 6 species.

Matuska et al. [9] propose a novel system for monitoring animals, consisting of a computing unit, for extracting features of animals, and a separate module to track movements. SIFT [10] and SURF [11] are used for feature extraction, and are classified using an SVM classifier [12]. Use of SIFT descriptors achieved an accuracy of 94% for animal species classification.

Sharma et al. [13] describe a system for animal detection that uses cross-correlation filters for template matching. The training data is used as a baseline for classification, and new images are matched with images in the database to detect presence of animals. This system obtains an overall accuracy of 86.25%.

2.2 Animal Intrusion Detection Systems

Suganthi et al. [14] present a system to detect intrusion by elephants to reduce agricultural losses. The proposed system uses multiple vibration sensors, and the number of triggered sensors is used to detect if an elephant is close by. If the sensors are triggered, a photograph is captured and Google Vision API is used to detect the presence of an elephant in the image. On confirmation of detection, alerts are sent to local authorities for further action.

Pooja et al. [15] use multiple PIR sensors to detect animal movement. The sensors are set so that when triggered, the number of sensors set off provides an indication of the species of animal. Based on the species, suitable actions are taken, such as playing an audio clip and alerting local sentries.

Several challenges still exist with all the surveyed methods. Most notably, the systems built for animal intrusion detection have to ensure the correct species identification. Animals react differently to stimulus—playing loud sounds might scare away wild boar, but might startle an elephant causing it to go on a rampage. It is not sufficient to detect the animal but also necessary to ascertain its intentions before creating alerts to ensure fewer false positives. All of this needs to be performed in real or near-real time. Since the solution also needs to be cost-effective, small compute devices such as embedded systems would need to be used, which can be deployed on site. The solution proposed in this paper attempts to address all of these challenges, and an overview of the design is presented in the next section.

3 Proposed Solution

The system proposed in this paper uses a network of cameras, connected to PIR motion sensors, so that image capture is triggered only when some movement is detected. This enables power conservation. The images captured through these cameras are processed to detect presence of wild animals, and if an animal is found, identify the species. Once identified, the animals are tracked for a suitable time in order to determine their intent—such as to find whether they are moving across the village, or into it. In the latter case, alerts are generated and local authorities are notified through proper channels. Understanding the intent goes a long way to reduce false positives, either due to a false detection or when there is no actual threat posed due to presence of the animal.

3.1 Object Detection and Tracking

YOLO object detection [8] model is used to detect the presence of animals in the captured images. YOLO is a DCNN object detection model which has good performance

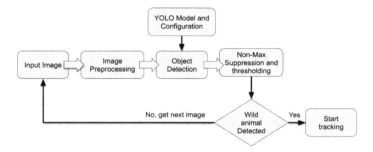

Fig. 1 Wild animal object detection framework

both in terms of accuracy as well as speed of inference. For the prototype version of the system proposed in this paper, five different species of animals—elephant, zebra, giraffe, lion, cheetah—are considered. Images of humans are also included, so there are a total of six different categories in the training data. The DCNN is fine tuned for better accuracy over these six categories.

Training data is obtained from publicly available wild animal videos, including those from YouTube channels and National Geographic videos. Frames extracted from these videos are manually annotated in the required format for training. The model is trained using images of dimensions 448×448. The learning rate was initialized to 0.001 with a decay rate of 0.995, and momentum is set to 0.9. The model converged after running it for 135 epochs. The average accuracy over detecting five species of animals is 98.8%, and for human detection the accuracy obtained is around 99.8%.

Figure 1 illustrates a flow chart of the animal detection module. The YOLO object detection model weights and configuration is loaded and the image is fed into the model. The outputs of the object detection network are tested against a threshold value and underwent non-max suppression to remove low confidence and overlapping predictions. If wild animals are detected in the frame, then the object tracking phase is triggered.

Once the animals are detected, identified, and localized in the frame, object tracking is used to determine intents or actions of the animal being monitored. The CSRT tracker [16] is used to track animals effectively, since it is both fast and accurate. The input to the tracker is the centroid of the detected bounding box. It is not necessary to provide the bounding box in every frame, however; once a tracker is initialized using one bounding box centroid, visual features from the marked area of the image are used to infer location of the animal in each subsequent frame. The tracker maintains state from previous frames, thus allowing for identifying the direction of movement.

Each track can be monitored separately, and can be used to obtain the direction of movement. Appropriate actions are taken if the intent of the animal matches a set of predefined rules, such as sending a user notification if the animal is found to be moving in a particular direction, or to flash lights, play audio clips if the animal moves close to the village.

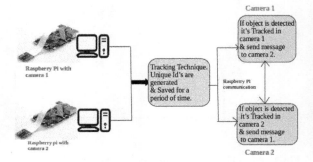

Fig. 2 System operation description using two Raspberry Pi devices, each equipped with a camera module

3.2 Prototype Design

This section presents the design of the overall system, integrating object detection, tracking, and notifications.[1] Raspberry Pi[2] is an embedded system which is capable of interfacing with a variety of peripheral devices through various protocols. It is also quite powerful for its size, housing a quad-core processor and 1GB RAM. It supports a camera module which is capable of recording video at 30 fps and 5 MP resolution.[3] Both integrate out-of-the box, and are used as the endpoint devices.

Figure 2 depicts the operation of the system, using two Raspberry Pi devices for illustration. Inter-device communication is restricted to the hand-off procedure, where a device indicates to its neighbours that it can no longer see an animal that it was previously monitoring. Tracking is implemented on device; however, object detection is slow given the processors constrained size. A lite version of the object detection model is run instead, which has lower accuracy but is capable of achieving up to 1 fps for detection.

If the animal moves out of range of the camera, a notification of the tracked person is sent to the other Raspberry Pi devices. The transmitted message contains information about class of object detected and identifies the sender, through which a monitoring program finds the location of the camera, and an alert notification is sent containing information about the type of animal spotted, and its last known location.

Unique identities are generated each time an object is detected, and a tracker is initialized over it. The detected object is tracked until it is visible in the camera's field of vision (FoV). If the system is no longer able to detect the object, a notification is sent to the other devices to signal an object tracking hand-off. The identities assigned to the animal remain unique only until the object is tracked in the given sequence, and could be assigned different ids if it re-appears in the camera FoV. In case, multiple wild animals are detected, each group of animals is assigned as a single entity, for example, group of lions are assigned a single identity. This is necessary because it is difficult to differentiate one animal from other within the same species.

---

[1] https://www.raspberrypi.org/products/raspberry-pi-3-model-b-plus/.
[2] https://www.raspberrypi.org/products/camera-module-v2/.
[3] https://www.pjreddie.com/darknet/yolo/.

4 Results

Table 1 shows the performance of the pre-trained model described in this paper against some of the surveyed object detection models. The pre-trained model presented in this paper achieves 98.8% accuracy for animal detection and 99.8% for humans. The table contains the average of these two values. Table 2 illustrates the time taken for processing an image for a few state-of-the-art object detectors, when using a GPU. Figure 3 depicts a few qualitative results of running the YOLO object detection model for animal detection. The bounding boxes for animals in the frame are annotated with a confidence score and the type of animal as predicted by the model. The figure shows results of running inference on elephants, humans, giraffe, and zebra.

Table 1 Summary of detection accuracy of surveyed papers

Model	# Species	Accuracy (%)
[5] (2018)	48	99.4
[7] (2018)	6	76.6
[3] (2017)	2	82.0
[4] (2017)	–	60.0
[2] (2016)	23	82.1
[9] (2014)	5	94.0
Ours	6	99.3

Table 2 Object detection performance compared to YOLO

Model	mAP	FPS
SSD300 [17]	41.2	46
SSD500 [17]	46.5	19
YOLOv2 [18]	48.1	40
TinyYOLO [18]	23.7	244

mAP is reported over the COCO dataset, and FPS is measured over GPU
Source [15]

Fig. 3 Qualitative results of object detection using YOLO

5 Discussion and Future Work

The object detection module is highly accurate. DCNN models for image classification and object detection are widespread in use, and it is evident that given sufficient training data, the models can generalize well in most domains. Similarly, the CSRT tracker is robust and reduces the need for continuous object detection, which is costly and compute intensive. This is especially advantageous, given the use of embedded devices like the Raspberry Pi. The notification system can be customized to dispatch messages using multiple protocols, such as SMS or e-mail. The action taken on animal detection can vary, and could include use of deterrents such as flashing bright lights or playing loud sounds, based on the animal species.

The YOLO object detection model is known for its accuracy and ease of use. However, running object detection on embedded devices remains a challenge. A faster and more resource optimal alternative for object detection could be explored. Recent developments to create networks specific to mobile devices, such as the MobileNet architecture [19] holds promise, and is a potential candidate to be used for object detection. Another alternative is to use a GPU device, but this would reduce cost-effectiveness of the solution.

One of the drawbacks of the approach presented here is when multiple cameras detect the same individual animal—it might result in multiple notifications being sent, and would appear as though more than one animal is detected, when in reality, there is only one. In order to circumvent this, a centralized server could monitor detections from each unit, and determine if there actually is just a single animal or several.

In addition, the CSRT tracker is a single object tracker—it bears no semantic notation of the object being tracked, and uses visual features to keep the tracklets continuous. It is thus, prone to failure if the background closely resembles the appearance of the animal. A more robust tracking mechanism is required, which considers not only visual features but also temporal and spatial features and can effectively track the animal under various conditions.

Use of infrared imagery is yet another area that offers room for improvement. In the proposed system, if the ambient light is not sufficient to capture a reliable image, object detection would fail. Since animal movement generally occurs during the night, use of IR images to detect animals would make the intrusion detection system more potent, offering a round-the-clock monitoring mechanism.

6 Conclusion

The proposed system attempts to reduce human–animal conflicts by continuous and automatic monitoring of vulnerable areas using computer vision to detect animal intrusions. The intrusion detection pipeline consists of three stages—animal detection, animal tracking and user alerts, and notifications. The proposed system is cost-effective and highly efficient, with an average accuracy of 98.8% in detecting and identifying animals in images. Although the prototype described in this paper is trained to recognize five different species of animals, it is easily extendable to detect and track other types of animals with sufficient training data. The choice of species can also be region specific, thereby providing a unique edge over other existing solutions. Such a system if implemented on a large scale has potential to largely reduce causalities due to animal intrusions.

References

1. Krizhevsky, A., Sutskever, I., Hinton, G.E.: Imagenet classification with deep convolutional neural networks. In: Pereira, F., Burges, C.J.C., Bottou, L., Weinberger, K.Q. (eds.) Advances in Neural Information Processing Systems, vol. 25, pp. 1097–1105. Curran Associates, Inc. (2012)
2. Zhang, Z., He, Z., Cao, G., Cao, W.: Animal detection from highly cluttered natural scenes using spatiotemporal object region proposals and patch verification. IEEE Trans. Multimedia **18**(10), 2079–2092 (2016)
3. Yousif, H., Yuan, J., Kays, R., He, Z.: Fast human-animal detection from highly cluttered camera-trap images using joint background modeling and deep learning classification. In: 2017 IEEE International Symposium on Circuits and Systems (ISCAS), pp. 1–4 (2017)
4. Kellenberger, B., Volpi, M., Tuia, D.: Fast animal detection in UAV images using convolutional neural networks. In: 2017 IEEE International Geoscience and Remote Sensing Symposium (IGARSS), pp. 866–869 (2017)
5. Norouzzadeh, M.S., Nguyen, A., Kosmala, M., Swanson, A., Palmer, M.S., Packer, C., Clune, J.: Automatically identifying, counting, and describing wild animals in camera-trap images with deep learning. Proc. Natl. Acad. Sci. **115**(25), E5716–E5725 (2018)

6. Swanson, A., Kosmala, M., Lintott, C., Simpson, R., Smith, A., Packer, C.: Snapshot Serengeti, high-frequency annotated camera trap images of 40 Mammalian species in an African Savanna. Sci. Data **2**, 150026 (2015)
7. Parham, J., Stewart, C., Crall, J., Rubenstein, D., Holmberg, J., Berger-Wolf, T.: An animal detection pipeline for identification. In: 2018 IEEE Winter Conference on Applications of Computer Vision (WACV), pp. 1075–1083 (2018)
8. Redmon, J., Divvala, S., Girshick, R., Farhadi, A.: You only look once: Unified, real-time object detection. In: 2016 IEEE Conference on Computer Vision and Pattern Recognition (CVPR), pp. 779–788 (2016)
9. Matuska, S., Hudec, R., Benco, M., Kamencay, P., Zachariasova, M.: A novel system for automatic detection and classification of animal. In: 2014 ELEKTRO, pp. 76–80 (2014)
10. Lowe, D.G.: Distinctive image features from scale-invariant keypoints. Int. J. Comput. Vis. **60**(2), 91–110 (2004)
11. Bay, H., Tuytelaars, T., Van Gool, L.: Surf: speeded up robust features. In: Leonardis, A., Bischof, H., Pinz, A. (eds.) Computer Vision - ECCV 2006, pp. 404–417. Springer, Berlin (2006)
12. Cortes, C., Vapnik, V.: Support-vector networks. Mach. Learn. **20**(3), 273–297 (1995)
13. Sharma, S., Shah, D., Bhavsar, R., Jaiswal, B., Bamniya, K.: Automated detection of animals in context to Indian scenario. In: 2014 5th International Conference on Intelligent Systems, Modelling and Simulation, pp. 334–338 (2014)
14. Suganthi, N., Rajathi, N., M, F.I.: Elephant intrusion detection and repulsive system. Int. J. Recent Technol. Eng. **7**(4S), 307–310 (2018)
15. Pooja, G., Bagal, M.U.: A smart farmland using Raspberry Pi crop vandalization prevention & intrusion detection system. Int. J. Adv. Res. Innov. Ideas Educ. **1**(S), 62–68 (2016)
16. Lukežič, A., Vojíř, T., Cehovin Zajc, L., Matas, J., Kristan, M.: Discriminative correlation filter tracker with channel and spatial reliability. Int. J. Comput. Vis. **126**(7), 671–688 (2018)
17. Liu, W., Anguelov, D., Erhan, D., Szegedy, C., Reed, S., Fu, C.Y., Berg, A.C.: SSD: Single shot multibox detector. Lecture Notes in Computer Science, pp. 21–37 (2016)
18. Redmon, J., Farhadi, A.: Yolo9000: Better, faster, stronger. In: 2017 IEEE Conference on Computer Vision and Pattern Recognition (CVPR), pp. 6517–6525 (2017)
19. Sandler, M., Howard, A., Zhu, M., Zhmoginov, A., Chen, L.: Mobilenetv2: Inverted residuals and linear bottlenecks. In: 2018 IEEE/CVF Conference on Computer Vision and Pattern Recognition, pp. 4510–4520 (2018)

VR Based Underwater Museum of Andaman and Nicobar Islands

T. Manju, Allen Francis, Nikil Sankar, Tharick Azzarudin, and B. Magesh

Abstract Nowadays, world is full of technology and inventions. Many computer applications are developed in the favour of mankind and to ease human work. As a top trending, technology acts a human organ. Every day wakes up using technology and doze off using a technology. In short, technology is moving along with us in our day-to-day life. So it is all about how we are going to use the technology. Speaking of technology, we can see many technologies ruling the world. Like Big Data, Internet of Things, etc., virtual reality is also one of the promising fields. The term virtual reality is used to describe a 3D virtual environment created by sensors which allows interaction. The environment develops an immersive feel that the person is actually in a real environment and he can manipulate objects or perform some actions [1]. Using this virtual reality technology, our aim is to develop an underwater museum to experience the feel of being under the sea.

1 Introduction

Andaman and Nicobar Islands is a place to most spectacular species of marine life in the world. Over 560 different species of corals had been recorded until date, and variety of the sights underwater can leave you spellbound. One of the few tourist attractions in India for underwater activities, the corals in Andaman Islands, makes it a must visit for everyone. Apart from being mesmerized by the beauty of the world within your world, what do you really see underwater? This project takes you through some of the marine life and corals you may see during your dive or underwater walk.

T. Manju (✉) · A. Francis · N. Sankar · T. Azzarudin
Thiagarajar College of Engineering, Madurai, India
e-mail: tmanju@tce.edu

B. Magesh
PERI Institute of Technology, Chennai, India

© The Editor(s) (if applicable) and The Author(s), under exclusive license to Springer Nature Singapore Pte Ltd. 2021
T. Senjyu et al. (eds.), *Information and Communication Technology for Intelligent Systems*, Smart Innovation, Systems and Technologies 196,
https://doi.org/10.1007/978-981-15-7062-9_63

"Virtual" has had the meaning of "being one thing in essence or impact, though not truly or in fact." Virtual reality (VR) is a simulated environment by several technologies. Instead of visualizing a scene or watching a displayed environment, users are immersed that they can feel their visualization with 3D worlds. This project implies that the immersive real-life experience through technology like virtual reality of the scenic extravaganza and underwater wonders of vivid flora and fauna which are displayed in the museums would certainly give a new dimension to the tourists in realizing the beauty of these islands. Applications of virtual reality include entertainment and education. Other distinct types of VR style technology include augmented reality and mixed reality [2]. We can see many applications of virtual reality in medical and army field. One of the civilian applications of the virtual reality is virtual tour [3].

2 Background Study

A virtual reality tour is a virtual simulation of an existing location usually comprising pre-recorded or edited movie, 3D based models of the real location as well as sound effects and narration through text. Substituting travelling, virtual tour evokes an experience of moving through the represented space. Regardless of technology used, virtual tour adds a view that graphically explains a story or a space. Thus, we can learn information of objects or locations through a virtual view. Initial stages of virtual tour included concepts like constructing real-time environment using photo-stitching method and using electric models with 3D modelling kit.

2.1 Photo-Stitching Method

The photo-stitching process can be divided into three main components: image registration, calibration, and blending. Image registration is a process of transforming data from multiple sources to one [4]. Calibration aims to minimize the differences between the ideal and the actual. Blending algorithms assign more weights near the centre of the image. The photo-stitching method stitches the photos of an actual environment to construct a virtual environment [5]. From the construction of an environment, we end up with a 360° cylindrical panoramic virtual space. The observer is covered with the environment in all directions. The observer is given an option to move the environment to a certain limit.

2.2 Video-Based Virtual Tour

The next generation in the virtual tourism is video-based virtual tour. The advantage of using video-based method is to show that the environment is constantly changing throughout the period. The virtual environment is created using 360° videos. We can also create the 360 video with hotspots. Veer Web site offers service to create 360 virtual video with hotspots (traceable content which shows the description of an object). Some of the applications of videos with hotspots are fully interactive video virtual tour for real estate, 360 video virtual tours for construction showing how the project evolves over time, or to compare before and after photos and videos.

2.3 Virtual Walkthrough

Virtual walk videos are documentary motion footage shot because the camera unceasingly moves forward through natural space. The impact is to permit viewers to expertise the sights they would see and also the sounds they would hear, and they will actually travel along a selected route as same pace as the camera. Virtual walks do not require the use of virtual reality headsets.

2.4 Head-Mounted Devices

Head-mount devices (HMD) were used to track the location of the user inside the virtual environment. Components of a virtual reality device comprises of a tracking device, sensors, and controllers. Sensors create a virtual environment around a limited space through projections. Head-mount devices reflect the projected images and allow the user to see through it. Controllers allow user to interact with the environment

3 Methodologies

In this project, we are going to create an underwater virtual environment using 3D objects (fishes, corals, and underwater predators). The 3D objects are created using video to mesh construction method. In order to give more immense feeling, the fishes in this environment are animated using rig animation. Considering the natural phenomena, flocking of fishes is implemented using flocking algorithm (Fig. 1).

3.1 3D Model Construction

The objects created using blender are not clear to visualize and take more time. Moreover, the objects available in Internet may not add the natural feel to the object. So in our project, we have created the objects using video to mesh construction method.

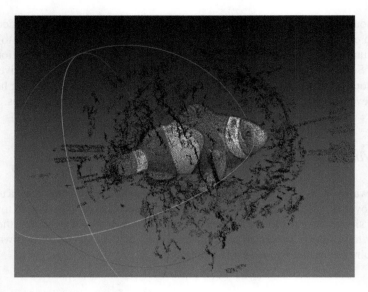

Fig. 1 Point cloud construction of an object

First of all, we need to record the object that we are going to create. So, we have to record the 360° video of that object. The lighting and the sharpness play a vital role in the result of the object. So, it is recommended to use best digital camera to record the video. Then, we have to convert this video into sequence of images. This can be done using any video to jpeg convertor software. The projection of the object in every frame will be different. Connecting every projection in the frames will give a point cloud of the object. Using Poisson mesh reconstruction and some functions, we can create the mesh from that point cloud. These methods can be implemented using the tool, Meshroom (Fig. 2).

3.2 Rig Animation

Rig animation is a technique used in skeletal animation for representing a 3D model using a series of interconnected digital bones. Every fish in our proposed work has different flocking style and different body movement, so we have to animate this for every fish using rig animation. First, we have to create rigs (bones) for every fish. Then, we have to embed the rigs to the flesh of the fish. So, every rig has random weights to control the fish. Now, we have to assign weights to every bone such that each bone can control specific body part of the fish. Rig animation can be implemented using Blender.

VR Based Underwater Museum of Andaman and Nicobar Islands 631

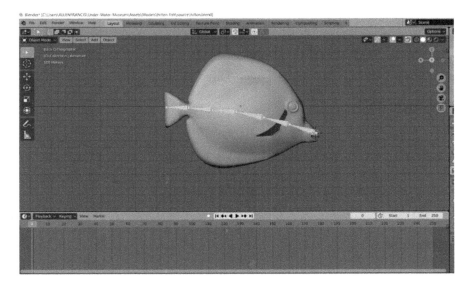

Fig. 2 Rig animation

3.3 Flocking Algorithm

Flocking of fishes include three natural behaviour.

- Separation: Movement of fish to avoid crowding while flocking in group
- Alignment: Movement to align with the group
- Cohesion: Behaviour to join the group

From Fig. 3, we can see the behaviours of the fishes which can be implemented using AI algorithm. We have to calculate the average heading direction of all the fishes. If a fish is about to collide with other fish, it should deviate from the average

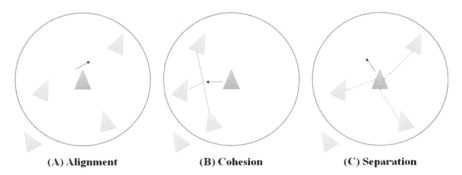

Fig. 3 Flocking algorithm

heading direction of the other fishes (separation). To join into a group ,the fish should join itself to the average position of all the fishes in the group (cohesion). The fish should always move towards the average direction of all the fishes (alignment).

4 Conclusion

Underwater museum using virtual reality has been developed using point cloud and rig animation. This is made automated using flocking algorithm. It will entertain kids as well as it also makes it good for diver training. It is also proposed to create offshore environment and made it as the diver option.

References

1. Maines, C., Tang, S.: An application of games theory to virtual university campus tour and interior navigation. In: International Conference on Development of E-System Engineering (DeSE) (2015)
2. Slater, M., Wilbur, S.: A framework for immersive virtual environments: speculations on the role of presence in virtual environments (1997)
3. Schuemie, M.J., van der Straaten, P., Krijn, M., vander Mast, C.: Research on presence in virtual reality: a survey. J. Cyber Psychol. Behavior (2001)
4. Liu, J., Cui, D.: Multiple vision point virtual tour technique based on panorama. Comput. Eng. **1** (2004) (in Chinese)
5. Fu, H., Chen, Z.: Panorama stitching based on vertical edge processing. Comput. Eng. **2** (2004) (in Chinese)

Network Performance Evaluation in Software-Defined Networking

Shilpa M. Satre, Nitin S. Patil, Shubham V. Khot, and Ashish A. Saroj

Abstract Software-defined networking (SDN) is a network architecture technique that enables the network to be managed or "programmed" intelligently and centrally utilizing software applications. The smart control plane is responsible for providing switching flow paths and optimizing network performance. Typically, the SDN controller runs on a server and is using protocols to tell switches where to send packets. Therefore, the controller's performance is extremely important. There are various parameters, such as Throughput, Bandwidth, Jitter, Bidirectional bandwidth, Round Trip Time (RTT), Missing UDP packet datagram, which is responsible for network performance. In addition, each parameter has different behaviors on different topologies, such as mesh topology, ring topology and tree topology. Present literature includes dozens of controller proposals. But, they do have not any comparative quantitative evaluation. In this article, we present a detailed qualitative overview of various SDN controllers, together with a quantitative analysis of their output throughout different network topologies with different parameters: throughput, bidirectional server-side bandwidth, bidirectional client-side bandwidth both individual and simultaneous, delay and security. More precisely, we will give the best controller according to user requirements for a particular topology. We also discuss the algorithm we built to find the best controller on the results of the experiment. This work compares five controllers to three topologies against various criteria. Finally, we discuss detailed results of the research on the efficiency, parameters and evaluation of SDN controllers.

S. M. Satre (✉) · N. S. Patil · S. V. Khot · A. A. Saroj
Department of Information Technology, Bharati Vidyapeeth College of Engineering
Navi-Mumbai, University of Mumbai, Mumbai, India
e-mail: shilpa.m.shelar@gmail.com

N. S. Patil
e-mail: nitin.patil1799@gmail.com

S. V. Khot
e-mail: shubhamkhot78631@gmail.com

A. A. Saroj
e-mail: ashish.saroj159@gmail.com

© The Editor(s) (if applicable) and The Author(s), under exclusive license to Springer Nature Singapore Pte Ltd. 2021
T. Senjyu et al. (eds.), *Information and Communication Technology for Intelligent Systems*, Smart Innovation, Systems and Technologies 196,
https://doi.org/10.1007/978-981-15-7062-9_64

1 Introduction

The network has been an important infrastructure of modern social development. Speed, communication and reliability are given more importance in networking. The growing technology in the networking area is software-defined networking (SDN). So it is miles crucial to expect the network performance based at the various factors inclusive of topology, hardware, controller and network parameters. Our venture will do the important examine of situation before implementation of network in software-defined networking.

1.1 Network Topologies

Network topologies are the arrangement of the devices in the networks. This arrangement of the device in the network can change the performance of the network [1, 2]. It is the network structure. There are different types of network topologies used by organizations at various layers of deployments, i.e., core layer, distribution layer and access layer. Such topologies consist mainly of the mesh, ring and tree. Each topology has its own advantages and inconveniences and is used according to the requirements of the organization.

(1) MESH TOPOLOGY: In this type of topology, all the devices are connected to each other directly.
(2) RING TOPOLOGY: The devices are connected in a circular form to the adjective devices.
(3) TREE TOPOLOGY: In these topology types, the devices are bound in a special type of structure like the tree branches.

There are a number of factors for calculating network performance. Such variables give the information on how the network works. We will use five network parameters to test the network performance.

(1) THROUGHPUT: The amount of material or items passing through a system or process.
(2) BIDIRECTIONAL BANDWIDTH SEQUENTIAL: In this type, the client and the server communicate together.
(3) BIDIRECTIONAL BANDWIDTH INDIVIDUAL: In these forms, one serves as the sender sending data and another acts as the receiver receiving data for a specific time frame and vice versa.
(4) JITTER: The inconsistent delay in delivery of the packets is Jitter.
(5) SECURITY: Because of the isolation of control and data planes an SDN could be very vulnerable to attacks [3]. A gap in the communication route between the two planes may theoretically lead to a major hole that can endanger the attackers.

2 Software-Defined Networking

SDN is the evolution in the networking sector. It has become the highlighted technology by the approach of separating the data plane and controller plane. To know better about the SDN concept, we should study the difference between traditional networks and SDN.

2.1 Traditional Network

There are different types of traditional networks that the organizations used. The "peers" are computer systems that are connected to each other via the Internet were used at first peer-to-peer networks. In other words, each device on a P2P network becomes both a file server and a client. An Internet connection and P2P software are the only requirements for a computer to join a peer-to-peer network. A computer network in which one centralized, powerful computer (called the server) is a gateway to which many less powerful personal computers or workstations (called clients) are connected after this client–server networks are used. The clients run programs and access data that are stored on the server. First cloud-based networking was cloud infrastructure, where some or all of the networking capabilities of a company are operating in the cloud. That can refer to either a private or a public cloud. It is based on cloud computing, which is the centralization of computing resources shared by users. In these types of networks, there are several disadvantages.

2.2 SDN

SDN stands for "software-defined networking" technology. SDN is a networking approach which uses open-source protocols such as OpenFlow to control software at the edge of the network. It is used for managing switch and router access. Having a common concept of SDN is almost impossible, since its architecture can vary considerably from one organization to another. The biggest difference between a conventional network and SDN is that conventional networks depend on physical infrastructure such as switches and routers to make and run properly connections. In comparison, a software-based network allows the user to monitor the allocation of resources via the control plane at a virtual level [4]. Rather than interacting with physical infrastructure, the user is interacting with software to provision new devices [5].

(1) **Control plane**: Control plane packets are processed by the router to update the routing table information
(2) **Data plane**: The forwarding occurs on the data plane. The data plane is a path through a router or switch for the packet forwarding.

(3) **Controllers**: An SDN controller is an application within architecture of software-defined networking (SDN) that manages flow control to enhance network management and application performance.

List of controllers used in the project:

A. Open daylight [6]
B. ONOS [7]
C. Floodlight [8]
D. RYU [9]
E. POX [10]

2.3 Traditional Network Versus SDN

From this viewpoint, an administrator will actively configure network services as certain network routes. The SDN also has more ability to communicate over the network with apps than a conventional switch. One can sum up the core difference between the two as virtualization. The SDN virtualizes all of your network. Virtualization produces an abstract version of your physical network that allows the delivery of services from a centralized location [4]

3 Implementation

To compare various SDN controllers, we conducted an extensive search of proposals not only in academic literature but also in the commercial domain [11, 12]. Here we first present the topology structure followed by the comparative analysis with parameters on various controllers and then the algorithm to find optimized results [13]. This method allows us to get appropriate results of controller behavior on different topologies and extensively map them in graphical format (Fig. 1).

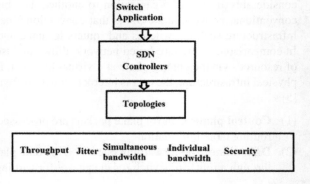

Fig. 1 Adapted experiment architecture

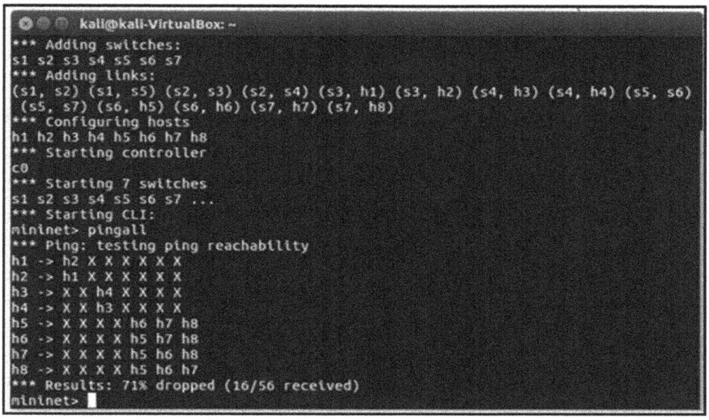

Fig. 2 Mininet console output

Tools for experimentation

Mininet: A network emulator that creates a virtual host, switch, controller and link network. Mininet hosts run regular Linux network software, and its switches support OpenFlow [14] for highly flexible custom routing and software-defined networking (Fig. 2).

Gnuplot: It a command-line program capable of generating two- and three-dimensional plots of suitable structures, data and results.

System Configuration

- Machine Vendor—HP
- Processor Chip—Intel i5
- Graphics—2 GB AMD RADEON
- Virtual Machine Client Vendor—VMware WorkStation 14

Controller Version

- Open Daylight—Oxygen
- ONOS—Sparrow
- FloodLight—v1.2
- POX—v0.2.7 [10]
- OVS Switches = 6
- Host Per Switch = 10

3.1 Experimentation

After an overview of five controllers, we conclude that the functioning, roles, and responsibilities of most of them do not provide a classification base. The findings of this review are more or less consistent in all the proposals mentioned in [12]. SDN's initial goal was to centralize the controller, and therefore, most of the controllers used a single controller, but this produced single-failure points and issues with scalability.

This work uses the Python code in Mininet to construct topologies and link them via commands to the controller on computer. The traffic in the network is overwhelmed when the controller is linked to check the connectivity of the links in the network [15]. Afterward, tests are carried out in generated surroundings to see the following controller results for the given topology and the following analysis parameter. To the best of our knowledge, there is no other work that collects and compares such controllers with topologies.

Experiment Phases:
(A) Testing Phase:
Stage 1: Run created topologies using mininet

sudomn –custom ~/Topologies/customtopo.py –topo ring –controller remote

Stage 2: Now flood network

Pingall

Stage 3: Open console of two hosts

Xterm h1 h30

Stage 4: Make one host as transmitter and another host as transreceiver

iperf -c server_ip -b 10mb -t 10 -i 1 – For Server
iperf -s -t 10 -i 1

Stage 5: Capture all results in file which look as following (Fig. 3)
Stage 6: Now we will pre-process this data to get time interval and bandwidth

cat results | grep sec | head -15 | tr - " " | awk '{print $4,$8}' >Output_File

Stage 7: Now for representation will use GNUPLOT

plot "Processed_Output" title "UDP Flow" with linespoints

We followed above stages repeatedly for all controllers and Topologies

```
Server listening on UDP port 5001
Receiving 1470 byte datagrams
UDP buffer size:  208 KByte (default)
------------------------------------------------------------
[119] local 10.0.0.43 port 5001 connected with 10.0.0.1 port 59162
[ ID] Interval       Transfer     Bandwidth        Jitter   Lost/Total Datagrams
[119]  0.0- 1.0 sec  1.23 MBytes  10.3 Mbits/sec   0.039 ms  13/  878 (1.5%)
[119]  0.0- 1.0 sec  13 datagrams received out-of-order
[119]  1.0- 2.0 sec  1.19 MBytes  10.0 Mbits/sec   0.008 ms   0/  850 (0%)
[119]  2.0- 3.0 sec  1.19 MBytes  10.0 Mbits/sec   0.018 ms   0/  850 (0%)
```

Fig. 3 Captured data output

3.2 Performance Analysis Phase

(1) Throughput: We observed interesting results here for mesh, ring and tree topologies. In OpenDayLight for mesh topology, it is consistent for 7 s and after that there is variation. While ring and tree topologies have linear graph. Ring topology has slight decrease in throughput at 9th second and resumed it at 10th second. After seeing graphs, we observed that Java-based controllers like OpenDayLight [6], ONOS [7] and floodlight [8] have multiple variation, but in Python-based controllers, it is more often consistent (Fig. 4).

(2) Individual bandwidth for client and server: We observed results for Flood-Light controller which is having lower link utilization of about 15 GB/Sec. While OpenDayLight controller is having inconsistent results per second (Fig. 5).

At same time, if we see results of Python-based controllers such as POX and RYU, these nearly have same results for ring and tree topologies. This results having performance measure from 25–30 GB with slight variation per second (Figs. 6 and 7). As we are having multiple results here, it is difficult to choose best controller. So we implemented algorithm which does this task for us (Sect. 4).

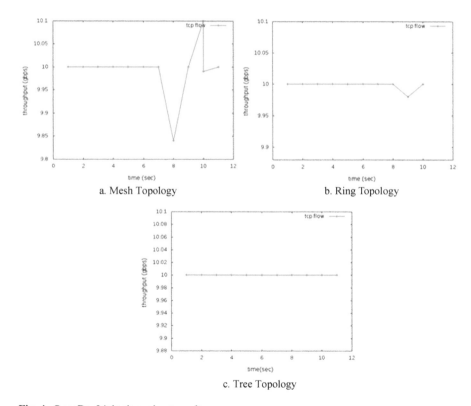

a. Mesh Topology
b. Ring Topology
c. Tree Topology

Fig. 4 OpenDayLight throughput results

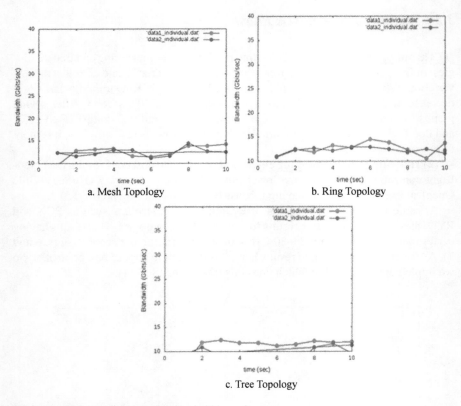

Fig. 5 FloodLight individual client bandwidth results

4 Algorithm

While building the network, the user is given the input of the value which they require in their network in the percentage format and mention the parameter priority. In algorithm, the user percentage which satisfies the user requirements will be marked one or else the value will be filled by zero and the value.

Stage 1: User needs to input parameter values in % and their priorities.

Throughput, Jitter, BW Individual Client, BW Individual Server, BW Simultaneous client, BW Simultaneous server, Security.

Stage 2: We build matrix of M x N where M is row which contains parameters and N is column which contains topologies with controllers (Tables 1 and 2).

Stage 3: Now multiply each row with respective parameter priority

Priority * Bits (0/1)	Priority1 * 0	Priority1 * 0	Priority1 * 1

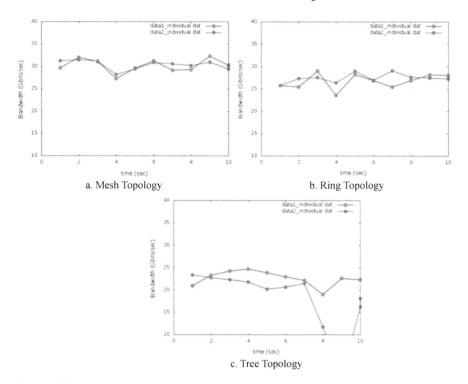

Fig. 6 RYU individual client bandwidth results

Stage 4: Now calculate total of each column and result having highest total value will be best possible matching controller as per customer requirements (Fig. 8).

Case 1: If all the values satisfy the condition, then that parameter is excluded from the calculations. The value will not affect the sum value since all the values are same.

Case 2: If any value in the table does not satisfy the value, the maximum value in the column is set as 1 as it is the best value in the parameter and closer to the customer input.

Then, the priorities will be multiplied with it, and the best result will be provided as the output to the user according to user inputs. If the two or more controllers have the same result, then there will be one more round and after which single result will be provided.

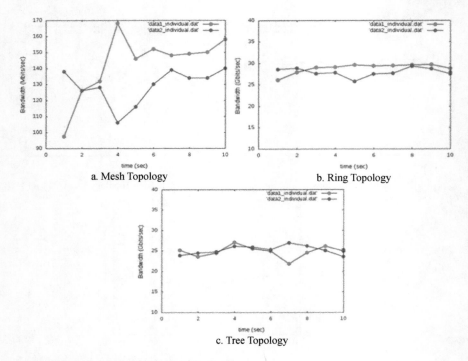

Fig. 7 POX individual client bandwidth results

Table 1 Algorithm matrix

No. of parameters	No. of controllers			
	C1	C2	C3 …	Cn
Parameter 1	0	0	1	1
Parameter 2	0	0	0	1
Parameter 3	1	0	0	1
⋮				
Parameter n	1	1	1	1

Where C1, C2, C3…, Cn are controllers
0 = Not satisfy user entered conditions
1 = Satisfies user entered conditions

5 Results

Based on the qualitative analysis of controllers and capabilities of traffic tools, and the evaluation of controllers using them, we have summarized the main findings below.

Table 2 Comparison table for all controllers

Used topologies	Mesh topology					Ring topology					Tree topology				
	POX (%)	ODL (%)	RYU (%)	ONOS (%)	Floodlight (%)	POX (%)	ODL (%)	RYU (%)	ONOS (%)	FloodLight (%)	POX (%)	ODL (%)	RYU (%)	ONOS (%)	FloodLight (%)
Throughput	75.966	83.983	84.014	76	84	75.798	84.014	84.033	76	84	84.033	84.033	83.991	75.806	86
Jitter	1.6	32.27	5.5	5.52	7.51	1.21	9.255	5.755	4.34	5.97	0.47	10.61	4.79	9.69	1.158
BW Individual client	84.821	83.47	85.795	86.064	84.208	84.945	83.674	85.536	85.934	94.421	85.402	83.911	86.045	85.681	94.679
BW Individual Server	93.966	79.672	86.022	91.614	79.966	88.676	81.935	85.833	87.808	97.507	84.982	89.222	10.776	91.8902	10.034
BW Simultaneous Client	94.894	79.916	95.675	84.502	85.869	82.682	76.774	99.212	85.904	85.89	95.112	85.493	96.692	85.933	85.841
BW Simultaneous Server	93.528	83.654	93.125	85.954	78.013	85.629	83.742	81.432	85.306	77.141	93.539	95.752	92.008	88.766	75.733
Security	3	1	2	2	2	3	1	2	2	2	3	1	2	2	2

Security parameter having values from 1 to 3, where, 1 = Low Security, 2 = Moderate Security, 3 = High Security

Fig. 8 Algorithm output

```
enter precentage of  throughput - 99
enter priority of throughput - 7
enter precentage of  jitter - 1
enter priority of jitter - 6
enter precentage of  bidirectional_individual_client BW - 75
enter priority of bidirectional_individual_client BW - 5
enter precentage of  bidirectional_individual_server BW - 80
enter priority of bidirectional_individual_server BW - 4
enter precentage of  bidirectional_simultaneous_client BW - 76
enter priority of bidirectional_simultaneous_client BW - 3
enter precentage of  bidirectional_simultaneous_server BW - 81
enter priority of bidirectional_simultaneous_server BW - 2
enter security level from(1-3) - 2
enter priority of security - 1
0 0 0 0 0 0 0 0 0 0 0 0 0 0
6 0 0 0 0 6 0 0 0 0 6 0 0 6
5 5 5 5 5 5 5 5 5 5 5 5 5 5
4 0 4 4 0 4 4 4 4 4 4 0 4 0
3 3 3 3 3 3 3 3 3 3 3 3 3 3
2 2 2 2 0 2 2 2 2 0 2 2 2 0
1 0 1 1 1 1 0 1 1 1 1 0 1 1 1
after checking case 1 and case 2
0 0 0 0 0 0 0 0 0 0 0 0 0 1
6 0 0 0 0 6 0 0 0 0 6 0 0 6
0 0 0 0 0 0 0 0 0 0 0 0 0 1
4 0 4 4 0 4 4 4 4 4 4 0 4 0
0 0 0 0 0 0 1 0 0 0 0 0 0 0
2 2 2 2 0 2 2 2 2 0 2 2 2 0
1 0 1 1 1 1 0 1 1 1 1 0 1 1 1
addition of column values
[13, 2, 7, 7, 1, 13, 6, 8, 7, 5, 13, 6, 3, 7, 9]
your conditions are satisfied by this controller - mesh_pox
```

- Considering overall performance then multi-threaded controller (FloodLight) significantly better than centralized and single-threaded controllers like NOX, POX, and Ryu. However, in order to perform better, they also need more physical resources.
- In case of jitter Python-based controller POX is having better performance as its result value is less as compared to others.
- Individual bandwidth of client is not having significant effect of controllers and topologies.
- In the physical topology of controllers, a number of performance parameters are directly affected. In this regard, we expect to perform an extension study in order to equate the distributed controllers with specific thematic settings (datacenter, WAN, web, etc).
- We also noticed that machine configuration and number of nodes affect the performance of controller. Isolating the performance of controller from the results is not possible.

6 Conclusion

We have successfully studied five controllers, namely OpenDay light, FloodLight, ONOS, RYU, Pox on different topologies, namely mesh topology, ring topology and tree topology, and on studying these five controllers, we saw that all these five are showing variations to each other. But the main variation is that the looping topologies show better results in Python-based controllers than the multi-threaded Java-based controllers.

In our research, we have generated 75 graphs, and for analyzing these graphs, we have developed our own algorithm as the manual analysis of these graphs was quite difficult.

The algorithm that we have used is giving optimized results as per user requirement. Thus, we can say that our research will be helpful for other researchers researching in the same field.

References

1. Beshley, M., Seliuchenko, M., Kahalo, I., Panchecko, O.: Experiment Performance Analysis on SDN Switch and Controller. 978-1-5386-2556-9/18/$31.00 2018 IEEE
2. Khondoker, R., Zaalouk, A., Marx, R., Bayarou, K.: Feature-Based Comparison and Selection of Software Defined Networking (SDN) Controllers. 978-1-4799-3351-8/14/$31.00c 2014 IEEE
3. Satre, S.M., Jadhav, V.P.: Analysis on security in cloud and sky computing. Recent Advances in Computer Science, E-Learning, Information & Communication Technology International Conference (CSIT-2016), Jawaharlal Nehru University New Delhi-110067, p-ISSN: 2393-9907; e-ISSN: 2393-9915; vol. 3, pp. 42–46, Jan. 2016
4. Zerrik, S., El Ouadghiri, D., Bakhouya, M.: Performance Evaluation of Software-Defined Networking Architectures Using Network Calculus. 978-1-5090-5146-5/16/$31.00 ©2016 IEEE
5. Satre, S.M., Jadhav, V.P.: QoS-oriented adaptive expedient broadcast routing for hybrid wireless network. IEEE Adv. Electr. Electron. Inf. Commun. Bioinform. In: 2016 International Conference (AEEICB - 2016), ISBN- 978-1-4673-9745-2, Feb. 2016
6. OpenDaylight: A Linux Foundation Collaborative Project. [Online] Available: https://www.opendaylight.org/
7. ONOS: **O**pen **N**etwork **O**perating **S**ystem. [Online] Available: https://wiki.onosproject.org/
8. FloodLight: The Floodlight Open SDN Controller is an enterprise-class, Apache-licensed, Java-based OpenFlow Controller. [Online] Available: http://www.projectfloodlight.org/floodlight/
9. Ryu SDN Framework Community, "Ryu Controller." [Online] Available: https://osrg.github.io/ryu/index.html
10. POX Controller Manual Current Documentation. [Online] Available: Available: https://noxrepo.github.io/pox-doc/html/
11. Yamei, F., Qing, L., Qi, H.: Research and Comparative Analysis of Performance Test on SDN Controller. 978-1-4673-8515-2/16/$31.00 ©2016 IEEE
12. Salman, O., Elhajj, I.H., Kayssi, A.: SDN Controllers: A Comparative Study. 978-1-5090-0058-6/16/$31.00 ©2016 IEEE
13. Patil, S.M., Malik, A.K.: Correlation based real-time data analysis of graduate students behaviour. In: Communications in Computer and Information Science book series (CCIS), vol. 1037, Springer Nature Singapore Pte Ltd. June 2019. https://doi.org/10.1007/978-981-13-9187-3_62
14. Darekar Sachin, M.Z.S., Kondke, H.B.: Network Behaviour of Open vSwitch as per the anticipated functionality of network application programmed over SDN using pox controller. In: 2nd International Conference on Computer Networks and Inventive Communication Technologies (ICCNCT - 2019) Springer Lecture Notes on Data Engineering and Communications Technologies
15. Darekar Sachin, M.Z.S.: Performance analysis of various open flow controllers by performing scalability experiment on software defined networks. In: 3rd International Conference on Inventive Computation Technologies (3rd ICICT 2018) conference proceedings and IEEE Xplore. www.icicts.com

Performance Improvement in Web Mining Using Classification to Investigate Web Log Access Patterns

Charul Nigam and Arvind Kumar Sharma

Abstract Data mining having one of the applications termed as Web mining which helps to eradicate particular information from Web data like Web content, Web structure, and Web usage data. Web service is the technique of detecting suitable service as per requirements and it is applicable in enterprises, industry, and government. Web service is a software arrangement considered to maintain interoperable machine wise interface over a grid. The purpose of this research study is to investigate an efficient methodology of the Web services which plays crucial role for acquiring different services. From this research study, the classification method in Web usage mining can be utilized to find visitor activity information form Web server log which helps the company for getting supporting information and used to specify the evaluation of Web application. The accuracy of our work is acquired better outcome than the existing classification, so it can be used in Web usage mining.

1 Introduction

Today, due to rapid expansion of the WWW a wealth of information on many awesome subjects and topics are reachable online. Web sites are playing important role in a vital characteristic no longer totally for groups however moreover for private people trying to detect a range of information. Web sites are desirable conversation channels to supply recommended data to the end-users. Several Internet Web sites have been growing and now not directly lead to expand the complexity of Web site design [1]. To simplify the complexity of Web site design we have to apprehend the form of Web site and their online working. This task is completed via way of the use of one of the techniques, acknowledged as Web mining. The objects of Web

C. Nigam (✉) · A. K. Sharma
Career Point University, Kota, India
e-mail: charul.nigam@gmail.com

A. K. Sharma
e-mail: drarvindkumarsharma@gmail.com

© The Editor(s) (if applicable) and The Author(s), under exclusive license to Springer Nature Singapore Pte Ltd. 2021
T. Senjyu et al. (eds.), *Information and Communication Technology for Intelligent Systems*, Smart Innovation, Systems and Technologies 196,
https://doi.org/10.1007/978-981-15-7062-9_65

usage mining are vast, heterogeneous and distributing documents. The logistic structure of Web is a plan structured through archives and hyperlinks, and mining results may additionally be on Web contents or Web structures [1]. The matters to do such as Web designing, developing, captivating Web sites are moreover a part of Web usage mining techniques. Because of its direct utility in e-commerce, Web analytics, e-learning, information retrieval, etc. [2]. Since Web has risen in 1991, the measure of Web site pages exponentially has developed. Web service is process of statement among two electronic devices over grid. Most of the requests of Web facilities are actually utilized in occupational and huge scale initiatives. Currently, in businesses these requests have raised from enormous request to vigorous setup of commercial developments. In present situation, net facilities are extensively utilized in business and university. It is a group of exposed etiquettes and applied for substituting data among requests or organizations [3].

AIM: We propose a Web mining model that illustrates how to utilize performance improvement in Web mining using classification to investigate Web log access patterns collected from a Web server of a Web site. Moreover, we have also analyzed the Web server access logs of user's behavior activity.

Organization: This paper is organized into different sections: Sect. 2 presents related work of Web mining in user behavior access patterns, Sect. 3 describes about Web mining and its taxonomy, Sect. 4 shows proposed methodology, Sect. 5 concludes the paper while references are mentioned at the last.

2 Related Work

In this section, some of the recent related works in the domain of Web mining are discussed as under.

The authors in [4] has said clustering as properly as classification primarily based records mining methodology in banking area which has utilized to limit the chance of choice making charge and analyze the non-public and attainable mortgage customers. The overall performance evaluation of K-means clustering approach has evaluated. They used K-means clustering to tightly closed statistics in banking to keep away from missing of inaccurate data. The authors in [5] have presented an analysis in banking sector by utilizing data mining has proposed to predict users using simple K-means algorithm that has utilized for clustering phase. The results have obtained to determine patterns of customers and to identify the customer retention modalities. In this work they have prepared banking data of users to remove inconsistencies and redundancies. The scaling and standardizing of data causes reduction in dispersion level between variables in bank datasets have been analyzed. The authors in [6] have adopted novel methods and effective procedure for closed pattern mining has proposed to detect meaningful and non-overlapping patterns. The performance analysis of proposed work for both synthetic and real dataset in banking process has tested rigorously. They propose that banking data has visualized based on closed pattern. The closed pattern of data has aggregated to avoid replication of bank datasets.

The authors in [7] presented a hybrid method of recommendation through extended matrix based on collaborative filtering with demographics information has proposed to generate more accurate prediction. The collaborative model of demographic information has applied to measure the performance that has related with conventional bias-based factorization model shows precision and recall without demographic data is evaluated. The authors in [8] reported the collaborative filtering to improve the accuracy of the recommender systems. This has achieved by accuracy from algorithms of memory based methods were optimized and formulated under the classification of k-NN. The Initial stage has to recover the quality of the recommendation system by hybrid techniques. It mainly used for content filtering and primarily collaborative filtering. At the second stage, the algorithms have admitted as social information with former hybrid recommendation techniques. It has developed and accommodated by the social adaptive techniques, social network analysis, and trust-aware algorithms. The authors in [9] adopted many classification approaches utilized in data mining concepts. Data mining is a wide field, which integrates the methods from multiple fields like pattern recognition, machine learning, statistics, and AI. To analyze the large number of data, this classification algorithm could be implemented on dissimilar kinds of datasets such as customers in life, data of patients, customer in telecom industry, health insurance industry, customers in banking, customer in sales industry, students in education, Web-related text files, files, image files, video files, audio files and voice files. The authors in [10] have proposed the classification methods for enhancing user access through Web data. This performed as classifiers and experiment of user access behavior from huge volume of databases. The vital role of Internet plays an impact in ordinary lifestyles. So, it is essential to review similarity growth of the data in the help of machine learning algorithms. For this purpose, by using the Naïve Bayes classification methods are categorized to identify frequent access patterns. The aim of this work diagnosed the shopping conduct of the person based totally on their situation. The effect of this proposed method shows greater high quality and effectively of machine through lowering the looking time with the sorts of users. The obstacles of the proposed classifier of Naïve Bayes ought to now not classify the undefined class. Authors in [11] have proposed a new feature selection algorithm by using minimal variance approaches. To decide the proposed approach which decreased the computational complexity, and reduced the quantity of preliminary buildings and raises the classification accuracy of the specific characteristic subsets. The clusters are designed to make use of the minimal variance techniques.

3 Web Mining—Background

Web mining is one of the data mining techniques to discover meaningful outline as Web information [12]. The data mining service called Web mining is utilized for extracting data from Web documents, Web log data, Web hyperlinks between Web documents and Web logs. It is a process of analyzing data from several angles and

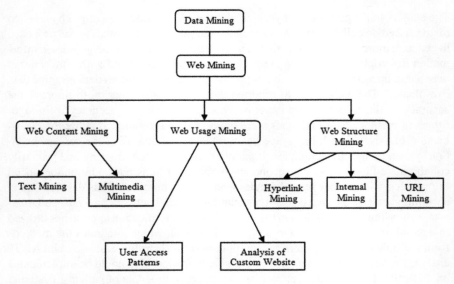

Fig. 1 Classification of Web mining [18]

encapsulating with valuable information that helps to develop revenue, cost of cuts, or both. It allows users to examine the information from various dimensions, sort it and capture relations. Precisely, data mining is utilized for discovering relationships among lots of fields in huge relational databases. It has one of the applications termed as Web mining that help to eradicate particular information from Web data [13]. Web mining is useful to e-commerce sites and e-services. It is mostly applied based on Web users through their communication with Web sites and usage of Web site to accept information from the Web. Web mining contains three wide categories—Web content mining, Web usage mining and Web structure mining which are shown in Fig. 1.

In our work the Web usage mining is considered.

3.1 Web Usage Mining

Web usage mining is a stream of Web mining, which in turn, is a part of data mining. It is widely used to extract the usage characteristics of Web users [14]. The extracted information could be used for several purposes like improving the Web applications, identifying the user's behavior, identifying fraudulent elements, etc. Some of the benefits of Web usage mining application are such as—improvement in Web design and layout, analyzing system performance to understand user's reaction and motivation, creating adaptive Web sites and predicting user's behavior, etc. Web usage mining includes three main phases which are shown in Fig. 2.

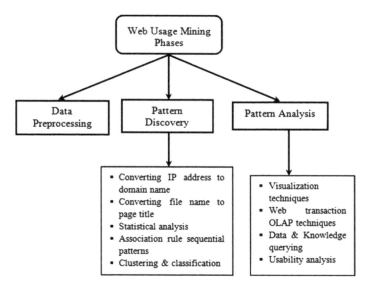

Fig. 2 Web usage mining phases [19]

From above figure, in first phase, Web logs could be pre-processed in order to preserve only useful and relevant information. In second phase, several Web usage mining approaches can be applied to identify interesting Web access patterns from preserved Web data [14]. Finally, in third phase, the Web usage patterns can be presented for Web log analysis. These three necessary phases are discussed as under.

3.1.1 Data Preprocessing

Data are generally imperfect with missing quality standards, missing assured attributes of importance, or having lone collective data in among hefty lowlife wrong or overview with shaggy overlaying inconsistencies into code and destinations. Data preprocessing responsibilities in data cleaning covers the absent of values, smooth noisy data, remove or identify outliners and resolution discrepancies. Data integration expending several databases, along file or data cubes with standardization plus combination in transformation of data [1]. Dropping the capacity but manufacturing a similar or related analytics results in decrease of data. Rare data is extremely subject to sound, loss of standards and discrepancy. Preprocessing of Web usage data is shown in Fig. 3.

Data preprocessing is significant phases in data mining method along contracts through planning besides alteration of the early dataset. Data pre-processing approaches [15] at various levels which are discussed as under:

- *Data Cleaning*

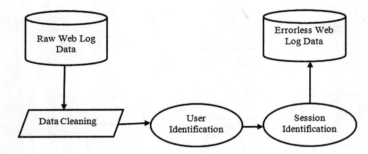

Fig. 3 Preprocessing of Web usage data

Data which is to be examined by data mining approaches can be imperfect, noisy and inconsistent. Unfinished, noisy and uneven data are mutual residence assets of huge, actual creation of databases and data warehouses. Due to several reasons data would be lacking attributes of attention could not be continuously be obtained such as client data [15]. Additional data may not be comprised basically as it has not measured significantly. Applicable data could not be noted unpaid to a mistake or since of apparatus faults. Data container has noisy consuming improper feature values, due in conformity with next there may have been mortal yet mainframe mistakes occurring at statistics penetration. Faults in data broadcast can also occur.

- *Data Integration*

It is in all like hood of data evaluation resolve incorporate the discarding over data, which chains data from several resources into a logical data stock as in data warehousing. This source might comprise several data base flat files, or data cubes. There are large amounts of disputes to reflect throughout combined of data. Representation of mixing can be complicated. Decision making with data mining is depicted in Fig. 4.

- *Data Transformation*

The data transformation converts the information or combined into approaches suitable for data mining conversion can include as follows:

- Normalization
- Smoothing
- Aggregation
- Generalization.

- *Data Reduction*

Complex information investigation and mining on immense of data can gross quite a while, create such examination unfeasible or infeasible. The idea of data decrease is ordinarily comprehended as any lessening the capacity or lessening the measurements. There are various techniques that have encouraged in examining

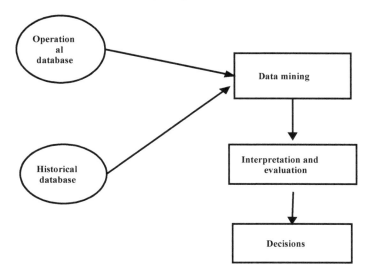

Fig. 4 Decision making with data mining

diminished aggregation then measure of statistics however yield beneficial information certain share put together techniques assignment including integrity in limitation of phase regarding information tuples. Mining over the lowered statistics index need to according to keep steadily educated yet consign the equivalent results.

3.1.2 Pattern Discovery

Web usage mining is utilized to uncover patterns in Web logs but often carried out only samples of data. The Web mining process would be in effective if the samples are not good representation of huge volume of data.

3.1.3 Pattern Analysis

This is one of the major phases of Web usage mining. In pattern analysis, the obtained Web usage patterns are analyzed to discover unknown information and useful information [16].

4 Proposed Methodology

Our proposed methodology tells how the development work should be carried out in the form of action. It is identified as a tool which is always used to discover some area for which data is composed and analyzed.

On the basis of data analysis, conclusions are drawn. Here, following issues are identified as under:

- Remove noisy data from Web logs
- Filter useful information from Web logs
- Transformation of data into numerical format
- Identify the IP address and session conditions of the user from Web logs.

4.1 Data Collection

Data collection is conducted from different locations, i.e., server side data, client side data and proxy server side data and cookies data [17]. Our proposed methodology is shown in Fig. 5.

Here the input data is taken from the dataset which consist of Web pages. Initially the input data is preprocessed using tag based preprocessing technique. Then modification process is done in which the user can alter, change or modify the contents in the Web page. Also the ratings for the Web pages are given and stored in the database. Then these ratings are classified using K-NN classifier and the priority for the Web page contents are obtained based on these ratings. Finally the user can access the Web pages and recommend it to others based on the prioritized data.

4.2 Proposed Algorithm

In this section, an algorithm is proposed for Web log preprocessing. This proposed algorithm is used to remove accessorial entries from Web log tables. Web usage data cleaning algorithm removes the irrelevant entries from Web logs and filtering algorithm discards the uninterested attributes from Web log files.

- *Tag Based Pre-processing Algorithm*

Generally contents of the Web pages are represented in head, body and tail. In the body part all contents of the Web pages are presented in which user can access information. Initially the pre-processing step is included for accessing Web contents by users. For this purpose, the process of tag based pre-processing is done. Here contents of Web pages are represented in tag format. i.e. </.....>.

The algorithm used for tag based pre-processing is as follows:

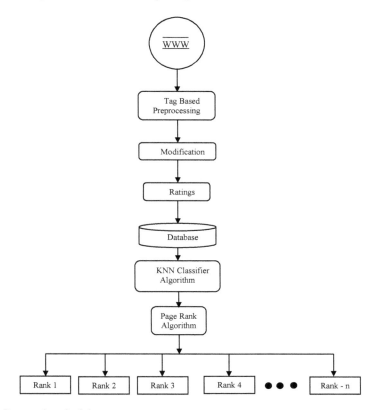

Fig. 5 Proposed methodology

```
Data as Dt_in
Words as W_d
Pre processed data as Pp_dt
While (Dt_in = !null)
W_d = Dt_in . Tokenize ("<" || ">")
End while
Pp_dt = W_d[all]
```

- **K-NN Classification Algorithm**

    Pseudo code of an algorithm for K-NN Classification is as follows:

```
Input data as Pp_dt, Weight as W_t, splits as S_p, Segments
as S_t, Function as knn_op(), Output data as Op_dt
Create knn_op()
knn_op().W_t → maximum,
For I to row-1
Initialize (S_t,N_u)
For each S_t ∈ S_ts do
Create d_k
Create knn_op()'
Host → device (In,d_k,knn_op()')
Initialize (S_p,S_t)
Create M_k
For each S_p ∈ S_ps do
Initialize M_k
Host → device (M_k)
Initialize (C_k,S_p)
For each C_k ∈ C_k do
Call Distance (In,S_p,C_k)
Call KNN (d_k,S_p,C_k,M_k)
End For each
End for Each
Device → host(knn_op()')
End For Each
Return knn_op()
```

This algorithm can be used in pattern discovery process of Web usage mining. K-NN is a classification algorithm and simple one and yields high competitive outcomes. It applied for both classification and regression predictive issues. The k-means classifier is also one of the partitioning techniques that analyze the classifiers. The nearest mean cluster is derived from the partition of n observations into k-classifiers. In each predefined category, the cluster center is extracted from k-means clustering. The classifiers are used to evaluate either Euclidean distance or Cosine similarity among the test and train tuples.

Here, evaluate the three significant aspects as under:

- Calculation time
- Predictive power
- Ease to interpret output.

5 Conclusion

Generally, the classification is utilized to classify every item in a number of data into one of predefined set of groups or classes. Classification algorithms make usage of mathematical methods such as neural network, decision trees, linear programming, and statistics. This paper proposes the classification algorithms in Web usage mining can be utilized to investigate Web log access patterns gathered from the Web server of a Web site. It also predicts user behavior patterns and visitor activity information from Web log data which helps the company to produce supporting important information and used to specify the evaluation of Web applications to improve the performance and convenience of the Web site. In future we propose and implement a Web service recommendation model which repeatedly recognizes the utility of facilities and proactively determining and endorsing facilities to the end consumers.

References

1. Chaudhary, J., et al.: Evaluating similarity of websites using genetic algorithm for web design reorganisation. In: Advances in Intelligent Systems and Computing, International Conference, vol. 555. Springer, Berlin (2017)
2. Mughal, M.J.H.: Data mining: web data mining techniques, tools and algorithms: an overview. Int. J. Adv. Comput. Sci. Appl. (IJACSA) **9**(6) (2018)
3. Sharma, A.K., et al.: Enhancing the performance of the websites through web log analysis and improvement. Int. J. Comput. Sci. Technol. **3**(4) (2012)
4. Çaliş, A., Boyaci, A., Baynal, K.: Data mining application in banking sector with clustering and classification methods. In: International Conference on Industrial Engineering and Operations Management (IEOM), pp. 1–8 (2015)
5. Oyeniyi, A., Adeyemo, A., et al.: Customer churn analysis in banking sector using data mining techniques. Afr. J. Comput. ICT **8**, 165–174 (2015)
6. Király, A., Laiho, A., Abonyi, J., Gyenesei, A.: Novel techniques and an efficient algorithm for closed pattern mining. Expert Syst. Appl. **41**, 5105–5114 (2014)
7. Valdiviezo Diaz, P., Bobadilla, J.: A hybrid approach of recommendation via extended matrix based on collaborative filtering with demographics information. In: International Conference on Technology Trends, pp. 384–398 (2018)
8. Thorat, P.B., et al.: Survey on collaborative filtering, content-based filtering and hybrid recommendation system. Int. J. Comput. Appl. **110** (2015)
9. Phyu, T.N.: Survey of classification techniques in data mining. In: Proceedings of the International Multi Conference of Engineers and Computer Scientists, pp. 18–20 (2009)
10. Raju, P.S., et al.: Data mining: Techniques for enhancing customer relationship management in banking and retail industries. Int. J. Innov. Res. Comput. Commun. Eng. **2**, 2650–2657 (2014)
11. Venkataraman, S., et al.: A novel clustering based feature subset selection framework for effective data classification. Ind. J. Sci. Technol. **9** (2016)
12. Parekh, A, et al.: Web usage mining: frequent pattern generation using association rule mining and clustering. Int. J. Eng. Res. Technol. (IJERT) **04**(04) (2015)
13. Fatma, M., Rahman, Z.: Building a corporate identity using corporate social responsibility: a website based study of Indian banks. Soc. Respons. J. **10**, 591–601 (2014)
14. Meghwal, A.R., et al.: Identifying system errors through web server log files in web log mining. Int. J. Comput. Sci. Technol. (IJCST) **7**(1) (2016)
15. Bell, D., Mgbemena, C.: Data-driven agent-based exploration of customer behavior. Simulation **94**(3), 195–212 (2018)
16. Parekh, A., et al.: Web usage mining: frequent pattern generation using association rule mining and clustering. Int. J. Eng. Res. Technol. (IJERT) **04**(4) (2015)
17. Chitraa, V., Davamani, A.S.: A survey on pre-processing methods for web usage data. Int. J. Comput. Sci. Inf. Secur. **7**(3), 78–83 ((2010))
18. Sharma, S., et al.: Explorative study of web data mining techniques and tools: a review. Int. J. Comput. Sci. Technol. (IJCST) **8**(1) (2017)

19. Cooley, R., et al.: Web mining: information and pattern discovery on the World Wide Web. In: International Conference on Tools with Artificial Intelligence, Newport Beach, pp. 558–567. IEEE, New York (1997)

Array Password Authentication Using Inhibitor Arc Generating Array Language and Colored Petri Net Generating Square Arrays of Side 2^k

S. Vaithyasubramanian, D. Lalitha, A. Christy, and M. I. Mary Metilda

Abstract To secure the information provided, Web sites ask the users to create a password. We have seen in many instances such passwords are hacked and information gets leaked. Security of login-password system is the foremost requirement in the field of information technology. In this paper a new password authentication system is proposed. Authentication using Array Password procedure is proposed, Inhibitor Arc generating Array language and Colored Petri net generating square arrays of side 2^k are discussed. These Petri net model generates Arrays, Authentication and implementation process of the these Arrays into Array Password methodology is described with illustrations.

1 Introduction

Petri nets have been introduced by Carl Adam Petri in the year 1962, Later it has several extensions as Colored Petri nets, Timed Petri nets, Timed colored Petri nets, Interval timed Petri nets, Stochastic Petri nets, Fizzy Petri nets, array languages in picture recognition [1–3] etc. Petri nets has been extensively used in varies fields. The concept of Petri nets and colored Petri net has been used in array generation. Petri nets and colored Petri nets are also used in tile pasting and kolam generation. Timed

---

S. Vaithyasubramanian (✉) · D. Lalitha · M. I. Mary Metilda
Department of Mathematics, Sathyabama Institute of Science and Technology, Chennai, India
e-mail: discretevs@gmail.com

D. Lalitha
e-mail: lalkrish2007@gmail.com

M. I. Mary Metilda
e-mail: metilda81@gmail.com

A. Christy
Department of Computer Science Engineering, Sathyabama Institute of Science and Technology, Chennai, India
e-mail: ac.christy@gmail.com

© The Editor(s) (if applicable) and The Author(s), under exclusive license to Springer Nature Singapore Pte Ltd. 2021
T. Senjyu et al. (eds.), *Information and Communication Technology for Intelligent Systems*, Smart Innovation, Systems and Technologies 196,
https://doi.org/10.1007/978-981-15-7062-9_66

colored Petri nets are used in Traffic management and modeling bank ATM. Petri nets were also used in modeling, analysis and execution of Robotic tasks. Recently they are playing a big role in Password generation [4, 5].

Validating Web login accounts, official mails, personal mails, online ticketing accounts, online banking accounts and so on of an user, users login account, Password and PIN are essential. Credentials plays a major role in authenticity of information security. Passwords strength and usability is the key factor in preserving the authenticity, integrity, accountability of user accounts [6, 7]. Hackers and Password crackers make use of vulnerabilities in authentication process and in password formation to gain access to the user accounts causing loss of information [8–10]. In view of this authentication procedure numerous practices have been proposed. In spite of the fact the conventional login and Password based process are anything but difficult to execute, they have been exposed to attacks with the help of available digital technology [11]. Solution to this authentication issue can be overcome by the implementation of multifactor authentication and Array passwords without changing the existing authentication process.

Password is a string of letterings made of alphabets, numbers and special characters available in normal computer key board. String languages are associated with Petri nets by attaching a label to every transition. This gives a word of the language generated [12]. Similarly an array language can also be generated by a Petri net [13, 14]. The language generated by the Petri net is defined as the set of all arrays reaching the final set of places. A subset of the set of places is defined as the final set of places. The tokens in the places of the basic Petri net contain only black dots. The transition sequence which takes an initial marking into a final marking is considered, in that sequence every transition is replaced by the corresponding label. Arrays over a given alphabet are taken as tokens in Petri nets defined for array generation. Firing rules associated with transitions are defined in such a way that when they are fired the arrays grow in size and move from one place to another. To bring in a better control over the firing sequences, inhibitor arcs can be used. Petri nets with inhibitor arcs have more control over the array that gets generated, than a Petri net without inhibitor arcs [15]. In the array generating model some of the transitions are labeled. If a transition does not have a label, then firing the transition just moves the array in the input place to all its output places. If a transition has a label, then the result of firing the transition will depend on the label.

In this paper we discuss Inhibitor Arc generating Array language and Colored Petri net generating square arrays of side 2^k. And how these Array languages can be implemented as authentication process is discussed.

2 PnAL: Petri Net Array Languages

Petri net is a dynamic model used to generate Array languages using catenation rules. The arrays from all the input places are consumed, but only the array from the specified input place is put in the output places. Starting from the initial assignment of arrays, all enabled transitions are fired. The arrays move and also may grow in size. The arrays that reach the final set of places are collected to form the array language generated by the Petri net. These languages are denoted by PnAL. The Catenation rules, Array generating Petri nets, Inhibitor Arc and Colored Petri nets are discussed in this section with illustrations.

2.1 Catenation Rule

Catenation rules are of the form $A \otimes B$ or $A \oplus B$. The array from the input place is joined/catenated according to the rule specified and put in all the output places. In this notation A always denotes the array coming from the input place. The number of rows of A is represented by m and the number of columns of A is represented by n. B is an array language defined in the net. An array language B, involved in column catenation, will have fixed number of columns and a variable number of rows. The number of rows of B will take the same value as m. Similarly an array language B involved in row catenation, will have fixed number of rows and a variable number of columns. The number of columns of B will take the same value as n.

Definition 2.1.1 (*Array Generating Petri Net (AGPn)*)

An Array Generating Petri Net (AGPn) is an eight tuple $(\Sigma, P, T, I, O, M_0, \varphi, F)$ where

Σ is an alphabet of characters, P is the set of places, T is the set of transitions such that $P \cap T$ is empty, I is the input function from T to bags of places, O is the output function from T to bags of places, M_0 is the initial assignment of arrays to places, φ is a partial function which assigns labels to selected transitions, F is a subset of P, called the final set of places.

2.2 Inhibitor Arc

An inhibitor arc is differentiated from the normal directed arc by replacing the arrowhead by a small circle. The corresponding place which are arrow headed by small circle is called an inhibitor input. A transition is disabled if the inhibitor input place has tokens. So an inhibitor arc places the condition that the transition would fire only if the place is empty. If inhibitor arcs are used in the Petri net to generate array languages, then the model has more generative power. The language Pn_1AL generated by the net in Fig. 1 cannot be generated if the inhibitor arcs are not used.

Example 2.2.1 The array generating Petri net $AGPn_1 = (\Sigma, P, T, I, O, M_0, \varphi, F)$, Where $P = \{P_1, P_2, \ldots, P_8\}$, $T = \{t_1, t_2, \ldots, t_8\}$, All the arrays used in the catenation rules, φ, Σ, input and output functions are given below. The initial marking M_0, is the array A in P_1. $F = \{P_3\}$.

Fig. 1 Net using inhibitor arcs

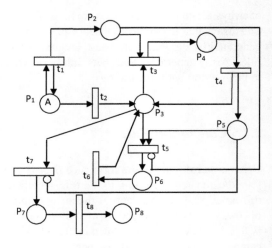

$$\Sigma = \{\bullet, x\}, A = \begin{matrix} \bullet \ x \\ x \ x \end{matrix}, A_1 = \bullet \ x, A_2 = x \ x,$$

$$A_3 = \bullet \bullet, A_4 = (\bullet)_m, \varphi(t_1) = A_1 \otimes A, \varphi(t_3) = P_3,$$

$$\varphi(t_4) = A \otimes A_2, \varphi(t_5) = P_3, \varphi(t_6) = A \otimes A_3,$$

$$\varphi(t_7) = A \otimes A_3, \varphi(t_8) = A \oplus A_4$$

The firing sequence that moves the start array to P_8 will be of the form t_1 n $t_2(t_3t_4)$ n (t_5t_6) n t_7t_8. Both the sequences t_3t_4 and t_5t_6 are made to fire exactly the same number of times as t_1. This control is brought in by the two inhibitor input places P_2 and P_5. Unless P_2 is empty t_5 cannot fire. If t_1 fires two times, then P_2 will have two tokens and so t_3t_4 is also fired two times. Now two arrays are pushed into P_5. The transition t_7 cannot fire until P_5 is empty. This makes sure that the sequence t_5t_6 is also fired two times. The firing sequence $t_1^2t_2 \ (t_3t_4)^2(t_5t_6)^2t_7t_8$ generates the third array of the language. The derivation is shown in Fig. 2.

$$\text{If} \quad B_1 = \bullet \ x \ \bullet, B_2 = x \ x \ \bullet, B_3 = \bullet \bullet \bullet,$$

$$\text{then} \quad Pn_1AL = B_1^k \otimes B_2^k \otimes B_3^k / k \geq 1$$

2.3 Colored Petri Net

Example 2.3.1 AGCPn1 $= (\Sigma, P, T, I, O, V, TA, \varphi, F)$ where $\Sigma = \{a\}, P = \{P_1, P_2, P_3, P_4, P_5\}, T = \{t_1, t_2\}, V$—The tokens used are arrays and integers. The declarations are given in Fig. 3. A is a two dimensional array whose initial value is assigned. I and J are integers whose initial value is also fixed. Initial Token Attribute

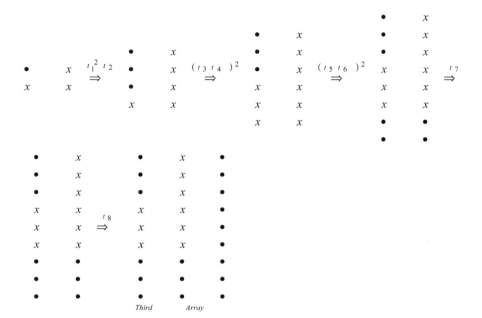

Fig. 2 Derivation up to second array of the language

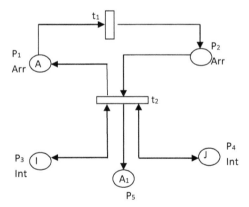

Fig. 3 CPn generating square arrays of side 2^k

is the set $\{\langle A, P_1, A\rangle, \langle I, P_2, 3\rangle, \langle J, P_3, 2\rangle, \langle A_1, P_5, A_1\rangle\}$, the function φ is defined in the net and $F = \{P_6\}$.

Fig. 4 Derivation of array of size 4

$$\begin{matrix} a & a \\ a & a \\ a & a \end{matrix} \xrightarrow{(t_1 t_2)^2} \begin{matrix} a & a & a & a \\ a & a & a & a \\ a & a & a & a \\ a & a & a & a \end{matrix}$$

Array in P_1 → Array in P_1 & P_5

$\phi_{t_1}[\langle A, P_1, A \rangle] = [\langle A, P_2, A \oplus a_m \rangle]$.

if $I < 2^J$, then

$\phi_{t_2}[\langle A, P_2, A \rangle, \langle I, P_3, I \rangle, \langle J, P_4, J \rangle] = [\langle A, P_1, A \otimes a^n \rangle, \langle I, P_3, I+1 \rangle, \langle J, P_4, J \rangle]$.

if $I = 2^J$, then

$\phi_{t_2}[\langle A, P_2, A \rangle, \langle I, P_3, I \rangle, \langle J, P_4, J \rangle] = [\langle A, P_1, A \otimes a_n \rangle, \langle I, P_3, I+1 \rangle,$
$\langle J, P_4, J+1 \rangle, \langle A, P_5, A \otimes a_n \rangle]$

The function ϕ_{t_1} and ϕ_{t_2} describe the actions of the transactions. At any stage either t_1 or t_2 is enabled and they fire alternately. Whenever t_1 fires a column of a's is joined to the array in P_1 and is put in P_2. Then t_2 gets enabled to fire. Since t_2 has three input places, the result of firing t_2 will depend on the I, J values. If $I < 2^J$, then a row of a's is joined to A and it is put in P_1. In P_3, the value of I is replaced by $I + 1$ and in P_4, the value J is unaltered. If $I > 2^J$, then a row of a's is joined to A and it is put in both P_1 and P_5. In P_3, the value of I is replaced by $I + 1$ and in P_4, the value of J is replaced by $J + 1$. The square arrays are built till it reaches the size 2^J. When it reaches the size of 2^J it is put in the final place P_5. All the arrays that reach the final place P_5 are collected to form the array language generated by the net.

Color Int = int;
Color C = string;
Color S = string;
Color Arr = ARRAY2;
Var i, j, m, n :Int;
Val I = 0;
Val j = 1;
Val S = λ ;(Empty String)

Color Int = int;
Color Arr = ARRAY2;
Var I, J: Int;
Val I =3, J = 2;
Var A: Arr;
Val
$A_1 = A = \begin{matrix} a & a \\ a & a \end{matrix}$;

The derivation of the array of size 2^2 starting from the initial marking is given in Fig. 4. The language generated by CPn_1AL consists of square arrays of size 2^k, $k > 1$.

3 Creating and Authenticating Access Code

Primary requirement in the field of computer technology is security of information. Hackers eagerness in knowing about others individual and official data to

Table 1 Array password authentication process

Step	Authentication process
1	Create preferred Login ID—for new user
2	Array Password creation or generation phase Instruction and guidelines from service providers to user
3	Users choice of: array password from arrays generated from inhibitor arc's array language or colored petri nets square arrays
4	Array size determination depending on step 3
5	Provision of array as preferred by user in step 4
6	Input entries in places of arrays by users Users need to remember array size and input given by them
7	Reconfirmation for password array size and inputs For password confirmation along with mobile number
8	If array size and password entered in step 6 and 7 matches process ends and user can proceed for accessing the service. If step 6 and 7 doesn't matches go to step 4
9	For existing users educating about array password and provision to change into new authentication process

threaten them, misusing others identity in making money forges this digital technology shaky. To enhance the effective security, As a prevention technique the proposed methodology take measures by implementing two factor authentication on Web login-password validation process. In existing login-password authentication process once the user enters user ID it will be directed to enter password for validation. The proposed Array Password authentication process enhances security by two factor authentication first factor identification of order of array second factor password input in array type. The proposed authentication process exploits multifactor authentication and robust array password to strengthen the validation process. The step by step authentication and implementation process are portrayed in Table 1 and 2 respectively. Sample Array Passwords are illustrated in Table 3.

In discussing the advantages it has multifactor authentication and new array password technique which will be difficult to crack. On the other hand as limitations such as it requires size modification to impart array passwords, educating users how to create, how to use, time complexity and remembering ability.

Example 3.1 (*Sample Array Passwords*)

4 Conclusion

Petri net models to generate array languages are defined. To improve the generative capacity of the model, inhibitor arc and colored Petri nets are also used for array generation. As a new technique this methodology can be used for information

Table 2 Array password implementation process

Step	Implementation Process
1	Visit login page, enter user name
2	Display to enter array size
3	User needs to enter their preferred array provided at the time of login account creation. If array entered by user matches proceeds to array password
4	If Array entered doesn't matches second last chance for users if not matched access denied and alert to users registered mobile number
5	Array password: user needs to deliver their preferred inputs in array displayed in the login screen
6	If input matches access will be granted to the user if not another last chance to deliver the correct input else access denied and intimation to users registered mobile
7	If user forgets password reset option with OTP authentication to users registered mobile number

Table 3 Sample array passwords

Array size: 4 × 4	Array size: 4 × 2	Array size: 2 × 4
Array password	$P\ a$	$P\ @\ 5\ 5$
$a\ a\ a\ 2$	$s\ s$	$w\ 0\ 6\ D$
$b\ b\ d\ 1$	$w\ o$	
$b\ w\ o\ r$	$r\ d$	
$P\ a\ s\ s$		

security. It requires educating users about creating array password and implementation process. Difficulty for Hackers and Password cracking analysts are they need to recognize the array size then input characters provided by users to get access. As multifactor authentication scheme this methodology enhances security features of login-password authentication services. In future analysis like user usability, remembrance ability, studies on service provider aspects can be carried out.

References

1. Rosenfeld, A.: Picture Languages—Formal Models of Picture Recognition. Academic Press, New York (1979)
2. Peterson, James L.: A Note on colored petri nets. Inf. Process. Lett. **11**(1), 40–43 (1980)
3. Balbo, G.: Introduction to generalized stochastic Petri nets. In: International school on formal methods for the design of computer, communication and software systems, pp. 83–131. Springer, Berlin (2007)
4. Jensen, K.: Coloured Petri nets: a high level language for system design and analysis. In: High-Level Petri Nets, pp. 44–119. Springer, Berlin (1991)

5. Sifakis, J.: Performance evaluation of systems using nets, Net Theory and Applications. Lect. Notes Comput. Sci. **84**, 307–319 (1980)
6. Walters, M., Matulich, E.: Assessing password threats: Implications for formulating university password policies. J. Technol. Res. **2**, 1 (2011)
7. Shen, J., Du, Y.: Improving the password-based authentication against smart card security breach. J. Software **8**(4), 979–987 (2013)
8. Thangavel, T.S., Krishnan, A.: Efficient secured hash based password authentication in multiple websites. Int. J. Comput. Sci. Eng. **02**(05), 1846–1851 (2010)
9. Malempati, S., Mogalla, S.: Simplified native language passwords for intrusion prevention. Int. J. Comput. Appl. **24**(4), 45–49 (2011)
10. Cheswick, W.: Rethinking passwords. Commun. ACM **56**(2), 40–44 (2013)
11. Mary Posonia, A., Vigneshwari, S., Albert Mayan, J., Jamunarani, D.: Service direct: platform that incorporates service providers and consumers directly. Int. J. Eng. Adv. Technol. IJEAT **8**(6), 3301–3304 (2019)
12. Peterson, J.L.: Petri Net Theory and the Modeling of Systems. Prentice Hall, Englewood Cliffs (1981)
13. Lalitha, D.: Rectangular array languages generated by a Colored Petri net. In: Proceedings of IEEE International Conference on Electrical, Computer and Communication Technologies, ICECCT 2015, pp. 1–5 (2015)
14. Lalitha, D.: Rectangular array languages generated by a Petri net. Adv. Intell. Syst. Comput. Comput. Vis. Robot. **332**, 17–27 (2015)
15. Lalitha, D., Rangarajan, K., Thomas, D.G.: Rectangular arrays and Petri nets. In: International Workshop on Combinatorial Image Analysis, Lecture Notes in Computer Science, vol. 7655, pp. 166–180 (2012)

sVana—The Sound of Silence

Nilesh Rijhwani, Pallavi Saindane, Janhvi Patil, Aishwarya Goythale, and Sartha Tambe

Abstract When it comes to living in a society, it is important to communicate with people around us for a better living and to survive in the human race and when it comes to communication with people, the one who are hearing or speech impaired are always left behind, in other words, they have a problem communicating with other people and when it comes to video calling, they always had to use the usual text chat to communicate. Our aim is to remove this barrier and create a platform where hearing and speech impaired people can communicate even via video calls. The proposed system translates Indian sign language (ISL) into text for the hearing-impaired user while for the normal user, and it converts the speech into text. It also helps hearing-impaired people to communicate with Google Voice Assistants without having voice making it a smart assistant.

1 Introduction

Communication is the only means to access the information and thereby get different opportunities. While we see that we are completely surrounded by technology, can

---

N. Rijhwani (✉) · P. Saindane · J. Patil · A. Goythale · S. Tambe
Department of Computer Engineering, Vivekanand Education Society's Institute of Technology, Chembur 400074, India
e-mail: rijhwaninilesh@gmail.com

P. Saindane
e-mail: pallavi.saindane@ves.ac.in

J. Patil
e-mail: janhvipatil1810@gmail.com

A. Goythale
e-mail: ashgoythale123@gmail.com

S. Tambe
e-mail: tambesartha@gmail.com

© The Editor(s) (if applicable) and The Author(s), under exclusive license to Springer Nature Singapore Pte Ltd. 2021
T. Senjyu et al. (eds.), *Information and Communication Technology for Intelligent Systems*, Smart Innovation, Systems and Technologies 196, https://doi.org/10.1007/978-981-15-7062-9_67

we use it for betterment of people including the one also who cannot speak or hear. People who are hearing or speech impaired are left behind during communication.

According to WHO, the count of people having disabilities has reached one billion and it is still counting [1]. The records of 2011 census of India states that 103.8 million people off the total population in India are senior citizens. Out of which 27 million people carry some sort of disabilities [2, 3]. Technology has always helped people in their difficult times. But when it comes to hearing-impaired people, they always fall short of the services that technology provides. To make technology available to hearing-impaired people like the way it is to all. To help them not only to communicate with their loved ones but also with the doctors and other people in case of emergencies. We have often seen that the disabled people are treated differently in society, e.g.: they have to go to different schools and they have different studies accordingly. These disabled people have to use a different kind of sign language which they have to learn in order to express effectively. There are different Indian sign languages depending on the person using it. It can be classified as image-based language and sensor-based language. Still a lot of research is yet to be done depending on the advantages and disadvantages of the different regional sign languages. The society's demanding support for hearing-impaired people became the Easter egg for the researchers in this field. Creation of sign language and a working system that helps such people to communicate using their own language without any hesitation without any other interpreter will be a golden gift to society [4]. The project helps the deaf and dumb users to be able to communicate efficiently and make them realize that they are one of us.

2 Related Work

Ashish S. Nikam et al. discussed various studies that are already done on the sign language. There are two approaches to use sign language firstly the image and gesture-based and secondly the sensor and gesture-based. But there is a lot of work pending in gesture-based in upcoming years. The different approaches for the sign hand recognition which are as follows: HMM, i.e., hidden Markov model and ANN, i.e., artificial neural network [5].

Eriglen G. et al. discussed a real-time Alabanian sign language detection using both the hands of the user. Depth map was constructed using the Kinect device [6] and also the use of k-means clustering for the classification of the signers for the partitioning of the pixels into groups. An accuracy of 91% was achieved by using this. Different work on the sensors used in the sign language detection was done by Almasre and Al-Nuaim [7].

G. Sabaresh et al. discussed focusing on the details that how the disabled or here the deaf and dumb people speak with the vocal individuals or here the normal people with the use of less demanding route. The communication was done using the gestures interpreter with the microcontroller and content showing gadgets [8].

A sign to speech/text system for hearing impaired people implemented by Patil et al. [9], this system consists of the collector at the transmitter side, and for the binding of each finger, five sensors were used for the detection. The user received the signs from the transmitter. Arduino was interfaced with the RF transmitter on the transmitter side, whereas, on the collector side, the RF is used to get the signs. EMIC 2 text-to-speech is used to create the sound where the mic is used for generation of two letters namely English and Spanish.

Conversion of specific hand gestures into audio using a VR ATMEGA32L by Ahmed et al. by developing a hand glove is discussed in [10]. Playing a recorded audio was done using the PIC18F4620 by developing a data glove. Software for editing the automatic speech was developed by Wald. Conversion of hand gestures into speech was also done using MATLAB. Five-fingered prosthetic hand system in which wireless glove that can translate sign language into speech was developed.

Feature extractions were used to extract the features which were later stored in the database in the form of Hindi text and compared with the given input video of the sign, which was developed by Paulo Trigueiros et al. for the recognition of Portugal sign language [11]. SVM was used for the comparison which gave an accuracy of 99.4% for vowels, whereas 99.6% for consonants. K. Sangeetha et al. discussed the android application where the gestures were used to translate the signs of the hearing-impaired people [12]. Vajjarapu L. et al. discussed the system, where the image was captured by the mobile camera and then was converted into the speech for this many skin detection and image processing techniques were used [13]. Numbers from 0 to 9 were detected using the dynamic hand gesture technique which was proposed by M. M. Grasuie and H. Seyedarabi. They are continuously trying to develop an automated speech recognition system with advancements [14].

Surbhi R. et al. described a system that dynamically recognizes a dynamic gesture of words in the form of Indian sign language and converts those words into text and speech. We got motivated by their lacuna and future scope that is a system for Indian sign language words that work in real time [15].

3 Proposed System

This paper talks about developing a Web application named "*sVana-The Sound Of Silence*" using Node.js that uses a computer's webcam to capture a person signing the word and alphabet, and translate it in real time. This app lets the user make a video call as well as talk to Google Assistant via sign language. The app would convert the sign language to text and convert speech-to-text at the time of video call. Figure 1 shows the overall flow of the proposed system.

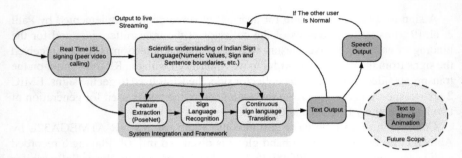

Fig. 1 System flow diagram

3.1 System Description

We aim toward analyzing and recognizing various alphabets and words from a database of sign images. The proposed system is designed in such a way that when the user visits our Web app, first the user enters the username, then the user has to select the type of user either normal user or hearing impaired. Figure 2 gives a user the information about sVana and what is sVana and it takes username from the user and the type of the user, i.e., either normal user or hearing-impaired user.

If the user selects the option of normal user, then directly the video calling canvas is opened. Skipping the complete dataset and training models and the speech-to-text conversion model is loaded which converts the real-time speech-to-text including the language Hinglish as well, considering the scenario being used in India.

While for hearing-impaired users, the user has two options either to load the pretrained model which consists of 26 alphabets of ISL or if a user wants to train the model with his/her own dataset, they can create their own dataset by training the model providing the complete words and training it with their unique gestures.

After the user completes the training model session, then the user can proceed to video calling, where during the live real-time corpus streaming video call, the signs shown by user are converted into the words or alphabets and the a sign which is predefined for sending the message or user can train it with its own sign is used to send the message and converse with the other user.

Fig. 2 Home page

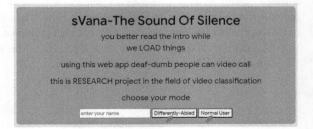

3.2 Dataset Collection

Using our system's functionality of saving the trained dataset model, we created our own variable dataset by asking the students of our college to sign the English alphabets according to the ISL, and took 50 sample images for each alphabet varying the background and the position of person.

We investigated different machine learning techniques like PoseNet, K-nearest neighbors (KNN). openCv Python library for training the dataset and models. *PoseNet* is a vision model that traces the different body joints and gives the different IDs to different parts and saves the pose of the body. As PoseNet uses CV techniques just to detect the human figures from the image or video, it does not stores any of the images reducing the size of the dataset.

If we had created a dataset first instead of the system, then we had to store the image for each alphabet and such 50 images per alphabet could have sized more than 100 MB of .json file. But because of PoseNet we were able to shrink that size to 8 MB only (Figs. 3 and 4).

Database consists of 50 * 26 images (50 samples of each alphabet in English language) with each image and video frames clicked in different light conditions with different hand orientation which were signed by college students. Images were not stored for each alphabet in the dataset, and instead PoseNet was used to reduce our dataset size.

Training model is saving the sign which our students are signing for alphabets S in Fig. 5 and L in Fig. 6, respectively, according to ISL.

As observed in the above figures, although the user signing the sign is different then too the system predicts the alphabet signing 's' in Fig. 7. and 'l' in Fig. 8 which shows the accuracy and diversity of our created dataset.

Fig. 3 Sign of the alphabets in ISL [16]

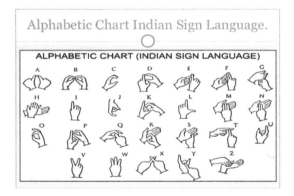

Fig. 4 PoseNet saving the pose of body in form of ID given to the joints [17]

Fig. 5 Dataset collection signing alphabet 's'

3.3 Selection of Dataset by User

After log in on the Web app, the user has two options either to load the data model or create the new one. It is based on the user's ease of selection of the dataset. If the user selects the load model option, then our own created model is loaded and preprocessed using MobileNet and further WebRtc is loaded for video calling and

Fig. 6 Dataset collection signing alphabet 'l'

Fig. 7 sVana predicted alphabet 's'

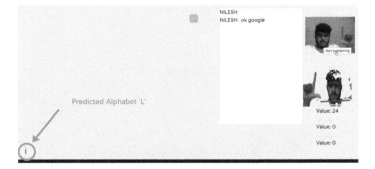

Fig. 8 sVana predicted alphabet 'l'

data be kept ready for live extraction of the signs. While if the user chooses to create his own dataset, then the user first enters the word to store and train the model by user-defined signs according to their ease KNN is used in front of the PoseNet for training the model and extracting the data both.

As observed in Fig. 9. There are three predefined functionalities that are '**Ok Google**' for activating the smart Google Assistant, '**send**' for sending the message, '**Add Word**' to train the model with his/her own sign for the word entered in the text field.

Figure 10 shows the pretrained model which can be converted into user-defined model by clearing the dataset for a particular alphabet or word and adding a new image for it based on a ease of the user and it also shows the implementation of PoseNet below the live video frame which plots the live pose graph of an user done via skin detection using KNN.

After entering the words which a particular user wants to train, now starts adding images of the sign of its his/her own choice and ease. Figure 11 shows training for '*Ok Google*' where 50 samples are processed via PoseNet.

Fig. 9 Provision to a user either to load the pretrained model or create user-specific data model

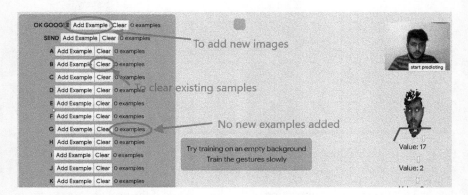

Fig. 10 Pretrained data model convertible to user specific

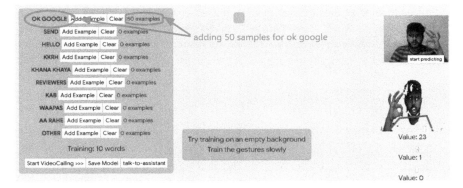

Fig. 11 Training of a user-specific data model for 'Ok Google'

3.4 Starting Video Call

After the data model is ready and the user gets connected on the peer connection, the sign language recognition can be activated by hitting the button 'click to start predicting.' PoseNet performs the task of image preprocessing and feature extraction. And the user signs are interpreted the same way they were used to train the data model. Lastly, they are converted into text as shown in the chat box.

Figure 12 shows a live video call during the testing of the application. Where two user objects were created, main user and secondary user, the user at the right in a small box is the main user calling, whereas the other one is the second user in current scenario the main user was hearing-impaired user and the secondary user was normal user.

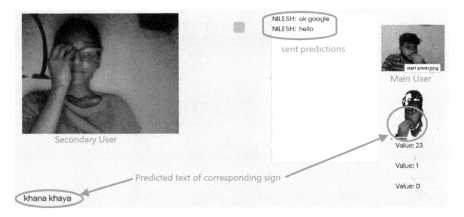

Fig. 12 Ongoing video call between two users

For the main user, the sign that the user signed on the webcam got converted into the corresponding word '*khana khaya*' and the '*send*' sign was used to send the predicted text '*Ok Google*' and '*hello*'. While for the secondary user, the speech is converted into text using simple speech-to-text inbuilt node peer API and the message is automatically sent.

3.5 Evaluation

For the testing purpose of our Web application, node appmetrics dash was used, where it was tested under the different test scenarios, different pairs of users, viz. Both normal users, both hearing-impaired users and normal-hearing-impaired users, datasets with different sizes and different working conditions such as network bandwidth and different configuration of systems.

Figure 13 shows the response time to the initial http request when tested on the localhost server. It took an average response time of 1 μs where the user was a normal user and the requirement did not include datasets or any training models, just the speech-to-text API was loaded for the normal user. On the other hand, if the user was hearing impaired, the system took an average response time of 1.2 ms as it involved loading of all models required for sign language recognition. The initial dataset of users contained zero words until an user chose the pretrained ISL dataset of 26 alphabets. The average response time when recorded Onlive Server with different scenarios and more than 100 server requests including both types of users came to be 54 ms and the Heroku CLI was used to host our node app online.

Figure 14 shows the request-response time considering all possible types of requests on the localhost server, e.g., load data model took 0.5 ms where the loaded data model contained zero words but when the dataset of 26 alphabets was loaded it took 740 μs to load the model and prepare it for the video calling. When considering the different test cases scenarios Average response time was approximately 634 μs according to Heroku for different variants of datasets and constant net speed of 10 mbps.

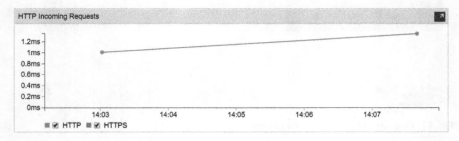

Fig. 13 Response time to server requests

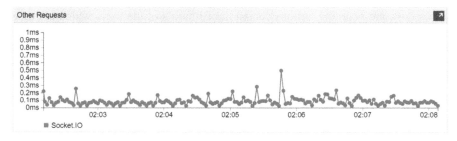

Fig. 14 Request-response time

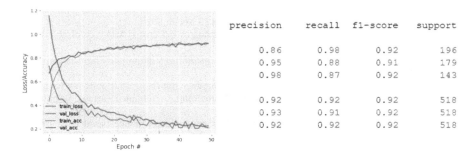

Fig. 15 Accuracy and training loss

Figure 15 shows the training loss and accuracy of the pretrained data model when it was tested under the different test scenarios for 50 epochs. The dataset showed the accuracy of 94%. But when the user-created datasets were considered including the different variants of sizes, it shows the accuracy of 97% for that particular user when worked in good environmental conditions and net connectivity speed.

4 Conclusion and Future Work

This publication is about an application that supports the complete dual-way communication between hearing-impaired and normal users. The application consists of two parts. Firstly, the speech-to-text conversion for the normal users which uses speech processing methods. Secondly, the motion recognition part for hearing-impaired users which use image processing methods and ISL as a data model. Both the parts were successfully implemented along with the smart Google Assistant which converts the image ISL to speech output to activate Google Assistant on an user's mobile. The development of the Bitmoji animation is under progress where the Bitmoji animation is created from the user-specified dataset during the video call, with the reference of user image which is stored differently for a particular user on the local storage of the user.

Also, we are planning on introducing the Bitmoji's of the users which are used for the purpose of viewing the stored sign for a word and the live feedback is shown for another user if both the users are the hearing impaired. The Bitmoji will be shown for the received message from another user but according to the dataset of that particular user. Also, we are focusing on implementing the Morse code digitized communication which will be beneficial to the society for the deaf-blind users.

References

1. WHO. World report on disability (2011). Available at http://whqlibdoc.who.int/publications/2011/9789240685215_eng.pdf?ua=1. Accessed on 23 March 2020
2. 2011 Census India: Retrieved from https://censusindia.gov.in/Census_And_You/disabled_population.aspx. Accessed on 23 March 2020
3. Census Data on Disabled Population, India. https://pib.gov.in/newsite/PrintRelease.aspx?relid=122878. Accessed on 23 March 2020
4. Mahesh, M., Jayaprakash, A., Geetha, M.: Sign language translator for mobile platforms. In: 2017 International Conference on Advances in Computing, Communications and Informatics (ICACCI), Udupi, 2017, pp. 1176–1181
5. Nikam, A.S., Ambekar, A.G.: Sign language recognition using image based hand gesture recognition techniques. In: 2016 Online International Conference on Green Engineering and Technologies (IC-GET), Coimbatore, 2016, pp. 1–5
6. Gani, E., Kika, A.: Albanian Sign Language (AlbSL) number recognition from both hand's gestures acquired by kinect sensors. Int. J. Adv. Comput. Sci. Appl. (IJACSA) 7(7) (2016)
7. Al-Nuaim, H., Almasre, M.A.: Recognizing Arabic sign language gestures using depth sensors and a KSVM classifier. In: 2016 8th Computer Science and Electronic Engineering (CEEC)
8. Sabaresh, G., Karthi, A.: Design and implementation of a sign-to-speech/text system for deaf and dumb people. In: 2017 IEEE International Conference on Power, Control, Signals and Instrumentation Engineering (ICPCSI), Chennai, 2017, pp. 1840–1844
9. Patil, P., Prajapat, J.: Implementation of a real time communication system for deaf people using Internet of Things. In: 2017 International Conference on Trends in Electronics and Informatics (ICEI), Tirunelveli, 2017, pp. 313–316
10. Ahmed, S., Islam, R., Zishan, M.S.R., Hasan, M.R., Islam, M.N.: Electronic speaking system for speech impaired people: Speak up. In: 2015 International Conference on Electrical Engineering and Information Communication Technology (ICEEICT), Dhaka, 2015, pp. 1–4
11. Trigueiros, P., Ribeiro, F., Reis, L.P.: Vision-based Portuguese sign language recognition system, from book New Perspectives in Information Systems and Technologies, April 2014
12. Sangeetha, K., Barathi Krishna, L.: Gesture detection for deaf and dumb people. Int. J. Dev. Res. 4(3), 749–752 (2014)
13. Lavanya, V., Praveen, A., Madhan Mohan, M.S.: Hand gesture recognition and voice conversion system using sign language transcription system. Natl. J. Electron. Commun. Technol. (2014)
14. Gharasuie, M.M., Seyedarabi, H.: Real-time dynamic hand gesture recognition using hidden Markov models. 978-1-4673-6184-2/13/$31.00 ©2013 IEEE
15. Surbhi, R., Gawande, U.: Development of full duplex intelligent communication system for deaf and dumb people. In: 2017 7th International Conference on Cloud Computing, Data Science & Engineering—Confluence (2017)
16. ISL image referred from Examrace. https://www.examrace.com/Current-Affairs/NEWS-First-Indian-Sign-Language-Dictionary-of-3000-Words-What-is-ISLR-TC.htm. Accessed on 12 Feb 2020
17. Pose Estimation-By tensorFlow. https://www.tensorflow.org/lite/models/pose_estimation/overview. Accessed on date 11 Jan 2020

Coreveillance—Making Our World a "SAFER" Place

C. S. Lifna, Akash Narang, Dhiren Chotwani, Priyanka Lalchandani, and Chirag Raghani

Abstract The culture in our civic society has undergone an abrupt change due to the extensive use of CCTV surveillance. Currently, our urban society is exclusively dependent upon CCTV footage for divulging any abnormal situation. This scenario has given rise to new openings for the researchers to judiciously utilize video analytics techniques to focus on many sensitive issues in society such as contravention of human rights (HR). The purpose of this paper is to design a surveillance platform equipped with situation intelligence to aid government and non-government working departments in taking corrective action against unfavorable intruders and destructive mishaps.

1 Introduction

In this modern era of the smart world, as discussed in paper [1], we have achieved great heights by implementing various technical features in various aspects of our daily life. Security is the key for a person's safety. The most commonly used security

---

The authors gratefully acknowledge the support extended by University of Mumbai as Minor Research Project Grant No. 960 (Circular No. APD/ICD/2019-20/762 dated. March 17, 2020).

C. S. Lifna (✉) · A. Narang · D. Chotwani · P. Lalchandani · C. Raghani
Department of Computer Engineering, VES Institute of Technology, Mumbai, India
e-mail: lifna.cs@ves.ac.in

A. Narang
e-mail: 2016.akash.narang@ves.ac.in

D. Chotwani
e-mail: 2016.dhiren.chotwani@ves.ac.in

P. Lalchandani
e-mail: 2016.priyanka.lalchandani@ves.ac.in

C. Raghani
e-mail: 2016.chirag.raghani@ves.ac.in

© The Editor(s) (if applicable) and The Author(s), under exclusive license to Springer Nature Singapore Pte Ltd. 2021
T. Senjyu et al. (eds.), *Information and Communication Technology for Intelligent Systems*, Smart Innovation, Systems and Technologies 196,
https://doi.org/10.1007/978-981-15-7062-9_68

Table 1 Comparison of the proposed coreveillance system with the existing system

No.	Evaluation parameter	Existing system	Proposed system
1	Type of monitoring	Manual	Digital
2	Visitor management and tracking	Manually entered	Automatically
3	Real-time reporting of vandalism	No	Yes
4	Intrusion detection	May or may not be caught	Real-time reporting with alerts and notifications

measures adopted in society are the security guards and the cameras placed in the society. The hour-long CCTV footage is recorded and stored for any future use. But, the time required to search the obligatory footage is wasted in time-sensitive cases. There is neither real-time processing of the footage nor alerts being provided during alarming situations. Till date, intrusion detection notification and vandalism actions are not being reported in real time. These are the lacunas in the existing security service that led to the proposition of the paper. Table 1 depicts the comparison with the existing system and the proposed coreveillance system.

To create the smart society [2–8] vision into reality, it is crucial to develop the technology infrastructure backbone that takes into consideration the challenges within social communities. This paper focuses on resolving the issues faced by societies. The goal is to build not just a smart society, but a smart and sustainable society. The objective of the paper is to incorporate video analytics along with the use of optimization methodology to tackle the shortcomings in buildings and societies. The important thing is that each anomaly requires a unique solution dependent upon local conditions. The proposed system can also be fine-tuned to address the issues in the car parking system.

Detection of real-world anomalies is often difficult, as they are very diverse and complicated. Hence, the approach discussed in this paper does not depend on any prior knowledge or supervised learning; rather sparse coding methods are adopted to get the desirable results. The normal event dictionary is created using the initial parts of the video that have normal events. The main thrust of the methodology is that anomalous events are not regenerable from normal events; even though there is a marginal possibility for false alarms when such a method is used.

2 Literature Survey

This section briefs about the work done in the respective domain which is related to our system and also explains the state-of-art techniques in video analytics which has been key to implement the proposed system.

In paper [9], the author discusses the human motion detection and notification system which is implemented using the contemporary digital image processing. By adopting a wider range of algorithms, the noise and signal distortions are highly minimized during the processing. Paper [11] uses live video streaming and analytics upon which a whole CCTV camera is connected to the surveillance system online. Hence, the direct feed of real-time video stream is used as straight input for further analysis in the system. This methodology was incorporated in the paper, to perform a multiple analysis on the video stream received based on the different criteria provided by the users to the proposed system. Based on the incidents reported, various corrective measures are incorporated such as notifications and alert to Local Governing Bodies.

Paper [12] is focused on event detection when processing is performed upon videos captured from different angles. As the images that are captured by the camera can be distorted, because of the optical effects that come into picture due to the weird angles of the captured object. With the help of innovative and various self-driven analysis tools, there is an opportunity to differentiate between normal and unusual movements in videos using Markovian model to track object trajectories in motion. Authors in paper [13] are more engrossed in differentiating the boundary between anomalous and normal behaviors. The behavior of normal and anomalous events varies from situation to situation. Hence, it is difficult to judge in different conditions. The paper describes an algorithm which uses weakly labeled videos to train the model, but no information is provided about where the anomaly exactly lies in the video. The system learns to find the anomaly by a machine instance learning (MIL) framework by keeping normal and anomalous videos as containers of each video labeled with segments of each video clip as instances of the bags. While testing, a full length video is used which is segmented to find the features of the video which is fed as input in the model and generates the anomaly score of the video.

3 Proposed System Architecture

This section explains the procedures, methodologies and techniques learnt and implemented during the phases of observational study. Figure 1 depicts the proposed system architecture for coreveillance system.

The data is collected using the available CCTVs fixed within the society and its nearby places. This data is further processed by cleansing it and processed using the video analytics techniques on the local server to identify the anomaly. Once an anomaly is detected, its summary is uploaded onto the cloud platform and relevant information is provided to the society immediately to alarm them about the suspicious activity happening around. Also, alerts are sent to the Local Governing Bodies which need to be notified to take preventive and corrective actions.

There are three types of actors that will be entering into the society:

Actor 1: A person who is already registered with the system and is treated as a valid user which lives in.

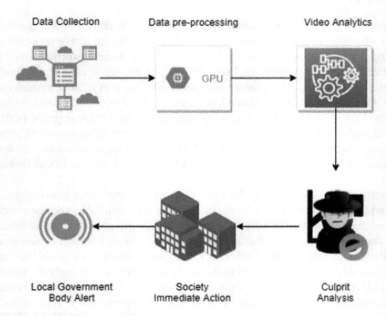

Fig. 1 Proposed system architecture for coreveillance

Actor 2: A person who enters the society for providing services and is treated as a visitor and is registered with the system. This actor is treated as a visitor.

Actor 3: The person who tries to enter into society by some malpractice. This actor is treated as an intruder.

The proposed system is directly connected to the CCTVs installed all over the area. In the case of Actor 2, each society member will be provided an application where he/she can login and track the activity of the visitor and can send/receive the alert messages. Whereas in the case of Actor 3 (intruder), the system continuously tries to track the activities. If the intruder creates any kind of threat or issue, the system will immediately send an alert message to the registered society members and also to Local Governing Bodies (like the Police/Fire Station) depending upon the anomaly detected. So, these respective authorities will take prompt action. Thus, our proposed system can be used to catch the culprit in real time.

4 Implementation Details

This section summarizes the implementation details of the proposed coreveillance system. The two stakeholders of the proposed system are the system administrator and the end users can access this system via the User App.

Initially, CCTV footage is captured from cameras installed in the society. Then, these real-time videos are segmented into frames and their features are extracted

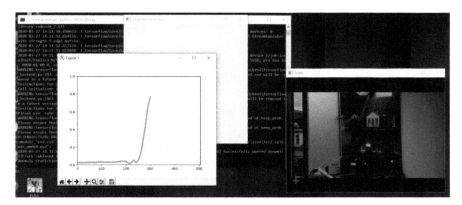

Fig. 2 Administrator dashboard when an explosion has been detected in the input video

using the C3D convolutional neural network model. The features extracted from frames are stored in text files. Then, the frames and their extracted features are fed to the multiple instance learning (MIL) Python module is used to detect the anomalies in the input frames. Based on the events in the given input frame, a graph is generated with X-axis determining the frame number and Y-axis determining the intensity of the anomaly on a scale of [0–1]. A threshold (which is set as 0.6) has been set to declare an event to be anomalous and to raise an alert. Figure 2 depicts the scenario when an explosion has been detected in the input video.

Once the whole video is covered, then the corresponding video frame where the anomaly has been detected is captured and summarized with the following details: Anomaly type, description, frames where the event intensity is above threshold (0.6). All the summarized data is then sent to the registered users (via the User App as depicted in Fig. 3a, b immediately by the system giving the description about the anomalous event with the help of notification module. The users can view the captured frames to know more about the anomalous event, ubiquitously as shown in Fig. 4a, b. These details can also forward the Local Governing Bodies to take corrective measures.

5 Conclusion and Future Scope

The motivation behind experimenting upon this paper was the incidents reported in the Indian newspapers [14–17]. Previous societal methodologies have proved to be working to some extent. But, there is a huge scope of improvement by using new technologies. So, this paper is a contribution to the future society and it is a step to make our society smarter than ever. The most important thing in a smart society [3] is that the people will experience the benefits of new technologies and better security.

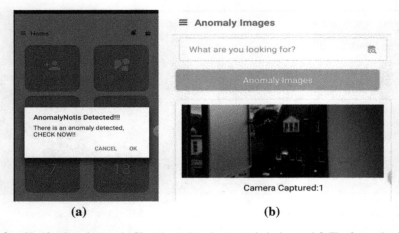

Fig. 3 **a** Notification alert on the User App when the anomaly is detected. **b** The frame details are available in the respective folder in User App

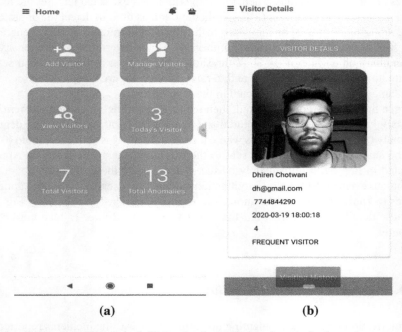

Fig. 4 **a** Dashboard of User App. **b** Visitor details are displayed as per the demand

Fig. 5 Future of smart society

Thus, the objective of this paper is to help society within their present living ecosystem and to build a smarter and safer society for the future as depicted in Fig. 5. The intention of implementing the proposed system was to reduce the crime investigation time in various residential and nonresidential places. As for the future scope, the authors are looking forward to merging the proposed system with the present Government Surveillance Systems. So that, an extra layer of security can be added to the current systems making them secure using the current latest upcoming technologies for the future.

Acknowledgements The authors gratefully acknowledge the support extended by University of Mumbai as Minor Research Project Grant No. 960 (Circular No. APD/ICD/2019-20/762 dated March 17, 2020)

References

1. Intelligent Buildings International Journal, Taylor and Francis. [Online] www.ibmbigdatahub.com. Accessed on 23 Sep 2019)
2. For smart cities, we need smart societies [Online] www.thehindubusinessline.com. Accessed on 23 Sep 2019
3. What is a Smart Society? [Online] https://link.medium.com/D6uCtZ2OPZ. Accessed on 23 Sep 2019
4. Campbell, T.: Beyond Smart Cities: How Cities Network. Learn and Innovate. Earthscan, Abingdon (2012)
5. Carta, M.: Creative city 3.0: new scenarios and projects, monograph it, no. 1 (2009). http://issuu.com/mcarta/docs/181_creative_city_3.0_monograph. Accessed on 23 Sep 2019
6. Castells, M.: The Rise of the Network Society: The Information Age: Economy, Society and Culture, vol. 1. Wiley Blackwell, Oxford (1997)
7. Kearney, A.T.: Global cities index and emerging cities outlook (2012). [Online] http://www.atkearney.com/documents/10192/dfedfc4c-8a62-4162-90e5-2a3f14f0da3a. Accessed on 23 Sep 2019
8. Cohen, B.: What Exactly is a Smart City? Co.Exist, 19 September 2012. [Online] http://www.fastcoexist.com/1680538/what-exactly-is-a-smart-city. Accessed on 23 Sep 2019
9. Mehta, R., Marodia, S.: Human motion detection and notification system (2019)
10. Facial recognition technology: fundamental rights considerations in the context of law enforcement, European Union Agency for Fundamental Rights, Published in Nov 2019

11. Mariadoss, P.: Performing real-time analytics using a network processing solution able to directly ingest IP camera video streams. U.S. Patent No. 8,325,228. 4 Dec. 2012
12. Cetin, A.E., Noyan, T., Davey, M.K.: Unusual event detection in wide-angle video (based on moving object trajectories). U.S. Patent No. 10,339,386. 2 Jul. 2019
13. Sultani, W., Chen, C., Shah, M.: Real-world anomaly detection in surveillance videos. In: Proceedings of the IEEE Conference on Computer Vision and Pattern Recognition (2018)
14. Duo arrested for stealing cars and bikes 'for fun' (14 Sep 2019 in Mumbai Mirror)
15. Sessions court convicts seven men of stealing Rs. 1.95 crore (31 Jul 2019 in Mumbai Mirror)
16. held for stealing cash, jewellery from 40 Noida flats over six months (17 Aug 2019 in India Today)
17. Burglar enters doctor's house in Blossom-II, beats her up (25 Jul 2017 in Times of India)

Large-Scale Video Classification with Convolutional Neural Networks

Bh. SravyaPranati, D. Suma, Ch. ManjuLatha, and Sudhakar Putheti

Abstract Convolutional neural networks have been established as an unbelievable class of models for picture confirmation issues. Enabled by these results, we give CNN's extensive trial evaluation a large degree of video-action syllabus using another dataset of 8M YouTube accounts. To get the Chronicles and its effects, we've used a YouTube video specification framework, which gives the names of the accounts they focus on. While the names are machine-generated, they are high-precision and are derived from a group of human-based icons, including metadata and question click signals. We have filtered the video names (Knowledge Graph Components) using both modern and manual curation strategies, including curiosity regarding whether the print is clearly indisputable. After that, we decode each video at one-layout per-second and use the deep CNN adjusted to ImageNet to remove the cover depicted immediately before the course of the action layer. Finally, we've stuffed the packaging features and made available both features and video level names for download. We train unique (ambiguous) game plan models on the dataset, survey them using significant evaluation estimates, and report them as baseline. Regardless of the size of the dataset, a portion of our models train the connection in less than a day on a singular machine using VGG. CNN our course release code for setting up model deals and generating predictions.

1 Introduction

Understanding and viewing video content is a great test for a variety of applications, including pregnancy, single help, elegant homes, free driving, stock film search, and sports video evaluation. Currently, fix the issue of multi-mark video filming for user-made chronicles on the Internet. There are two problems with the verification

Bh. SravyaPranati (✉) · D. Suma · Ch. ManjuLatha · S. Putheti
Computer Science and Engineering, Vasireddy Venkatadri Institute of Technology, Guntur, Andhra Pradesh, India
e-mail: sravyapranathi333@gmail.com

© The Editor(s) (if applicable) and The Author(s), under exclusive license to Springer Nature Singapore Pte Ltd. 2021
T. Senjyu et al. (eds.), *Information and Communication Technology for Intelligent Systems*, Smart Innovation, Systems and Technologies 196,
https://doi.org/10.1007/978-981-15-7062-9_69

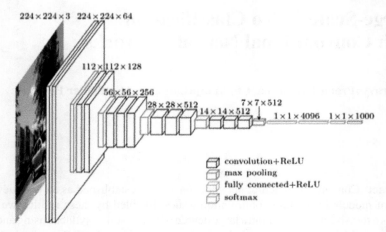

Fig. 1 Architecture of VGG16

of such data. Web-number accounts for material and quality are of great capricious value (see Fig. 1). Additionally, customer-generated marks are often low, ambiguous, and may contain bots. Current methods for video validation usually resolve accounts with features removed from successive house singles, which follow shortly after consolidation. Characteristic extraction model approaches arrive at important classical neural frameworks (CNN) pre-arranged in static images [1]. Handheld video features [2–4] as well as video housing and small video cuts [5, 6] pre-arranged CNNs reflect the development and appearance. Other more advanced models use different levels of spatiotransient convolutional plans [5, 7–11] to simultaneously integrate and integrate full-scale video. The temporal segment's standard approaches logically consolidate or enhance final aggregate or worse pooling, similar to today's pooling frameworks (LSTM [12], GRU [13]). These strategies can make mistakes in any situation. Basic systems, for example, may lead to a more or less frequent grouping of common or more important pooling. Continuous models are a significant part of the time used for short-term collection of variable-length groups [14, 15] but fail to gradually clear integer systems, although their abundance is large.

2 Related Work

Image benchmarks are expected to play an important role in moving PC Vision calculations for image comprehension. Beginning with different degrees of stamp datasets, for example, image comprehension query will soon use more datasets, for example, ImageNet's class time [6] view estimation. Using a significant number of parameters, ImageNet has been able to improve significant segment learning techniques, for example using the use of the Alexa Net [16] and Inception [11] architectures for class size (21,841) and various classes (25). High-level characters) and many verified images are available. The Do-Basic Relative effort develops fast video awareness of the system everywhere in stamp datasets, for example, KTH [17], Hollywood 2 [18], Wiseman [5] an99d thousands of video cuts for medium datasets with more than 50 movement commands. At this time, the biggest video benchmarks available are Sports-1 M [15], 487 Games Activities and 1 M Chronicles, YFCC-100 M, [34] 800 K Accounts, and some unadulterated metadata (titles, illustrations, names), 91 And 223 Chronicles 239,200 Human Development Classes and Orders and Thousands of Chronicles [9] Dataset on Operational Net. However, for all intents and purposes, all existing video benchmarks are limited to viewing motion and activity classes and are limited to groups of less than 500. YouTube-8M completes the opening of video benchmarks as follows: Hang benchmark achieves advanced video resolution and illustration that reflects the basic themes of the video. This is better than the average collection of rate classes—less than 500 classes for 4800 knowledge graph elements and special datasets. A notable addition to the volume of the eponymous Chronicles is 8 million accounts, 500,000 h of video.

3 YouTube-8M Dataset

YouTube-8M is a benchmark dataset for video comprehension, where standard assignment of video content is selected. We start with traditionally (albeit trapped) YouTube accounts of data for a variety of classes, including various games, workouts, animals, livelihoods, things, places to play, games, and more. We use the YouTube Video Classification Framework to obtain content for video and to retrieve accounts for a given item [19]. Knowledge graphs are materials [20] (formerly freebase topics [21]). They are linked to each video in accordance with the video's metadata, configuration, and material codes [19]. We use knowledge graph materials to briefly describe the needs of the video. For example, the video of biking on dirt roads and peacocks includes the mounting/subject of mountain biking, not the dirt, the road, the person, or the school. Then, the purpose of the dataset is not to appreciate what is accessible on each side of the video, but to know the two main elements that best describe the video. Note that, this is not the same as a simple event or scene detection attempt, where everyone has the same event or scene [21, 22]. This is not the same as most essay confirmation efforts, which is to label everything that is clearly in the picture. It is important to make countless names in each video without paying attention to the video. The purpose of this benchmark is to understand what is in the video and sketch it into two or three main points. Going with the subsections, we describe our language and video decision planning by following a brief summary of the dataset measurements.

4 VGG16

VGG16 is a convolution neural net (CNN) design that was utilized to win ILSVR(Imagenet) rivalry in 2014. It is viewed as one of the brilliant vision model engineering to date. The most exceptional thing about VGG16 is that as opposed to having an enormous number of hyperparameter they concentrated on having convolution layers of 3 × 3 channel with a walk 1 and constantly utilized same cushioning and maxpool layer of 2 × 2 channel of walk. It follows this plan of convolution and max pool layers reliably all through the entire engineering. At last, it has 2 FC(fully associated layers) trailed by a softmax for yield. The 16 in VGG16 alludes to it has 16 layers that have loads. This system is a truly huge system and it has around 138 million (approx) parameters.

4.1 VGG16 Architecture

Contributing to our ConvNets is a 224 × 224 RGB image of a certain size. The main preprocessing we do is to reduce the average RGB estimate found in the manufacture from each pixel. The image passes through responsive layers, where we use channels with less resonant fields: 3 × 3 (which is the smallest size to capture the idea of left/right, up/down, and vision). In one of the settings, we use 1 × 1 convolution channels, which can be seen as a direct change of information channels (nonlinear follow-up). Convention wak 1 fixed to pixel; The spatial cushion of consciousness. The layer input corresponds to the final target where spatial targets are saved after transformation, for example, 1 pixel per cushioning 3 × 3 assembly. Layers. Spatial pooling consists of five max-pooling layers that follow some part of the converter. Layers (not all conversions. Layers follow max-pooling). Max-pooling is done by a 2 × 2 pixel window, walk2. A heap (alternating between different models) is followed by three fully interconnected (FC) layers: the initial two have 4096 channels, the third one has 1000-way ILSVRC ordering, and 1000 channels (one for each class). The last layer is a delicate max layer. The arrangement of completely related layers is the same across all systems.

5 Results

We first split the videos into 8 million YouTube videos of train and test data. Now, we have trained the videos with the 0.50 VGG 16 model. Fine-tuning is the process of taking the weight of a previously trained network model and performing a new model of the second task. If the size of the dataset is inadequate, it can be used to speed up the training process and reduce the risk of high competition.

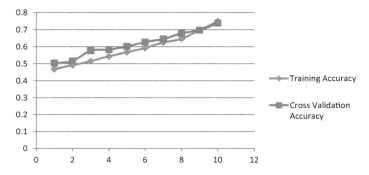

Fig. 2 Accuracy with epochs

The best tuning in our experiment is to use a pre-trained VGG16 to receive a 224 × 224 × 3 sized input image and replace the top convolutional layer with two or three thick layers with relay activation functions. Since the model requires some time to train or work, there is no complicated calculation of the rays. It blends in quickly and is rarely activated. In this experiment, we increase the number of original datasets by multiplying each leaflet image by 32 batch sizes, although the pre-trained model can reduce overfitting.

Using pre-treated VGG16 as a classifier with fully integrated membranes has proven to yield excellent results. It has low execution time and memory usage while maintaining high classification accuracy. A well-tuned pretrain model can achieve 74.85%+ evaluation accuracy for 25 of the 10 classes. A precision metric is used to measure the performance of the algorithm in a meaningful way. The accuracy of the model is usually determined after the model parameters, which are calculated in the form of percentages. It is a measure of how accurate your model's forecast is compared to actual data. Figure 2 represents the accuracy of the epochs. Figure 3 represents the loss with the epochs.

6 Conclusion

In this paper, we present YouTube-8M, a large-scale video benchmark for video classification and representation practice. Similar to making large-scale image datasets for image comprehension, our goal with YouTube-8M is to advance in the field of video comprehension. In particular, we address two major challenges with large-scale video comprehension: (1) collecting a large labeled video dataset with reasonable quality labels, and (2) eliminating computational barriers by pre-processing the dataset. Art frame-level features. As a side effect, one of the largest and most diverse types of general visual interpretation vocabulary (consisting of 4800 visual knowledge graph entities) is made from popular signals and manual curation on YouTube and is organized into 25 high-level categories. Finally, we describe the use

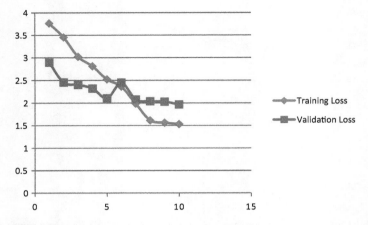

Fig. 3 Loss with epochs

of the dataset by performing transfer learning experiments on existing video benchmarks—VGG16. Our experiments show that the features learned in this dataset can be generalized to these standards, including the new advanced configuration in VGG16. Our experiments show that features learned on this dataset generalize well on these benchmarks, including setting a new state-of-the-art on VGG16.

References

1. Scheidegger, F., Cavigelli, L., Schaffner, M., Malossi, A.C.I., Bekas, C., Benini, L.: Impact of temporal subsampling on accuracy and performance in practical video classification
2. Everingham, M., Gool, L.V., Williams, C.K.I., Winn, J., Zisserman, A.: The pascal visual object classes (voc) challenge (2009)
3. Fei-fei, L., Fergus, R., Perona, P.: One-shot learning of object categories. IEEE Trans. Pattern Anal. Mach. Intell. **28** (2006)
4. Girshick, R.: Fast R-CNN. In: Proceedings of the International Conference on Computer Vision (ICCV) (2015)
5. Blank, M., Gorelick, L., Shechtman, E., Irani, M., Basri, R.: Actions as space-time shapes. In: Proceedings of the International Conference on Computer Vision (ICCV) (2005)
6. Deng, J., Dong, W., Socher, R., Lija, L., Li, K., Fei-fei, L.: Imagenet: A large-scale hierarchical image database. In: Proceedings of the IEEE Conference on Computer Vision and Pattern Recognition (2009)
7. Griffin, G., Holub, A., Perona, P.: Caltech-256 object category dataset. Technical Report 7694, California Institute of Technology (2007)
8. He, K., Zhang, X., Ren, S., Sun, J.: Deep residual learning for image recognition. CoRR, abs/1512.03385 (2015)
9. Heilbron, F.C., Escorcia, V., Ghanem, B., Niebles, J.C.: Activitynet: A large-scale video benchmark for human activity understanding. In: IEEE Conference on Computer Vision and Pattern Recognition (CVPR), pp. 961–970 (2015)
10. Hochreiter, S., Schmidhuber, J.: Long short-term memory. Neural Comput. **9**(8) (1997)

11. Ioffe, S., Szegedy, C.: Batch normalization: Accelerating deep network training by reducing internal covariate shift. In: Proceedings of the International Conference on Machine Learning (ICML), pp. 448–456 (2015)
12. Jiang, Y., Liu, J., Roshan Zamir, A., Toderici, G., Laptev, I., Shah, M., Sukthankar, R.: THUMOS challenge: Action recognition with a large number of classes. http://crcv.ucf.edu/THUMOS14 (2014)
13. Jiang, Y.-G., Wu, Z., Wang, J., Xue, X., Chang, S.-F.: Exploiting feature and class relationships in video categorization with regularized deep neural networks. arXiv preprint arXiv:1502.07209 (2015)
14. Jordan, M.I.: Hierarchical mixtures of experts and the algorithm. Neural Comput. **6** (1994)
15. Karpathy, A., Toderici, G., Shetty, S., Leung, T., Sukthankar, R., Fei-Fei, L.: Large-scale video classification with convolutional neural networks. In: IEEE Conference on Computer Vision and Pattern Recognition (CVPR), pp. 1725–1732, Columbus, Ohio, USA (2014)
16. Krizhevsky, A., Sutskever, I., Hinton, G.E.: ImageNet classification with deep convolutional neural networks. In: Advances in Neural Information Processing Systems (NIPS), pp. 1097–1105 (2012)
17. Laptev, I., Lindeberg, T.: Space-time interest points. In: Proceedings of the International Conference on Computer Vision (ICCV) (2003)
18. Laptev, I., Marszalek, M., Schmid, C., Rozenfeld, B.: Learning realistic human actions from movies. In: IEEE Conference on Computer Vision and Pattern Recognition (CVPR) (2008)
19. Google I/O 2013—semantic video annotations in the Youtube Topics
20. Knowledge Graph Search API
21. Tensorflow: Image recognition. https://www.tensorflow.org/tutorials/image_recognition
22. Jegou, H., Perronnin, F., Douze, M., Sanchez, J., Perez, P., Schmid, C.: Aggregating local image descriptors into compact codes. IEEE Trans. Pattern Anal. Mach. Intell. **34**(9), (2012)

Saathi—A Smart IoT-Based Pill Reminder for IVF Patients

Pratiksha Wadibhasme, Anjali Amin, Pragya Choudhary, and Pallavi Saindane

Abstract Women undergoing In Vitro Fertilization (IVF) treatment have to strictly administer the stringent schedule of the entire process, which leaves them physically and emotionally exhausted. Saathi is a smart IoT-based pill reminder which aims to help the women opting for the IVF. Saathi is specially designed for IVF undergoing women, giving them the facilities of setting the reminder of their daily medications and injections, having real-time tracking of medicine consumption, maintaining their prescriptions, generating reports from real-time tracking of medicine consumption, and also allowing them to communicate with their doctor. Thus, it helps the patient to adhere to their strict schedule and monitor their intake.

1 Introduction

Infertility or a couple being unable to conceive a child can cause significant stress and unhappiness, which can lead to psychological consequences as well. Around 12-15% couples in the USA are unable to conceive. Nowadays, the most used method of assisted reproductive technologies is In Vitro Fertilization (IVF). The process itself is expensive and has to follow strict rules for medication use [1]. Even if you follow the process strictly, there are only 30–40% chances of success. There are many medicine reminder applications, even nowadays we can set reminders in our phone too. There is no medicine reminder application that is linked with the pillbox itself through the cloud.

Our smart pillbox along with a software application makes it easier for users to remind them to take pills and keep track of their medications. Now, the question arises as to how many people will adapt to this trend of using a smart pillbox replacing the traditional method of manually remembering all medications. To check the acceptance level of the people to use a smart pillbox, a study has been carried

P. Wadibhasme · A. Amin · P. Choudhary · P. Saindane (✉)
Vivekanand Education Society's Institute of Technology, Chembur, Mumbai 400074, India
e-mail: pallavi.saindane@ves.ac.in

© The Editor(s) (if applicable) and The Author(s), under exclusive license to Springer Nature Singapore Pte Ltd. 2021
T. Senjyu et al. (eds.), *Information and Communication Technology for Intelligent Systems*, Smart Innovation, Systems and Technologies 196,
https://doi.org/10.1007/978-981-15-7062-9_70

out [1]. Out of a random 500 people chosen for the study, nearly 70% of them were willing to use smart pillboxes to remind them to take medication [2]. Similar studies on older patients [3], HIV-positive patients [4], and patients with coronary heart disease [5] have found a high acceptability of using new technology to remind them to take medication. It can be concluded from these that more people are now turning toward and adapting smart ways in exchange for a healthy mind and body. Saathi is specially designed for women undergoing IVF treatment. It will have seven temperature-controlled compartments: six to store the pills and one large compartment to store the injections. It will be connected to our android application 'SaathiCal.' The women undergoing IVF take a maximum of six pills a day, for which they can set reminders through the application. Saathi tracks the real-time medicine consumption of the patient and stores its history to further generate the report on the intake of medicine. This report can also be sent to the patient's doctor whom the patient has registered in the application. The patient can also communicate with the doctor through the application and also set a schedule for their next appointment. The doctor thus gets weekly updates regarding the patient through the application. Using Saathi, one not only saves time but also ensures that the schedule is followed properly.

2 Related Work

In recent years, many systems have been developed, both software and hardware, that help in drug management, sending reminders to the patient, and keeping a record of the pill consumption of the patient as well [6]. These dispensers are integrated with smartphones or smart watches as well [7]. These come in variants such as pill boxes with timers or pill boxes that have seven-day supply [8]. But the main problem with these pill dispensers is that these cannot be integrated with our smartphones. With the advent in cloud technology, the new generation prefers having information over the cloud. There is no integration of the schedule on hardware devices; it cannot be accessed conveniently. The memory systems used by these devices are limited only to the memory chip available. There is no means to set the schedule of pills in one go for a month recursively such as weekly/monthly. Hence, it becomes difficult to set reminders or medication schedules easily at one's comfort anywhere. Also, the pill reminders available in the market currently which are comparatively advanced than others are very huge and impossible to carry in a bag on the go [9]. Also, the reminder systems such as a software app integrated with the device are accessible only to the person consuming the medicines. For an IVF patient, the family, closest relatives and friends, play a major role in taking care of the patient both emotionally and physically. The currently available systems do not provide any means to provide real-time information to the patient's close ones about the patient's well-being. There is no provision either for the close ones to remind the patient from anywhere, through a button click, so that it sets a reminder on the device and software as well, rings at scheduled time, and ensures that the patient has taken the medicine properly. IVF schedule does not only have pills as medication, but hormonal injections are to be taken regularly as well. There is no system available currently that can provide a separate slot for the injectional liquid which is temperature controlled. There is no temperature-controlled system available for the pills either.

3 Existing Systems

3.1 Pill Sense

It uses a 3D printed pill bottle that has a magnetic switch sensor, an accelerometer, load cell, and PIP-Tag mote [10] for collecting data from sensors and transmitting it wirelessly to a base station attached to a nearby computer. It works on the idea of collaborative sensing where the switch sensor is utilized for monitoring cap removal, the accelerometer for monitoring pill pickup events, and the load cell for bottle weight sensing.

3.2 Neck-Worn Sensors

The piezoelectric sensor is used for sensing the mechanic stress resulting from skin motion during pill swallowing and generating voltage as a response. That data is then sent to mobile phones via bluetooth that runs algorithms for further analysis. A Bayesian network classifier was used for classifying the data received. The achieved precision and recall for the capsule were 87.09% and 90%, respectively. Major challenges are comfort and social acceptance [11] of the user, since it needs to be worn by the patient.

3.3 Proximity Visual Systems

RFID sensors and video cameras have been used in [12, 13] to characterize the medication taking activity in an in-home environment. The medicine bottles or pill straps are equipped with RFID tags and stored in a storage that embeds an RFID reader. The RFID technology identifies the medication bottles placed in the cabinet. However, once a bottle is out of the coverage of the reader's antenna, the identification process using RFID technology fails. The camera is used for tracking the occurrence of medication taking based on moving object detection and color model of the bottle, thus keeping a track of medications taken and their time.

Fig. 1 Proposed system architecture

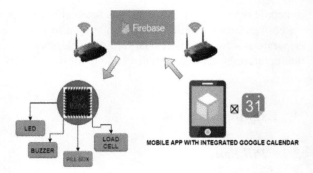

4 Proposed System

4.1 System Architecture

Firebase is used integrate software app and hardware. The hardware consists of a buzzer that is connected to the reminder module of the software app; a load sensor made up of HX711 and load cell with a precision of 0.001 mg, fixed between two plates, one that would act as a base, and one that would carry the pills for consumption and a TFT display which is connected to the hardware as well as the app (Fig. 1).

4.2 Modular Diagram and Reminder-Alarm Flowchart

The pill box consists of software and hardware integration. The patient registers with the app and sets reminder for the pill that has to be taken for an entire month. Whenever the reminder buzzer goes off on the hardware and the patient removes pill from the pill box, the weight of the pill box decreases giving a confirmation that the pill has been taken. When the weight goes below the threshold, a reminder is sent to user (Figs. 2 and 3).

4.3 Implementation of the Software App

The SaathiCal app aims at providing ease to its users. The home page opens as soon as the application is opened. The main activity directs the users to the first screen of the app. Since the process of IVF is very taxing, the app provides different ways to set alarms, remainders in the google calendar as well as offline on the device.

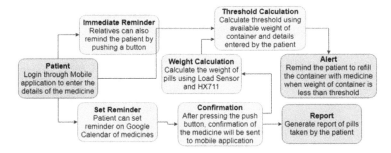

Fig. 2 Flow and linking of system modules

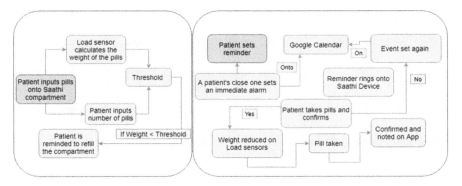

Fig. 3 Flowchart of reminder and weight alarm system

These methods are used since it is not convenient for the user to carry the pill box everywhere along.

The Alarm
A normal alarm or a task reminder can be set with special notes. Google api is used to connect to google calendar to interface the hardware to the application. In case the user sets the alarm for pill in google calendar, a buzzer will ring on the pill box, which can be put off from the app or the pill box indicating that the pill has been taken.

Customized User Login Activities
Next, we have login for users using firebase connectivity. The new user first registers using email and password, then after logging in the user will provide the pill details which will be stored in the profile of the user. The details such as pill name, number of pills, and user id are taken as input. The weight of the pill box is calculated, if it is below the threshold set then a notification is sent to the user to refill the pill box, else the weight of the pill box will be displayed. Using the details of the pills taken by the users, the data will be retrieved and compared with the actual prescription

prescribed by the personal doctor taken as image input and manual data entry from the user, thus generating the report of the user that will further help the doctor for analysis on next visit.

Additional Features
The app also has features such as listing of all the probable medicines taken during this process along with the description of the medicines and their side effects (Figs. 4 and 5).

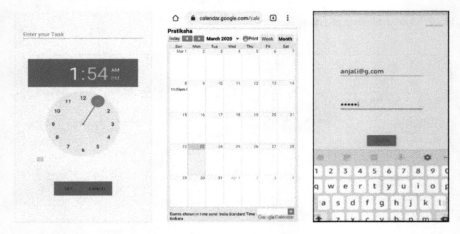

Fig. 4 Alarm, google calendar, and login system on the app

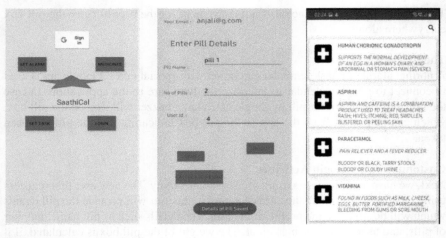

Fig. 5 User interface, pill input and medication on app

4.4 Hardware Prototype

The Saathi hardware consists of a temperature-controlled six compartment box, with a touch screen display, speaker, LED indicators, and weight sensors for each compartment.

The Pill Box

The pill box would have seven compartments. Every compartment would have its own load sensor to detect the weight of the pills, a buzzer, and an LED indicator to indicate the exact pill that has to be taken when the alarm rings.

The Load Sensor

The load sensor consists of a HX711 module with a load cell, fixed between two plates, one that would act as a base, and one that would carry the pills for consumption. The load cell would have a precision of 0.001 mg. There would be a load sensor module present for every compartment and every load cell in the pill box would have its own unique ID for integration with the software app.

The Temperature Control Module

The temperature control module would be made up of a DHT11---temperature and humidity sensor and Peltier modules for cooling, a relay module and a fan to maintain air circulation and regulate the temperature.

Touch Screen Display

The input on hardware would be through a TFT display. The input from the display would be connected to the app through firebase

Speakers and LED Indicators

The speakers would let the user know when alarm rings and LED indicators would be helpful to know the compartment concerned with the alarm (Figs. 6 and 7).

Fig. 6 ESP8266 node MCU connected to google calendar for reminders

Fig. 7 Load cell and LED display connected to node MCU

5 Future Scope

The proposed idea can further be improved by customizing it to the different types of protocols that exist for this treatment according to required medications. This treatment has many intrusive procedures and hence has a very long procedure. Thus, we can provide a chat room where women who have gone through the treatment can share their experiences, talk to each other, and provide emotional support to each other. It can also calculate the blood pressure, measure heartbeat, and generate a report and send it to the doctor. It can also provide the option of ordering the medicine after the appointment with the doctor.

6 Conclusion

The Saathi hardware and application was built keeping in mind the requirements of the women undergoing IVF treatment. For the fulfillment of this purpose, we gathered details required by interviewing a gynecologist. The main purpose of this device is to help the patient follow the schedule and keep track of all its medicine consumption and appointments with the doctor, generating reports for the same, and also allowing them to be in contact with their doctor. It is an aid that will act as assistance to the patient in the way of her becoming a mother.

References

1. Choi.: A pilot study to evaluate the acceptability of using a smart pillbox to enhance medication adherence among primary care patients. Int. J. Environ. Res. Public Health **16**(20), 3964 (2019)
2. A review of medication adherence monitoring technologies. Appl. Syst. Innov. (2018)
3. Reeder, B., Demiris, G., Marek, K.D.: Older adults' satisfaction with a medication dispensing device in home care. Inform. Health Soc. Care **38**, 211–222 (2013)
4. Miller, C.W., Himelhoch, S.: Acceptability of mobile phone technology for medication adherence interventions among hiv-positive patients at an urban clinic. AIDS Res. Treat. **2013**, 670525 (2013)

5. Santo, K., Singleton, A., Chow, C.K., Redfern, J.: Evaluating reach, acceptability, utility, and engagement with an app-based intervention to improve medication adherence in patients with coronary heart disease in the medapp-chd study: a mixed-methods evaluation. Med. Sci. **7**, 68 (2019)
6. Boon Nuddar, N., Wuttidittachotti, P.: Mobile application. In: Proceedings of the International Conference on Big Data and Internet of Things (2017)
7. Othman, N.B., Ek, O.P.: Pill dispenser with alarm via smart phone notification. In: IEEE 5th Global Conference on Consumer Electronics, Kyoto, pp. 1–2 (2016)
8. Hayes, T.L., Hunt, J.M., Adami, A., Kaye, J.A.: An electronic pillbox for continuous monitoring of medication adherence. In: Conference of the IEEE Engineering in Medicine and Biology Society (2006)
9. Wu, H.K., Wong, C.M., Liu, P.H., Peng, S.P., Wang, X.C., Lin, C.H., Tu, K.H.: A smart pill box with remind and consumption confirmation functions. In: Conference Proceedings IEEE Consumer Electronics, pp. 658–659 (2015)
10. Aldeer, M.M.N., Martin, R.P., Howard, R.E.: Tackling the fidelity-energy trade-off in wireless body sensor networks. In: Proceedings of the IEEE/ACM International Conference on Connected Health: Applications, Systems and Engineering Technologies (CHASE), Philadelphia, PA, USA, pp. 7–12 (2017)
11. Kalantarian, H., Alshurafa, N., Sarrafzadeh, M.: A survey of diet monitoring technology. IEEE Pervasive Comput. **16**, 57–65 (2017)
12. Hasanuzzaman, F.M., Tian, Y., Liu, Q.: Identifying medicine bottles by incorporating RFID and video analysis. In: Proceedings of the IEEE International Conference on Bioinformatics and Biomedicine Workshops (BIBMW), pp. 528–529 (2017)
13. Hasanuzzaman, F.M., Yang, X., Tian, Y., Liu, Q., Capezuti, E.: Monitoring activity of taking medicine by incorporating RFID and video analysis. Netw. Model. Anal. Health Inf. Bioinform. **2**, 61–70 (2013)

Predictive Analysis of Alzheimer's Disease Based on Wrapper Approach Using SVM and KNN

Bali Devi, Sumit Srivastava, and Vivek Kumar Verma

Abstract Alzheimer's is a neurodegenerative disorder, and early prediction of Alzheimer's disease is important for its treatment and detection. This paper proposes machine learning-based method for Alzheimer's patient detection using SVM and KNN. This proposed method uses the concept of wrapper method for feature selection to improve the accuracy of the model. This framework focuses on neurological brain disease with analytics and prediction over it. We are using machine learning with supervised classifiers for model training, testing, and evaluation. Many neurological diseases related to the brain, spine, and nerve are dementia, Alzheimer's, Parkinson's, and brain tumors. We are focusing on the most popular neurological disease like Alzheimer's and do all the prediction and execution over the machine learning platform. We have also evaluated the prediction level of the framework with many assessment and evaluation metrics for checking the accuracy and ability of our model. We have reported our model accuracy as 92% on average, which is higher within its dataset.

1 Inroduction

Alzheimer's dementia [2] is a deteriorating neurological disorder that is the most popular type of dementia usually beginning in late medieval or old age. Alzheimer's disorder is an advanced, persistent disease that influences brain cells and induces an intellectual functioning disability. It is a brain dysfunction that burns out the ability to think, recall, visualize, and understand. Dr. Alois Alzheimer is credited for the discovery of this disorder and he found differences in a woman's brain tissue that had expired from a severe mental condition in 1906. The symptom involved lack of memory, communication problems, and bizarre behavior [3]. He studied her brain after she died and found several odd clumps (today named amyloid plaques) and

---

B. Devi (✉) · S. Srivastava · V. K. Verma
School of Computing and IT, Manipal University Jaipur, Jaipur, Rajasthan, India
e-mail: baligupta03@gmail.com

© The Editor(s) (if applicable) and The Author(s), under exclusive license to Springer Nature Singapore Pte Ltd. 2021
T. Senjyu et al. (eds.), *Information and Communication Technology for Intelligent Systems*, Smart Innovation, Systems and Technologies 196,
https://doi.org/10.1007/978-981-15-7062-9_71

twisted packets of fibers. Another characteristic is the impairment of the associations inside the brain among cells in the brain (neurons). Neurons relay signals from various parts of the brain to the organs and muscles in the body, as well as from the brain. Additional parts of the brain are damaged when neurons die. Damage is extensive in the final phase of Alzheimer's and the brain tissue has shrunk substantially [11]. Alzheimer's dementia is ranked as the sixth major risk factor in the USA, although news reports suggest that the condition could be ranked third as just a cause of death for older adults, just behind cancer and heart disease. Alzheimer's is an overarching concept that had to describe the signs that affect memory, everyday activity performance, and language skills. Alzheimer's disease is the commonest type of dementia [12], which is getting worse over time and affects thinking, vocabulary, and reasoning.

We are focusing on the most popular neurological disease like Alzheimer's and do all the prediction and execution over the machine learning platform. This proposed method uses the concept of wrapper method [13] for feature selection to improve the accuracy of the model. We have also evaluated the prediction level of the framework with many assessment and evaluation metrics for checking the accuracy and ability of our model. We have reported our model accuracy as 92% on average, which is higher within its dataset.

2 State of Art: SVM and KNN

2.1 Support Vector Machine (SVM)

A support vector machine (SVM) [6] with a separate hyperplane is a discriminating classifier. In another way, given that the labeled train data (supervised learning) [5], this produces an optimal hyperplane that classifies new patterns. This hyperplane is a two-dimensional line space that divides a plane into two portions in which it lays at either side in each portion. We are using SVM [5, 6] in our model as a classifier to classify Alzheimer's dementia patient and cognitive normal person. The weight function between Alzheimer's dementia patient and cognitive normal person is given below in Eq. 1. Classification-based class separation with Alzheimer's dementia patient and cognitive normal person is presented in Fig. 1.

$$F(s) = \pm(w^T s + b) \quad (1)$$

Fig. 1 SVM classifier with AD and CN

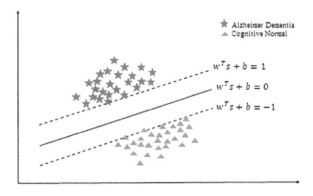

2.2 K-Nearest Neighbors (KNN)

K-nearest neighbors (KNN) [9] classifier is a technique for supervised machine learning model that could be used both for classification and for predictive regression problems. It is, however, primarily used in research for the classification of predictive problems. We are using KNN ($k = 20$ as **k** = square root $(n)/2$)) [9, 10] for searching the k-nearest neighbors to find the actual class of Alzheimer's dementia patient and cognitive normal person. The function which we have used for finding the distance is Euclidian distance that is given in Eq. 2.

$$E_d = \sqrt{\sum_{n=1}^{i}(x_n - y_n)^2} \qquad (2)$$

3 Related Work

Basia et al. [12] suggested CNN-based model for Alzheimer's disease prediction and detection. They have applied CNN-based model over MRI data by performing the data augmentation approach deemed for image deformation, flipping, scaling, cropping, and rotation. They had taken ADNI datasets with 352 images and Milan datasets with 55 images. The main concern with this paper is that they have not used any feature selection method to produce results with high accuracy. Razavi et al. [3] showed an idea using unsupervised feature engineering, which uses softmax regression using the ADNI dataset. They have not used any selection method in his model to produce the selected features that have been really used in Alzheimer's detection and prediction. Segovia et al. [11] suggested a partial least squares and support vector machine-based framework for AD detection and prediction. PLS used to create orthogonal score vectors by maximizing the covariance between different sets of variables. They have also calculated Fisher's discriminant ratio (FDR), which is the class separability for a function that can be calculated by mean in numerical terms. They have used mean-based reparability features which provide a basic average between feature selection and sometimes due to this there is less accuracy as model considered average selected features. Soheil et al. [4] proposed a framework using computer vision to predict and detect Alzheimer's disease. He has used the concept of the convolution neural network to produce the end-to-end architecture of the diagnosis of Alzheimer's patients. Based on the learning model, they have developed the biomarkers to predict Alzheimer's disease, but they have not used any feature selection method to produce the selected and relevant features for Alzheimer's disease prediction. Lu et al. [7] described an MRI-based model using deep learning multimodal and multiscale feature extraction method for metabolism and regional volume. They have worked out on ADNI MRI images using deep neural networks for model development, but they have not used any features selection model to produce the best model with selected features. Gunawardena et al. [8] showed a convolutional neural network-based model on the MRI dataset of ADNI [1]. They have used different image segmentation and image preprocessing method for image feature extraction. They have not used any feature selection method to produce better results with the relevant feature set. In our proposed work, we have used the concept of wrapper method for feature selection as well as for correlation between the features. After feature selection, we have applied two supervised classifier support vector machine and k-nearest neighbors as for classification between Alzheimer's dementia patient and cognitive normal person.

4 Proposed Framework and Algorithm

We have formulated the problem of feature selection using the wrapper method [13] with the comparison of two images data at a time. We have taken two images as I1, 2; both the images data are taken as functions $\Omega \rightarrow ZI$ where Ω is a region of the image data with function Z and I denotes the number of image data input. We are taking the region Ω in the form of centralized storage of MRI images data. We are also taking image data value $I = 1$ if image data is grayscale. If we map the field of image I_1, I_2 then these are the form of $f_{12}: \Omega \rightarrow ZI$, in the given format that in Eq. 3.

$$I_1(f_{12}) \approx I_2(f_{12}) + f(x) \tag{3}$$

where $\approx$ is a precise operation between flows from Image-1 to the Image-2 with N image data region to identify the features of all the images data in the same format. If we are using the wrapper method for feature selection, then wrapping of the two methods creates a specific feature set with high optimal results. If an image data is wrapping twice to create multiple features, we have multiple procedures with the symbol as W_1 and W_2, and Wrap as W_r then the formulated Eq. 4 is given below:

$$W_r(W_r(W_1; I_1(f_{12})); W_2)(f_{12})$$
$$\rightarrow W_r(I_1; W_1)(f_{12} + W_2(f_{12}))$$
$$\rightarrow I_1((f_{12}) + W_2(f_{12}) + W_1((f_{12}) + W_2(f_{12})))$$
$$\rightarrow W_r(I_1; W_2 + W_r(W_1; W_2))(f_{12}) \qquad (4)$$

Wrapping of multiple image data, we can form using the given Eq. 5.

$$(W_1 * W_2) \rightarrow W_2 + \text{wrap}(W_1, W_2) \qquad (5)$$

For the n image extraction data, we write Eq. 5 as in general way in Eq. 6:

$$(W_1^n * W_2^n) \rightarrow (W_2 + \text{wrap}(W_1, W_2))^n \qquad (6)$$

We can modify Eq. 4 with the help of wrapping of multiple image data given below:

$$W_r(W_r(I_1; W_1); W_2) = W_r(I_1, W_1 * W_2) \qquad (7)$$

For the general way of wrapper method [13], we have used the whole model with the correlation of two wrapper method [13] which is given below in Eq. 8.

$$\text{Cov}[W_1 * W_2] \rightarrow \frac{1}{\Omega} \sum_{X \in \Omega}^{\infty} (I_1(f_{12}) * I_2(f_{12}))$$
$$- \frac{1}{\Omega^2} \sum_{X \in \Omega}^{\infty} I_1(f_{12}) \sum_{X \in \Omega}^{\infty} I_2(f_{12}) \qquad (8)$$

Correlation between two wrapper methods with Alzheimer's dementia patient and cognitive normal person data is detected with the help of covariance between these two as in Eq. 9.

$$\text{Cor}[W_1 * W_2] \rightarrow \frac{\text{Cov}[W_1, W_2]}{\sqrt{\text{Cov}[W_1, W_1]\text{Cov}[W_2, W_2]}} \qquad (9)$$

For the overall framework process, we have taken preprocessed dataset from ADNI (MRI data) as a form of extracted features clinical data, after that we have processed that data with cleaning and feature selection by wrapper method [13]. After that we have applied supervised machine learning algorithms as support vector machine (SVM) [6] and k-nearest neighbors (KNN) [9] used for classification between Alzheimer's dementia patient and cognitive normal person. With the continuation of the whole process, we have training data with the help of the same above classification algorithms. On the other hand, we have applied cross-validation using test and train data for prediction and evaluation of our results. The proposed process diagram is given below in Fig. 2.

Fig. 2 Proposed process framework

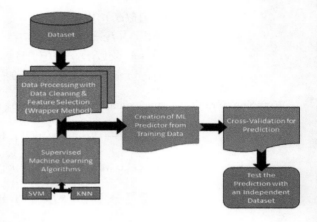

4.1 Process Algorithm

S-1: Import the preprocessed clinical MRI data from ADNI.
S-2: Perform data preprocessing and feature selection using wrapper method

$$\text{Cov}[W_1 * W_2] \to \frac{1}{\Omega} \sum_{X \in \Omega}^{\infty} I_1(f_{12}) * I_2(f_{12}))$$

$$- \frac{1}{\Omega^2} \sum_{X \in \Omega}^{\infty} I_1(f_{12}) \sum_{X \in \Omega}^{\infty} I_2(f_{12})$$

S-3: After applying wrapper method finding the correlation between the correlated features.

$$\text{Cor}[W_1 * W_2] \to \frac{\text{Cov}[W_1, W_2]}{\sqrt{\text{Cov}[W_1, W_1]\text{Cov}[W_2, W_2]}}$$

S-4: Split the dataset in train set and test set as 70:30.
S-5: Apply the SVM and KNN using the selected features from dataset (step S-3).
S-6: Evaluate the performance using evaluation metrics on test set.

5 Dataset

We have collected dataset from Alzheimer's Disease Neuroimaging Initiative [1] as a part of the Laboratory of Neuro Imaging, University of Southern California. We have collected already feature extracted preprocessed data (1510 person data: 196

AD and 1314 CN) for feature selection, analysis, and prediction. We have divided this data in the train set and test set as 70:30.

6 Result and Discussion

We have applied our proposed process model over the test data and find the result with the help of area under the curve, evaluation metrics, and comparative analysis with other works in the same field. The area under the curve is the curve between the false positive rate (FPR) and the true positive rate (TPR). It is defined as a proposed model ability in terms of separable quality of classifiers. We have maintained area under the curve for SVM and KNN with the AUC score, respectively, 0.916 and 0.901. As per our result, it shows that the AUC results are good and classifier having class separable quality greater than 0.908 on average. AUC scores of both SVM and KNN are given below in Fig. 3.

We have also found the evaluation metrics with the help of the confusion matrix from the test data. For the same, we have evaluated accuracy, precision, recall, and $F1$ score form the confusion matrix with respect to given test data. Precision is the positive predicted value; recall is the true positive rate or the ability of the model, whereas the $F1$ score is the harmonic mean between precision and recall. All the same evaluation metrics are given in Table 1.

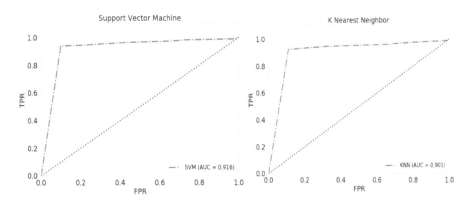

Fig. 3 AUC score: SVM and KNN

Table 1 Evaluation metrics

Classifier	Accuracy	Precision	Recall	$F1$ score
SVM	92.3	92.1	92.7	92.4
KNN	91.7	91.2	91.9	91.55
Average	92.0	91.65	92.3	91.98

Table 2 Comparative analysis

Comparative work	Methodology	Accuracy (%)
Basia et al. [12]	CNN-based model	75–85
Razavi et al. [3]	Unsupervised, softmax regression	87.7
Segovia et al. [11]	PLS + SVM	91.6
Lu et al. [7]	DNN-based model	82.93
Gunawardena et al. [8]	CNN-based model	84.4
Soheil et al. [4]	Biomarker's and CNN	88.3 (complex part)
Our work	SVM + KNN + wrapper method	92

As per the above evaluation, our model gives on average accuracy 92%, precision 91.65%, recall 92.3%, and $F1$ score 91.98% which is higher and given below in the comparison Table 2.

7 Conclusion

This work presents machine learning-based method using SVM and KNN classifier to predict a person Alzheimer's dementia patient and cognitive normal person. We have showed our analysis on preprocessed clinical data of MRI images. We have also used wrapper method for feature selection to improve the accuracy of our model. Sometimes some features are not relevant for training the model and then discriminant factors for feature selection play an important role over it. Wrapper method produces discriminant factors and wrapping it. We have evaluated accuracy and ability of our proposed model using area under the curve, evaluation metrics, and comparative analysis between existing researches. Our model reported higher accuracy as 92% with the higher $F1$ score as 91.98%. We will include in future the concept of deep CNN or any other features extraction technique to extract real-time features from the MRI image and after that we will perform feature selection and model development over it.

References

1. ADNI, http://adni.loni.usc.edu/study-design/. Accessed Online January 12, 2020
2. Bondi, M.W., et al.: Alzheimer's disease: past, present, and future. J. Int. Neuropsychol. Soc. (JINS) **23**(9–10), 818–831 (2017). https://doi.org/10.1017/s135561771700100x
3. Razavi, F., Tarokh, M.J., Alborzi, M.: An intelligent Alzheimer's disease diagnosis method using unsupervised feature learning. J. Big Data **6**, 32 (2019). https://doi.org/10.1186/s40537-

019-0190-7
4. Esmaeilzadeh, S., Belivanis, D.I., Pohl, K.M., Adeli, E.: End-To-end Alzheimer's disease diagnosis and biomarker identification. MLMI@MICCAI (2018). https://doi.org/10.1007/978-3-030-00919-9_39
5. Goel, V., Jangir, V., Shankar, V.G.: DataCan: Robust approach for genome cancer data analysis. In: Sharma, N., Chakrabarti, A., Balas, V. (eds.) Data Management, Analytics and Innovation. Advances in Intelligent Systems and Computing, vol. 1016. Springer, Singapore (2020). https://doi.org/10.1007/978-981-13-9364-8_15
6. Devi, B., Kumar, S., Anuradha, Shankar, V.G.: AnaData: a novel approach for data analytics using random forest tree and SVM. In: Iyer, B., Nalbalwar, S., Pathak, N. (eds.) Computing, Communication and Signal Processing. Advances in Intelligent Systems and Computing, vol. 810. Springer, Singapore (2019). https://doi.org/10.1007/978-981-13-1513-8_53
7. Lu, D., Popuri, K., Ding, G.W., Balachandar, R., Beg, M.F.: Alzheimer's disease neuroimaging initiative. Multimodal and multiscale deep neural networks for the early diagnosis of Alzheimer's Disease using structural MR and FDG-PET images. Sci Rep. **8**(1), 5697 (2018). https://doi.org/10.1038/s41598-018-22871-z
8. Gunawardena, K., et al.: Applying convolutional neural networks for pre-detection of alzheimer's disease from structural MRI data.. In: 24th International Conference on Mechatronics and Machine Vision in Practice (M2VIP), pp. 1–7 (2017)
9. Shankar, V.G., Devi, B., Srivastava, S.: DataSpeak: data extraction, aggregation, and classification using big data novel algorithm. In: Iyer, B., Nalbalwar, S., Pathak, N. (eds.) Computing, Communication and Signal Processing. Advances in Intelligent Systems and Computing, vol. 810. Springer, Singapore (2019). https://doi.org/10.1007/978-981-13-1513-8_16
10. Devi, B., Shankar, V.G., Srivastava, S., Srivastava, D.K.: AnaBus: a proposed sampling retrieval model for business and historical data analytics. In: Sharma, N., Chakrabarti, A., Balas, V. (eds.) Data Management, Analytics and Innovation. Advances in Intelligent Systems and Computing, vol. 1016. Springer, Singapore (2020). https://doi.org/10.1007/978-981-13-9364-8_14
11. Segovia, F., Górriz, J.M., Ramírez, J., Salas-González, D., Álvarez, I.: Early diagnosis of Alzheimer's disease based on partial least squares and support vector machine. Expert Syst. Appl. **40**(2), 677–683 (2013). ISSN 0957-4174. https://doi.org/10.1016/j.eswa.2012.07.071
12. Basaia, S., Agosta, F., Wagner, L., Canu, E., Magnani, G., Santangelo, R., Filippi, M.: Automated classification of Alzheimer's disease and mild cognitive impairment using a single MRI and deep neural networks. NeuroImage Clin. **21**, 101645 (2019). ISSN 2213-1582. https://doi.org/10.1016/j.nicl.2018.101645
13. Modi, M., Patel, S.: An evaluation of filter and wrapper methods for feature selection in classification (2014)

A Novel Method for Enabling Wireless Communication Technology in Smart Cities

Vijay A. Kanade

Abstract The paper discloses an innovative system and approach to harness wireless Wi-Fi technology by enabling trees to generate wireless signal via biodegradable chips. The wireless signals undergo beamforming in order to focus the signals in a desired direction. The signals apart from beamformed ones are exposed to a tiny membrane that essentially converts the wireless signals (form of radio waves) incident on it into visible light. The radio waves hence are converted into photosynthetically active form of radiation and therefore the converted radiation is further distributed in the tree habitat. The radiation is used by the trees for photosynthesis purpose. Hence, the adverse impact of wireless signals on tree growth is avoided through beamforming and conversion of wireless radio waves to visible spectrum.

1 Introduction

Wi-Fi is a wireless LAN technology that has become an indispensable part of our lives today. This computer networking technology enables Internet connection over various wireless media such as wireless routers and wireless modem. Wi-Fi hotspots are observed in restaurants, cafes, coffee outlets, etc., that provide free Internet services to their users. One of the advantages of these hotspots is that it keeps people connected, and thereby is useful in case of an emergencies.

Further to the wireless routers, Wi-Fi has allowed the dumb devices to communicate over the Internet. Wi-Fi thus has led to the emergence of 'Internet of Everything (IoE).' IoE connects devices such as light bulbs, thermostats, smart watches, smart fridges, and various other smart appliances [1].

Wi-Fi uses IEEE 802.11 wireless LAN and mesh standards for their deployment, usage, and security protections, which ensure that the consumers can rely on the technology [4].

V. A. Kanade (✉)
Intellectual Property Research, Pune, India
e-mail: kanade.science@gmail.com

As discussed above, Wi-Fi can be termed as oxygen of the modern world, which allows seamless Internet connection to its customers. Various commercial giants like IBM, Samsung, Philips, etc., are in the race for developing the smart devices that are Internet enabled. As per the recent statistics, every second about 127 new devices get connected over the Internet. It is also predicted that by 2027, about 41 billion devices would be connected over the Internet and by 2030, each user would own 15 connected devices [2, 3]. However, this large-scale development and manufacturing of smart devices is leading to the generation of high volume of 'electronic waste.' Since Internet is of paramount importance for these devices to function, there has also been substantial upsurge in production of wireless hotspots (i.e., routers, nanochips) that are used at various commercial and public places. Such electronic development is only adding to the existing burden of electronic wastage. Hence, there seems to be a requirement for more sustainable development of wireless communication media, that allows the devices to connect over the Internet, yet ensure that the 'electronic waste' is kept in check.

The present research proposal discloses an innovative approach for providing the sustainable deployment of Wi-Fi technology without any load on 'electronic waste.' We intend to use biodegradable chips, that can be fixed onto the trees and would act as Wi-Fi connection media for the community that is in its vicinity.

2 Modules

The various modules employed for harnessing the inventive communication technology are discussed in the following section.

2.1 Biodegradable Chip

As per the recent statistics, about 70–75% of the discarded electronics piles up in a landfill, wherein the value of these discarded items worldwide is about $62.5 billion. As per world economic forum, about 50 million tons of e-waste are produced every year. If the current trend is followed, wherein we do not take measures to control the e-waste generation, then this number could amount to 120 million tons by 2050 [6]. Further, the discarded e-waste is taking up space for its disposal and releasing toxic chemicals in the ecosystem at the same time, thereby polluting the environment substantially.

To overcome these challenges, biodegradable semi-conductor chips have been developed that are made from wood. This implies, on their usage in forests or green vegetation (those dominated by trees), the chips are degraded by the fungus present in such vegetative habitat [10, 11]. The chips on degradation become a part of the topsoil and are as safe as fertilizers.

The chips are made up of a biodegradable cellulose nanofibril (CNF) layer. This layer is flexible and transparent derivative of wood composed of plurality of nanofibers [5].

2.2 Photosynthetically Active Radiation

Photosynthetically active radiation (PAR) is the light range (400–700 nm) in the electromagnetic spectrum at which photosynthesis occurs. Photosynthesis is a chemical process wherein light energy during daytime is converted into chemical energy that is used by the plants as a source of food.

The visible segment of the EM spectrum involves seven wavelength ranges that correspond to seven colors. The colors in the context of increasing wavelength and decreasing energy are as follows: VIBGYOR, wherein each alphabet defines colors as violet, indigo, blue, green, yellow, orange, and red [7]. Hence, during daytime, the plants use visible light section of the EM spectrum for photosynthesis.

2.3 Wi-Fi Beamforming

Generally, a wireless signal from a broadcast antenna is dispersed in all directions. However, with beamforming the wireless signal is focused on a device having a receiving antenna, rather than spreading the signal in omni-directions. Such an arrangement establishes a faster and optimal connection that is more secure and reliable in comparison to conventional EM devices.

In beamforming, multiple antennas are positioned in close vicinity that broadcast the same signal at different time periods. At some instants, the waves from various antennas overlap each other to produce constructive interference, wherein the signal magnitude is enhanced. Such constructive interference is observed in some areas, while in other areas destructive interference occurs, wherein the waves cancel each other, and the signal gets weaker for the receiver to detect. Therefore, in beamforming, the signals form a targeted beam of EM energy that allows the signal to focus in one specific direction [8, 12].

2.4 Conversion of Radio Waves to Visible Light

Radio waves or the wireless signals given out by the wireless media such as Wi-Fi routers and modem can be converted into light. According to the recent research at University of Copenhagen's Niels Bohr Institute, a sensor (or a detector) has been developed that can convert radio signals into light signals at room temperature [9]. In the present research, we use a nanomembrane (approx. 200 nm thick) of silicon

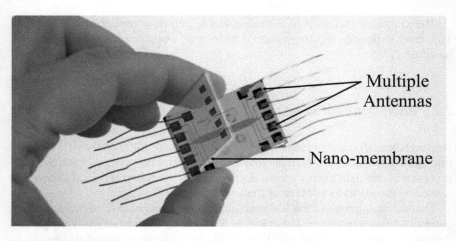

Fig. 1 Biodegradable chip with silicon nitride nanomembrane

nitride attached to an electronic circuit of the biodegradable chip. The membrane detects the EM waves given out by the antennas in its vicinity. On detection of the EM waves, the membrane starts flicking back and forth, i.e., mechanical vibrations are produced in the nanomembrane. As the membrane is connected to the electrical circuit, the mechanical vibrations of the membrane are converted to electrical pulses that are transmitted to the attached voltage sensor. The EM fields of the electrical pulses are detected by a voltage sensor attached to the circuit and in turn produce optical signals via artificial light generating unit due to voltage fluctuations produced by the electrical circuit. Hence, the generated wireless signals are converted into visible light form (Fig. 1).

3 Inventive Communication Technology

In the proposed communication technology, biodegradable Wi-Fi chips with multiple antennas are deployed on multiple trees/plants or vegetation occurring in a habitable natural environment, wherein trees are found in abundance. The deployable spaces for this innovative communication technology include cities, towns, offices, banks, households, rural areas, etc. These chips receive Internet service via wired or wireless connections from various Internet service providers. Here, the Wi-Fi chips are enabled with the silicon nitride nanomembrane along with the attached electronic circuit.

The Wi-Fi chips are configured in such a way that signals undergo beamforming and focus them only in specific directions, wherein the receiver device is functional. This implies as a city or a rural resident navigates through the city or village, the resident would encounter number of Internet hotspots on the way. These hotspots are essentially trees, that are necessary to keep our environment green and clean. The tree hotspots continuously emit Wi-Fi signals in its surrounding area, hence the residents can connect to the very tree hotspot to access the Internet service. Further, the remaining portion of the wireless signals (i.e., signal that do not undergo beamforming) hit the surface of the nanomembrane. The nanomembrane then undergoes back and forth mechanical movement. These mechanical vibrations of the nanomembrane are converted into electrical signals and transmitted to the coupled voltage sensor. The sensor detects the altering electromagnetic field of the received signal. The variation in the EM fields are recorded and the artificial light generating unit (i.e., LED) is used for producing light in response to the changing EM fields sensed by the voltage sensor. The LED light bulb (or any other artificial light source) is externally attached to the nanomembrane for producing optical signals that may be used by the tree hotspots for photosynthesis. Hence, the initially generated wireless Wi-Fi signals are partly converted to visible light in the disclosed communication technology. The generated artificial light is used by the trees/plants or nearby vegetation for the photosynthesis purpose. The artificial light produced in the above process acts as a steady source light, wherein the photons are generated without any interruptions. Further, the produced artificial light is of wavelength of the visible light, which is essential for photosynthesis. These properties of the artificial light make it favorable for the chlorophyll to trigger photosynthesis.

The inventive Wi-Fi technology is diagrammatically represented in Fig. 2.

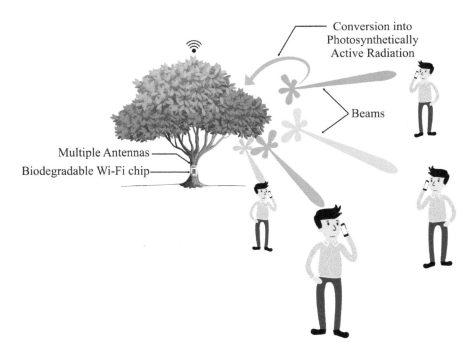

Fig. 2 Inventive wireless communication technology

3.1 Beamforming Mathematics

One of the critical aspects for the inventive Wi-Fi technique to operate is that of beamforming. Hence, developing a computationally simple method to exhibit such technical feature is of paramount importance [15]. We used a mathematical model as a beamforming technique to calculate the beam pattern output for two beams generated by the Wi-Fi-enabled tree. Individual beams (2 beams) were computed by using step-by-step procedure, and beam pattern for two beams was evaluated based on the directional requirement.

Beamforming refers to a linear combination of the output produced by each antenna element, wherein a single beam is computed by using the below equation:

$$\varsigma(\theta) = \sum_{k=0}^{k=N} a_k(\theta) \cdot x_k \quad (1)$$

where

N number of antenna elements
K index variable
a_k complex coefficient of the kth element
x_k voltage response from the kth element
ς beam response
θ angle of the beam main lobe (out of 'n' number of lobes).

Equation 1 above is evaluated by segregating it into following three quantities:

(a) Generating vector of beam coefficients.
(b) Computing a matrix of the element responses over transmitter angles.
(c) Evaluating matrix product of the beam coefficients and the element responses.

Further, multiple beams (2 beams) are generated in parallel by generating beam coefficients for each beam as per below procedure:

$beam(\theta) = $ Generate a vector of beam coefficients for an arbitary beam

$\Upsilon \leftarrow -\pi \cdot \sin(\theta)$

for $k \in 0 \ldots N - 1$

$b_k \leftarrow e^{i \cdot \Upsilon \cdot k}$

$$\frac{b}{\sqrt{b \cdot b}}$$

$b_1 = beam(45°)$
$b_2 = beam(-30°)$

(a): Beam coefficient for Generating a

Beam in A Specific direction

Furthermore, the matrix of element responses computed for a transmitter placed along a range of 0° to 180° is as disclosed below:

$$dst(\theta, r) = \text{//Forms the vector product//}$$
$$\text{for } k \in 0 \ldots N - 1$$
$$d_k \leftarrow e^{\text{dist}(k, \theta, r, N) \cdot 2 \cdot \pi \cdot j}$$
$$\frac{d}{\sqrt{d \cdot d}}$$

(*b*): Element Responses for a Range of Angles from 0° *to* 180°

Then, we plot the output generated by computing Eq. 1. Hence, the beam pattern formed for two beams, at 45° and −30° are as below:

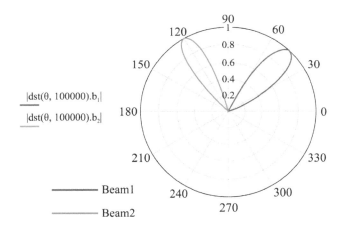

(c): Two beam patterns for antenna element and transmitter range = 1000 λ

4 Advantages

The proposed inventive communication technology has following advantages:

- Plants usually undergo photosynthesis during daytime due to light received from the Sun. However, during nighttime, due to absence of light energy, plants cannot proceed with the photosynthesis process. The disclosed technology allows the plants to undergo photosynthesis at night as the certain section of the wireless

Fig. 3 Operative inventive Wi-Fi communication technology in a city

signals are converted into artificial light necessary for the chlorophyll to carry out photosynthesis.
- The biodegradable computer chips have overall environmental benefits, wherein such chips do not allow generation of any electronic waste. These chips are eco-friendly form of electronics that can become part of nature when the biodegradable material is no longer intact to carry out the purpose of broadcasting signals.
- The technology allows the entire city, town or a geographic area to be fully Wi-Fi enabled. The disclosed inventive communication technology provides an ecosystem, wherein the individual users need not opt for Internet service facility on their personal gadgets. Since the communication technology can be deployed in public places, the users can coexist and use one common service instead of individual Internet service subsets (Fig. 3).
- In one of the research studies, it was identified that radiation from Wi-Fi significantly causes a drop in root and shoot growth. The research study also concluded that the Wi-Fi radiation alters and changes the size and shape of the leaves and further also reduces hairs of the roots. Another important finding of the study mentions that radiation levels below the international standard guidelines for microwave radiation cause the plant growth to stunt and thereby the plant loses its ability to manifest any defensive mechanism to protect itself [13]. Hence, the presented communication technology ensures that no plants or trees or green vegetation are affected by the Wi-Fi radiation, since the proposed approach utilizes the beamforming technique for focusing the wireless signals to its intended users. Further, the later segment of the unused wireless signals are utilized by the biodegradable chips to convert them into visible light spectrum. This ensures that the wireless signals are not allowed to reach or hit the vegetation or trees or plants. Thus, the technology possesses an eco-friendly edge over the conventionally available Wi-Fi technologies.

- The communication system uses the multi-user multiple-input and multiple-output (MU-MIMO) principle for wireless communication, wherein multiple antennas are used for sending and receiving multiple set of streams [14]. Hence, the Wi-Fi technology supports multiple users at one go.

5 Conclusion

The proposed innovative approach presents a futuristic model for harnessing Wi-Fi technology by enabling trees to serve as hotspot that broadcast wireless signals in their vicinity. The Wi-Fi service is developed by using a set of biodegradable chips fixed on the group of trees. The biodegradable chips possess multiple antennas that undergo wireless beamforming for providing the Internet service to the intended users in its vicinity. Apart from the beamformed signals, the remaining portion of the wireless signals are converted into visible light spectrum by using the silicon nitride-based nanomembrane. The visible light spectra can then be used by the trees for photosynthesis purpose. Further, the inventive technology can be installed at various public places in the cities, rural areas, etc. The disclosed Wi-Fi service also provides flexibility to its users if they wish to avail the common Internet service operating within their geography rather than indulge into choosing independent Internet service from Internet service providers on their personal devices. Hence, the research proposes a novel wireless communication technology for smart cities of the future.

Acknowledgements I would like to extend my sincere gratitude to Dr. A. S. Kanade for his relentless support during my research work.

References

1. Soffar, H.: The importance and uses of Wi-Fi technology, 27 Mar 2015
2. Gyarmathy, K.: Comprehensive guide to IoT statistics you need to know in 2020, 26 Mar 2020
3. Heslop, B.: By 2030, Each person will own 15 connected devices—here's what that means for your business and content, 04 Mar 2019
4. Beal, V.: 802.11 IEEE wireless LAN standards
5. Crew, B.: Scientists develop biodegradable computer chips made from wood, 27 May 2015
6. World Economic Forum: The world's e-waste is a huge problem. It's also a golden opportunity, 24 Jan 2019
7. Fondriest Environmental, Inc.: Solar radiation and photosynethically active radiation, fundamentals of environmental measurements, 21 Mar 2014
8. Fruhlinger, J.: Beamforming explained: how it makes wireless communication faster, 15 Oct 2019
9. Gibney, E.: Sensor turns faintest radio waves into laser signals. Nat. Mag. 6 Mar 2014
10. Orcuttarchive, M.: A biodegradable computer chip that performs surprisingly well, 14 July 2015
11. Cai, Z.: Biodegradable computer chips made from wood. Forest Products Laboratory (FPL), Madison, WI, Research Station (2015)

12. Introduction to 802.11ax High-Efficiency Wireless, 5 Mar 2019
13. Havas, M. et al.: Effects of wi-fi radiation on germination and growth of broccoli, pea, red clover and garden cress seedlings: a partial replication study. Curr. Chem. Biol. **10**(1) (2016)
14. Geier, E.: How MU-MIMO Wi-Fi works to improve the speed and capacity of home networks, 01 June 2015
15. Mathscinotes: Beamforming math, 20 Jan 2012

Electronic Aid Design of Fruits Image Classification for Visually Impaired People

V. Srividhya, K. Sujatha, M. Aruna, and D. Sangeetha

Abstract Deep learning has attained immense achievement in many fields, such as computer vision and usual speech processing in recent years. It has a burly erudition capability and can create enhanced use of datasets for characteristic extraction when compared to traditional machine learning systems. The researchers start doing research work in this deep learning due to its popularity. In this manuscript, we chiefly initiate several superior neural networks of deep learning and their functions. Deep learning technologies are becoming the major approaches for natural signal and information processing, like image classification, speech recognition. Deep learning is a technology inspired by the functioning of human brain. Convolutional neural networks (CNN) become very popular for image classification in deep learning. CNN's perform better than human subjects on many of the image classification datasets. In this project, a deep learning convolutional network based on Tensorflow is deployed using python for binary image classification.

1 Introduction

Globally, the Blindness and vision impairment have 2.2 billion people by the fact sheet of October 2019. Out of above 82% of sightless people are 50 years aged or more than, who have been in this world with an incapability to do jobs like study,

---

V. Srividhya (✉) · K. Sujatha
EEE Department, Dr. MGR Educational and Research Institute, Chennai, India
e-mail: sripranav2007@gmail.com

K. Sujatha
e-mail: drksujatha23@gmail.com

V. Srividhya · M. Aruna · D. Sangeetha
EEE Department, Meenakshi College of Engineering, Chennai, India
e-mail: arunaraj_m@yahoo.co.in

D. Sangeetha
e-mail: sangitadevarajan@gmail.com

© The Editor(s) (if applicable) and The Author(s), under exclusive license to Springer Nature Singapore Pte Ltd. 2021
T. Senjyu et al. (eds.), *Information and Communication Technology for Intelligent Systems*, Smart Innovation, Systems and Technologies 196,
https://doi.org/10.1007/978-981-15-7062-9_73

inscribe, or walk with no help. In this manuscript, royal projected the finger reader in the new design that is similar to the projected method but with extra costs of elegant mobile phone which is necessary for the device to the occupation. If we use the finger reader we need to be aware with the habit of touch screens and mobile phones that includes the trouble. A new profit-making invention known as a camera that allows the customer to study paper wordings beside several extra tasks but the tools used increase its market place price exponentially [1]. The involvement of this manuscript is threefold over. While bearing out chores on a daily basis, an attempt to diminish the reliance of the customer on the public around him. First, the thought of a wearable gadget that maintains the common human affinity of peaking at things to act together with the surroundings. Second is a model with little price solution to the crisis faced by the visually impaired whereas cooperating with their surroundings. Conventionally white cane and guide dogs have been worned by blind people to evade barriers in their walking pathway. Resolutions can also inform the customer of a probable barrier 3–4 m ahead like Smart cane and Ultra cane can recover the recognition expanse for cane and however, they are not able to organize barriers based on probable hazard and also do not retain routing. Third, we built smart solutions recently using Google Glass or Microsoft's seeing AI use a cloud server to execute all the processing, warning their usability in regions without connectivity [2].

2 Materials and Methods

2.1 *Tensorflow and Execution Model*

Tensorflow is a mechanism learning scheme that functions at huge range and in various surroundings. Tensorflow uses statistics flow charts to symbolize calculation, shared state, and the procedures that transform that state. It maps the nodules of a statistics flow chart diagonally various mechanisms in a cluster, and within a mechanism across multiple computational devices, together with multicore Central processing units, common-purpose General processing units, and custom-designed ASICs identified as Tensor Processing Units (TPUs). Several Google services use Tensorflow in fabrication and it has to turn into broadly used for mechanism learning research and we have released it as an open-source project. In this manuscript, we illustrate the Tensorflow statistics flow model and display the gripping presentation attains for numerous factual-world applications [3]. To signify all calculations and status in a mechanism learning algorithm the Tensorflow uses a single statistics flow graph, together with the individual numerical functions, the factors and their revised rules, and the input preprocessing (Fig. 1). The statistics flow graph conveys the message among subcomputations explicitly, thus making it trouble-free to perform autonomous computations in parallel and to division computations diagonally several devices [4–9].

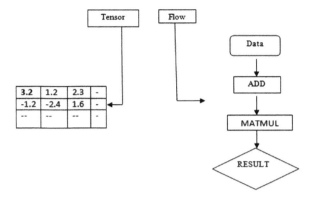

Fig. 1 Working of tensorflow

For training, set up to high precision involves a huge quantity of computation and we use Tensorflow to range out this computation diagonally a cluster of GPU-enabled servers. Convolutional neural networks (CNN) become very popular for image classification in deep learning.

2.2 Proposed Method

To reduce the rate of the loss function mostly stochastic gradient descent (SGD) are being used by several neural networks that iteratively renovates the constraints of the system by moving them in the direction. A number of refinements to SGD accelerate union by shifting the revised rule. Based on the new optimization technique the researchers often want to do testing, but doing that in mistrust engages transforming the constraint server realization. An open-source (Object Detection API) structure work is built on peak of Tensorflow which aspires to make it easy to create, instruct and set up object detection molds. To accomplish this, Tensorflow Object detection application programming interface is a computing interface to a software component or a system, offers the customer with multiple pretrained object detection models with teachings and example codes for finetuning for object detection tasks. The run time of Tensorflow is a cross-platform library. Multiple client languages are supported by Tensorflow. Our interior customers are mainly well-known with Python and C++ languages. In this project, a deep learning convolutional network based on Tensorflow is deployed using python for binary image classification. We analyzed for different fruits such as Papaya, Banana, Mango, Guava, etc. (Fig. 2, 3, 4, 5, and 6).

Fig. 2 Papaya fruit maturity with unripe

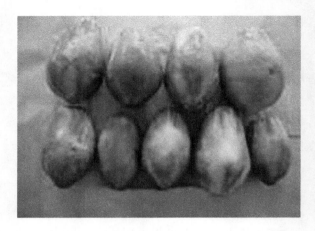

Fig. 3 Papaya fruit maturity with full ripe

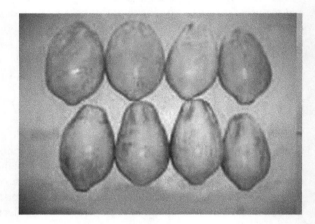

2.3 Block Diagram Description

In the proposed system, ultrasonic sensor is used to monitor the distance of the object. It has three main parts a raspberry pi 3, camera, and an ultrasonic sensor. When the object is captured by the camera and analyze the image using tensor flow and detect what is the picture is about, then by using a speaker or headphone, the voice will assist the person about that picture. The function of the Bluetooth module is used to transfer sound data with telephones or handheld computers for communication between the impaired people and the system (people's mobile phone) (Fig. 7) [10].

Electronic Aid Design of Fruits Image ... 731

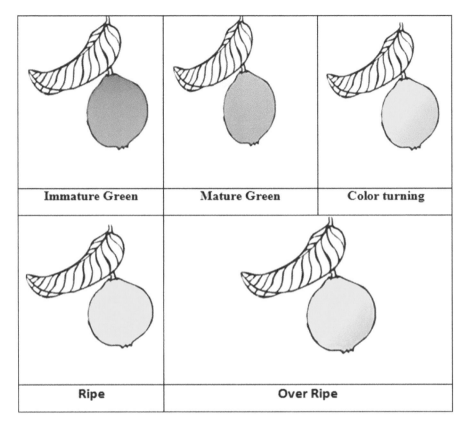

Fig. 4 Guava fruit maturity stage by stage

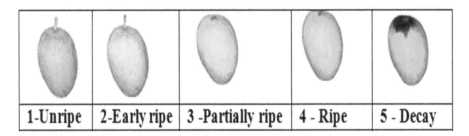

Fig. 5 Mango fruit maturity stage by stage

3 Result and Discussion

The objectives of the system are to detect obstacles that white canes or guide dogs they cannot, expanding their detection range. Vision is one of the most important

Fig. 6 Banana fruit maturity stage by stage

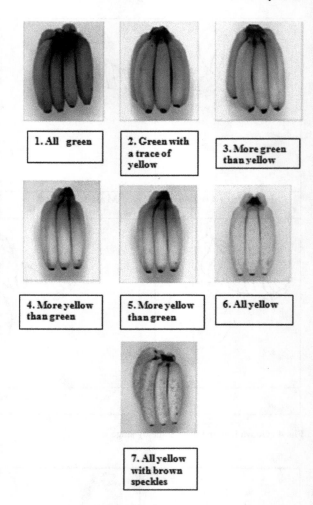

senses that helps people interact with the real world. Commonly, visually impaired develop their other senses to feel their surroundings, but in certain cases, this is not enough. Our prototype is wireless with the purpose to be comfortable for the user to classify the image with good quality. For example if we want to purchase the fruits with good quality, with the use of wearable device and with the help of tensor flow analysis of image classification we can able to purchase a good quality of fruits.

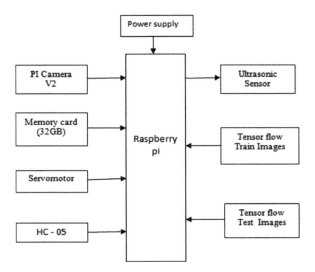

Fig. 7 Block diagram of proposed system

4 Conclusion

Our project helps sightless to move around in indoor or outdoor environments. The objectives of the system are to detect obstacles that white dogs guide dogs cannot, expanding their range of detection. This article proposes the use of a system that detects and recognizes nearby obstacles or objects, giving an audible warning in order to avoid a collision or predictable. The prototype is based on deep learning capabilities. In this system, we find the obstacle name by using tensor flow to find the obstacle name. It is very reliable and easy to implement the fruits image classification for visually impaired people with good quality.

References

1. Dou, L., Wang, X., Ma, Z., Ribarsky, W.: Hierarchicaltopics: visually exploring large text collections using topic hierarchies. IEEE Trans. Vis. Comput. Graph. **19**(12), 2002–2011 (2013)
2. Bosch, H., Thom, D., Heimerl, F., Puttmann, E., Koch, S., Krüger, R., Worner, M., Ertl, T.: Scatterblogs2: real-time monitoring of microblog messages through user-guided filtering. IEEE Trans. Vis. Comput. Graph. **19**(12), 2022–2031 (2013)
3. Maceachren, A.M., Jaiswal, A., Robinson, A.C., Pozanowski, S., Savelyev, A., Mitra, P., Zhang, X., Blanford, J.: Senseplace2. GeoTwitter analytics support for situational awareness. In: IEEE Conference on Visual Analytics Science and Technology, 181–190 (2011)
4. Abadi, M., Barham, P., Chen, J.: TensorFlow: a system for large-scale machine learning. In: **I**: 12th USENIX Symposium on Operating Systems Design and Implementation (OSDI'16), Savannah, GA, USA (2016)
5. Jaswal, D., Sowmya, V., Soman, K.P.: Image classification using convolutional neural networks. Int. J. Advance. Res. Technol. **3**(6), 1661–1668 (2014)

6. Fatma, M.A., Mazenand Ahmed, A., Nashat.: Ripeness classification of Bananas using an artificial neural network. Arab. J. Sci. Eng. **44**(8), 6901–6910 (2018)
7. Maheswaran, S., Sathesh, S., Priyadharshini, P., Vivek, B.: Identification of artificially ripening fruits using smart phones. In: International Conference on Intelligent Computing and Control (I2C2). IEEE Robotics & Automation Society, 23–24 June 2017. IEEE society (2017)
8. Ahmad, Z., Seemab Gul, E.: Brain tumor detection & features extraction from MR images using segmentation, image optimization & classification techniques. Int. J. Eng. Res. Technol. **7**(5), 182–187 (2018)
9. Sujatha, K., Kumaresan, M., Ponmagal, R.S., Vidhushini, P.: Vision based automation for flame image analysis in power station boilers. Aust. J. Basic Appl. Sci. **9**(2), 40–45 (2015)
10. Reshma, R., Sreekumar, K.: Literature survey on methodologies for classification, maturity detection, defect identification and grading of fruits. Int. J. Comput. Appl. **180**(36), 18–22 (2017)

Temperature Regulation Based on Occupancy

Rajesh Kr. Yadav, Shanya Verma, and Prishita Singh

Abstract Expanding demand and cost of energy has driven numerous organizations to discover smart ways for controlling, monitoring and preserving energy. The rising technology of the Internet of Things (IoT) (Xu et al. in IEEE Trans Ind Inform 10(4), 2233–2243, 2014 [1], Shah and Yaqoob in A survey: internet of things (IoT) technologies, applications and challenges. In: IEEE Smart Energy Grid Engineering (SEGE), Canada, 2016, [2]) can be used to manage energy utilization in various sectors. Such devices help in making the interactions between people and their home duties easy. The paper proposes the use of motion sensors, temperature and humidity sensors in order to perform temperature regulation of the room. A major problem with conventional techniques for reducing the energy consumption of air conditioners is the inability of such systems to accurately determine the occupancy of the room. However, we have considered various other factors as well, both internal and external like energy dissipated by electronics, sunlight, humidity, etc. In the end, a function of these parameters is made in order to predict the temperature of the room. Uses of IoT are expanding quickly. This project is also one of the applications of IoT. This project has a wide scope as it can be used in any room or party halls and can help in conservation of energy.

1 Introduction

Every household requires different amount of power for the functioning, but usage of sensor leads to the decrease in the energy consumption which in return saves the

---

R. Kr. Yadav (✉) · S. Verma · P. Singh
Delhi Technological University, New Delhi 110042, India
e-mail: rkyadav@dtu.ac.in

S. Verma
e-mail: shanyaverma03@gmail.com

P. Singh
e-mail: prishitasingh25@gmail.com

© The Editor(s) (if applicable) and The Author(s), under exclusive license to Springer Nature Singapore Pte Ltd. 2021
T. Senjyu et al. (eds.), *Information and Communication Technology for Intelligent Systems*, Smart Innovation, Systems and Technologies 196,
https://doi.org/10.1007/978-981-15-7062-9_74

environment as well as the cost incurred. Due to more controlled and better use of the appliances, the devices even tend to have a longer life span. This paper deals with the temperature regulation of a room based on the occupancy and environmental conditions using digital motion sensors and temperature and humidity sensors. We have compiled different sensors to function together in order to obtain better results. Also, two different cases have been considered, one consisting of two different gates for the entry and exit and the other for similar gate for entering and leaving the room. Moreover, the system becomes adaptive with respect to the usage of the users as a record of the data used is also maintained.

The motion sensors are placed at the gateway of the room. At the entry gateway, sensors detect if one enters the room. Similarly, at the exit gateway, they detect if one leaves the room. If there is only one gate, then one sensor is placed outside the room and one inside the room. The order of activation tells us about the entry or exit. Both the sensors are placed near the gate. The motion sensor detects the change in heat when a person walks into the sensor's range, triggering the device.

This project predicts the temperature of the room according to the number of people present in the room on the basis of the signals received by the motion sensors integrated with the further different parameters such as the presence of mobile phones which produce 1 W of heat. The factor of sunlight is ignored because the temperature in the shade is not cooler than the temperature in the sunlight. Shade only feels cooler because of avoiding solar radiation. Aggregate heat produced by appliances is also considered using the specific heat capacity C, the mass of the substance m and the change in temperature ΔT in the equation:

$$q = m * c * T \quad \text{(Thermodynamic Equation)} \tag{1}$$

Humidity is also an important parameter. Relative humidity is inversely proportional to temperature. The following equation can be used to calculate the heat index or apparent temperature, where Tf represents the air temperature in degrees Fahrenheit and RH denotes the relative humidity expressed as a whole number.

$$\text{HI} = a - b + c - d - e + f - g + h - 42.379 \tag{2}$$

where:

$$a = 2.04901523 * Tf \tag{3}$$

$$b = 0.22475541 * Tf * RH \tag{4}$$

$$c = 10.14333127 * RH \tag{5}$$

$$d = 6.83783 \times 10^{-3} * (Tf^2) \tag{6}$$

$$e = 5.481717 \times 10^{-2} * (RH^2) \qquad (7)$$

$$f = 1.22874 \times 10^{-3} * (Tf^2) * (RH) \qquad (8)$$

$$g = 1.99 \times 10^{-6} * (Tf^2) * (RH^2) \qquad (9)$$

$$h = 8.5282 \times 10^{-4} * (Tf) * (RH^2) \qquad (10)$$

Other parameters considered are the presence of fans. Each and every device is a 100% efficient heater. So, the room temperature is going up with the fans on, even though it may feel like it is cooler. However, the effect is very slight and is almost impossible to measure. The last parameter considered is the type and the number of lights present.

2 Related Work

In [3], researchers proposed the blueprint of the room temperature and humidity controller using fuzzy logic. It proposes a model which consists of two fuzzy logic controllers that control temperature and humidity, respectively. In [4], researchers proposed a method of temperature regulation inside a hospital ward using a computer remote control system. However, the paper fails to consider the factor of occupancy. In [5], researchers studied controlling the heating, ventilation, air-conditioning systems on the basis of the changing location of the occupied. A temperature control system that minimizes energy was created in [6].

3 Proposed Work

3.1 Overview

A typical smart room is stocked with a lot of sensors for measuring house conditions, such as temperature, humidity, light. Temperature and humidity can be measured by a sensor and motion sensors find the number of persons in a room. Here, we have even added different parameters to this project to make it even more useful.

3.2 Methodology

Case of two gates. When two different gates are used for entering and leaving the room.

Brief Description of the Drawings. Figure 1 illustrates how the system functions at the entry gate of the hall. Following steps are followed in one cycle.

1. Firstly, the number of people in the room, represented by n, is initialized to 0.
2. Then, the motion detector detects if there is any movement at the entry gate.
3. If there has been a movement at the entry gate, the motion sensor is monitored. The variable "x" is calculated, which represents the number of people that have just entered the room.

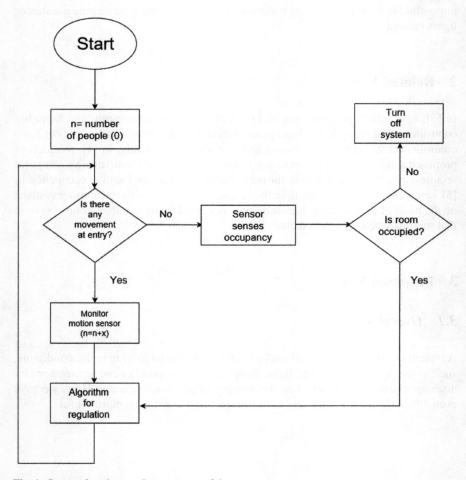

Fig. 1 System functions at the entry gate of the room

4. "*x*" is then added to "*n*" to find the total number of people in the room.
5. If the room is occupied, then the control is passed on to the algorithm for regulation. In case the room is not occupied, then the system is switched off.

The explanation of Fig. 2 is similar to that of Fig. 1. The only difference is that it is a flowchart describing the functioning at the exit gate due to which the number of people leaving the room is detected and reduced. There is a possibility that a person enters from the exit gate and vice versa. To counter this problem, one sensor can be placed outside the room and one inside the room near both the gates. If the outer sensor of the exit gate is activated first, it means that a person is entering from that gate. Similarly, if the inner sensor of the entry is activated first, it means a person is exiting from it.

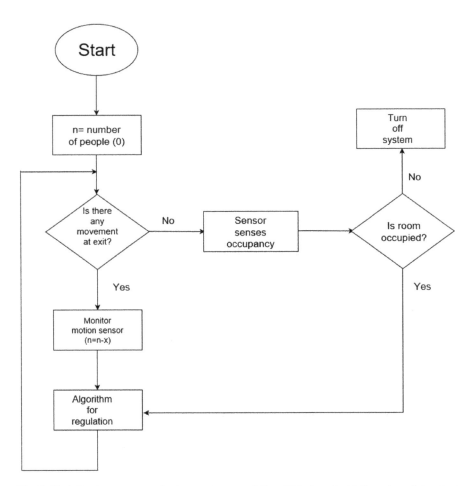

Fig. 2 Variation in occupancy for temperature regulation. This flowchart tells us how the system functions at the exit gate of the room

Case of only one gate. There might be cases when there is only one gate in the room. In such a case, one sensor, let S1, is placed outside of the room and the other; let S2 is placed inside the room near the gate. If S1 is activated before S2, it clearly means that a person has entered the room and if S2 is activated before S1, it means a person is exiting the room. Hence, movement of each person can be traced effectively to keep a record of occupancy.

Algorithm for Regulation. The working of the system is based on energy efficiency. If the number of people occupying the room reduces to zero, then the system is switched off. Whenever a motion is detected, the data is updated but the regulation of temperature is performed five minutes after the observation of the data record. As stated above, other parameters like humidity, etc., will also be considered.

3.3 Implementation

The room considered consisted of ten 20 W LED lights. Other electrical appliances in the room included five fans.

Implementation for occupancy. HC-SR501 PIR motion sensors are placed at the entry and the exit gates of the room. In this way, we can calculate the number of people entering and leaving the room. In the case of one gate, two sensors are used near the gate. Following connections were made:

- The GND of the Arduino Uno is connected to the GND pin of the sensor.
- The digital pin, 3 of the Uno is connected to the OUT pin of the sensor.
- The voltage, 5 V, of the Arduino Uno is connected to the VCC pin of the sensor.

Implementation for temperature and humidity. We have made use of DHT11 Digital Relative Temperature and Humidity Sensor in order to calculate the temperature and humidity. Following connections were made:

- The GND of the Arduino Uno is connected to the negative pin or the GND pin of the sensor.
- The Analog0 or A0 of the Uno is connected to the Data pin of the sensor.
- The voltage, 5 V, of the Arduino Uno is connected to the VCC pin of the sensor.

Implementation for phones, other electronic appliances and lights. It is considered that the number of phones in the room is equal to the number of people in the room. A phone dissipates around 1 W. The heat produced by the appliances in the room is calculated by using the thermodynamic equation, that is, Eq. (1).

4 Result and Discussion

Smart devices installed in the house are beneficial in terms of control. Due to more controlled and better use of these devices, they also have a longer life span. After enabling the smart home system, the differences between the total cost, as well as, savings over a period of time, respectively, will surely decrease.

The motion sensors are placed at the entry and exit gates. Figures 3 and 4 are the snippets of the output obtained at the entry and exit gates, respectively. In this way, the number of people present in the room can be found. Figure 5 is the output in the case of one gate. Sensor 2 is placed outside the room, and sensor 1 is placed inside the room. In both cases, one person is currently occupying the room. So the heat dissipated by mobile phones will be 1 W which is ignored as it is negligible. Also, the room considered consisting of ten 20 W LED lights. Other electrical appliances in the room included five fans. So, the heat dissipated by the lights is ((80/100) * 20) * 10, which is 160 J.

The heat dissipated by the fans is very slight and thus is ignored. Since on an average a human being produces 100–120 J of heat, we take the amount of heat

```
motion detected at 46 sec. Num of people entered are
1
motion ended at 49 sec
---
motion detected at 59 sec. Num of people entered are
2
motion ended at 62 sec
---
motion detected at 75 sec. Num of people entered are
3
motion ended at 78 sec
```

Fig. 3 Snippet for the output obtained from the motion sensor placed at the entry gate

```
motion detected at 32 sec. Num of people exited are
1
motion ended at 35 sec
---
motion detected at 44 sec. Num of people exited are
2
motion ended at 46 sec
```

Fig. 4 Snippet for the output obtained from the motion sensor placed at the exit gate

```
motion detected at2 117 sec
motion ended at2 117 sec
---
motion detected at2 122 sec
---
motion detected at1 127 sec
Num
1
```

```
motion detected at2 150 sec
---
motion detected at1 157 sec
Num
2
motion ended at1 201 sec
motion ended at2 207 sec
---
motion detected at1 210 sec
---
motion detected at2 214 sec
Num
1
```

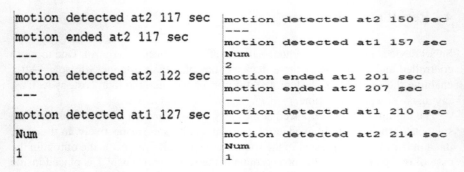

Fig. 5 Output obtained in the case of one gate where motion from 2 to 1 means entry

dissipated to be approximately 100 J. Since, in the given case, when occupancy varies, then the heat dissipated is also recalculated if any change in the data record is found after five minutes of triggering of the motion sensor. Here, the heat dissipated due to occupancy will remain 100 J since occupancy remains constant after a span of five minutes.

Figure 6 is the output obtained from the DHT11 temperature and humidity sensor. It shows the current temperature and humidity of the room. These values can then be used to calculate the final temperature of the room. In this case, the current temperature of the room is 27 °C and humidity is 38.

Fig. 6 Output obtained from the DHT11 temperature and humidity sensor

```
Temperature and Humidity Data
27 *C, 38 H
27 *C, 38 H
27 *C, 38 H
27 *C, 38 H
27 *C, 38 H
27 *C, 38 H
27 *C, 38 H
27 *C, 38 H
27 *C, 38 H
27 *C, 38 H
27 *C, 38 H
27 *C, 38 H
```

An advisable room temperature ranges from 16 to 30 °C. In the further extension of this paper, if the temperature exceeds the range, then a warning or prompt can be sent to the administrator. We know that if we know the relative humidity and the air temperature, then the Eq. (2) can be used to calculate the heat index. So, according to (2), the heat index is 26.8 °C. This means that our body feels as if the temperature is 26.8 °C.

Now, 4184 Watts of energy leads to a temperature rise of 1 °C. In the above results, the total energy produced is 160 W + 100 W, that is, 260 W. So, this leads to a total temperature rise of (260/4184), that is, 0.06 °C.

4.1 Future Scope

In the further extension of the paper, a system can be designed that does not turn off the system when no one is present in the room in order as it might lead to uncomfortable temperature when an occupant enters into the room again after some time. Secondly, identification of the groups entering in parallel can be considered. Thirdly, we can make use of IoT that will offer exemplary benefits. [1, 2, 7].

5 Conclusion

We defined a method to successfully regulate the temperature of the room using motion sensors and temperature and humidity sensors, along with taking other factors into consideration. An algorithm that helps to decide what should be the temperature of the room based on the number of people present was specified.

Smart houses depend on the basis of requiring the least human usage as conceivable, along with keeping the required comfort level. They are without a doubt a promising innovation for the coming time. If the features of a home automation system are extended, it also leads to making the technology significantly more powerful or robust than it is as of now in terms of cost effectiveness. It also leads to the reduction of greenhouse gases, even in the cases of security systems, tracking, remote monitoring and control and more.

References

1. Xu, L., He, W., Li, S.: Internet of things in industries: a survey. In: IEEE Trans. Ind. Inform. **10**(4), 2233–2243 (2014)
2. Shah, S., Yaqoob, I.: A survey: internet of things (IoT) technologies, applications and challenges. IEEE Smart Energy Grid Engineering (SEGE), Canada. In (2016)
3. Das, T., Das, Y.: Design of A room temperature and humidity controller using fuzzy logic. Am. J. Eng. Res. West Bengal, India **2**(11), 86–97 (2013)

4. Lai, X., Zhong, J.: The schematic design of the ward temperature regulation based on computer remote control. In: International Conference on Electronic Information Technology and Intellectualization (ICEITI) (2016)
5. Vosughi, A., Xue, M., Roy, S.: Occupant-location-catered control of IoT-enabled building HVAC systems. IEEE Trans. Control Syst. Technol. 1–9 (2019)
6. Chianese, A., Piccialli, F.: Designing a smart museum when cultural heritage joins IoT. In: 8th International Conference on Next Generation Mobile Apps Services and Technologies, , UK, pp. 300–306 (2014)
7. Malche, T., Maheshwary, P.: Internet of things (IoT) for building smart home system. In: International Conference on I-SMAC (IoT in Social, Mobile, Analytics and Cloud) (I-SMAC), India, pp. 65–70 (2017)

Score Prediction Model for Sentiment Classification Using Machine Learning Algorithms

Priti Sharma and Arvind Kumar Sharma

Abstract With the explosive boom of social media content over the Internet in the recent few years, users now present their views on nearly something in the discussion. There are many microblogging web sites like Twitter, Facebook, LinkedIn, Tumbler, etc. Twitter has grown to be a very famous verbal information exchange tool amongst the users and is one of the most open and simplest platforms to share their sentiments on extraordinary topics. In this context, we acquire the tweets from twitter data and preprocess the tweet that removes irrelevant inappropriate phrases such as name, symbols, etc. It additionally compares every tweet to a database of positive, negative, and impartial words. This paper presents an enhanced score prediction model for sentiment analysis using machine learning as a real-time data processing system for evaluating the public sentiments based on the data of social media Twitter.

1 Introduction

Today, with the emergence of social media, the focus is shifted to this wealth of knowledge where sentiments of people can help in making expert decisions for business growth. Social feedback is the term that refers to the sentiments or opinions of people exchanges over OSNs. Human beings utilized to assess their reviews and opinions of the purchaser earlier than buying any product. Thus, we can say social networks contain hidden opinions of the users. Finding the opinions and sentiments from social media and monitoring them over the Internet is a critical task. Therefore, it needs automated a real-time data processing model for evaluating the public sentiments. Sentiment is the emotion, feeling, opinion, attitude, ideas that show the behaviour of the users. Opinion and sentimental mining are one of the important

P. Sharma (✉) · A. K. Sharma
Career Point University, Kota, India
e-mail: priti.shr18@gmail.com

A. K. Sharma
e-mail: drarvindkumarsharma@gmail.com

research areas due to huge daily posts over social networks, extracting people's opinions is a difficult task. About 90% of today's data is supplied during two recent years and getting perception into this massive-scale statistics is not trivial [1, 2]. The fast increase of Twitter and the public get right of entry of tweets have made twitter a famous research subject. For example, researchers have examined the usage of Twitter in advertising merchandise and sharing consumer's opinions. Through machine learning phenomenon, it has been around and became an established area for helping organizations in making accurate decisions. In this context, the research in this paper is purely based on the data of social media Twitter.

AIM: We propose a methodology for sentiment classification prediction using machine learning as a real-time data processing system for evaluating the public sentiments purely based on the data of social media Twitter. Moreover, we acquire the tweets from twitter data and preprocess the tweets which remove irrelevant inappropriate phrases such as name, symbols, etc. In addition, it compares every tweet to database of positive, negative, and impartial words.

Organization: The remainder of the paper is structured as follows: Sect. 2 presents related work based on sentiment classification methods including machine learning, Sect. 3 describes briefly machine learning approaches including sentiment analysis, Sect. 4 covers proposed methodology and Sect. 5 concludes our paper while references are mentioned at the last.

2 Related Work

In previous years, many researchers are carried out their research works in sentiments analysis using social media. This section reviews literature related to sentiment classification methods including machine learning.

In [3], Tang et al. explored Twitter sentiment classification with the approach known as sentiment-specific word embedding. Recurrent Neural Network (RNN) is employed in [4] for sentiment classification. On the other hand, n-gram machine learning phenomenon is used in [5] for sentiment analysis. Merger of offline and online feedback is the main focus in [6] for analyzing public sentiments. Deep learning based analysis is described in [7]. Sentiment classification with context-sensitive approach is studied in [8]. Sentiment patterns are explored in [9] based on the dependency-based rules. Twitter corpus is used in [10] for opinion mining. Other sentiment-based approaches found in the literature include the dependency graph-based approach [11]. Lexical based approaches are followed in [12].

3 Machine Learning in Sentiment Analysis

Machine learning algorithms are broadly classified into two categories such as- supervised and unsupervised learning. The supervised approach needs training phase while the other category does not need training and uses some sort of similarity measure to learn and perform the intended operation. The most popular supervised learning approaches are employed for sentiment classification of tweets. The algorithms are known as Naive Bayes, SVM, C4.5, and Random Forest. Naive Bayes is a probabilistic classifier that is based on Bayes' theorem. SVM is the classification algorithm that is broadly utilized in data extraction. SVM is binary classifier but supports kernels for multi-class classification as well. C4.5 is an extension of well knows algorithm known as ID3 and it generates decision trees that are used for classification of tweets. Random forest is another algorithm that can be used for classification in machine learning [13]. Moreover, sentiment analysis is incredibly helpful in social media monitoring observing as it enables us to pick up an outline of the more extensive general sentiment behind specific points. Social media monitoring tools are like brand analytics that make that procedure simpler and speedier than any time in recent memory, on account of real-time monitoring abilities. The capacity to disentangle bits of knowledge from social information is a training that is broadly embraced by connotations over the world. Computational investigation of sentiments, opinions, evaluations, attitudes, affects, appraisal, views, emotions, subjectivity, etc., expressed in text. At some point, it is called opinion mining. It has numerous applications extending from ecommerce, marketing to governmental issues, and some other research. Figure 1 shows sentiment analysis of twitter data extracted by the features as corpus building to the model has analyzed the sentiment data by the emoticons of negative, positive, and neutral are categorized.

Fig. 1 Process of Twitter sentiment analysis

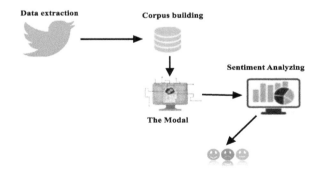

4 Methodology for Sentiment Classification

The proposed research study will help in evaluating the currently available sentiment analysis systems, and machine learning techniques, it will help to find the best possible robust model which can analyze data efficiency and performance of the system. This section provides methodology used to have sentiment classification with the tweets dataset collection from Twitter. Twitter API is used to have live connectivity to twitter and collected tweets. Afterward, the tweets are pre-processed to have the training and testing sets. The following subsections provide more details on the methodology proposed.

4.1 Problem Formulation

Supervised learning methods for sentiment classification are discussed in [3, 5, 14, 10]. There are many specific methods employed for sentiment analysis. They include sentiment-based word embedding [3], N-gram machine learning [5], deep learning [7], dependency-based rules [9], and graph-based approach [11], and lexicon-based method [12]. From the past literature, it is concluded that the methods employed are useful in sentiment classification. However, it is an open problem to have further optimizations and domain-specific investigations to exploit sentiment classification in a better way. It's a challenge to collect social media content and use it for discovering business intelligence. Towards this end, in this paper, we proposed a framework for garnering sentiments for purely based on the data of social media Twitter.

4.2 Proposed Framework

A framework is proposed to guide the research on sentiment classification. Since sentiments are opinions in social media, tweets of social media are collected from Twitter website. Thus, the tweets collected have wealth of knowledge in the form of sentiments of people. Mining such intelligence can boost decision-making process of any organization. Our proposed framework is aimed at helping in sentiment analysis with different machine learning methods. The work flow of proposed framework is shown in Fig. 2.

Fig. 2 Framework for Sentiment classification of Tweets

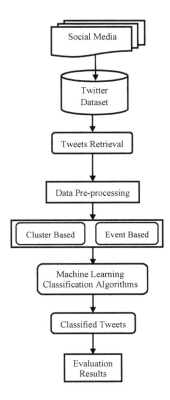

4.2.1 Dataset

Numerous assessments and marked sentiment datasets have been made, particularly for twitter posts and Amazon item surveys. The most popular and widespread are as under:

- Sanders corpus
- Sentiment strength Twitter dataset
- SemEval (Semantic Evaluation) dataset
- Large movie review dataset
- Stanford Twitter sentiment
- Amazon reviews.

4.2.2 Data Pre-processing

The manipulation of the data involves the cleaning process in which the data which are mined from Twitter API streaming is processed to remove the unwanted words from the tweets. As the technique aims tweets in English, the tweets which are transcribed in other patterns are thrown out. The characters which are non-English in the tweets are also deleted. Because of the familiar language type on Tweet, the words that are misspelled can habitually happen. The method cleans the files by eradicating words that do not occur in the emotion bilingual dictionary. Hashtags and consumer references though are retained in the dataset. In preprocessing method, it eradicates undesirable information in the dataset and authenticates the fault in the dataset [15]. Now in preprocess with any undesirable or codes are transformed into insignificant standards and eliminating valueless information. It is so-called data filtering or data pre-processing.

The proposed algorithm for the Pre-processing and Pos-tagging is as follows:

Algorithm 1: Pre-processing and Pos-tagging

```
Input as Ip_dt
Features as Fe_dt
Sentiment value as Se_vl
Initialize:
de_se[] = [0]*len(Fe_dt)
For i←0 to len(Ip_dt)
For j←0 to len(Ip_dt.sentence)
For k←0 to len(Fe_dt)
If
Fe_dt[k] is Ip_dt.sentence[j]
de_se[k] = Se_vl (Ip_dt.sentence[j])
```

4.2.3 Classification of Emotion Prediction

The sentimental analysis is then classified as per the score prediction meter. The score is generated according to the positive, negative, and neutral emotions which are analyzed from the particular tweets and are categorized in this classification process.

The proposed algorithm for the classification of emotion prediction is represented as follows:

Algorithm 2: Classification Emotion Prediction

```
Document level SC(D, r)
D ←Text Cleaner (D)
{s1, s2, sk } ←Get Sentences (D)
for i <- 1, k do
s_score i ←Sentence Level SC(si)
end for
APS ←(∑_{i=1}^{k} s_scorei)_{/k}
if APS >= r then
res ← POS
else
res ←NEC
end if
return res
end procedure
```

Finally, the score for the classified emotions are predicted by means of score predictor to estimate the positive or happy feelings of tweets among the whole tweets which are extracted.

4.3 Evaluation Metrics

The parameters used for performance evaluation of the proposed system are to be taken precision, recall, accuracy, and F-measure. Classification evaluation provides the required business intelligence to twitter social data. This social know-how can help the organization to make strategic decisions. These values are used to have metrics like precision, recall, accuracy, and F-measure. A model is proposed to explore sentiment classification with machine learning methods like Naive Bayes, SVM, C4.5 and Random forest, etc. The model guides the operations in terms of training and testing phases.

Precision
Precision metric helps us to measure how many tweets are estimated correctly of a certain category. It measures the correctness of the classifier. It is larger precision means as low false positives, when the lower precision means high false positives. It is computed using the equation [14]:

$$\text{Precision} = \text{TP}/(\text{TP} + \text{FP}) \qquad (1)$$

Recall
Recall helps us to measure how many tweets are predicted correctly as belongs to given category. The recall is also known as true positive rate (TP) is a proportion of

positive cases which are correctly identified. It is computed using the equation [14]:

$$\text{Recall} = \frac{\text{True Positive}}{\text{True Positive} + \text{False Negative}} \qquad (2)$$

$F1$ Score

F1 score also called as F-Score or F-Measure. $F1$ score is the measure of calculating the weighted average of precision and recall. It is the weighted harmonic mean of the evaluation metric as precision and recall. $F1$ score ranges from 0 to 1 and $F1$ score is considered perfect when it is 1 which means that the model as low false positives and low false negatives. $F1$ score is calculated using the equation [14]:

$$F1 \text{ Score} = 2 \times (\text{Precision} \times \text{Recall})/(\text{Precision} + \text{Recall}) \qquad (3)$$

Accuracy

Accuracy helps to find how many tweets are predicted correctly out of all tweets in datasets. Classifier accuracy of a test set is a percentage of test set that is correctly classified by classifier. Accuracy is measured by identifying the correctness of classified results. It is computed using the equation [14]:

$$\text{Accuracy} = \frac{Tp + Tn}{Tp + Fn + Tn + Fp} * 100\% \qquad (4)$$

5 Conclusion

Sentimental analysis is a way towards deciding if a bit of composing viz products, movie reviews, tweets, etc. is positive, negative, or neutral. Marketers can utilize sentimental analysis to explore the popular assessment of their organization to analyze the satisfactory behaviour of customers. This paper offers score predictor, simultaneous processing of data structure to estimate the sentiment changes which occurs at the perspective of Twitter. Thus, the twitter streaming API dataset is loaded and cleaned by performing some pre-processing algorithms. Then the sentimental analysis has been estimated and emotions of the public are categorized as positive, negative, and neutral. The purpose of this method utilized to take a look at proposed systems with present datasets to be in a position as future work with real-world data.

References

1. Bagheri, H., et al.: Big Data: challenges, opportunities and Cloud based solutions. Int. J. Electr. Comput. Eng. **5**(2), p. 340 (2015)
2. Bagheri, H., et al.: Multi-agent approach for facing challenges in ultra-large scale systems. Int. J. Electr. Comput. Eng. **4**(2), 151 (2014)
3. Tang, D., et al., Learning sentiment-specific word embedding for twitter sentiment classification, pp 1555–1565 (2014)
4. Tang, D., Qin, B., Liu, T.,: Document modeling with gated recurrent neural network for sentiment classification, pp. 1422–1432 (2015)
5. Tripathy, A. et al.: Classification of Sentiment Reviews Using n-Gram Machine Learning Approach. Elsevie, pp. 117–126 (2016)
6. Su, L.Y.F., et al.: Analyzing Public Sentiments Online Combining Human and Computer-Based Content Analysis. Information, Communication and Society, pp. 1–24 (2016)
7. Majumder, N. et al.: Deep learning-based document modeling for personality detection from text. IEEE Intell. Syst. pp. 74–79 (2017)
8. Ren, Y., et al.: Context-sensitive Twitter sentiment classification using neural network, pp. 1–7 (2016)
9. Poria, S.: Sentic patterns: dependency-based rules for concept-level sentiment analysis, pp. 45–63. Elsevier, Amsterdam (2014)
10. Pak, A., Paroubek, P.: Twitter as a corpus for sentiment analysis and opinion mining, p 1320–1326 (2013)
11. Wang, X.: Topic sentiment analysis in Twitter: a graph-based hashtag sentiment classification approach, pp 1–10. ACM (2011)
12. Dang, Y., et al.: A lexicon-enhanced method for sentiment classification: an experiment on online. Prod. Rev. IEEE (2010), pp 1–8
13. Madhuri, K.: A machine learning based framework for sentiment classification: Indian Railways case study. Int. J. Innov. Technol. Explor. Eng. (IJITEE) **8**(4) (2019)
14. Visa, S., Ramsay, B., Ralescu, A., Knaap, E.: Confusion matrix-based feature selection. In: CEUR Workshop Proceedings, pp. 120–127 (2011)
15. Jadon, M. et al.: Sentiment analysis for movies prediction using machine leaning techniques. In: ICICI, LNDECT 38, 2019. Springer Nature, Berlin

CFD Analysis Applied to Hydrodynamic Journal Bearing

Mihir H. Amin, Monil M. Bhamare, Ayush V. Patel, Darsh P. Pandya, Rutvik M. Bhavsar, and Snehal N. Patel

Abstract Finite difference method of computational fluid dynamics has been applied to hydrodynamic journal bearing application. Discretization of momentum and energy equations are done and applied to application considered. The non-dimensional results obtained from the analysis showed temperature variation and pressure distribution for changing different input parameters, speed of bearing, dynamic viscosity, coupled parameter, etc. Focus of current research was to add magnetic parameter to existing in energy and momentum, non-dimensionalize it and convert that equation into finite difference form after applying suitable boundary conditions. Then final equations were coded in commercially available software MATLAB to obtain quantitative results. Suitable validation has been done quantitatively. This has been accompanied by undergoing parametric analysis to input parameters to understand in detail analysis of the present application.

M. H. Amin (✉) · M. M. Bhamare · D. P. Pandya · R. M. Bhavsar
CHARUSAT University, Nadiad-Petlad Road, Changa, Gujarat 388421, India
e-mail: mihiramin219@gmail.com

M. M. Bhamare
e-mail: bhamaremonil@gmail.com

D. P. Pandya
e-mail: darshpandya077@gmail.com

R. M. Bhavsar
e-mail: bhavsarrutvik247@gmail.com

A. V. Patel
BVM Engineering College, Near Bhaikaka Library, Vallabh Vidhyanagar, Anand 388120, India
e-mail: ayushvpatel1510@gmail.com

S. N. Patel
Alumnus, IIT Kharagpur, Kharagpur, India
e-mail: snehalpatel619@gmail.com

© The Editor(s) (if applicable) and The Author(s), under exclusive license to Springer Nature Singapore Pte Ltd. 2021
T. Senjyu et al. (eds.), *Information and Communication Technology for Intelligent Systems*, Smart Innovation, Systems and Technologies 196,
https://doi.org/10.1007/978-981-15-7062-9_76

Nomenclature

C	Clearance in bearing
e	Eccentricity(distance between centres of journal bearing)
F_j	Frictional force appearing in the surface of journal
f	Friction coefficient $f = \frac{F_j}{w} = (C/R)\frac{F}{W}$
F_m	Induced magnetic forces for unit value
h	Thickness of film(Lubricant)
h_m	Intensity of magnetic field
h_{mo}	Peculiar value of intensity of magnetic field
l	Variable of couple stress(CS) $l = \left(\frac{\eta}{\mu}\right)^{1/2}$
L	Dimensionless parameter of CS $L = l/C$
L_b	Bearing length in axial direction
p	Pressure of lubricant
P	Dimensionless pressure
R	Bearing or journal radius
u	Circumferential component of velocity
v	Radial component of velocity
w	Axial component of velocity
w	Load-carrying capacity
W	Load-carrying capacity (dimensionless) $W = \frac{w(C/R)^2}{\mu \omega L_b R}$
X_m	Magnetic fluid susceptibility
(x, y, z)	Cartesian coordinates
φ	Attitude angle
η	Innovative constant of material
μ	Dynamic fluid Vis
θ	X/R
ρ	Lubricant density
ω	Angular velocity
MR	Magnetorheological fluid

1 Introduction

Computational mechanics is the discipline involved with the help of computational methods to understand the principles of fundamental mechanics correlation with applied mathematics computer programming. It is widely used to solve computationally challenging difficulties in science and engineering.

Magnetic fluid contains three basic elements mainly carrier fluid or base fluid, magnetic fragments and a layer of coating on each and every molecules. The agglomeration of the particles is prevented by coating on the particles. Apart from liquid property, governing by an external applied magnetic force plays a very major role in the application of ferrofluid. Ferrofluid can widely resolve many difficult dynamic sealing, lubrication, heat transfer and problems regarding damping as well as broad area cover of high technology like automobile, aeronautical, defence, space and vacuum technology and medical science compare to conventional fluid [1–4].

With the help of momentum and continuity equation, magnetic force and modified Reynolds' equation are acquired, and simultaneously, energy equation is also derived to get temperature and distribution of pressure renders performance of bearing, load-carrying capacity, attribute angle, friction coefficient, friction force. Here, micro-continuum theory of Stokes, a study of hydrodynamic loaded journal bearing with couple stresses, is conducted. It is found that fluids with microstructure are efficient and shows better properties as lubricant compare to Newtonian fluid. Fluids having coupled stresses are considered to be better than the Newtonian fluids [5–10].

Bingham lubricated journal bearing performance characteristics is studied by computational fluid dynamics analysis using Raimondi and Boyd charts.

Simulation with computational fluid dynamics study of MR fluid journal bearing is done. A tool is developed to solve the coupled magnetic rheological flow problem with the presence of magnetic field increase load-carrying capacity but also increases some amount of friction coefficient. Also, suitable meshing techniques are performed and simulated at high eccentricities [11, 12].

Analysis of finite dynamic journal bearing is made with different material (ferrous–ferrous material pair and ferrous non-ferrous material pair combination). Results prove that for ferrous non-ferrous combination, there is less pressure and temperature rise but relatively high wear loss and also included with a view to tribological performance on the experimental aspects of ferrofluid lubrication [13].

Undergoing numerical analysis with writing a code for solving governing differential equations, using finite difference method gave a challenging task to be employed during solving the complex equations. In the present research, focus was more on solving energy equation and getting non-dimensional temperature distribution in journal bearing. This issue was not addressed comprehensively in previous research from the best knowledge of authors, which motivated the authors to solve present problem.

To the best of knowledge, the authors in the present analysis have found out after referring to the literature survey that the majority of previous research has focussed on pressure distribution and that too majority of the literature focussed only on effects of eccentricity on pressure distribution. But authors in present research have given the best try to find out the effects of various parameters aforesaid on not only pressure distribution but also temperature distribution. In addition to this, effect of considering magnetic fluid on pressure and temperature distributions has also been addressed quantitatively. Also, effect of using different types of working fluid on the performance of bearing has been discussed in detail. This novelty in present research has been addressed in detail quantitatively with governing equations explained in detail and method of implementation of the same explained in the form of algorithm developed.

2 Theoretical Analysis

$$F_m = (\nabla \times h_m) \times \bar{B} + \mu_0 M_g (\nabla h_m) \quad (1)$$

The equation indicates the magnetic force due to the sum of force due to induced current and due to the magnetization of magnetic material by applied magnetic field.

$$F_m = \mu_0 M_g (\nabla h_m) \quad (2)$$

The term for force due to induced current from above equation is taken zero as for non-magnetic fluid.

$$F_m = \mu_0 X_m h_m (\nabla h_m) \quad (3)$$

Here, magnetization (M_g) is substituted as product of magnetic intensity (H) and magnetic susceptibility in Eq. (2).

$$\rho \frac{dV}{dt} = -\nabla p + F_m + \frac{1}{2} \nabla \times B + (\mu - \eta \nabla^2) \nabla^2 V \quad (4)$$

$\rho \frac{dV}{dt}$	is inertia term which is relation to motion.
$-\nabla p$	is pressure force.
F_m	is force due to magnetic field.
$\frac{1}{2} \nabla \times B$	is related to body force which includes weight.
$(\mu - \eta \nabla^2) \nabla^2 V$	is combination of viscous force and coupled stress

$$\nabla \cdot V = 0 \quad (5)$$

It indicates the divergence of V and is derived from continuity equation considering density as constant.

This incompressible isothermal fluid is distinctive by **two** things—shear viscosity (μ) and innovative constant—of material (η) which are directly affected by property of couple stress fluid. After neglecting the effect of couples due to body and inertia forces, assuming the hydrodynamic lubrication which is applied to thin film and Eqs. (4) and (5) becomes

$$\frac{\partial p}{\partial x} = F_{mx} + \mu \frac{\partial^2 y}{\partial y^2} - \eta \frac{\partial^4 u}{\partial y^2} \quad (6a)$$

$$\frac{\partial p}{\partial y} = 0 \quad (6b)$$

$$\frac{\partial p}{\partial z} = F_{mz} + \mu \frac{\partial^2 w}{\partial y^2} - \eta \frac{\partial^4 w}{\partial y^2} \quad (6c)$$

where

F_{mx} Magnetic force in circumferential (x) direction
F_{mz} Magnetic force in axial (z) direction

$$\frac{\partial u}{\partial x} + \frac{\partial v}{\partial y} + \frac{\partial w}{\partial z} = 0 \tag{7}$$

at bearing surface,

$$u(x, 0, z) = v(x, 0, z) = w(x, 0, z) = 0 \tag{8a}$$

$$\left.\frac{\partial^2 u}{\partial y^2}\right|_{y=0} = \left.\frac{\partial^2 w}{\partial y^2}\right|_{y=0} = 0 \tag{8b}$$

$$u(x, h, z) = v(x, h, z) = w(x, h, z) = 0 \tag{9a}$$

$$\left.\frac{\partial^2 u}{\partial y^2}\right|_{y=h} = \left.\frac{\partial^2 w}{\partial y^2}\right|_{y=h} = 0 \tag{9b}$$

Equations (8a) and (9a) are the boundary condition of no slip. And Eqs. (8b) and (9b) derived from CS at the surface of solid. After integrating Eqs. (6a) and (6c) having boundary conditions, the velocity circumferential velocity and radial velocity can be obtained as:

$$u = \omega R \frac{y}{h} + \frac{1}{2\mu}\left(\frac{\partial p}{\partial x} - F_{mx}\right) \times \left[y(y-h) + 2l^2\left\{1 - \frac{\cosh\left(\frac{2y-h}{2l}\right)}{\cosh\left(\frac{h}{2l}\right)}\right\}\right] \tag{10}$$

where $l = \left(\frac{\eta}{\mu}\right)^{\frac{1}{2}}$

$$w = \frac{1}{2\mu}\left(\frac{\partial p}{\partial z} - F_{mz}\right)\left[y(y-h) + 2l^2\left\{1 - \frac{\cosh\left(\frac{2y-h}{2l}\right)}{\cosh\left(\frac{h}{2l}\right)}\right\}\right] \tag{11}$$

3 Numerical Analysis

Integrating Equation (7) from 0 to h, we get,

$$\int_0^h \left(\frac{\partial u}{\partial x} + \frac{\partial v}{\partial y} + \frac{\partial w}{\partial z}\right) dy = 0 \tag{12}$$

When Eqs. (10) and (11) are substituted in (12), we get,

$$\frac{1}{12\mu}\frac{\partial}{\partial x}\left[F_{mx}\left\{-12l^2 h + h^3 + 24l^3 \tanh\left(\frac{h}{2l}\right)\right\}\right]$$
$$+ \frac{\partial h}{\partial x} \times \frac{\omega R}{2} - \frac{1}{12\mu}\frac{\partial}{\partial z}\left[\left\{-12l^2 h + h^3 + 24l^3 \tanh\left(\frac{h}{2l}\right)\right\}\frac{\partial P}{\partial z}\right]$$
$$+ \frac{1}{12\mu}\frac{\partial}{\partial z}\left[F_{mz}\left\{-12l^2 h + h^3 + 24l^3 \tanh\left(\frac{h}{2l}\right)\right\}\right]$$
$$- \frac{1}{12\mu}\frac{\partial}{\partial x}\left[\frac{\partial P}{\partial x}\left\{-12l^2 h + h^3 + 24l^3 \tanh\left(\frac{h}{2l}\right)\right\}\right] = 0 \tag{13}$$

$$\frac{\partial}{\partial x}\left(g(h,l)\frac{\partial p}{\partial x}\right) + \frac{\partial}{\partial z}\left(g(h,l)\frac{\partial p}{\partial z}\right) = 6\mu\omega R\frac{\partial h}{\partial x} + \frac{\partial}{\partial x}[g(h,l)F_{mx}] + \frac{\partial}{\partial z}[g(h,l)F_{mz}] \tag{14}$$

$$F_{mx} = \mu_o X_m h_m \frac{\partial h_m}{\partial x}, \quad F_{mz} = \mu_o X_m h_m \frac{\partial h_m}{\partial z} \tag{15}$$

Substituting (15) in (14)

$$\frac{\partial}{\partial x}\left\{\frac{g(h,l)}{\mu}\mu_0 X_m h_m \frac{\partial h_m}{\partial x}\right\} - \frac{\partial}{\partial x}\left\{\frac{g(h,l)}{\mu}\frac{\partial P}{\partial x}\right\} + \frac{\partial}{\partial z}\left\{\frac{g(h,l)}{\mu}\mu_0 X_m h_m \frac{\partial h_m}{\partial z}\right\}$$
$$- \frac{\partial}{\partial z}\left\{\frac{g(h,l)}{\mu}\frac{\partial P}{\partial z}\right\} + 6\omega R\frac{\partial h}{\partial x} = 0 \tag{16}$$

Formulation of energy equation,

$$\frac{DT}{Dt} = \frac{\partial T}{\partial t} + u\frac{\partial T}{\partial x} + v\frac{\partial T}{\partial y} + w\frac{\partial T}{\partial z}$$

where

$\frac{D}{Dt} = \frac{\partial}{\partial t} + u\frac{\partial}{\partial x} + v\frac{\partial}{\partial y} + w\frac{\partial}{\partial z}$ is the total derivative

C_p form of the equation is,

$$\rho C_p \frac{DT}{Dt} = k\nabla^2 T + \mu\emptyset + \frac{J^2}{\sigma} \tag{17}$$

Where

$\rho C_p \frac{DT}{Dt}$ is transient or time dependent temperature.
$k\nabla^2 T$ indicates temperature as function of position.

$\mu\emptyset$ is (heat source) due to viscous dissipation.
$\frac{J^2}{\sigma}$ is (heat source) due to magnetic field which may be attributed as Joule's heating.

Here,

$$\nabla^2 T = \frac{\partial^2 T}{\partial x^2} + \frac{\partial^2 T}{\partial y^2} + \frac{\partial^2 T}{\partial z^2}$$

And,

$$\emptyset = \left\{ \left(\frac{\partial u}{\partial y} + \frac{\partial v}{\partial x}\right)^2 + \left(\frac{\partial w}{\partial x} + \frac{\partial u}{\partial z}\right)^2 + \left(\frac{\partial v}{\partial z} + \frac{\partial w}{\partial y}\right)^2 \right\}$$
$$- 0.67 \left\{ \frac{\partial u}{\partial x} + \frac{\partial v}{\partial y} + \frac{\partial w}{\partial z} \right\}^2 + 2 \left\{ \left(\frac{\partial v}{\partial y}\right)^2 + \left(\frac{\partial w}{\partial z}\right)^2 + \left(\frac{\partial u}{\partial x}\right)^2 \right\} \quad (18)$$

Substituting $\left(\frac{\partial u}{\partial x} + \frac{\partial v}{\partial y} + \frac{\partial w}{\partial z}\right) = 0$ and $\frac{\partial u}{\partial x} = \frac{\partial w}{\partial x} = 0$ (as x-axis is parallel to bearing axis) and assuming $\frac{\partial v}{\partial x} = \frac{\partial v}{\partial y} = \frac{\partial v}{\partial z} = 0$, we get

$$\emptyset = 2 \left\{ \left(\frac{\partial w}{\partial y}\right)^2 + \left(\frac{\partial u}{\partial y}\right)^2 \right\} \quad (19)$$

Now, J (current density) is given as,

$$J = \sigma \left(E + \vec{V} \times \vec{B} \right) \quad (20)$$

where

$$\vec{V} = u\hat{i} + v\hat{j} + w\hat{k} \quad (21a)$$

$$\vec{B} = B_1\hat{i} + B_2\hat{j} + B_3\hat{k} \quad (21b)$$

Here, $B_x = B_z = 0$ and $B_y =$ constant which is assumed with $E = 0$, we get

$$\therefore J = \sigma \left(\vec{V} \times \vec{B} \right) \quad (22)$$

Resolving the bracket term of Eq. (22),

$$\therefore \vec{V} \times \vec{B} = \begin{vmatrix} \hat{i} & \hat{j} & \hat{k} \\ u & v & w \\ 0 & B_0 & 0 \end{vmatrix} = u B_0 \hat{k} - w B_0 \hat{i} \quad (23)$$

$$J^2 = J \cdot J \tag{24}$$

With $J = \left(uB_0\hat{k} - wB_0\hat{i}\right)\sigma$

Substituting Eq. (22) and (23) in (24),

$$\therefore J^2 = \sigma^2\left[u^2 B_o^2 + w^2 B_o^2\right] \tag{25}$$

$$\therefore J^2 = \sigma^2 B_0^2\left[u^2 + w^2\right] \tag{26}$$

$$\therefore \frac{J^2}{\sigma} = \sigma B_0^2\left[u^2 + w^2\right] \tag{27}$$

Substituting Eqs. (19) and (27) in (17), energy equation is given as,

$$\rho C_p \frac{DT}{Dt} = k\nabla^2 T + 2\mu\left\{\left(\frac{\partial u}{\partial y}\right)^2 + \left(\frac{\partial w}{\partial y}\right)^2\right\} + \sigma B_0^2\left[u^2 + w^2\right] \tag{28}$$

where

$$\frac{\partial T}{\partial t} = \frac{\partial T}{\partial t} + u\frac{\partial T}{\partial x} + w\frac{\partial T}{\partial z};$$
$$\nabla^2 T = \frac{\partial^2 T}{\partial x^2} + \frac{\partial^2 T}{\partial z^2} \tag{29}$$

The modified Reynolds equation is given as,

$$\frac{\partial}{\partial \theta}\left(G(H,L)\frac{\partial P}{\partial \theta}\right) + \frac{1}{4\lambda^2}\frac{\partial}{\partial Z}\left(G(H,L)\frac{\partial P}{\partial Z}\right) = 6\frac{\partial H}{\partial \theta} + 4\lambda^2\gamma\frac{\partial}{\partial \theta}\left[G(H,L)H_m\frac{\partial H_m}{\partial \theta}\right]$$
$$+ \gamma\frac{\partial}{\partial Z}\left[G(H,L)H_m\frac{\partial H_m}{\partial Z}\right] \tag{30}$$

where

$$\frac{\partial}{\partial \theta}\left(G(H,L)\frac{\partial P}{\partial \theta}\right) + \frac{1}{4\lambda^2}\frac{\partial}{\partial Z}\left(G(H,L)\frac{\partial P}{\partial Z}\right) = 6\frac{\partial H}{\partial \theta}$$
$$+ 4\lambda^2\gamma\frac{\partial}{\partial \theta}\left[G(H,L)H_m\frac{\partial H_m}{\partial \theta}\right] + \gamma\frac{\partial}{\partial Z}\left[G(H,L)H_m\frac{\partial H_m}{\partial Z}\right]$$

H non-dimensional thickness of film $H = h/C$
H_m non-dimensional intensity of magnetic field

$$\left(H_m = \frac{h_m}{h_{mo}}\right)$$

where L becomes 0

$$\frac{\partial}{\partial \theta}\left(H^3 \frac{\partial P}{\partial \theta}\right) + \frac{1}{4\lambda^2}\frac{\partial}{\partial Z}\left(H^3 \frac{\partial P}{\partial Z}\right) = 6\frac{\partial H}{\partial \theta}$$
$$+ 4\lambda^2 \gamma \frac{\partial}{\partial \theta}\left[H^3 H_m \frac{\partial H_m}{\partial \theta}\right] + \gamma \frac{\partial}{\partial Z}\left[H^3 H_m \frac{\partial H_m}{\partial Z}\right] \quad (31)$$

Now,

$$\bar{y} = \frac{y}{c} \quad (32a)$$

$$\bar{T} = \frac{T - T_a}{T_w - T_a} \quad (32b)$$

where

T_a ambient temperature
T_w wall temperature

$$\bar{t} = \frac{Ut}{C} \quad (32c)$$

U velocity (maximum)
C distance.

Similarly, velocity of fluid in z-direction is,

$$\bar{w} = \frac{w}{U}; \quad (33)$$

where ω = Velocity of fluid in z-direction.
Position in z is given by,

$$\bar{Z} = \frac{z}{L} \quad (34)$$

where

L maximum length in Z-direction.

$$\bar{\theta} = \frac{x}{R} \quad (35)$$

Obtaining Eq. (32b) in the form of T and taking partial derivative gives,

$$\therefore \frac{\partial T}{\partial t} = \frac{\partial \bar{T}}{\partial t}(T_\omega - T_a) \qquad (36)$$

From Eq. (32c) and then substituting it in Eq. (36), we get

$$\therefore \frac{\partial T}{\partial t} = \frac{\partial \bar{T}}{\partial \bar{t}}\frac{U}{C}(T_\omega - T_a) \qquad (37)$$

Similarly, partial derivative of T with respect to x is,

$$\therefore \frac{\partial T}{\partial x} = \frac{\partial \bar{T}}{\partial \bar{\theta}}\frac{(T_\omega - T_a)}{R} \qquad (38)$$

Similarly, partial derivative of T with respect to z is,

$$\frac{\partial T}{\partial Z} = \frac{\partial \bar{T}}{\partial Z}(T_\omega - T_a) = \frac{\partial \bar{T}}{\partial \bar{Z}}\frac{(T_\omega - T_a)}{L} \qquad (39)$$

Partially differentiating Eq. (38) with respect to x we get,

$$\frac{\partial^2 T}{\partial x^2} = \frac{(T_\omega - T_a)}{R^2}\frac{\partial^2 \bar{T}}{\partial \bar{\theta}^2} \qquad (40)$$

Similarly, partially differentiating Eq. (39) with respect to z,

$$\frac{\partial^2 T}{\partial z^2} = \frac{(T_\omega - T_a)}{L^2}\frac{\partial^2 \bar{T}}{\partial \bar{z}^2} \qquad (41)$$

Now, variation in velocity is needed to be non-dimensional

$$\frac{\partial u}{\partial y} = \frac{\partial \bar{u}}{\partial y} \times U = \frac{\partial \bar{u}}{\partial \bar{y} \times C} \times U = \frac{\partial \bar{u}}{\partial \bar{y}} \times \left(\frac{U}{C}\right) \qquad (42)$$

Similarly,

$$\frac{\partial w}{\partial y} = \frac{\partial \bar{w}}{\partial \bar{y}} \times \left(\frac{U}{C}\right) \qquad (43)$$

Substituting Eqs. (36), (37), (38), (39), (40), (41), (42) in Eq. (28),

$$\rho c_p \frac{DT}{Dt} = \rho c_p \left(\frac{\partial \bar{T}}{\partial \bar{t}}\frac{U}{C}(T_\omega - T_a) + \frac{\partial \bar{T}}{\partial \bar{\theta}}\frac{(T_\omega - T_a)}{R}U\bar{u} + \frac{\partial \bar{T}}{\partial \bar{Z}}\frac{(T_\omega - T_a)}{L}U\bar{w}\right) \qquad (44)$$

Also,

$$k\nabla^2 T = k\left(\frac{(T_\omega - T_a)}{R^2}\frac{\partial^2 \bar{T}}{\partial \bar{\theta}^2} + \frac{(T_\omega - T_a)}{L^2}\frac{\partial^2 \bar{T}}{\partial \bar{z}^2}\right) \quad (45)$$

With,

$$\mu\left\{\left(\frac{\partial u}{\partial y}\right)^2 + \left(\frac{\partial w}{\partial y}\right)^2\right\} = 2\mu\left\{\left(\frac{\partial \bar{u}}{\partial \bar{y}}\right)^2 \frac{U^2}{C^2} + \left(\frac{\partial \bar{w}}{\partial \bar{y}}\right)^2 \frac{U^2}{C^2}\right\} \quad (46)$$

And

$$\sigma B_0^2(u^2 + w^2) = U^2 \sigma B_0(\bar{u}^2 + \bar{w}^2) \quad (47)$$

Adding Eqs. (44), (45), (46), (47) and rearranging,

$$\left\{\frac{\partial \bar{T}}{\partial \bar{\theta}}\frac{\bar{u}}{R} + \frac{\partial \bar{T}}{\partial \bar{Z}}\frac{\bar{\omega}}{L} + \frac{\partial \bar{T}}{\partial \bar{t}}\frac{1}{C}\right\} = \frac{\sigma B_0^2 U}{\rho c_p (T_\omega - T_a)}(\bar{u}^2 + \bar{w}^2)$$
$$+ \frac{k}{\rho c_p}\left[\frac{1}{UR^2}\frac{\partial^2 \bar{T}}{\partial \bar{\theta}^2} + \frac{\partial^2 \bar{T}}{\partial \bar{z}^2}\frac{1}{L^2}\right]$$
$$+ \frac{U \times 2\mu}{C^2 \rho c_p (T_\omega - T_a)}\left\{\left(\frac{\partial \bar{u}}{\partial \bar{y}}\right)^2 + \left(\frac{\partial \bar{w}}{\partial \bar{y}}\right)^2\right\} \quad (48)$$

H is only a function of θ. While H_m is not a function of θ. Using these to reduce Eq. (31), nonlinear PDE, cannot be solved analytically so finite difference technique used to discretize above momentum conservation

$$3H^2\frac{\partial H}{\partial \theta}\frac{\partial P}{\partial \theta} + \frac{H^3}{4\lambda^2} \cdot \frac{\partial^3 P}{\partial \theta^2} + H^3\frac{\partial^2 P}{\partial \theta^2} = 6\frac{\partial H}{\partial \theta}$$
$$+ 4\lambda^2 \gamma \times 0 + \gamma H^3\left[\left(\frac{\partial H_m}{\partial z}\right)\left(\frac{\partial H_m}{\partial z}\right) + H_m\frac{\partial^2 H_m}{\partial z^2}\right] \quad (49)$$

By Taylor Series expansion,

$$f(x + \Delta x) = f(x) + \frac{\partial f}{\partial x}\Delta x + \frac{\partial^2 f}{\partial x^2}\frac{(\Delta x)^2}{2!} + \frac{\partial^3 f}{\partial x^3}\frac{(\Delta x)^3}{3!} + \cdots. \quad (50)$$

$$f(x - \Delta x) = f(x) - \frac{\partial f}{\partial x}\Delta x + \frac{\partial^2 f}{\partial x^2}\frac{(\Delta x)^2}{2!} - \frac{\partial^3 f}{\partial x^3}\frac{(\Delta x)^3}{3!} + \cdots. \quad (51)$$

$$f(x + \Delta x) - f(x - \Delta x) = 2\frac{\partial f}{\partial x}\Delta x + 2\frac{\partial^3 f}{\partial x^3}\frac{(\Delta x)^3}{3!} + \cdots. \quad (52)$$

Rearranging,

$$\therefore \frac{\partial f}{\partial x} = \frac{f(x + \Delta x) - f(x - \Delta x)}{2\Delta x}$$
$$- 2\frac{\partial^3 f}{\partial x^3}\frac{(\Delta x)^2}{3!} - 2\frac{\partial^5 f}{\partial x^5}\frac{(\Delta x)^4}{5!} + \cdots. \quad (53)$$

$$\frac{\partial f}{\partial x} = \frac{f_{k+1} - f_{k-1}}{2\Delta x} + O(\Delta x)^2 \quad (54)$$

where $O(\Delta x)^2$ is truncated series at power 2.

Adding (50) and (51),

$$f_{k+1} + f_{k-1} = 2f_k + 2\frac{\partial^2 f}{\partial x^2}\frac{(\Delta x)^2}{2!} + 2\frac{\partial^4 f}{\partial x^4}\frac{(\Delta x)^4}{4!} + \cdots. \quad (55)$$

$$\therefore \frac{f_{k+1} + f_{k-1} - 2f_k}{(\Delta x)^2} + O(\Delta x)^2 = \frac{\partial^2 f}{\partial x^2} \quad (56)$$

Now using Eq. (55) and (57),

$$\frac{\partial P}{\partial \theta} = \frac{P_{k+1,m} - P_{k-1,m}}{2\Delta \theta} \quad (57)$$

Similarly from other equations,

$$\frac{\partial P}{\partial z} = \frac{P_{k,m+1} - P_{k,m-1}}{2\Delta z} \quad (58)$$

$$\frac{\partial^2 P}{\partial z^2} = \frac{P_{k,m+1} + P_{k,m-1} - 2P_{k,m}}{(\Delta z)^2} \quad (59)$$

$$\frac{\partial^2 P}{\partial \theta^2} = \frac{P_{k+1,m} + P_{k-1,m} - 2P_{k,m}}{(\Delta \theta)^2} \quad (60)$$

Substituting Eq. (57), (58), (59) and (60) in Eq. (49),

$$3H_{km}^2 \left[\left(\frac{M_{k+1,m} - H_{k-1,m}}{\neg_1 a} \right) \left(\frac{P_{k+1,m} - P_{k-1,m}}{1, 10} \right) \right]$$
$$+ \frac{H_{k,m}^3}{H\lambda^2} \left[\frac{P_{k,m+1} + P_{k,m-1} - 2P_{km}}{(\Delta z)^2} \right]$$
$$+ H_{k,m}^2 \left[\frac{P_{k+1,m} + P_{k-1,m} - 2P_{k,m}}{(\Delta \theta)^2} \right] = 6 \times \frac{H_{k+1,m} - H_{k-1,m}}{2\Delta \theta}$$
$$+ \gamma H_{k,m}^3 \left[\left(\frac{M_{m_{k,1n+1}} - H_{m_{k,1n-1}}}{2\Delta z} \right)^2 + H_m \left\{ \frac{R_{m_{k,m+1}} + K_{m_{k,m-1}} - 2K_{m_{k,m}}}{(\Delta z)^2} \right\} \right]$$
$$(61)$$

Discretization of energy equation,

$$\frac{\partial \bar{T}}{\partial \bar{\theta}} = \frac{\bar{T}_{k+1,m} - \bar{T}_{k-1,m}}{2\Delta\bar{\theta}} \quad \text{(Central difference)} \tag{62}$$

$$\frac{\partial^2 \bar{T}}{\partial \bar{\theta}^2} = \frac{\bar{T}_{k+1,m} + \bar{T}_{k-1,m} - 2\bar{T}_{k,m}}{(\Delta\bar{\theta})^2} \quad \text{(Central difference scheme)} \tag{63}$$

$$\frac{\partial \bar{T}}{\partial \bar{z}} = \frac{\bar{T}_{k,m+1} - \bar{T}_{k,m-1}}{2\Delta\bar{z}} \tag{64}$$

$$\frac{\partial^2 \bar{T}}{\partial \bar{z}^2} = \frac{\bar{T}_{k,m+1} + \bar{T}_{k,m-1} - 2\bar{T}_{k,m}}{(\Delta\bar{z})^2} \tag{65}$$

Equation (48) can be written as,

$$\frac{\bar{u}}{R}\frac{\bar{T}_{k+1,m} - \bar{T}_{k-1,m}}{2(\Delta\bar{\theta})} + \frac{\bar{w}}{L}\frac{\bar{T}_{k,m+1} - \bar{T}_{k,m-1}}{2(\Delta\bar{z})} + \frac{T_{n+1} - T_{n-1}}{2(\Delta t)} = \frac{\mu\phi}{u(T_w - T_a)}$$
$$+ \frac{k}{\rho C_p}\left(\frac{1}{R^2 u}\frac{\bar{T}_{k+1,m} + \bar{T}_{k-1,m} - 2\bar{T}_{k,m}}{(\Delta\theta)^2}\right) + \frac{\sigma B_0^2 U}{\rho c_p(T_\omega - T_a)}(\bar{u}^2 + \bar{w}^2) \tag{66}$$

Equation (61) is discretized form of momentum equation which is in finite difference form, and Eq. (66) is discretized form of energy equation which is also done using finite difference methods of computational fluid dynamics. Suitable boundary conditions are also given in the above explanations. The following algorithm has been applied to solve Eqs. (61) and (66) to get the desired results.

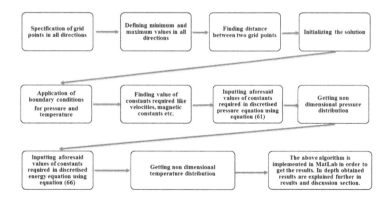

4 Validation

Reduced form of Reynolds equation is given as follows:

$$\frac{\partial}{\partial \bar{x}}\left(\bar{h}^3 \frac{\partial \bar{p}}{\partial \bar{x}}\right) + \frac{X^2}{Y^2}\frac{\partial}{\partial \bar{y}}\left(\bar{h}^3 \frac{\partial \bar{p}}{\partial \bar{y}}\right) = \frac{C}{X}\frac{\partial \bar{h}}{\partial \bar{x}} \quad (67)$$

This equation in finite difference method can be written as follows:

$$V\bar{p}_{i,j} = \left[\frac{\overline{h^2_{i+0.5,j}}}{\left(\bar{h}_{i+0.5,j} + \bar{h}_{i-0.5,j} + 2\bar{h}^3_{i,j}\frac{X^2}{Y^2}\frac{\Delta x^2}{\Delta y^2}\right)}\bar{p}_{i+1,j} \right.$$
$$+ \frac{\bar{h}^3_{i-0.5,j}}{\left(\bar{h}_{i+0.5,j} + \bar{h}_{i-0.5,j} + 2\bar{h}^3_{i,j}\frac{X^2}{Y^2}\frac{\Delta \bar{x}^2}{\Delta g^2}\right)}\bar{p}_{i-1,j}$$
$$+ \frac{\bar{h}^3_{i,j}\frac{X^2}{Y^2}\frac{\Delta x^2}{\Delta y^2}}{\left(+2\bar{h}^2_{i,j}\frac{X^2}{Y^2}\frac{\Delta \bar{x}^2}{\Delta y^2}\right)}(\bar{p}_{i,j+1} + \bar{p}_{i,j-1})$$
$$\left. - \frac{\Delta \bar{x} C}{2X}\frac{(\bar{h}_{i+1,j} + \bar{h}_{i-1,j})}{\left(\bar{h}_{i+0.5,j} + \bar{h}_{i-0.5,j} + 2\bar{h}^3_{i,j}\frac{X^2}{Y^2}\frac{\Delta \bar{x}^2}{\Delta y^2}\right)}\right] \quad (68)$$

Non-dimensional pressure distribution obtained by solving momentum equation by finite difference method has been matched exactly with analytical result as shown in Fig. 1.

This justifies the usage of finite difference method in solving governing equations in present analysis.

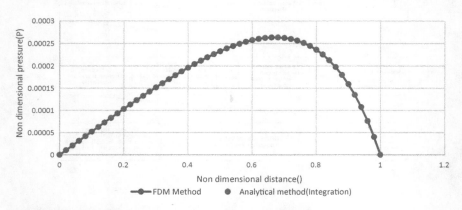

Fig. 1 Validation of finite difference method (FDM) versus analytical method

CFD Analysis Applied to Hydrodynamic Journal Bearing 769

5 Results and Discussion

The pressure and temperature distribution obtained from the above study for variation in quantities like coupled stress parameter, alpha, radius, axial length, theta, eccentricity, clearance and results obtained are shown below (Figs. 2 and 3).

Coupled stress theory was developed in 1966 wherein rotation field of a fluid is in terms of velocity field of the fluid [14]. Whenever complex fluids like the one used for present analysis which is magnetic fluid is used, then coupled stress theory developed by Stokes is used. Results in Fig. 4 suggest that with the increase in coupled stress parameter, the pressure distribution in the direction of the circumference at middle plane increases, when all other input parameters are kept constant. Actually, pressure

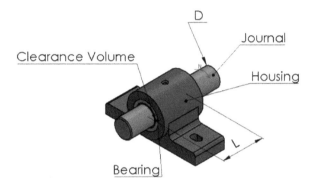

Fig. 2 Journal bearing

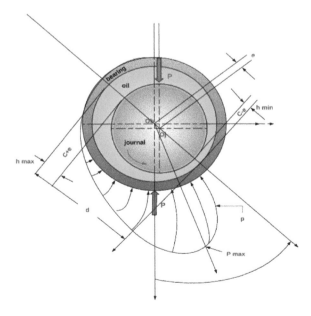

Fig. 3 Hydrodynamic journal bearing

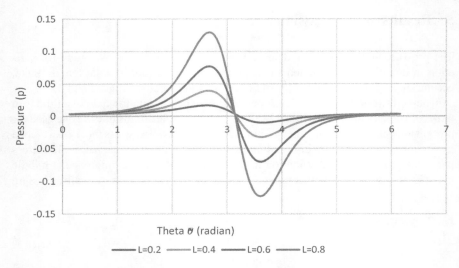

Fig. 4 Variation of pressure with angle (θ) for different lengths

is force per unit area, and it is also a type of stress in reality. Also, couple stress parameter is actually addition of stress/force to the working fluid inside the bearing. So, this is the reason of increase in pressure with increase in couple stress parameter. The trend of each curve is such that suction and compression of working fluid are observed from pressure distribution behaviour along the 360° angle of bearing.

It can be observed from Fig. 5 that with the increase in magnetic coefficient parameter, the pressure distribution in the direction of circumference at middle plane increases. If the value of alpha is 0, it means that fluid is non-magnetic. In addition

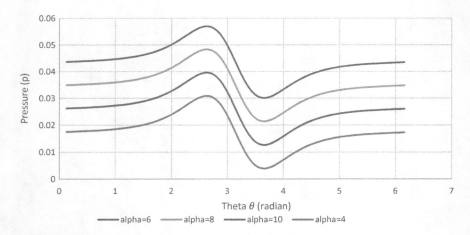

Fig. 5 Variation of pressure with angle (θ) for different value of alpha

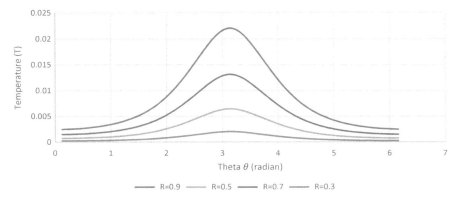

Fig. 6 Variation of temperature with angle (θ) for different radius

to this, there is a shift in maximum pressure or peak pressure towards left and right w.r.t angle, during compression and suction, respectively.

Figure 6 shows non-dimensional temperature variation with the angle theta. The trend of each curve shows that maximum non-dimensional temperature is achieved at a point where film thickness of working fluid is minimum. The point where film thickness is minimum, pressure is highest. This causes high temperature. It is very necessary to identify this place so that dissipation of heat from this point is done and temperature can be set in required limits. Also, with increase in radius of bearing, temperature increases. This is because increase in radius of bearing causes dissipation of heat difficult. So it is observed that with the increase in radius, the temperature distribution at central point increases, when all other input parameters are kept constant.

Increase in axial length increases non-dimensional temperature distribution which can be observed from Fig. 7. The trend in Fig. 7 is similar to Fig. 6 which has the same explanation of highest achievable temperature at minimum film thickness. Increase in axial length too increases the temperature at any point along angle theta. As heat dissipation reduces with increase in axial length of bearing, temperature increases with increase in axial length.

Figure 8 shows the effect of change in non-dimensional temperature with clearance of bearing. Trend in Fig. 8 shows maximum temperature achieved at minimum film thickness. With the reduction in clearance, pressure inside the bearing increases which causes an increase in temperature of working fluid. This qualitative observation is evident from Fig. 8 quantitatively. So it is observed that with the increase in bearing clearance, the temperature distribution at central point decreases.

Due to increase in eccentricity, film thickness varies considerably. Higher eccentricity causes reduction in film thickness in areas in working fluid where distance between journal and bearing is reduced. Simultaneously, distance between journal and bearing in other areas increases. Overall effect is larger pressure in areas of lower film thickness and lower pressure in areas in higher film thickness. This causes higher

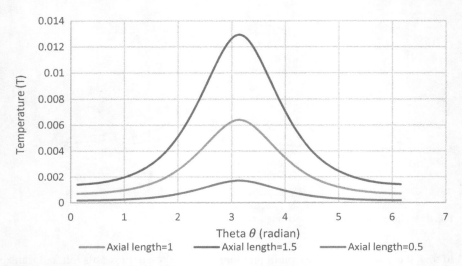

Fig. 7 Variation of temperature with angle (θ) for different axial lengths

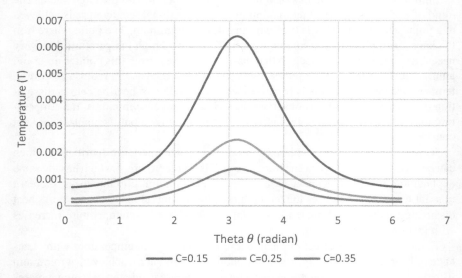

Fig. 8 Variation of temperature with angle (θ) for different clearance of bearing

temperature in areas in working fluid with larger pressure with lower film thickness and vice versa. This trend can be observed from Figs. 9, 10 and 11 clearly.

Increase in angular velocity of journal increases temperature inside working fluid. This is because velocity gradient in working fluid increase with increase in angular velocity of journal. This causes larger stresses in working fluid which increase the

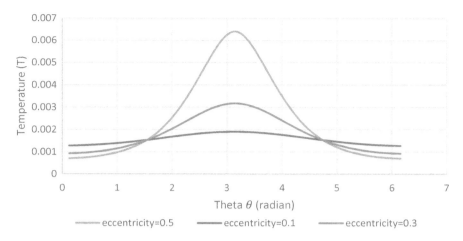

Fig. 9 Variation of temperature with angle (θ) for different eccentricities

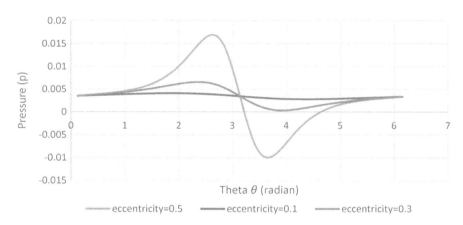

Fig. 10 Variation of pressure with angle (θ) for different eccentricities

temperature of working fluid. So, increase in angular velocity increase the temperature of working fluid along angle theta which is evident from Fig. 12. So, from the above figure it is observed that with the increase in angular velocity (omega), the temperature distribution at central point increases. This is because with the increase in angular velocity (omega), temperature distribution at central point increases.

Using higher viscous working fluid may increase life of bearing, but also increase the temperature as stresses increase with increase in fluid dynamic viscosity. This trend can be observed from Fig. 13. So, from the above figure it is observed that with the increase in fluid viscosity (μ), the temperature distribution at central point increases. This is because with the increase in fluid viscosity (μ), temperature distribution at central point increases.

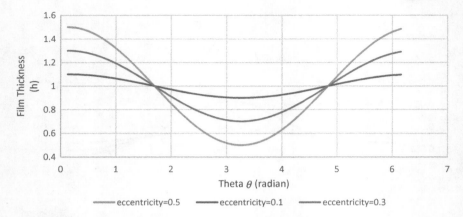

Fig. 11 Variation of film thickness with angle (θ) for different eccentricities

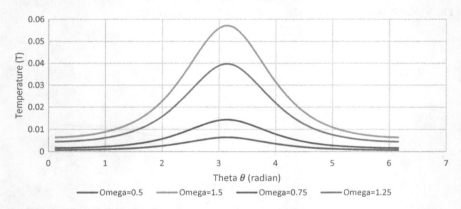

Fig. 12 Variation of temperature with angle (θ) for different values of omega

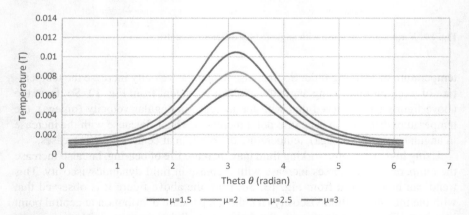

Fig. 13 Variation of temperature with angle (θ) for different values of fluid viscosity

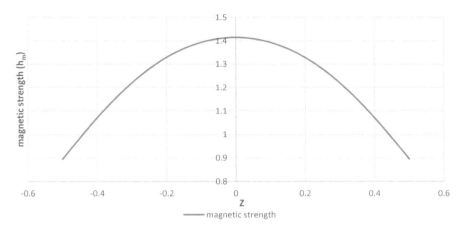

Fig. 14 Variation of magnetic strength (h_m) with axial distance (z)

It can be observed from Fig. 14 that magnetic strength along z-axis goes to maximum at centre. Since, from the nature of equation of magnetic strength, magnetic field is maximum at centre, this goes perfectly with result obtained in Fig. 14. So as the magnetic strength along z-axis goes to maximum at centre.

6 Conclusion

A complete analysis using techniques of computational fluid dynamics is applied to journal bearing application including magnetic fluid as a working fluid. Results obtained were in the form of temperature and pressure distributions using parametric analysis. Various input parameters were altered, and its effect on temperature and pressure distribution has been observed. Complete theoretical analysis applied to bearing application has been presented comprehensively. Effects of various input parameters on non-dimensional temperature distribution have been addressed quantitatively. Fluids having coupled stresses are considered to be better than the Newtonian fluids. Hydrodynamic journal bearings have shown better performance with non-Newtonian fluids than the Newtonian fluids as per several results and discussion. And so effects of using different types of working fluid affecting the bearing performance can be studied. Experimental work in addition to present analytical work can be an extension to present analysis. A useful experimental model to undergo parametric analysis can be developed, and desired results may be achieved.

References

1. Nada, G.S., Abdel-Jaber, G.T., Abdo, H.S.: Thermal effects on hydrodynamic journal bearings lubricated by magnetic fluids with couple stresses. In: *2nd International Conference on Energy Engineering*, pp. 1–11 (2009)
2. Hu, Z., Wang, Z., Huang, W., Wang, X.: Supporting and friction properties of magnetic fluids bearings. Tribol. Int. **130**, 334–338 (2019)
3. Patel, N.S., Vakharia, D., Deheri, G.: Hydrodynamic journal bearing lubricated with a ferrofluid. In: Industrial Lubrication and Tribology (2017)
4. Ohara, T., Yutaro, A.B.E., Globeride Inc.: Magnetic fluid sealing device and magnetic fluid sealed bearing. U.S. Patent 10,451,115 (2019)
5. Nada, G.S., Osman, T.A.: Static performance of finite hydrodynamic journal bearings lubricated by magnetic fluids with couple stresses. Tribol. Lett. **27**(3), 261–268 (2007)
6. Osman, T.A., Nada, G.S., Safar, Z.S.: Static and dynamic characteristics of magnetized journal bearings lubricated with ferrofluid. Tribol. Int. **34**(6), 369–380 (2001)
7. Manser, B., Belaidi, I., Hamrani, A., Khelladi, S., Bakir, F.: Performance of hydrodynamic journal bearing under the combined influence of textured surface and journal misalignment: a numerical survey. Comptes Rendus Mécanique **347**(2), 141–165 (2019)
8. Urreta, H., Leicht, Z., Sanchez, A., Agirre, A., Kuzhir, P., Magnac, G.: Hydrodynamic bearing lubricated with magnetic fluids. J. Phys. Conf. Ser. **149**, 012113 (2009). https://doi.org/10.1088/1742-6596/149/1/012113
9. Manojkumar, N.U., Jagadish, H., Kirankumar, B.: CFD analysis of hydro-dynamic lubrication journal bearing using castor oil. In: Recent Trends in Mechanical Engineering (pp. 671–683). Springer, Singapore (2020)
10. Sriram, G., Arumugam, S., Ramachandran, M.: Finite element analysis of a journal bearing lubricated with nano lubricants. FME Trans. **48**(2), 477 (2020)
11. Bompos, D.A., Nikolakopoulos, P.G.: CFD simulation of magnetorheological fluid journal bearings. Simul. Model. Pract. Theory **19**(4), 1035–1060 (2011)
12. N. S. Patel, D. P. Vakharia, and G. M. Deheri, "A studyon the performance of a magnetic-fluid-based hydrodynamic short journal bearing. ISRN Mech. Eng. **2012**, 7, Article ID 603460 (2012)
13. Patel, N.S., Vakharia, D.P., Deheri, G.M., Patel, H.C.: Experimental performance analysis of ferrofluid based hydrodynamic journal bearing with different combination of materials. Wear **376**, 1877–1884 (2017)
14. Stokes, V.K.:. Couple stresses in fluids. Phys. Fluids **9**(9), 1709–1715 (1966)

Author Index

A
Adaramola, Bernard Akindade, 377
Adhikari, Neerav, 97
Ahamad, Abaan, 167
Aithal, Himajit, 107
Akhund, Tajim Md. Niamat Ullah, 43
Akshay, S., 409
Amin, Anjali, 697
Amin, Mihir H., 755
Amritha, P. P., 463, 533
Angne, Hemali, 251
Anitha, H. M., 453
Anuradha, A., 33
Arjun, N., 573
Aruna, M., 727
Atkari, Aditya, 251
Azzarudin, Tharick, 627

B
Bachute, Shubham, 233
Badri Prasad, V. R., 241
Bansal, Jayshri, 53
Bansal, Pratosh, 53
Baral, Daya Sagar, 139
Bedekar, Mangesh, 127
Bera, Padmalochan, 291
Bhamare, Monil M., 755
Bharath, Vivith, 241
Bharati, Jeeva, 565
Bhatnagar, Shaleen, 473
Bhatnagar, Shrey, 523
Bhat, Prashant, 223
Bhatt, Jigar, 127
Bhavsar, Rutvik M., 755

Bhingarkar, Sukhada, 233
Bista, Umanga, 139
Borwankar, Saumya, 523

C
Chakrawarti, Rajesh Kumar, 53
Chaudhari, Anita, 367
Chaudhari, Kanchan, 347
Chavan, Dhiraj, 389
Chawla, Dimple, 67
Chhabda, Riya, 483
Chotwani, Dhiren, 681
Choudhary, Pragya, 697
Christy, A., 659
Colaco, John, 313

D
Dabadge, Aditya, 233
Dastoor, Sarosh, 483
Desai, Miral M., 77
Deshmukh, Amar, 347
Deshmukh, Rohan, 347
Deshpande, Dhananjay S., 377, 419
Deshpande, Sachin, 589
Devale, Indrajeet, 233
Devan, Amrutha S., 493
Devi, Bali, 707
Dhargalkar, Nishant, 251
Dholakiya, Dhruvkumar, 213
Dongare, Aniket, 233

F
Francis, Allen, 627

G
Gaddipati, Mohith Sai Subhash, 493
George, Jossy, 515
Gopalan, Karthik, 157
Gopan, Akhila, 493
Goythale, Aishwarya, 669
Gupta, Deepak, 303, 445, 507, 565, 573, 581, 599

H
Habeebullah, Abdulkadir, 377
Hegde, Prajna, 223
Hussain, Mohammed Mohsin, 543

I
Indu, S., 543

J
Jain, Abhishek, 203
Jain, Khushi, 523
Jain, Swati, 67
Jain, Vaibhav, 259
Jayarekha, P., 453
Jena, Debasish, 291
Jha, Yash, 523
Joseph, Solley, 515
Joshi, Basanta, 97, 139
Joshi, Suvrat Ram, 139

K
Kale, Dilip, 251
Kanade, Vijay A., 717
Kant, Shri, 323
Karn, Rupesh, 139
Kaurani, Venus, 107
Kaur, Gagandeep, 17, 25
Kaushik, Ajay, 543
Kaushik, Sriram, 157
Khot, Shubham V., 633
Kranjcec, Bojana, 337
Krishnaa, Gowtam, 599
Krishnaja, S., 493
Kshirsagar, Tanmay, 213
KumarGupta, Naveen, 203
Kumari, Sangeeta, 347
Kumar, Rakesh, 1

Kumar Singh, Alok, 543
Kundu, Goutam, 193

L
Lalchandani, Priyanka, 681
Lalitha, D., 659
Lifna, C. S., 681
Lohani, R. B., 313

M
Madhavan, Shobhana Palat, 573, 581
Magesh, B., 627
Mahapatra, Sudhir Kumar, 555
Mallik, Piyush Kumar, 445
Mane, Sonali J., 177
ManjuLatha, Ch., 689
Manju, T., 627
Manohar, N., 409
Mary Metilda, M. I., 659
Mathew, Punya, 259
Mehta, Kavish, 259
Mewada, Hiren K., 77
Mirsic, Leo, 337
Mishra, Aniket, 589
Mishra, Arun, 279
Mishra, Bharati, 291
Mishra, Nidhi, 473
Mittal, Kushagra, 269
Mohammed, Bilal Kedir, 555
Mondal, Safikureshi, 193
Mrunalini, M., 357
Mukherjee, Nandini, 193
Munde, Anjali, 149
Muthu Manikandan, M., 157
Muthuraj, S., 533

N
Naik, Rutajagruti, 347
Nair, Aswathy, 493
Narang, Akash, 681
Nauriyal, Vaibhav, 269
Nayak, Amit, 213
Newaz, Nishat Tasnim, 43
Nigam, Charul, 647
Nimbalkar, Komal, 389
Nirala, Aman Kumar, 419
Nwiah, Edward, 323

O
Oreskovic, Stjepan, 337

P

Painter, Vaidehi, 483
Pai, Shashidhar, 241
Panday, Sanjeeb Prasad, 117
Pandey, Shraddha, 523
Pandit, Hardik B., 33
Pandya, Darsh P., 755
Patel, Ayush V., 755
Patel, Ritesh, 399
Patel, Snehal N., 755
Patil, Aseem, 607
Patil, Chinmay, 389
Patil, Janhvi, 669
Patil, Nitin S., 633
Patil, Shankar M., 177
Patil, Sushant, 389
Patil, Vijaykumar N., 177
Patoliya, Jignesh J., 77
Pavan Kumar, D., 357
Pawar, Urvashi, 483
Potdar, Avinash M., 555
Pundir, Sumit, 269
Putheti, Sudhakar, 689

R

Rabadiya, Kinjal, 399
Raghani, Chirag, 681
Rajadhyax, Devesh, 389
Rajak, A. R. Abdul, 167
Rajaraman, S., 565
Rashel, Masud Rana, 43
Ravoor, Prashanth C., 617
Reza, Md. Sumon, 43
Richa, Er, 87
Rijhwani, Nilesh, 669
Routray, Kasturi, 291

S

Sahay, Milind, 543
Saifuzzaman, Mohd., 43
Saindane, Pallavi, 669, 697
Salau, Ayodeji Olalekan, 377, 419
Sangeetha, D., 727
Sankar, Nikil, 627
Santhya, R., 533
Sarancha, Vitaliy, 337
Saroj, Ashish A., 633
Satre, Shilpa M., 177, 633
Sayagavi, Ashwini V., 617
Sengupta, Sharmila, 389
Sethi, Kamalakanta, 291

Sethumadhavan, M., 463, 533
Shakya, Aman, 97, 139
Sharma, Arvind Kumar, 647, 745
Sharma, Priti, 745
Shobha Rani, N., 409
Shreyass, G., 303
Shroff, Naman, 303
Singh, D. P., 269
Singh, Prishita, 735
Singh, Rahul, 543
Singh, Sweta, 1
Snigdha, Shouvik Roy, 43
Sonar, Mayuresh, 367
Sowmyanarayanan, R., 599
SravyaPranati, Bh., 689
Sreekanth, N. P., 507
Srinivasan, Seshadhri, 463
Sriram, P. R., 157
Srivastava, Sumit, 707
Srividhya, V., 727
Subhashruthi, N. J., 157
Subhedar, Shiva, 203
Sudarshan, T. S. B., 617
Sujanani, Anish, 241
Sujatha, K., 727
Sulyma, Vadym, 337
Suma, D., 689

T

Tambe, Sartha, 669
Thapa, Sobit, 117
Thayyil, Ashiema G. A., 493
Tongia, Rasesh, 259
Tuan Van, Pham, 433

U

Udaykumar, Aniket, 241

V

Vaithyasubramanian, S., 659
Vakharwala, Parantap, 483
Vardhan, Anand, 573
Vardhan, R. Dhanush Shri, 581
Vartak, Raj, 367
Vedak, Jateen, 367
Verma, Shanya, 735
Verma, Vivek Kumar, 707
Vitale, Ksenija, 337

W

Wadibhasme, Pratiksha, 697
Wadkar, Harshad S., 279
Wazid, Mohammad, 269

Y

Yadav, Rajesh Kr., 735
Yogeshwar, B. R., 463

CPSIA information can be obtained
at www.ICGtesting.com
Printed in the USA
LVHW020557031120
670482LV00001B/1